D0908734

ANATOMY OF THE MONOCOTYLEDONS

IV. JUNCALES

ANATOMY OF THE MONOCOTYLEDONS

EDITED BY C. R. METCALFE
Keeper of the Jodrell Laboratory
Royal Botanic Gardens, Kew

IV. JUNCALES

BY

D. F. CUTLER
Jodrell Laboratory
Royal Botanic Gardens, Kew

OXFORD
AT THE CLARENDON PRESS
1969

Oxford University Press, Ely House, London W. 1
GLASGOW NEW YORK TORONTO MELBOURNE WELLINGTON
CAPE TOWN SALISBURY IBADAN NAIROBI LUSAKA ADDIS ABABA
BOMBAY CALCUTTA MADRAS KARACHI LAHORE DACCA
KUALA LUMPUR SINGAPORE HONG KONG TOKYO

PRINTED IN GREAT BRITAIN

EDITOR'S PREFACE

THE present volume of *Anatomy of the monocotyledons* is the first to be written by my colleague Dr. David F. Cutler. It is a great pleasure to welcome this valuable contribution to the series of volumes describing the systematic anatomy of the vegetative organs of the Monocotyledons which is being prepared from the Jodrell Laboratory at Kew, with the interest and collaboration of botanists at other centres. Since the subject matter of this volume covers the families included by Dr. J. Hutchinson in the Glumiflorae–Juncales in his system of classification it is most immediately complementary to my own previous volume on the Gramineae. A further volume of my own, covering the Cyperaceae, or family of the sedges, is now at an advanced stage of preparation, and, when this appears, we shall between us have dealt with the Glumiflorae as a whole.

The method of treatment adopted by Cutler is similar to my own for the Gramineae, but in dealing with all of the families except the Juncaceae he has given separate descriptions for a great many species. This is fully justified, especially with the Restionaceae, partly because the anatomy of this last family has been rather imperfectly covered in spite of the well-known writings of Gilg-Benedict and others. Furthermore the anatomy of this group has turned out to be of exceptional taxonomic interest and, in the future, it will be necessary for those who undertake revisions of this notoriously difficult family to pay attention to Dr. Cutler's findings and suggestions. It was indeed fortunate that the author has been able to enjoy the collaboration not only of colleagues on the staff of the Kew Herbarium but also of Mr. L. A. S. Johnson and Dr. S. T. Blake from Australia as well as Dr. O. Leistner from South Africa, all of whom have been liaison officers at Kew from their respective countries. Many other botanists have collaborated in various ways and the author gladly acknowledges all of the help that he has received.

The family Thurniaceae, consisting only of the small genus *Thurnia*, from Venezuela and Guyana, has also proved to be interesting because of the peculiar arrangement of the vascular bundles which is at present believed to be unique amongst the Monocotyledons. There does not appear to be any anatomical evidence to support the opinion, sometimes expressed, that *Thurnia* is closely related to the Rapateaceae or Juncaceae. The minute plants comprising the family Centrolepidaceae, which one might have expected would be too small to yield anatomical data of taxonomic importance, are, however, unusual in several ways, and the occurrence of the elaborate hairs that are to be seen in some of the species is most surprising.

Hitherto the anatomy of the Juncaceae has been more thoroughly investigated than that of the other families included in the Juncales, and, for this reason, they have been treated less exhaustively than the other families in this volume.

Nearly all of the plants that have been described are herbs, some of them minute. In dealing with them the author has not, therefore, had to meet the same challenge as that which confronted Dr. Tomlinson when he wrote about the Palms in Volume II of this series. In his preface to that volume he reminded his readers of the great sizes that are attained by many Palms when he quoted L. H. Bailey as having commented: 'To procure material of the great palms is like setting forth to collect a windmill except that one does not have the advantage of steps built on the derrick.' On the other hand Dr. Cutler, unlike Tomlinson, has not had the advantage of being able to study the plants that he has investigated in the field as well as in the laboratory, apart from British members of the family Juncaceae. Attempts to grow members of the Restionaceae at Kew have met with very limited success. Cutler has had, therefore, to rely on herbarium specimens and on material specially pickled for anatomical investigation by botanists in various countries who have been kind enough to collaborate with us. Nevertheless, partly because most of these plants include a well-developed skeletal system of sclerenchyma in their make-up, it has been possible to obtain first-class microscopical preparations in which the diagnostic characters have been clearly visible. This volume is, in fact, an excellent example of what can be achieved in favourable circumstances, even when the anatomist is reluctantly driven to make use of dried specimens. The story that is presented here constitutes a most fascinating and stimulating account of the anatomy of the Juncales. The author has done very well to produce such an interesting and well-informed contribution to his subject so early in his botanical career.

C. R. METCALFE
Keeper of the Jodrell Laboratory

Royal Botanic Gardens, Kew
December 1967

AUTHOR'S PREFACE

THE vegetative anatomy of the four families included by Hutchinson (1959) in his Glumiflorae–Juncales is described in this volume. These families are Juncaceae, Thurniaceae, Centrolepidaceae, and Restionaceae. Recent anatomical research has shown that the two new families, Anarthriaceae and Ecdeiocoleaceae, should be separated from the Restionaceae, but as they clearly belong to Hutchinson's Juncales they have been included in the present volume.

Certain representatives of this group have a world-wide distribution, for example Juncaceae. Others are almost entirely restricted to the southern hemisphere. Among these, Restionaceae and Centrolepidaceae have the widest distribution. Thurniaceae are restricted to S. America, and Anarthriaceae and Ecdeiocoleaceae to south-west Australia.

The smaller families Thurniaceae, Centrolepidaceae, Anarthriaceae, and Ecdeiocoleaceae have been studied in greater detail than the larger family Juncaceae. It is justifiable to give a fuller descriptive treatment to these smaller families because they have been relatively neglected until recently. This is in contrast to the Juncaceae, of which the anatomy is already more completely known, and the literature referring to this family has been fully cited, thereby avoiding needless repetition.

The Restionaceae have also been described in some detail, for two reasons. First, the family is a 'difficult' one taxonomically. The specialized xeromorphic adaptations of members of this group have rendered gross morphological distinction between both genera and species problematical. An examination of the extensive lists of synonyms given for practically every species bears witness to this (Pillans 1928). A study of the anatomy of this group has provided valuable new information for taxonomic application. Second, while working on this family for a Ph.D. thesis, I was able to devote considerably more time to it than to the other families. During these studies it became apparent that it was necessary to examine as many species as possible for a proper understanding of the family. It was not possible to rely on a small selection of species from each genus as being 'representative'. The proportion of first-hand information in this account is, therefore, much higher than the proportion of data reported from the literature.

There is a current trend to make statements about possible phylogeny or classification based on rather fragmentary evidence. For example, similarity quotients between families have been published by some authors using data from various sources. Undoubtedly the anatomical data published in this volume could be treated in this way but their use in such a manner at the present time can only be regarded as premature. Our knowledge of the systematic anatomy of the monocotyledons as a whole is still very incomplete and this is abundantly clear in all the volumes of *Anatomy of the monocotyledons* that have been published or are still being prepared. There is a temptation, and indeed one is often pressed to go into print, when taxonomic evidence afforded by a particular line of investigation seems to point clearly to certain inter-relationships or new classifications. However, restraint must be exercised if the literature is not to be cluttered with redundant and imperfectly conceived classifications.

The purpose of a volume of this present kind must surely be twofold. Primarily it should set out the current state of knowledge of the anatomy of the families included. Secondarily it should point out the possible taxonomic implications of the data it presents. In families such as the Restionaceae there is a great deal of new anatomical evidence for certain far-reaching revisions in the classification. For the reasons set out above, the author has not attempted to make such revisions here. It is hoped that those who see the imperfections of the current classifications will be stimulated to use the new information as an integral part of an over-all study of the taxonomy of the family concerned.

In spite of the limitations to which attention has been drawn there are some broad observations that can justifiably be made at the present time.

The family Juncaceae, with the possible exception of *Prionium*, seems to constitute a homogeneous anatomical unit. The South American representatives are somewhat distinctive, but the range of structure shown by them can be readily accommodated in the anatomical concept of the family. The possible interrelationships of individual genera are discussed after the descriptions of each genus.

There appears to be justification for maintaining the family status of Thurniaceae. Its species are anatomically unlike any seen in the Juncaceae. *Thurnia* is probably most like *Prionium* in general appearance but the two genera are anatomically distinct.

Centrolepidaceae are a most interesting and unusual family. Far from indicating possible affinities with other families, the anatomical evidence points to possible taxonomic isolation of very long standing. The Centrolepidaceae do not appear to represent a homogeneous assemblage of genera; there is some anatomical evidence that *Hydatella* and *Trithuria* should be excluded from the family. A remarkable feature of the Centrolepidaceae is that their members, although reduced in size and complexity, nevertheless exhibit enough variation in anatomical characters for these to be of taxonomic value.

It has emerged from the investigation of the Restionaceae that it may be easier to ascribe generic identity to certain specimens by examining transverse sections of their culms with a microscope than by an examination of their flowers. This principle can even be extended to the identification of the occasional species. This is of particular value with sterile or male material. Anatomy is also of value in matching male and female plants of the same species. Probably the most important conclusion which has emerged from the anatomical study of the family is that no single genus is represented in both Australia and South Africa. This undoubtedly indicates that *Restio* itself is in need of extensive revision. Details of taxonomic conclusions relating to all genera in the family occur in the main text.

Each of the new families Anarthriaceae and Ecdeiocoleaceae is distinguished by a combination of characters. There is, as yet, very little to indicate the families to which they are most closely related, but it seems reasonable to classify them in the Juncales.

Some years ago Dr. C. R. Metcalfe was describing the work of the anatomy section of the Jodrell Laboratory at Kew to a party of visiting students. I was a member of that group and can recall the enthusiasm with which he told us about the application of the study of plant anatomy to the identification of fragmentary plant material. Later, I joined the staff of the Jodrell Laboratory and found that my first impressions had been well founded. Dr. Metcalfe has been a constant help and guide to me during the preparation of this book. He has shown a keen interest at every stage and his constructive criticism of the text has been invaluable. I am glad to have this opportunity to thank him.

It should be stressed that this book is the product of a team, all the members of which have played an important part. Special thanks are due to Miss Carol Milne and Mrs. Avril Turner for their technical assistance in producing microscope slides. Miss Mary Gregory has both supplied me with bibliographical information and painstakingly read and reread the typescripts. Her keen eye has spotted many inconsistencies and it is largely due to her that the task of proof-reading has been made so much lighter for me. Nevertheless, the responsibility for any remaining errors rests firmly with me. Mr. T. Harwood has processed the photographs, and Miss V. Horwill has done the greater part of the typing.

Mr. J. F. Levy, senior lecturer in the Botany Department, Imperial College, London, deserves particular mention. As my supervisor during my Ph.D. studies at college he was always ready to discuss problems and share interest in the research into the Restionaceae.

Among the taxonomists who have afforded me assistance and advice, I am indebted to Dr. R. Melville, Mr. A. A. Bullock, and Mr. H. K. Airy Shaw at Kew, Mr. L. A. S. Johnson and Dr. S. T. Blake who have both been Australian liaison officers at Kew, Dr. O. Leistner, until recently South African liaison officer at Kew, and Mr. J. Lewis of the British Museum (Natural History). Many other colleagues in the Kew Herbarium have also helped me from time to time.

The following have supplied me with material for examination: the Director of the Berlin Botanic Garden and Museum, Professor V. I. Cheadle (California), Dr. P. B. Tomlinson (Florida), Dr. L. Moore (New Zealand), Dr. K. Dalby (London), Dr. Lund (Windermere), Dr. Wynn-Parry (Bangor, North Wales), Mr. L. A. S. Johnson and Dr. R. Carolin (Sydney), Dr. S. T. Blake (Brisbane), Mr. R. G. Strey (Durban), and Mr. R. D. Royce and Dr. N. Brittan (Perth).

I am indebted to the Crown Agents for permission to reproduce the figures of Anarthriaceae, Ecdeiocoleaceae, and Thurniaceae that first appeared in my papers on those families in the *Kew Bulletin* **19**, 3, 1965, and to the Council of the Linnean Society of London for permission to reproduce Figs. 1–24, from *Proceedings of the Linnean Society* **179**, 2, 1968.

D. F. C.

Jodrell Laboratory
Royal Botanic Gardens, Kew
December, 1967

CONTENTS

NOTE

Tables 1–3 are on pp. 225–6

Key to shading of line drawings is on Fig. 1, p. 22 and on front endpapers

GENERAL MORPHOLOGY OF THE JUNCALES

1. Introduction

In writing a volume devoted to anatomical descriptions of vegetative parts of plants it has to be assumed that readers will be familiar to some extent with the gross morphology of the families described. However, just as it is only too easy, from a habitual study of thin sections, to think of plant cells as two-dimensional structures, it is equally easy to forget that the sections themselves were prepared from what was originally a whole plant. Again, although it may be reasonable to expect that the reader will be familiar with the range of form exhibited by species of *Juncus*, it is perhaps unreasonable to assume that he will be equally aware of the morphology of *Centrolepis* or *Restio*. A brief description of the gross morphology of the families included in this volume is, therefore, probably not out of place. Fuller descriptions are readily available in floras and taxonomic works.

2. Morphology

Although members of the six families show a wide range in size, their general characteristics can be fairly closely defined. They are perennial or annual herbs, and are rarely shrubby. The leaves are linear, often grass-like, sometimes cylindrical, with open or closed basal sheaths. The leaf blade is either well developed (e.g. Thurniaceae) or very reduced (e.g. most Restionaceae). The smallest members of this order occur in the family Centrolepidaceae, where many species attain the height of only 2–4 cm. The largest member is probably *Prionium*, a South African genus of the Juncaceae. The single species of this genus is shrubby and aloe-like and sometimes exceeds 1–2 m in height. It is rivalled in size only by members of the Thurniaceae and a few of the Restionaceae.

Restionaceae

Of individual families described here, the Restionaceae have the largest number of genera, but these are quite similar to one another. Species vary in height from about 10 cm to 2 m. The aerial parts of the majority of species consist of several-noded, wiry, sterile or fertile shoots. These may be simple or branched, erect or flexuose, terete, quadrangular or flattened, solid or fistular. The shoot has taken over the physiological function of the leaf. Deciduous or persistent leaf bases occur at each node, each sheath being split to the base. The sheath is rarely produced at the apex into a reduced, linear, foliaceous blade that is frequently deciduous. A ligule is present in very few species. The rhizome is tufted or creeping, covered with brown scarious scales, tough, rarely above 1 cm in diameter but occasionally reaching 2 cm. The adventitious roots are wiry or relatively fleshy, rarely exceeding 2 mm in diameter.

Anarthriaceae

Anarthriaceae have cauline leaves and some species also have basal leaves. The leaves are laterally compressed or more or less cylindrical and are sometimes equitant at the base. The culms are also laterally compressed, or terete and grooved.

Ecdeiocoleaceae

The Ecdeiocoleaceae lack basal leaves and the cauline leaves are reduced to sheaths. The culms, which are restionaceous in appearance, arise from a creeping rhizome.

Centrolepidaceae

The Centrolepidaceae are small, tufted, perennial or annual herbs. The leaves are linear or thread-like and are usually crowded or closely imbricate. The leaf bases are inflated and partially sheathing. Ligules are present in several species, these often bearing fine hairs. The culm is short and is involved to a limited extent in photosynthesis. The basal part of the stem may develop the appearance of a small, upright rhizome. The roots are very fine.

Juncaceae

The Juncaceae include plants with a wide range of form, from the large *Prionium* to small, distichous species in such genera as *Distichia* and *Oxychloë*. The leaves occur mostly in a basal tuft and are cylindrical to dorsiventrally or, rarely, laterally flattened, mostly linear or filiform, sheathing at the base or entirely reduced to a sheath; sheaths are open or closed and are sometimes ciliate at the top. A median ligule is present in *Oxychloë*. The inflorescence axis is short in *Oxychloë* and *Distichia*, but is a well-developed organ with photosynthetic tissues in other genera. The rhizome is creeping or upright; it is particularly large in *Prionium*. The roots are mostly fine; they are particularly large and fleshy in *Prionium*. The southern hemisphere genera *Rostkovia* and *Marsippospermum* are very like *Juncus* in general appearance. *Distichia* and *Oxychloë* are cushion plants.

Thurniaceae

The Thurniaceae are perennial herbs with elongated, leathery leaves that are sheathing at the base. The leaf margins are smooth or spinulose-serrate. The inflorescence axis is stout, obtusely 3- or 4-angled. The rhizome is robust and upright and the adventitious roots reach 6 mm in diameter.

3. Microscopical Structure

(a) The leaf

As in Gramineae (Vol. I of this series) and Cyperaceae the leaf provides a large number of anatomical characters that are of taxonomic importance. However, it must be borne in mind that since Restionaceae in general lack leaf blades, it is not possible to make strict comparisons between all members of the Juncales.

(b) The leaf epidermis in surface view

This order contains members that have leaves with both an abaxial and adaxial epidermis, and others that have an abaxial epidermis only because of their cylindrical form. Cells of each epidermis are normally arranged with their long axes parallel with that of the leaf. There are exceptional members in which the long axes of the cells run across the width of the leaf. In leaves that have subepidermal fibre strands or girders, the epidermal cells above these (the costal cells) are usually narrower and frequently longer than those (intercostal) cells between them. Zones of cells over the midrib and next to the margins also often differ from the others. Adaxial cells, when present, are normally wider and often shorter than the abaxial cells.

There is no distinction in the Juncales between long and short epidermal cells of the type found in Gramineae. Frequently the cells surrounding the stomata are shorter and less regular in outline than others. Certain epidermal cells of some families contain silica-bodies.

The Thurniaceae are unusual in that in one species cells of the hypodermis become superficial between longitudinal files of epidermal cells (see p. 78).

Stomata normally occur only in the abaxial surface in species that have large differences between abaxial and adaxial epidermal cells, but are often present in both surfaces of other species.

The paracytic type of stoma is almost universal in these families. The pair of subsidiary cells lies parallel to the long axis of the guard cells, and the whole stoma is arranged with its long axis parallel to that of the leaf. There is a recognizable range of types of subsidiary-cell outline. These types are often sufficiently constant in a species to be of taxonomic significance. Sometimes several types may occur in one species, and intermediate forms between the main types can often be recognized. It is important to note that the stomata from the same relative position in each species have been described whenever possible. There is some slight variation in subsidiary-cell shape in stomata from near the base or apex of the leaf.

Anomocytic stomata, i.e. stomata lacking subsidiary cells, have been recorded for one genus of Centrolepidaceae (p. 112).

Some stomata in Thurniaceae have a tendency to be tetracytic, that is, with 2 lateral and 2 terminal subsidiary cells.

Hairs and **papillae** are not a common feature of the leaves of the Juncales. Hairs occur sporadically, and are absent from the Thurniaceae, Ecdeiocoleaceae, and Anarthriaceae. The hairs show a range of form, but are usually of a simple kind, being unicellular or filamentous, but some very remarkable types are to be found in a few genera, see pp. 200, 244. Papillae are somewhat more frequent in occurrence. They normally take the form of a single perpendicular or curved protuberance from a cell, but in some species there are several smaller papillae to each cell. Because of the limited distribution of hairs and papillae, their presence and type are of value in the identification of species.

Cuticular marks are conspicuous in some species. They take the form of either longitudinal striations or granular and irregular marks. In some species a granular cuticle can have superimposed striations.

The term **'tannin'** is used in this book to describe dark amorphous substances that occur as cell inclusions. Epidermal cells containing tannin are characteristic of some species.

(*c*) *The lamina in transverse section*

Leaves of the Juncales exhibit a wide range of form in transverse section. Those of the Centrolepidaceae are oval or circular, rarely reaching 1 mm in width. The small blades of the Restionaceae, when they persist, are dorsiventrally flattened with rounded margins; they are usually 5–6 times broader than thick. Thurniaceae have leaves that are broad and dorsiventrally flattened, with or without a V-shaped section. Juncaceae have a wider range of form than other Juncales. The sections may be circular, circular with an adaxial groove, or circular with both adaxial and abaxial grooves, kidney-shaped, arc-shaped, V-shaped, broad and flat or narrow and flat, conspicuously thickened at the midrib or narrowing on either side of the midrib, with rounded or acute margins. The lamina may be of equal or unequal widths on either side of the midrib. The lamina is very reduced in Ecdeiocoleaceae. In Anarthriaceae the lamina is laterally flattened with acute or rounded angles, or is circular and grooved.

The appearance of the section varies according to the level at which it is taken in many species and it is important to remember that for comparative purposes sections are described from about half-way down the length of the lamina.

The **epidermis** does not often exhibit many characters of importance in transverse section, but the proportions of height to width of individual cells may be of taxonomic significance in rather extreme cases. Cells above sclerenchyma girders or strands may sometimes differ in some way from the others. They may, for example, be taller or shorter, narrower or wider. The outer wall of epidermal cells shows some variation in thickness in different species, but this is an unstable character and should be used for taxonomic purposes with due caution.

Of more significance is the position of **stomata** in relation to the leaf surface. Certain species can be distinguished from other closely similar relatives by looking to see whether the stomata are sunken or superficial. There is variation in the sectional outline of subsidiary cells; some have pronounced ridges, some are the same size as the guard cells and others are markedly larger. The guard cell may be mounted centrally or towards one end or the other of such large subsidiary cells. The guard cells themselves, besides exhibiting a wide range of wall thickness and outline of lumen, may possess a marked outer lip or ridge next to the aperture. Some may also have an inner lip, or an inwardly directed ridge.

The transverse section also gives a fuller understanding of the nature of **hairs** or **papillae,** and in particular the mode of attachment of the hairs.

Some species have a well-developed **hypodermis** composed of colourless cells or fibres, or both. In others there is no trace of hypodermis and the tissues next to the epidermis consist of chlorenchyma. **Chlorenchymatous cells** range in form from more or less rounded and isodiametric to palisade-like. They either fit closely together or are separated to varying degrees by air-spaces. Some cells have well-developed projections or lobes that separate them from one another. There is often variation in the number of layers of abaxial and adaxial chlorenchyma from specimen to specimen of the same species. Information given about the number of layers must, therefore, be taken as an approximation. Only when there are widely differing numbers of layers, e.g. 1–2 as opposed to 6–7, does this appear to be of any significance taxonomically; even then this character would need to be used in conjunction with others for maximum reliability. Protective cells, which are fully described for Restionaceae (p. 120), also occur in certain *Juncus* and *Thurnia* species. Mesophyll of the leaf is usually described under the heading 'chlorenchyma' in this work. Other, colourless cells that are often present in leaves, particularly in the centre of wide cylindrical leaves, are termed ground tissue. Some authors treat such colourless cells as part of the mesophyll. In a number of species of Juncaceae and Anarthriaceae the leaves and culms in section are very similar to one another. The use of the term 'ground tissue' in the leaf emphasizes the similarity of this colourless tissue to the 'central ground tissue' of the culm. Stellate cells occur in diaphragms in the leaves of some species, for example, certain *Juncus* species.

The **vascular bundles** are usually arranged in a single row or arc in the flatter leaves, and in a circle in cylindrical leaves. In laterally flattened leaves the bundles may occasionally appear to be in two rows, but it is normally easy to locate a large keel bundle at one angle and two small 'marginal' bundles at the opposite angle (see *Anarthria*, p. 332). Thurniaceae are very unusual and perhaps unique in the monocotyledons in having vascular bundles of two orders of size in two distinct rows with the **phloem** pole of the smaller bundle opposite the **phloem** pole of the larger bundle (see p. 80 for a full description).

Large- or medium-sized vascular bundles usually alternate with others that are smaller. They are usually embedded more or less centrally in the flatter leaves, but are more peripheral in leaves that are circular or oval in section. The larger vascular bundles frequently have one or more conspicuous metaxylem vessels on either flank; smaller metaxylem vessels and tracheids occur between these; the protoxylem elements may persist, or they are broken down and a cavity remains in their place. The phloem pole either borders directly on the xylem, or is separated from it by 1–several layers of narrow, frequently thick-walled cells. The phloem pole outline can be a useful diagnostic character; it is usually circular, semicircular, or oval, but can be U-shaped, among other forms. The phloem is usually without sclereids but these do occur in occasional species. The xylem of smaller vascular bundles is usually composed of tracheids or vessels of more or less uniform, narrow size; the phloem pole is small and sometimes weakly developed.

Bundle sheaths; most vascular bundles are surrounded by one or two bundle sheaths. The inner sheath is sclerenchymatous, being composed of narrow or wide fibres; the fibres may have square and not pointed ends (they might properly be termed 'thick-walled parenchymatous cells'). The inner sheath may be complete, as it is around the majority of larger bundles, or represented by caps at the xylem and phloem poles, as with smaller vascular bundles. The outer sheath, if present, is usually composed of 1 layer of wide, thin-walled, parenchymatous cells. The outer sheath may be interrupted abaxially or adaxially or laterally by a sclerenchymatous extension of the inner bundle sheath.

Sclerenchyma is restricted to the inner sheaths of vascular bundles in some representatives of the Juncales, but many flat-leaved species also have marginal strands. The outline of marginal strands may have taxonomic application. Girders or strands of sclerenchyma may be present opposite some or all vascular bundles. The outline and extent of penetration of such girders or strands is usually of taxonomic significance. When referring to sclerenchyma a girder is defined as a connection between vascular bundle and the epidermis; an incomplete or partial girder is an extension from the bundle sheath which does not reach the epidermis; a strand is free from the bundle sheath and may be subepidermal or occur anywhere in the chlorenchyma or ground tissue. The amount of sclerenchyma in a given species is liable to vary with the habitat. The variation is, however, usually only one of quantity and does not affect the form in which the sclerenchyma is present. If, for example, in a particular species the sclerenchyma is represented by strands these may vary in size in plants from different habitats but the strands will be present in all specimens. In the experience of systematic anatomists, although the structure of a species can and does vary with the habitat, anatomical variation in response to ecological change is much less marked than is commonly believed and is more often quantitative than qualitative.

Sclereids are rare outside the phloem; they are recorded in the chlorenchyma of *Anarthria*, for example.

Air-canals are frequent between vascular bundles and occur in the chlorenchyma of *Prionium*. They are more fully developed in some species than others, and reach their greatest size in the leaf base.

Tannin is present in scattered cells in a number of species from all families; sometimes it may be of a pathological nature and if it appears so this is recorded in the description.

Crystals are very rare in the Juncales; they have been recorded for *Anarthria* and *Distichia* only.

Silica-bodies are present only in Restionaceae and Thurniaceae; granular silica is recorded for Restionaceae. The silica-bodies are characteristically spheroidal and nodular; sometimes they are less regular and conform to the shape of the lumen of the cell in which they occur. Since there is not the range in form of silica-bodies such as occurs in the Gramineae, the taxonomic value of this character is reduced, but nevertheless the presence or absence of silica-bodies remains of great significance.

(*d*) *The culm*

(i) *General morphology*

The culm shows a wide range of size and form in the Juncales. It is very short in Centrolepidaceae, and tall in Juncaceae, Restionaceae, and Thurniaceae. The culm of most species contains some chlorenchyma, but is most highly developed as a photosynthetic organ in Restionaceae, Juncaceae (except *Oxychloë* and *Distichia*), Ecdeiocoleaceae, and Anarthriaceae. The culms of Thurniaceae also contain some chlorenchyma.

In its simplest form the culm has several central vascular bundles, a cortex that is not easily distinguishable from the central ground tissue, and an epidermis without stomata (e.g. certain Centrolepidaceae). In its most complex form it is a highly organized structure performing all the functions of a leaf as well as providing mechanical support to the inflorescence (e.g. Restionaceae).

(ii) *Culm in surface view*

The culm surface does not appear to provide anatomical characters of any great diagnostic significance in Centrolepidaceae and some genera of Juncaceae among the Juncales. Among the other representatives of the Juncales, however, an understanding of the range of culm surface anatomy is as important as a knowledge of that of the leaf surface. In the Restionaceae its study is of great importance, since leaves are generally lacking.

The range of variation exhibited by culm surface anatomy is the same as that shown by leaf surface anatomy and the reader is referred to the accounts on pp. 2–3.

(iii) *Culm in transverse section*

The outline of culm sections is variable. The circular form is most commonly encountered, but oval, 3- or 4- to several-sided and even winged or kidney-shaped outlines are also to be found. The culm may be grooved or channelled, or smooth. The thickness of culms varies from over 1 cm (e.g. *Thurnia*) to less than 1 mm (e.g. *Centrolepis*).

The **epidermal cells** show the same range of characters as those in the leaf, with some more in addition. These further characters are found mainly in Restionaceae. Cells may be of irregular heights, areas of high or low cells either being unrelated to the distribution of sclerenchyma or corresponding closely to subepidermal strands or girders. Some cells, particularly those next to the stomata, are extended further into the chlorenchyma than others in certain species of Restionaceae. The anticlinal walls of the cells may be straight or wavy, and the thickening of the anticlinal walls may be even or wider at the outer end of the wall and gradually decreasing to the inner end. In cells of the last type, the lumen is flask-shaped and the neck of the flask is wavy.

Stomata, which may be superficial or sunken, are present in most species; guard and subsidiary cells exhibit the same range as in the leaves. In species with grooved culms, stomata are often confined to the sides and base of the groove. A more or less well-defined ring of **chlorenchyma** is present next to the epidermis in most species, although some, e.g. *Ecdeiocolea monostachya*, may have a more or less continuous **hypodermis** composed of sclerenchyma. Chlorenchyma is particularly well developed in Restionaceae, Juncaceae, Ecdeiocoleaceae, and Anarthriaceae. The outermost cells are palisade-like (or peg-cells, see p. 119) in representatives of all these families. In other members the cells are rounded and more or less lobed; they may be isodiametric or taller than wide. The number of cell layers in the chlorenchyma may be relatively constant (e.g. Restionaceae) or somewhat variable. The chlorenchyma may be divided into sectors by sclerenchyma strands or girders (or pillar cells, Restionaceae, p. 121). The chlorenchyma may contain protective cells around the substomatal cavity (see p. 120), e.g. Restionaceae, some Juncaceae, Thurniaceae. A **parenchyma sheath** separates the chlorenchyma from a **mechanical cylinder** in Restionaceae; this is not usually so even in those Juncaceae that have a mechanical cylinder.

Vascular bundles are similar in structure to those of the leaf and may be graded according to size. (In *Thurnia*, an additional xylem pole exists in some bundles, see p. 82.) The smaller bundles are usually nearer to the outside of the culm. All bundles may be free and scattered throughout the ground tissue, or all may be embedded in an outer, mechanical cylinder (in 1 or more rings), or some may be enclosed in the mechanical cylinder and some free in the central ground tissue. In Restionaceae no bundles are present to the outside of the sclerenchyma cylinder whereas in Juncaceae some small bundles may be (Pl. IV). **Bundle-pairs**—a large and small bundle on the same radius, and enclosed in a common sclerenchyma sheath—exist in *Ecdeiocolea*.

Bundle sheaths, if distinct from the sclerenchyma of a mechanical cylinder, are usually represented only by the inner sclerenchyma sheath (or caps) but in certain representatives, outer, parenchymatous sheaths also occur. **Sclerenchyma,** in addition to that of the bundle sheaths and the mechanical cylinder already referred to for certain species, also occurs in other species as strands either next to the epidermis or in the central ground tissue. Girders are found, for example, in some species of Juncaceae. The mechanical cylinder itself may have a ribbed outline; some of these ribs extend to the epidermis in certain species. As in the leaf, the distribution and outline of girders and strands are of taxonomic importance, usually at the species level. The ribs themselves also provide some additional characters, since they also show a range of outlines. Sclereids are uncommon, and are therefore conspicuous when they occur.

The **central ground tissue** is composed of parenchymatous cells. In species where the vascular bundles are scattered throughout the ground tissue there is usually no central cavity, but a cavity is frequently formed when the vascular bundles are confined to outer rings.

Air-canals are usually confined to the centre of the culm, and in such a situation they are interrupted by diaphragms or at the nodes. A few species have air-canals developed in the chlorenchyma opposite to vascular bundles, and in the genus *Thamnochortus* (Restionaceae) in the central, translucent ground tissue there are areas of thin-walled cells that sometimes break down to form canals.

Tannin is present in certain species in cells of the epidermis, chlorenchyma, or ground tissue.

Silica-bodies are present only in Restionaceae and Thurniaceae. They are spheroidal-nodular in shape. In Thurniaceae they occur in cells of the epidermis above sclerenchyma strands. In Restionaceae they are mostly confined to special cells of the parenchyma sheath or outer layer of the mechanical cylinder. Such cells with their silica-bodies are called stegmata. The special cells have thickened (lignified or suberized) inner and anticlinal walls and a thin outer wall. Certain genera of Restionaceae have silica-bodies in some epidermal cells (Pl. VIII B). Silica-sand occurs in Ecdeiocoleaceae.

Crystals have been found only in *Anarthria*, *Distichia*, and *Lyginia*, where they are of the rhombic type.

(*e*) *The rhizome*

Several of the families of the Juncales have rhizomes. It has not been possible to examine many rhizomes anatomically, but when this has been done transverse sections have proved to be the most informative. Even so, the range of variation in basic structure has proved slight, and it is difficult to assess the taxonomic significance of many of the characters that have been seen to be variable.

The outer part of the rhizome is composed of entrant leaf or scale bases and adventitious roots and their associated vascular bundles. The **cortex** is normally parenchymatous, the cells being polygonal or variously lobed. Sometimes air-canals develop in the cortex. There may be an **exodermis** of several layers of cells with slightly thickened walls. It is often difficult to discern an **endodermoid layer**, but sometimes cells have U-shaped wall thickening and mark off the cortex from the central region. The outermost **vascular bundles** are usually in a state of fusion and segregation, and may still show evidence of the collateral form which they have in the leaf base. The inner vascular bundles run in various directions throughout the centre of the rhizome and are amphivasal but each may have a well-defined protoxylem pole. Individual bundles are usually enclosed in a several-layered sclerenchymatous sheath. The **ground tissue** may be parenchymatous and thin-walled, the cells polygonal or aerenchymatous and lobed, or the cell walls may be lignified. Tannin and starch are common in the outer cells, but crystals have not been seen.

(f) The root

Roots in transverse section show a wider range of variation than is found in rhizomes. The roots within a genus or family may be quite similar, however, and there is usually as much root structure variation within the larger families as between them.

The piliferous layer is frequently persistent, but may be detached in older roots. The **root-hairs** show some variation; they may be as wide as or narrower than the cell from which they arise. In Centrolepidaceae they seem to develop from one side of the cell consistently. Beneath the piliferous layer, there may be a **hypodermis** composed of cells similar in size to the epidermal cells, but this is not common. One more frequently finds an **exodermis** of 1–3 layers of cells with slightly thickened, lignified walls. The **cortex** is either 'solid' and composed of several layers of rounded or polygonal parenchymatous cells with small intercellular spaces (e.g. many *Luzula* spp.) or composed of several distinct zones, the outer of which is made up of thin-walled cells, the middle of radiating plates of cells with air-cavities between them, and the inner of more layers of more closely packed cells. This second type of arrangement is similar to that of the grasses and sedges, etc. An **endodermis** separates the cortex from the vascular tissue. The cells of the endodermis are usually arranged in 1–(2) layer(s). All the walls are evenly thickened, or the inner and anticlinal walls are thick and the outer wall is thin, giving the cells a characteristic U-shaped appearance in transverse section. Casparian strips have been seen in some species. The **pericycle** is normally well developed as 1–2 layers of narrow cells with cellulose walls, but it could not be detected in Centrolepidaceae (see p. 95). **Phloem** strands alternate with **protoxylem** strands in a peripheral ring next to the pericycle in many species. In some Restionaceae, however, additional phloem strands are distributed in the central region of the root. In most of the families, the wide **metaxylem vessels** are arranged in a ring immediately to the inner side of the protoxylem–phloem ring, but in some species there may be 2 rings of metaxylem vessels or 1 ring and a central vessel. The Centrolepidaceae are very unusual, however, in having only 1 central metaxylem vessel, or 2 or 3 vessels evenly spaced and touching the endodermis; the phloem poles occupy all the space contained by the endodermis and not occupied by xylem.

4. Diagnostic Microscopical Characters

In the first volume in this series, *Gramineae*, the author, C. R. Metcalfe, was able to give an account of the anatomical characters that he had found to be of diagnostic value in that group. The Juncales are composed of families showing a wide range of morphology, making it impractical to follow the scheme used for Gramineae. For example, whereas the members of Restionaceae are separable on characters of culm anatomy, it is the leaf anatomy of Juncaceae which provides most characters of taxonomic value for that family. For this reason, anatomical characters of particular diagnostic value are described together with the text for each family. This has also meant that individual treatment could be given to each of them.

It is well known that it is difficult to classify some biological material because the characters chosen to distinguish between taxa may not be completely distinct, but show a continuous gradation between two extremes. Anatomical characters are not exempt from this principle. However, with care, certain diagnostic characters can be selected which are reasonably clear cut and can be used for the identification of most of the material for comparison. Intermediate forms of a character do still inevitably appear from time to time, and it has been necessary to qualify a number of statements with 'frequently' or 'usually'. This does not detract in any way from the value of the text. Indeed, instead of giving an artificially simple picture of the anatomy, this approach indicates the complexity of the problem which actually exists.

In certain instances it is difficult to convey in words the exact meaning of a particular descriptive term—a difficulty that is increased if a new term has to be coined. A number of descriptive terms is illustrated by line drawings to make their meaning clear. These drawings are given in Fig. 25, p. 120. Each drawing should be taken as a guide to the typical form of the character defined, but, in view of what has been said above, slight variations from the typical should be expected.

5. Material and Methods

Most of the material examined came from herbaria; the source is indicated in brackets after details of collector, location, etc. for each specimen. The voucher specimen for such material is, therefore, the herbarium sheet from which it was removed. Other specimens examined were from spirit material, collected by various botanists, but notably V. I. Cheadle for Restionaceae. Professor Cheadle holds the voucher specimens for his collections. Other spirit material is also either backed by herbarium specimens, or in itself constitutes the voucher. This spirit material is preserved at Kew for reference purposes. A small number of living plants was kindly sent from Sydney by O. Evans.

It is particularly important with families such as Restionaceae, where there can be considerable debate as to the correct identity of a specimen, to have not only voucher specimens, but reference to the person or persons responsible for the identification of the material. Most of the South African material of Restionaceae examined had been seen by N. S. Pillans and is mentioned in his authoritative monograph (1928). A. A. Bullock also assisted with identifications and authority citations. It was the author's good fortune to have the advice of L. A. S. Johnson (Australian liaison officer at Kew for 1962–3) and R. Melville of Kew on the identity of Australian Restionaceae.

Most of the species of Juncaceae that have been examined were obtained from the Kew Herbarium, but these were supplemented by spirit material from various sources; material of *Prionium* was collected specially by R. G. Strey.

All but one specimen of Centrolepidaceae came from the Kew Herbarium; a living plant was received from New Zealand. For the remaining 3 families, Anarthriaceae, Ecdeiocoleaceae, and Thurniaceae, it was necessary to rely entirely on herbarium material.

Revival of dried material

Herbarium material was placed in cold water which was brought to the boil. It was then boiled for between 2 and 10 minutes, until it had resumed its original shape and texture. It could then be treated as fresh material, and fixed in the manner described in the next paragraph.

Fixing

Both living and revived materials were cut up into convenient pieces, fixed in freshly made formalin acetic alcohol, and stored in screw-top jars until required. The minimum fixing period was 48 hours, but material can be stored indefinitely in this fixative without deterioration.

Standard levels

The need to take material from standard levels in the plant for anatomical description is now widely appreciated. For all families *leaf* transverse sections and epidermal preparations were made from material taken half way down the lamina, unless otherwise stated. The *culm* (inflorescence axis) was sectioned at about half-way down its length, and very occasionally also near the base or apex. When a culm has nodes (e.g. Restionaceae) sections were prepared from the centre of an internode. *Roots* selected for sectioning were mature and of average diameter. *Rhizomes* were sectioned, as far as possible, through an internode several nodes from the apex. The nature of rhizomes makes it difficult to prepare strictly comparable sections from specimen to specimen.

Anatomical methods

The material was always washed free of fixative before further treatment. Some members of Juncaceae were softened in 4 per cent hydrofluoric acid (overnight) prior to sectioning.

Surface preparations

Flattened leaves. The specimen was laid on a glazed tile, with the surface for investigation face down; it was irrigated with a hypochlorite bleach solution ('Parozone') and the tissues scraped from the back using a safety-razor blade, until the epidermis was reached; one end of the specimen was held down with a small bottle cork to keep hands free from 'Parozone'. The prepared surface was cleared for about 10 minutes in a dish of 'Parozone' and then transferred to a dish of water when careful brushing with a soft camel-hair brush removed any loose cells. After washing, the preparation was ready for staining.

Cylindrical leaves. Sometimes a portion of the leaf was split lengthwise and one half of it placed, with the epidermis facing downwards, on a tile. It was then scraped as described above. Alternatively a portion of the cylindrical leaf was placed lengthwise in the clamp of a Reichert sledge microtome where it was supported by pith or cork. The first cut was regulated so as to remove a thin sliver of epidermis. The prepared epidermis was cleared for staining.

Culms. These were treated as cylindrical leaves.

Sections

Leaves and culms. (i) *Transverse sections*. The specimens were clamped upright in pith or carefully selected piece of bottle cork in a Reichert sledge microtome. A channel was cut in the cork to accommodate the cylindrical stem or leaf to avoid too much crushing. The specimen was kept irrigated with 50 per cent alcohol by using a small camel-hair brush; sections were made at between about 18 to 25 μm thick; they were removed from the knife edge to a dish of 50 per cent alcohol with the brush. They were then cleared in preparation for staining. (ii) *Longitudinal sections*. Bottle cork was found to be the best supporting medium for sectioning and it was prepared in the following manner. A 3- to 4-mm thick disc about 1·5 cm in diameter was taken from the end of a lenticel-free cork, using a safety-razor blade. The disc was bisected with a sloping cut. The two halves were brought together so that the long edge had a V-shaped section. The specimen was arranged so that one edge of the long axis was level with the bottom of the V. When placed in the clamp, pressure caused the cork to bend outwards on either side, raising the specimen and presenting a flat surface for the knife blade. (If the cork is cut so that the long edge is flat before clamping, it will bend outwards on clamping and release a narrow, cylindrical specimen.)

Roots and rhizomes. These were treated as stems, except that no surface preparations were made.

Macerations

It was necessary to macerate certain specimens containing cells whose nature was difficult to interpret in sections. The chromic/nitric acid method was employed. The time required for separation of cells varied from specimen to specimen and had to be judged by inspection. The macerate was washed in several changes of water, until all traces of acid were removed. Water was replaced by 95 per cent alcohol (2 washes) and 98 per cent alcohol (2 washes). A hand centrifuge was used between washings to consolidate the macerate. Samples of the finally prepared pellet were mounted in Euparal containing 2–3 drops of fast green per 10 ml; cells took up the green stain as the slides dried.

Staining

Temporary preparations (both cleared and uncleared) were stained on the slide, using either methylene blue or chlor-zinc-iodide. Any excess stain was removed and the sections mounted in glycerine. Alternatively, sections were mounted in glycerine containing 5 ml of 1 per cent aqueous methylene blue to 45 ml of 50 per cent glycerine.

Carbolic acid solution was also used in the preparation of temporary mounts, since this causes silica-bodies to turn pink.

Permanent mounts were prepared in the following way. Sections were cleared in 'Parozone', washed in tap-water, transferred to 50 per cent alcohol and then placed in an excavated glass block containing the stain. A lid was placed over the stain to prevent evaporation. By experiment it was found that a stain mixture of 80 per cent safranin (1 per cent safranin in 50 per cent alcohol) and 20 per cent Delafield's haematoxylin would give consistently good results after immersion for 2–6 hours. When it was required to stain overnight, the well-tried safranin/haematoxylin 94 per cent/6 per cent mixture was used. The sections were removed from the stain with forceps and agitated in a petri dish containing 50 per cent alcohol and 2–3 drops of conc. HCl. The sections were watched and removed to 95 per cent alcohol as soon as the correct colour balance appeared. It is easy to remove excess stain very rapidly, and great care is needed at this stage. After 5 minutes in 95 per cent alcohol, the sections were transferred to a covered petri dish of absolute alcohol (5 min) and then to xylol, prior to mounting in neutral Canada balsam. Slides were then baked in an oven at 58 °C. for 10–14 days before use.

Wax embedding

It was necessary to embed only very few specimens, since nearly all the material studied could be sectioned as described above. Conventional embedding techniques were employed.

Methods of examining slides

Whilst the conventional light microscope was used as the primary instrument for the examination of the microscope slides, other optical methods were also used. Polarized light microscopy was an invaluable aid in crystal and starch examination. Elliptically polarized light made it possible to examine hairs on the epidermis of stem or leaf in unstained specimens. This is particularly useful, because in stained material it is not always easy to make out the structure of such complex hairs as those exhibited by *Meeboldina* (p. 241), for example. Unstained macerations can also be viewed successfully by this method. The technique consists in inserting a thin mica plate or phase plate between the analyser and specimen. Instead of the dark-grey-white effects obtained with polarized light, in elliptically polarized light the crystalline specimen appears coloured against a differently coloured background.

Anoptral contrast and phase contrast microscopy were also employed in the examination of unstained macerations.

Recording the data

Anatomical observations were recorded in several ways. First, marginally punched index cards were used to record the presence or absence of a large number of characters. A separate set of cards had to be used for each family, since the same characters were not of equal importance in all the families. The cards provide a permanent record of far more characters than can be conveniently included in a written account. They can also be used to assist in the determination of character correlations. The code and lists of numbers to enable additional sets of cards for Restionaceae to be prepared is given in a Ph.D. thesis housed in the library of the University of London and at the Jodrell Laboratory, Kew (Cutler 1965).

Written descriptions were first entered on specially prepared forms. These forms save time since one is constantly reminded to examine the specimen for all the characters on a particular list. In the author's opinion forms also enable generic descriptions to be written more quickly.

Illustrations were prepared using both a prismatic camera lucida and a camera.

Measurements

Measurements of cell dimensions are given in descriptions in terms of relative height to width of individual cells or the relative heights or thicknesses of cell layers. Because it was usually possible to examine only 1 or 2 specimens of each species, definite measurements could be misleading. It has to be borne in mind that anyone wishing to identify material often finds himself with only a single specimen at hand, so that a description is of practical use only when it is based on characters that show the least variation. For this reason a statistical analysis of sizes and dimensions was not undertaken.

Leaf, culm, rhizome, and root dimensions, where given, relate to the material examined. Measurements are either included with the individual species descriptions or, if the species are not described separately, with the list of material examined.

The following relative terms are used with particular meanings. (Note that the meaning of a term is conditioned by the type of cell to which it refers.)

Cell walls: very thick $> 8\ \mu m$; thick 6–8 μm; moderately thickened 2·5–6 μm; slightly thickened 1–2·5 μm; thin $< 1\ \mu m$.

Sclerenchyma fibres: wide 20–25 μm, medium-sized 15–20 μm, narrow $< 15\ \mu m$.

Tracheids in peripheral vascular bundles: wide $> 15\ \mu m$, medium-sized 8–15 μm; narrow $< 8\ \mu m$.

Metaxylem vessels in medullary vascular bundles and cells of central ground tissue: wide $> 30\ \mu m$; medium-sized 15–30 μm; narrow $< 15\ \mu m$.

Orientation

The following terms have particular meaning in this work:

height, high—refer to anticlinal walls,
width, wide—refer to the length of periclinal walls.

6. Abbreviations

Certain abbreviations are used in the text; their meanings are as follows:

c. = *circa*, about
I.S. = inner bundle sheath
O.S. = outer bundle sheath
L.S. = longitudinal section
mx = metaxylem
mxv = metaxylem vessel
px = protoxylem
pxv = protoxylem vessel
scl. = sclerenchyma
T.L.S. = tangential longitudinal section
T.S. = transverse section
vb (plural vb's) = vascular bundle(s)

v. = very
(K) = Kew Herbarium material
(M) = Kew Museums material
(S) = spirit material at Jodrell Laboratory

7. Taxonomic Considerations

In common with other volumes in this series which contain descriptions of more than one family, the sequence followed is that adopted by Hutchinson (1959). Hutchinson included the Juncales with Cyperales and Gramineae in the Glumiflorae.

It may be helpful to the reader to know the main ways in which several taxonomists have classified the members of Hutchinson's Juncales. The following brief account gives an indication of the fortunes of the various families.

Hamann (1961) published an interesting review of the history of various classifications involving the families recognized by Engler (1930) in his Farinosae. The review contains a very useful table summarizing classifications by twenty-six authors. The table also provides data on other families that have, from time to time, been classified with members of Engler's Farinosae. Certain information relating to Hutchinson's Juncales can be extracted from this table. It should be remembered that since Anarthriaceae and Ecdeiocoleaceae were not described until 1965 (Cutler and Shaw), when Restionaceae is mentioned it includes the genera making up these new families. Also, when some of the earlier authors wrote about Juncaceae they understood it to contain *Thurnia*, which was later elevated to family rank in Thurniaceae. The combinations of families are being stressed in the following analysis: no regard is given to the name of the combination, neither is any account taken of families other than Restionaceae, Centrolepidaceae, Thurniaceae, and Juncaceae. When a single family constitutes a 'combination' this means that one or more authors did not include it with any of the other three at present under consideration.

Each family is denoted by its initial letter; the number of authors recognizing each combination is given.

R+C+T+J : 5	R+C+T : 3	R+C : 14
T+J : 12	R+J : 1	R : 1
C : 2	T : 2	J : 1

It can be seen that most authors group R with C and nearly as many T with J (this last combination includes all citations for Juncaceae prior to the separation of Thurniaceae). The number of instances in which R and C are grouped together in all classifications is 22 (5+3+14), and in which T and J are grouped together in all classifications is 17.

Hamann's own analysis has produced a series of lists of similarity quotients. He found that both Restionaceae and Centrolepidaceae share more selected characters in common with Gramineae than they do with one another and that they are both much more similar to one another than either is to Juncaceae or Thurniaceae. Juncaceae and Thurniaceae themselves were found to be very similar.

When considering the various combinations of families and the similarity quotients it is well to remember that the groupings were mainly constructed using conventional characters of gross morphology, and that the similarity quotients were based on selected characters. It is the author's opinion that considerable changes in the similarity quotients could result if some of the additional anatomical characters described in the present volume were used for comparative purposes. Nevertheless, as stated in the author's preface, we are a long way from the day when sufficient data will be available for this type of analysis.

Hutchinson's Glumiflorae contain families that share certain basic anatomical characters in common (the type of vascular bundle, for example). Only when the survey of the other monocotyledonous families is complete will it be apparent whether additional families should be included in the division.

Every effort has been made to indicate the apparent degree of affinity between the constituent genera within each family described, on the basis of anatomical structure.

Literature Cited

CUTLER, D. F. (1965) The taxonomic significance of the anatomy of the Restionaceae. Thesis, University of London. 778 pp.

—— and SHAW, H. K. A. (1965) Anarthriaceae and Ecdeiocoleaceae: two new monocotyledonous families, separated from the Restionaceae. *Kew Bull.* **19**, 489–99.

ENGLER, A. and PRANTL, K. (1930) Farinosae. In *Die natürlichen Pflanzenfamilien* **15a**.

HAMANN, U. (1961) Merkmalbestand und Verwandtschaftsbeziehungen der Farinosae. Ein Beitrag zum System der Monokotyledonen. *Willdenowia* **2**, 639–768.

HUTCHINSON, J. (1959) *The Families of Flowering plants* II *Monocotyledons*, 2nd edn. Clarendon Press, Oxford.

METCALFE, C. R. (1960) *Anatomy of the monocotyledons* I *Gramineae*. Clarendon Press, Oxford.

PILLANS, N. S. (1928) The African genera and species of Restionaceae. *Trans. R. Soc. S. Afr.* **16**, 207–440.

FAMILY DESCRIPTIONS

JUNCACEAE

(Figs. 1–14, Pl. I, II, III. A, IV. A)

Introduction

THE family Juncaceae is composed mainly of perennial or annual herbs, but there is one shrub-like genus (*Prionium*). The rhizome is erect or horizontal or, rarely, absent. Stems are mostly leafy at the base. Leaves are cylindrical to flat, mostly linear or filiform, sheathing at the base or entirely reduced to a sheath; sheaths are open or closed and sometimes ciliate at the top. A median ligule has been recorded for *Oxychloë* (Guédès 1967).

The distribution of the family is world-wide but some genera are restricted to the southern hemisphere. The plants are most frequent in temperate and cold or montane regions, usually in wet or damp habitats.

The limits of the family used here are mainly those adopted by Buchenau edn. 1 and Vierhapper in edn. 2 of Engler's *Pflanzenfamilien*. The tribes *Xeroteae* and *Calectasieae* of Bentham and Hooker (*Genera Plantarum*) are not included. Barros's treatment of *Oxychloë* and *Andesia* is followed, and Benoist's genus *Voladeria* is described, although in Engler's *Syllabus* II (1964) this is not regarded as being well enough known to be included in the family. *Thurnia* is treated separately, under Thurniaceae.

Family Description

Leaf Surface

Leaves of most spp. have both abaxial and adaxial surfaces but in some *Juncus* spp. in which the leaf is cylindrical the adaxial surface is not present. Various intermediate stages also occur in certain spp. where the adaxial surface is reduced but not eliminated.

Abaxial surface

Epidermis: cells 4- or 6-sided, walls thin to moderately thickened, straight or wavy; cells ranging from as long as wide to 30 times longer than wide, but mainly in the ranges 2–5 or 6–12 times longer than wide; in *Prionium* 1½–3 times wider than long. **Hairs** present only in *Luzula* (q.v.). **Papillae** present on margins of *Luzula* spp. **Stomata** when present paracytic, usually as in culm of same sp. **Cuticular marks:** striations not normally conspicuous.

Adaxial surface

Epidermis. Cells 4- or 6-sided, anticlinal walls thin to moderately thickened, straight or wavy. Cells ranging from as long as wide to 15 or more times longer than wide, but mostly 1–3 or 6–10 times longer than wide; in *Prionium* 1½–3 times wider than long. Cells frequently wider than those of abaxial surface. **Stomata** absent, except in *Prionium*.

LEAF T.S.

Outline circular, oval, dorsiventrally flattened, arc-shaped, half-moon-shaped, kidney-shaped, or V-shaped, with or without ridges or grooves. **Epidermis.** Cells always in 1 layer. (i) Abaxial cells mostly between as high as wide and 1½ times higher than wide, some up to 1½ times wider than high; outer walls usually moderately thickened to thick, other walls less thickened. (ii) Adaxial cells indistinguishable from abaxial (*Prionium*), larger than abaxial (*Luzula*) or absent. Outer walls moderately thickened or thick, other walls usually thin or slightly thickened. Anticlinal walls of all epidermal cells straight, never wavy. Adaxial cells frequently diminishing in size towards leaf margins and becoming similar in size to abaxial cells. Cells of both surfaces frequently smaller than remainder where opposite vb's in certain *Luzula* spp. End walls of cells not perpendicular to leaf surface in some spp., e.g. *J. monanthus* (Haslinger 1914). **Stomata** superficial or sunken; subsidiary cells thin-walled; guard cells frequently with outer lip (v. pronounced in some spp.), sometimes also with lip at inner aperture. **Hypodermis** sclerenchymatous or parenchymatous, present in *Prionium* and *Rostkovia.* **Vascular bundles** in 1 (2 or 3) ring(s) in spp. with cylindrical leaves and in 1 layer in dorsiventrally flattened leaves. Larger and medium-sized bundles frequently alternating with smaller ones. Bundles in *Juncus, Oxychloë,* and *Rostkovia* with 1 (–4) wide mxv's per flank; flanking mxv's narrower in *Luzula* and v. narrow in *Distichia,* medium-sized in *Prionium*; *Marsippospermum* with several, narrow mxv's on either flank. Protoxylem canal present in some spp. Vessels showing same range of wall-pitting and perforation plates as in culm (see p. 20). Phloem pole frequently separated from xylem by 1–2 layers of narrow, thin, or moderately thick-walled parenchymatous cells (3–4 layers in *Marsippospermum*), or abutting directly on it (some *Luzula* spp.). Outline of phloem pole variable, from oval to triangular or circular. Small sclereids present in phloem of some spp. and structures resembling secretory-canals present in occasional *Juncus* spp.

Bundle sheaths: O.S. normally present as 1 continuous or interrupted layer of wide, parenchymatous cells; girder-like extensions of this sheath present at poles of bundles in *Prionium,* these reaching both abaxial and adaxial epidermis; I.S. sclerenchymatous, as complete sheath, or as bundle caps only; caps well developed in *Marsippospermum.* **Chlorenchyma:** cells palisade-like or more or less isodiametric, sometimes with pegs or lobes; 1–several layers in cylindrical leaves, layers interrupted by scl. strands or girders in some spp.; chlorenchyma cells in flattened leaves present next to both abaxial and adaxial surfaces; in *Prionium* arranged in shape of hourglass or figure of 8. Protective cells (see p. 120) present in *Rostkovia*; in some spp. of *Juncus* procumbent, lining substomatal cavity, walls next to cavity thickened, others thin. **Sclerenchyma.** Fibres elongated, with pointed ends and oblique pits; present in bundle sheaths or caps and in some spp. of *Juncus* and *Oxychloë* in girders next to epidermis. Strands 1–25 cells wide present in ground tissue of *Prionium*; fibres in several layers next to epidermis in *Rostkovia* except below stomatal strips. Marginal fibre strands present in some flat-leaved *Juncus* spp. and most *Luzula* spp. **Air-canals** present in some spp.; frequently alternating with vb's; central canal present in some spp. with cylindrical leaves and traversed by diaphragms of stellate or lobed cells in certain spp. In *Prionium*, air-canals present in chlorenchyma (Fig. 13 D, p. 67). **Tannin** present in some cells of ground tissue of certain spp. **Crystals**[1] present in *Oxychloë*. **Silica-bodies** absent.

CULM SURFACE

Hairs: absent. **Papillae** present in v. few spp., next to stomata. **Epidermis:** cells 4–6-sided, usually 4–5 times longer than wide but occasionally as long as wide or up to 10 times longer than wide; wider than long in *Prionium*; walls straight or wavy, thin to moderately thickened or v. thick. **Stomata** paracytic, with long axis parallel to that of culm; sometimes in files corresponding to shallow longitudinal grooves. Guard cells normally with rounded ends; subsidiary cells with acute, obtuse, perpendicular, or rounded ends; long outer wall (adjacent to epidermal cells) usually curved, sometimes sinuous. **Cuticular marks:** longitudinal striations present in v. few spp.

CULM T.S.

Outline circular, oval, sometimes grooved. **Cuticle** variable in thickness. **Epidermis.** Cells in 1 layer; variable in height; in most spp. slightly wider than high to twice as high as wide, exceptionally twice as wide as high or 3 times higher than wide. Those opposite ridges frequently wider and taller than others, those next to stomata occasionally smaller than remainder. Outer walls often thickest; anticlinal walls not normally wavy, with even thickening or thick at outer ends and thin at inner ends, tapering between; inner walls often thin. **Stomata** superficial or sunken; subsidiary cells variable, normally about twice as high as guard cells; guard cells frequently with lip at outer aperture, less frequently with lip at inner aperture also; walls of guard cell normally moderately thickened, thin or slightly thickened in some *Luzula* spp. and *Prionium*; lumina of guard cells triangular with apex directed towards aperture, except in *Prionium* where following outline of guard cell wall. **Chlorenchyma.** Palisade cells present in outer (or all) layers, cells sometimes with short pegs, or in some genera (*Luzula*, *Andesia*) and occasional spp. of other genera rounded and/or more or less lobed; cells of inner layers rounded in spp. with outer palisade cells. Number of layers of cells variable, from 1–2 to 6–10; inner layers sometimes distinct from central ground tissue. Cells irregularly lobed, in 6–8 layers in *Rostkovia*. Protective cells procumbent, lining substomatal cavity, each with thickened walls next to cavity, present in some spp. of *Juncus*. Areas of wide, thin-walled cells present in some spp. **Parenchyma sheath** 1-layered, frequently interrupted, present in some spp., particularly those with continuous scl. cylinder; probably morphologically equivalent to outer bundle sheaths. **Sclerenchyma** confined to I.S. in some spp.; present as small girders or strands next to epidermis e.g. some *Juncus* spp.; forming mechanical cylinder joining members of the ring of vb's in some spp.; sometimes present as strands in culm centre in *Juncus*.

Vascular bundles in most *Juncus* spp. with 1 (2) conspicuous mxv's on either flank, xylem separated from phloem pole by 1–2 layers of parenchymatous cells (these often with thickened, lignified walls); in *Luzula* spp. flanking mxv's often narrow. In *Rostkovia* vb's elongated radially with 1–2 files of

[1] Crystals observed in *Distichia* may be an artifact.

narrow mxv's on either flank; phloem widely separated from xylem by several layers of narrow, thick-walled cells; px elements exhibiting unusually tortuous arrangement, side walls frequently visible in T.S. In *Oxychloë* with narrow mxv's on flanks not to be confused with groups of 2–4 wider, parenchymatous cells to outer side of mxv's; phloem abutting directly on xylem. Protoxylem canal present in some spp. Vessel lateral wall-pitting in mxv's reticulate, scalariform, or alternate, perforation plates frequently oblique, simple, scalariform, or scalariform-reticulate.

Bundle sheaths. Vb's in some genera (*Distichia*, some *Luzula* spp.) with scl. sheaths only; sheath indistinct around bundles embedded in culm scl. cylinder. Most genera with 2 sheaths to each bundle: (i) O.S. parenchymatous, cells in 1 layer, sometimes interrupted; (ii) I.S. sclerenchymatous, fibres normally in 1–2 layers on flanks and more at xylem and phloem poles, occasionally as caps at poles only. Some spp. with 3 sheaths (i) and (ii) as above and a third, inner parenchymatous sheath composed of 1 layer of cells, these sometimes with U-shaped wall thickenings on inner and side walls next to phloem, e.g. *J. acutus*. **Central ground tissue** parenchymatous, continuous, or with a central cavity, sometimes with transverse diaphragms of stellate cells (some *Juncus* spp.). Cells of ground tissue normally wide and thin-walled. **Air-canals** present in culm centre of many spp., also noted in chlorenchyma of some *Juncus* and *Luzula* spp. and *Rostkovia*, canals formed by breakdown of wide parenchymatous cells. **Silica** absent. **Crystals** in *Distichia*. **Tannin** present in some cells of certain spp.

RHIZOME

Outer tissues often irregularly arranged because of entrant scale-leaf and culm bases. **Epidermis:** cells in 1 layer; mostly wider than high, walls slightly or moderately thickened. **Cortex.** (i) Outer part composed of 5–20 layers of more or less hexagonal cells all about as high as wide; walls frequently thin. (ii) Middle part composed of 5–8 layers of cells with moderately thickened walls (e.g. *Luzula* spp.) or made up of radiating plates of cells with air-spaces between them (e.g. many *Juncus* spp.). (iii) Inner part normally 1–4-layered, cells rounded, thin-walled. **Endodermoid layer** distinct in most spp. as single layer of cells with all walls or only radial and inner walls moderately thickened to thick. **Vascular bundles** amphivasal, scattered or arranged in 1–2 rings near to endodermoid layer; all except leaf traces contained within the endodermoid layer. Traces of px pole evident in vb's of some spp. Vessels angular, wall pitting scalariform or alternate; perforation plates oblique, simple or scalariform. **Bundle sheaths** sclerenchymatous, fibres in 1–4 layers. **Ground tissue:** cells narrow, parenchymatous, sometimes with thick walls. **Air-canals:** see cortex. **Crystals, silica, tannin:** none seen.

ROOT T.S.

Epidermis: cells in 1 layer, frequently slightly wider than high; walls slightly thickened. **Root-hairs** arising from cells similar to those of rest of epidermis, or from different, thicker-walled cells. **Hypodermis** present in some spp. as 1 layer of cells with thickened outer walls; other walls thin. **Cortex.** (i) Outer part sometimes definable as 3–9 layers of more or less hexagonal, isodiametric (longer than wide in L.S.) cells with slightly thickened walls;

otherwise composed of rounded cells and not readily distinguishable from remainder of cortex. (ii) Middle part either composed of rounded cells without wide intercellular spaces or air-cavities (e.g. *Luzula* spp.) or made up of radiating plates of cells separated by lysigenous air-cavities; plates frequently about 8–9 cells deep and 1–2 cells wide. (iii) Inner part distinct only in spp. with well-differentiated middle cortex, composed of about 2–4 or 7–8 layers of narrow, thin-walled, rounded cells. **Endodermis** 1 (–3)-layered, cells with characteristic, U-shaped wall thickenings. **Pericycle** normally composed of 1–2 layers of thin-walled cells. **Vascular system:** mxv's solitary or in multiples of 2–3, usually arranged in 1 ring to inside of px and phloem, sometimes also with 1 central vessel. Individual mxv's surrounded by 1 layer of flattened tracheids. **Ground tissue** composed of cells with thin or moderately thickened to thick walls.

DISTICHIA Nees *et* Meyen

No. of spp. 3; examined 2.

Genus Description

Leaf surface

Epidermis: abaxial cells 1–9 times longer than wide, 4-sided, walls moderately thickened, wavy. **Stomata:** subsidiary cells with obtuse ends; outline of pair of guard cells slightly lemon-shaped.

Leaf T.S. (sheathing base).

Outline oval with concave adaxial face, margins tapering to 2 layers of cells. **Epidermis:** (i) abaxial cells about half size of adaxial cells, outer walls v. thick, others thin; (ii) adaxial cells more or less square, outer walls slightly thickened, others thin. **Stomata** superficial, sparse; subsidiary cells slightly taller than epidermal cells, wide; guard cells as high as epidermal cells, thin-walled, except at small outer and inner lips. **Vascular bundles** arranged in 1-layered arc, all small and without conspicuous flanking mxv's; *c.* 15 in number, smallest (in sheathing edges) reduced, sometimes represented only by small scl. strands. **Bundle sheaths:** O.S. parenchymatous 1-layered; I.S. 1–3-layered, of narrow, thick-walled fibres. **Chlorenchyma** probably represented by 1–2 layers of broad palisade to isodiametric cells next to abaxial epidermis. **Ground tissue** parenchymatous, cells wide, thin-walled, separated by intercellular spaces. **Air-canals** present between vb's; formed by breakdown of v. thin-walled cells.

Lamina T.S.

Outline oval, with wide, shallow grooves (Fig. 1. A). **Epidermis** (only abaxial seen) as in leaf sheath. **Vascular bundles** in 1 ring, phloem in direct contact with xylem; no conspicuous mx vessels (Fig. 1. C). **Bundle sheaths:** O.S. parenchymatous, 1–2-layered on flanks, occasional cells at poles; I.S. scl. caps, 2–3-layered at phloem pole, 3–5-layered at xylem pole. **Chlorenchyma** *c.* 3–5-layered; cells of outer 3 layers palisade-like, with small pegs, those of inner layers more or less hexagonal, with more pronounced pegs. **Sclerenchyma** represented by bundle sheaths only. **Ground tissue** composed of isodiametric cells between vb's. Central cells breaking down to form cavity.

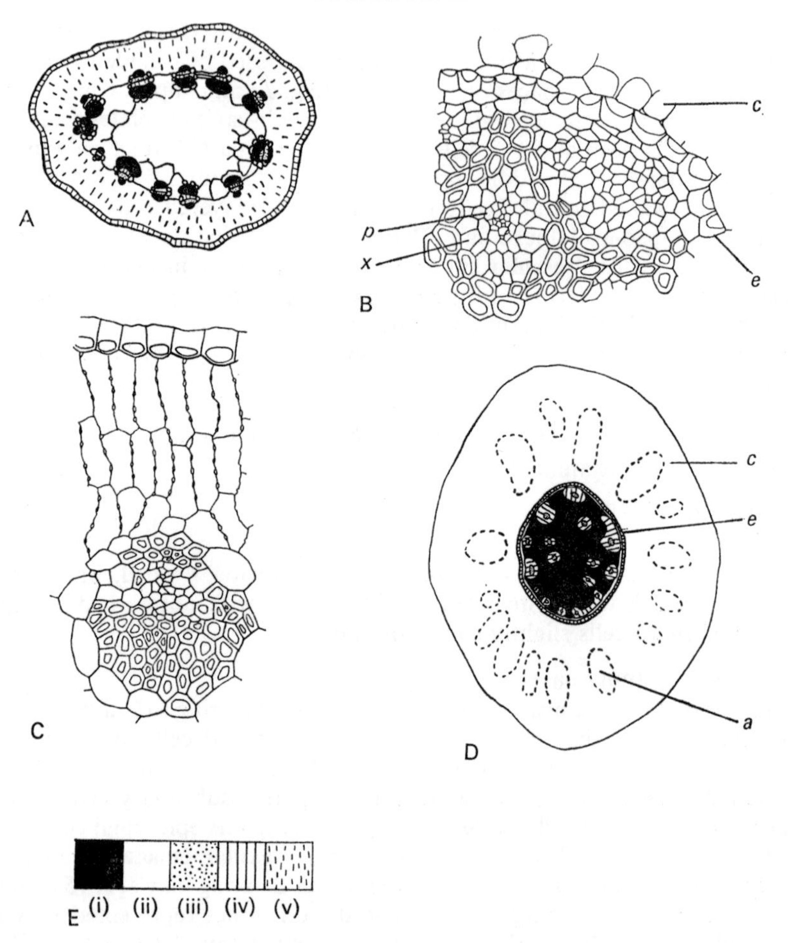

FIG. 1. Juncaceae. A–D, *Distichia muscoides*: A, diagram of leaf T.S. (×50); B, detail of culm vascular bundles (×200); C, detail of leaf T.S. (×200); D, diagram of culm T.S. (×50); E, key to shading in line drawings: (i) = sclerenchyma; (ii) = parenchyma; (iii) = phloem; (iv) =xylem; (v) = chlorenchyma.
a = air-space; c = cortex; e = endodermoid sheath; p = phloem; x = xylem.

Culm T.S.

Axis v. short, covered with congested leaf bases; only the distal 2–3 mm showing culm characters.

Outline oval (Fig. 1. D). **Epidermis:** cells as high as wide or slightly higher than wide; walls moderately thickened. **Cortex.** (i) Outer part 4–6-layered consisting of 4–6-sided cells with moderately thickened walls and numerous, conspicuous, simple pits. Most cells about as high as wide. (ii) Middle part consisting of radiating plates 2–4 cells wide, some with leaf traces embedded in them, cells often collapsed, thin-walled; wide air-spaces between plates. (iii) Inner part composed of 1–2 layers of rounded, thin-walled cells. **Vascular bundles:** 2 rings, both small and large vb's without conspicuous mxv's. Each vb composed of small, rounded phloem pole in an arc of narrow, angular tracheary elements (it is not possible to differentiate between vessels and tracheids in the material available). Many fused bundle pairs present (Fig. 1. B). **Bundle sheaths:** 1–several layers of narrow fibres indistinguishable on flanks from those of scl. cylinder. **Sclerenchyma** as 5–6-layered cylinder; cells of outer layer with U-shaped wall thickenings, particularly in more basal sections. **Central ground tissue** composed of medium-sized, hexagonal cells with thick walls and simple, more or less rounded pits. **Air-cavities** as described for middle cortex; smallest towards apex of axis. **Crystals** rectangular; rare (artifact?).

Rhizome T.S.

Outline circular. **Epidermis** and **stomata** not seen. **Cortex.** (i) Outer part consisting of 8–12 layers of rounded cells; cell walls of increasing thickness towards centre of rhizome. (ii) Middle part consisting of 5–6 layers of cells interrupted by a ring of air-canals of varying size; leaf traces embedded in plates of tissue. (iii) Inner part 1–2-layered, consisting of hexagonal cells, with intercellular spaces at angles between cells and also part way along walls. **Endodermoid layer** 1 cell thick, cells mostly with U-shaped wall thickenings but some cells thin-walled. One–2 cell layers to inside of endodermis apparently representing pericycle. **Vascular bundles** mostly amphivasal in culm centre; those next to pericyclic region fused. Leaf trace bundles more or less collateral. Tracheary tissues possibly containing vessels, but elements narrow and end walls not distinctly perforate. **Central ground tissue** indistinguishable from **bundle sheaths**; all cells between vb's thick-walled. **Air-cavities** as described for middle cortex; these arranged in net-like pattern as seen in T.L.S., each cavity being short, rhombic, and isolated from its neighbours.

SPECIES DESCRIPTION: as for genus.

MATERIAL EXAMINED

Distichia muscoides Nees *et* Meyen: (1) Pennell, F. W. 13492; Peru 1925 (K). (ii) Mandon, G. 1444; La Paz, Bolivia (K). (iii) Asphend, E. 11553, 1940 (K). Leaf T.S. 2×0·8 mm. Culm 1×0·8 mm.

D. tolimensis (Decne.) Buchenau: Pennell, F. W., and Hazen, T. E. 9860; Colombia 1922 (K). Leaf T.S. 3–4×1 mm. Culm T.S. 1×1·3 mm.

TAXONOMIC NOTES AND INFORMATION FROM THE LITERATURE

Haslinger (1914) recorded that *Distichia* resembles *Patosia* and *Oxychloë*. The root, like that of many *Juncus* species, has air-canals in the middle cortex.

Buchenau (1906) figured the leaf of *D. muscoides* in T.S. There are 3 layers of palisade chlorenchyma in his specimen. He also illustrated stomata in this species. Because the plant is so reduced, it is difficult to find anatomical characters that suggest possible lines of relationship. It is distinguishable from *Oxychloë* by the lack of scl. in the leaf apart from that surrounding the vb's.

JUNCUS L.

No. of spp. *c.* 300; examined 29.

GENUS DESCRIPTION

The genus was divided into 8 subgenera by Buchenau (1890). Representatives of all but one of these (*singulares*) were available for this present study.

Leaf surface

Hairs absent. **Papillae** seen in *J. subulatus*, extending over stomata, and as projections of certain epidermal cells in *J. mutabilis*. **Epidermis:** adaxial cells occurring only in some spp., when present usually wider and longer than abaxial cells. Walls of cells of both surfaces thin to moderately thickened or thick; sometimes straight, frequently wavy, particularly at the surface (Figs. 2. J; 3. G; 7. H, I). Outline of cells usually 4-sided, sometimes 4- or 6-sided. **Stomata:** abaxial only; types as for family (Fig. 2. C, G; 4. K). Stomata arranged in longitudinal bands in some subgenera, particularly those with subepidermal fibre girders or strands; in other subgenera, scattered. **Cuticular marks:** striations in general inconspicuous. **Tannin** v. infrequent.

Leaf T.S.

Outline v. variable; particular types frequently characteristic of one or several subgenera; circular, oval, kidney-shaped, with or without ridges, strap-shaped or V-shaped. **Epidermis:** abaxial cells frequently as high as wide, or occasionally twice as high as wide; adaxial cells (if present) frequently larger than abaxial, sometimes bulliform, up to 3 times higher than wide. Outer walls normally thicker than others, often convex, sometimes with irregular outer surface. **Stomata** more often superficial than sunken; guard cells frequently with lip at outer aperture, sometimes also with lip at inner aperture; subsidiary cells frequently taller than guard cells and extending further into leaf, but often not as tall as epidermal cells. **Hypodermis** absent. **Vascular bundles** in 1 row in dorsiventrally flattened leaves, normally in 1 ring (occasionally 2 or 3 rings) in cylindrical leaves. Larger bundles rarely with more than 2 medium-sized mxv's on either flank or none. Phloem pole frequently ensheathed by scl. or thick-walled parenchyma cells; sclereids in phloem rare. Protoxylem canal not always present. Some vb's with areas of narrow, thin-walled parenchyma cells flanking the px. pole and also to the inside of scl. bundle sheath.

Bundle sheaths: O.S. usually 1-layered, parenchymatous; I.S. sclerenchymatous, represented by caps to many smaller vb's and by complete sheaths to larger bundles; sheath normally thicker at phloem than xylem pole and narrowest on flanks. **Chlorenchyma:** cells next to epidermis normally palisade-like, 2–4 (–6) times higher than wide, often with few, short pegs; cells in 1–4 or 5(–7) layers, innermost cells frequently about as high as wide and with rounded corners; protective cells present in spp. of subgenus *thalassii* only. **Sclerenchyma** confined to I.S. in many spp., but also present as subepidermal girders in subgenera *genuini* and *thalassii*; represented by extensions from certain bundle sheaths to epidermis in 2 spp. of *poiophylli*, by marginal strands in some spp. of *poiophylli* and by a cylinder uniting the vb's in 1 sp. of *septati*. **Air-canals** present centrally in many spp. or situated between vb's in others. Canals formed by breakdown of parenchymatous ground tissue. Central cavities either continuous or variously partitioned by transverse or longitudinal diaphragms or septa. Many spp. lacking air-canals. **Tannin** present in some epidermal cells and in isolated cells of ground tissue of a v. few spp.

Culm surface

Papillae present in *J. mutabilis* and around stomata of v. few spp. **Epidermis**

and **stomata:** frequently as on abaxial leaf surface of same sp. but cells often smaller, e.g. Figs. 2. D, H; 3. H. **Cuticular marks:** none seen.

Culm T.S.

Outline circular or oval, with or without grooves or ridges. **Cuticle** thin or slightly thickened, occasionally moderately thickened or thick; infrequently ridged. **Epidermis:** cells usually resembling those of abaxial surface of leaf of same sp.; mostly about as high as wide, sometimes 1½–3 times higher than wide and occasionally wider than high. **Papillae** noted in *J. mutabilis* Lam. **Stomata** usually superficial, infrequently sunken; similar to those of leaf of same sp. **Chlorenchyma.** Cells in 2–6 layers; cells either all palisade-like, all isodiametric or outer cells palisade-like and inner cells isodiametric. Protective cells present in some spp. of subgenus *thalassii* only. Pegs present on palisade and isodiametric cells of many spp. Wide, thin-walled cells present in certain areas in some spp., these often breaking down to form air-cavities. **Parenchyma sheath:** one layer of parenchymatous cells (frequently distinguishable from chlorenchyma cells by their slightly thicker walls) sometimes present in spp. having scl. cylinder to inside of chlorenchyma (see below).

Sclerenchyma present most frequently as cylinder 1–4 (–6) cells wide, situated between larger vb's, joining them together by their bundle sheaths. Subepidermal girders, usually triangular in outline, present in subgenera *genuini* and *thalassii* (as well as cylinder in some spp.). Parallel-sided bundle-sheath extensions reaching to epidermis, present in *J. tenuis* (*poiophylli*). Scl. represented by bundle sheaths only in one member of *graminifolii* (*J. planifolius*). Strands of scl., each surrounded by a parenchymatous sheath, present in central ground tissue of some spp. of subgenus *thalassii*. Fibres of girders and bundle sheaths narrow, with v. thick walls; those forming cylinders sometimes similar, but more often wide, with moderately thickened walls and more strongly resembling modified parenchyma cells. **Vascular bundles** frequently arranged in 2 rings, the smallest bundles outermost; sometimes in more than 2 rings, some free in central ground tissue, or in cortical wings of tissue. Small bundles normally situated to outer side of scl. cylinder and not embedded in it. Both larger and smaller bundles frequently similar to those of leaves of same sp. Amphivasal bundles present at nodes of spp. with cauline leaves. Mxv wall pitting mostly scalariform or scalariform-reticulate; perforation plates slightly or steeply inclined, normally scalariform or scalariform-reticulate, sometimes simple. Laux (1887) described vessels with forked ends in *J. effusus*. **Bundle sheaths:** O.S. present round free vb's, e.g. in *J. acutiflorus*, *J. effusus*, absent from those in scl. cylinder; cells rounded in T.S. and 1½–3 times longer than wide in L.S.; I.S. sclerenchymatous, distinct in bundles situated on outer side of scl. cylinder. Fibres normally in 2–8 layers at phloem pole, in 1–4 layers at xylem pole and 1–2 layers on flanks; often distinguishable from fibres of cylinder by their smaller size and thicker walls. **Central ground tissue** consisting of continuous parenchyma in some spp. but interrupted by one or more **air-canals** in others. Central air-canals traversed by diaphragms of stellate or lobed cells in some spp. Air-canals sometimes present between outer vb's.

Rhizome T.S.

Epidermis: cells variable, wider than high or up to twice as high as wide, walls slightly or moderately thickened. **Cortex.** (i) Outer part frequently parenchymatous, cells in 1–9 layers; in some spp. these layers bounded on inner side by 5–9 layers of cells with slightly thickened walls. (ii) Middle part usually composed of radiating plates of rounded cells; plates 1–5 cells wide, and up to 50 cells deep and separated by air-cavities formed by breakdown of wide, thin-walled cells; plates replaced by continuous parenchyma in some spp. (iii) Inner part made up of 1–2 layers of narrow, rounded, parenchymatous cells. **Endodermoid layer:** well developed as single layer of cells with thickened inner and anticlinal walls and thinner outer walls in most spp.; not distinct in *J. articulatus* and with all walls of similar thickness in *J. acutus*. **Pericycle** 1–3-layered when apparent, individual cells normally less tall than those of endodermoid layer, and with cellulose wall thickening.

Vascular bundles. Small vb's immediately to inside of pericycle, v. tortuous, often fusing and dividing. *J. gerardii* unusual in having a more or less continuous cylinder of xylem punctuated by strands of phloem in this region. Principal vb's amphivasal, scattered through central ground tissue. Some vb's in certain spp. with definable px. pole. Wall pitting of mxv's scalariform or alternate; perforation plates slightly oblique, scalariform, scalariform-reticulate, or simple. **Bundle sheaths** frequently of 1–2 (–4) layers of narrow fibres. **Central ground tissue** usually parenchymatous, or occasionally composed entirely of narrow, thick-walled cells.

Root T.S.

Epidermis: cells thin-walled, about as high to twice as high as wide. **Root-hairs** arising from cells similar to or narrower than those of epidermis, but distinguishable from them in most spp. by having slightly thickened walls. **Cortex.** (i) Outer part composed of 3–6 (–9+) layers of hexagonal cells with slightly or moderately thickened walls ('exodermis' of some other authors). In *J. gerardii*, cells of outermost layer with thicker outer than anticlinal and inner walls. (ii) Middle part composed of radiating plates of cells separated by air-spaces in most spp. (iii) Inner part composed of 2–4 or 7–8 layers of rounded, narrow cells; walls thin, becoming thickened in older roots. **Endodermis** frequently 1-layered, sometimes 2–3-layered; cells with thick inner and anticlinal walls and less thickened outer walls. **Pericycle** not always developed; usually 1-layered, composed of narrow cells with moderately thickened walls. **Vascular tissue:** mxv's arranged in a ring and sometimes accompanied by a central mxv in large roots. **Central ground tissue:** cells narrow, with thick walls. Fig. 4. G illustrates part of the root of *J. conglomeratus* in T.S.

Subgenus Descriptions: Points of Difference

1. *Junci subulati* Fr. Buchen.

Leaf T.S.

Outline circular. **Stomata** sunken, overarched by small papillae from margins of surrounding epidermal cells. **Sclerenchyma:** I.S. only. **Air-canals** central, diaphragms of stellate cells.

Culm T.S.

Outline circular. **Stomata** as in leaf. **Chlorenchyma:** palisade cells in 2–3 layers. **Sclerenchyma** as a cylinder. **Vascular bundles** in 2 rings.

MATERIAL EXAMINED

Juncus subulatus Forssk.: Sandwith, N.Y., 5444; Berrow, N. Somerset 10.9.58 (K). Leaf diameter 2 mm. Culm diameter 3 mm.

2. *Junci poiophylli* Fr. Buchen.

Leaf T.S.

Outline either flat, grooved, or arc-shaped with abaxial grooves, or trough-shaped; conspicuous square margins in *J. compressus*. **Epidermis:** adaxial surface reduced in area in *J. bulbosus* (Fig. 3. B) and *J. squarrosus*, and extending full width of leaf in *J. bufonius* (Fig. 2. B), *J. compressus* and *J. trifidus*. **Stomata** superficial. **Sclerenchyma** as strands in margins (except in *J. bulbosus* and *J. trifidus*). **Bundle sheaths:** I.S. of some vb's extending as girders to adaxial or abaxial epidermis. **Vascular bundles** grouped in threes (1 large, central, with 1 small on either side) in *J. gerardii* (Fig. 2. E) and in part in *J. compressus*, as for genus in other spp. **Air-canals** present between bundle groups in *J. compressus* (Fig. 2. K) and *J. gerardii*, between portions of mesophyll enclosing vb's in *J. bulbosus* and frequently present between vb's in *J. bufonius*, *J. tenuis* (Fig. 3. F) and *J. trifidus*.

Culm T.S.

Outline circular (Figs. 2. A, F, I; 3. A; Pl. IV. A) or oval (Fig. 3. E) with or without ridges. **Chlorenchyma** frequently composed of rounded cells. **Sclerenchyma** as a cylinder (Pl. IV. A); parallel-sided girders extending from opposite some larger vb's to epidermis in *J. tenuis* (Fig. 3. E), v. wide opposite larger vb's in *J. compressus*. **Vascular bundles** usually arranged in 2 rings, smallest vb's outermost (Fig. 2. A), but vb's occasionally in 1 ring (Fig. 3. E). Occasional free medullary vb's present in some spp. **Air-canal:** central cavity present in some specimens, e.g. *J. trifidus*.

MATERIAL EXAMINED

Juncus bufonius L.: (i) Cutler, D. F. 963; Halesowen, Worcestershire 1963 (S). (ii) Cutler, D. F. 584B; Staines Moor, Middlesex 1965 (S). (iii) Cutler, D. F. 476; New Forest, Hampshire 1964 (S). (iv) Metcalfe, C. R.; Frensham near Little Pond 7. 1963 (S). Leaf width 1–1·2 mm, thickness 0·2 mm. Culm diameter 0·3–1·0 mm.

J. bulbosus L.: (i) Turrill, W. B., and Montford, M. M. S.138; Studland Heath 22.7.36 (K). (ii) Cutler, D. F.; Bedgebury, Kent 23.6.65 (S). (iii) Kerry, Eire (K). Leaf diameter 0·7×0·5 mm. Culm diameter 1·0 mm. Rhizome diameter 3·0 mm.

J. compressus Jacq.: Hubbard, C. E. 12104; Cassington, Oxfordshire 17.7. 1944 (K). Leaf width 1 mm, thickness *c.* 0·1 mm. Culm diameter 1·1× 0·9 mm.

J. gerardii Lois.: (i) Dalby, K.; Shingle Street, Essex 11.10.63 (S). (ii) Blakelock, R. H.; Wrabness Salt Marsh, Essex 1952 (K). (iii) Isle of Grain

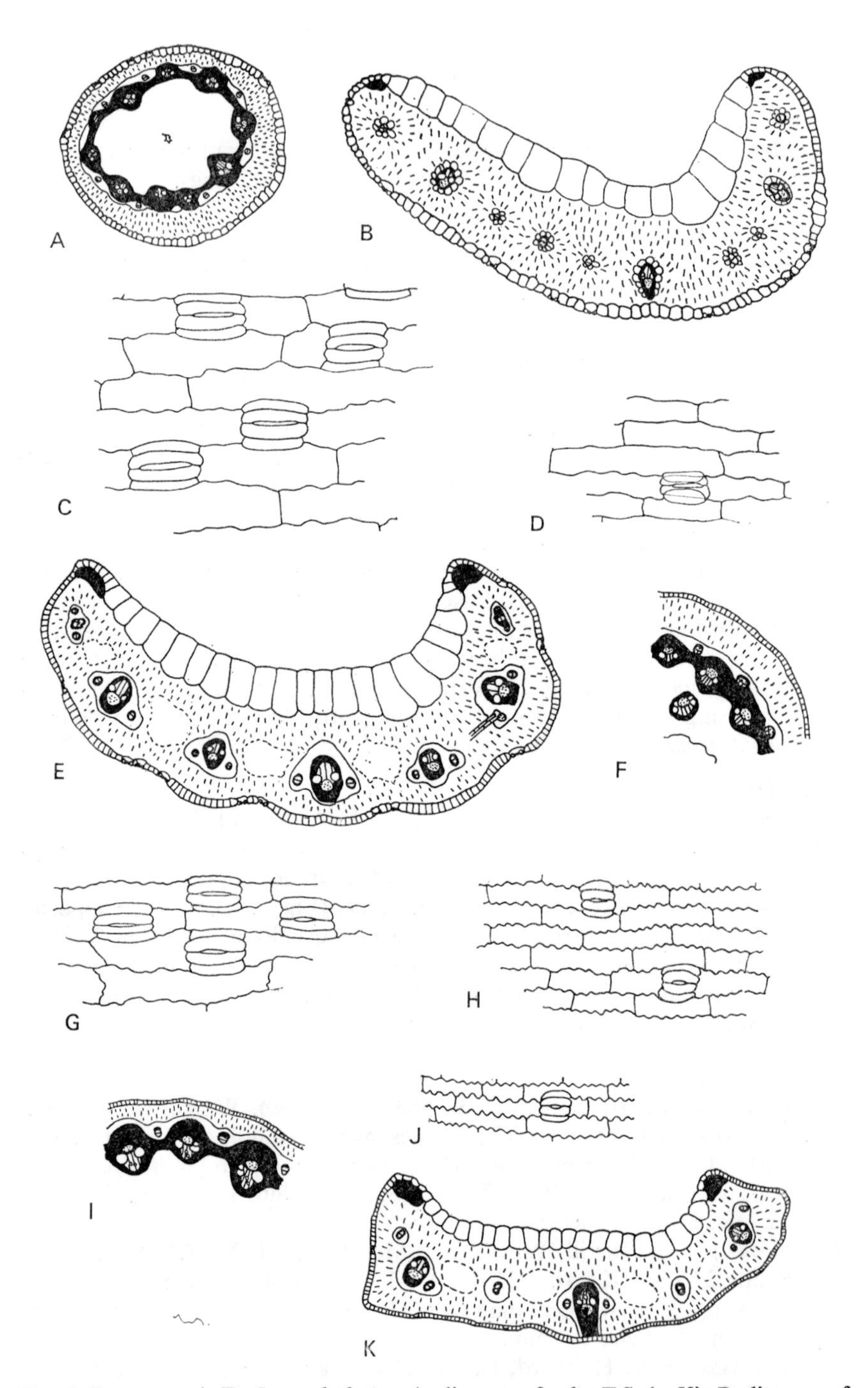

FIG. 2. Juncaceae. A–D, *Juncus bufonius*: A, diagram of culm T.S. (×50); B, diagram of leaf T.S. (×50); C, surface detail of leaf abaxial epidermis (×200); D, surface detail of culm epidermis (×200). E–H, *J. gerardii*: E, diagram of leaf T.S. (×50); F, diagram of sector of culm T.S. (×50); G, surface detail of leaf abaxial epidermis (×200); H, surface detail of culm epidermis (×200). I–K, *J. compressus*: I, diagram of sector of culm T.S. (×50); J, surface detail of leaf abaxial epidermis (×200); K, diagram of leaf T.S. (×50).

1951 (K). Leaf width 1–1·5 mm, thickness 0·2–0·4 mm. Culm diameter 1·3–1·5 mm. Rhizome diameter 1·5–2·5 mm. Root diameter 0·8 mm.

J. squarrosus L.: (i) Clarke, C. B. 1901 (K). (ii) Metcalfe, C. R.; Frensham Little Pond 7. 1963 (S). (iii) Cutler, D. F.; Scottish Lowlands 1.8.64 (S). Leaf width 0·8 mm, thickness 0·5 mm. Culm diameter 1·0×0·8 mm.

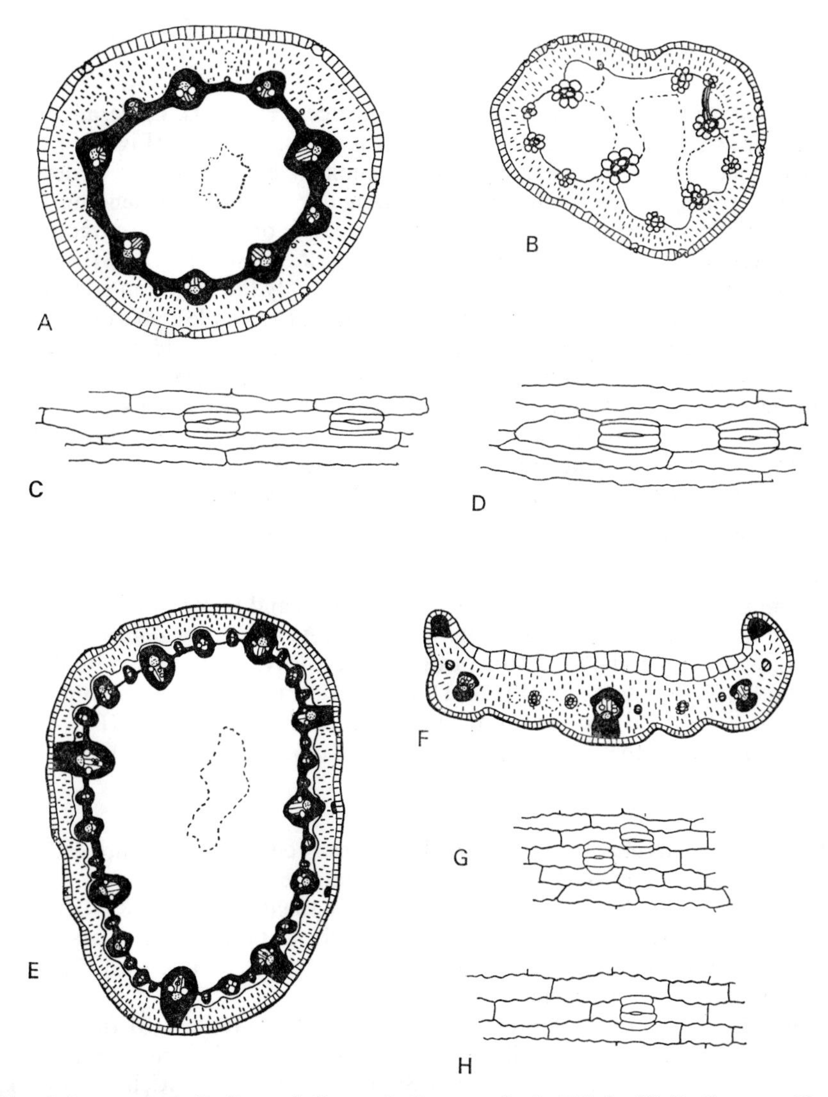

FIG. 3. Juncaceae. A–D, *Juncus bulbosus*: A, diagram of culm T.S. (×50); B, diagram of leaf T.S. (×50); C, surface detail of leaf adaxial epidermis (×200); D, surface detail of leaf abaxial epidermis (×200). E–H, *J. tenuis*: E, diagram of culm T.S. (×50); F, diagram of leaf T.S. (×50); G, surface detail of leaf abaxial epidermis (×200); H, surface detail of culm epidermis (×200).

J. tenuis Willd.: (i) Townshend, W.; Chester, Pennsylvania (K). (ii) Turrill, W. B., and Montford, M. M.; Richmond Park, Surrey 6.8.32 (K). (iii) Metcalfe, C. R.; Frensham Little Pond 7.1963 (S). Leaf diameter 1·0×0·8 mm. Culm diameter 1·5×1·0 mm.

J. trifidus L.: Townsend, C. C.; Glen Afric, Inverness 18.7.47 (K). Leaf width 0·6 mm, thickness 0·2 mm. Culm diameter 0·6 mm.

3. ***Junci genuini*** **Fr. Buchen.**

Leaf T.S.

Outline circular or oval, sometimes with ridges (Fig. 4. J). **Epidermis:** cells over girders taller than remainder in *J. inflexus*. **Stomata** (Fig. 4. E, K) usually superficial, occasionally slightly sunken. **Sclerenchyma:** girders or strands, triangular, subtriangular or rectangular in outline, extending from epidermis, penetrating chlorenchyma layers partly or completely; frequently opposite vb's and reaching the vb sheaths, e.g. *J. effusus* (Fig. 4. B); strands opposite most vb's in some spp. (Fig. 4. N) and only the larger ones in others (Fig. 4. J). **Vascular bundles** arranged in 2–3 rings, the smallest vb's outermost. **Chlorenchyma:** palisade cells in 2–8 layers and sometimes divided into sectors by scl. girders. **Parenchyma sheath** not developed. **Bundle sheaths:** O. S. parenchymatous, 1-layered; I.S. sclerenchymatous. **Ground tissue:** wide, rounded parenchyma cells surrounding vb's and vb sheaths. **Air-canals:** central air-canal present in many spp. **Diaphragms** transverse, composed of stellate cells, present in many spp., e.g. *J. conglomeratus*, *J. effusus*, and *J. inflexus*.

Culm T.S.

Similar to cylindrical leaf, but with more vb's and fewer stomata. **Outline** ridged in *J. inflexus* (markedly; Fig. 4. H; Pl. I. A) and *J. conglomeratus* (Fig. 4. A), with *c.* 14 and 40+ distinct ridges respectively. **Epidermis:** cells similar to those of leaf of same sp. (Fig. 4. D, I, M). **Air-canals:** in *J. inflexus* present in ring, 1 opposite and next to the inner side of each scl. strand (Fig. 4. H; Pl. I. A). Many other spp. with single, central cavity (Fig. 4. C, L).

MATERIAL EXAMINED

Juncus conglomeratus L.: Cutler, D. F.; Halesowen, Worcestershire 30.10.62 (S). Culm diameter 2×1·5 mm. Root diameter 1·1 mm.

J. effusus L.: (i) Milne-Redhead, E. 5645; Ampfield, S. Hants (K). (ii) Metcalfe, C. R.; Surrey, July 1951 (S). (iii) Cutler, D. F., Halesowen, Worcestershire 30.10.62 (S). (iv) Hort. Kew 17.2.66 (S). Leaf diameter 4·8 mm. Culm diameter 3–4·5 mm.

J. filiformis L.: (i) Lund; Waterhead Bay, Windermere 10.8.63 (S). (ii) Gregor, A. G. 12.7.19 (K). Leaf diameter 1×0·8 mm. Culm diameter 0·6–0·8 mm.

J. inflexus L.: (i) Turrill, W. B., and Summerhayes, V. S.; Richmond Park, Surrey 1.8.1930 (K). (ii) Cutler, D. F. 863; Halesowen, Worcestershire 1963 (S). (iv) Cutler, D. F. 581, 582; Staines Moor, Middlesex 1965 (S). (iv) Hort. Kew 1951 (S). Leaf diameter 1·0×0·6 mm. Culm diameter 2–2·2×1·0 mm.

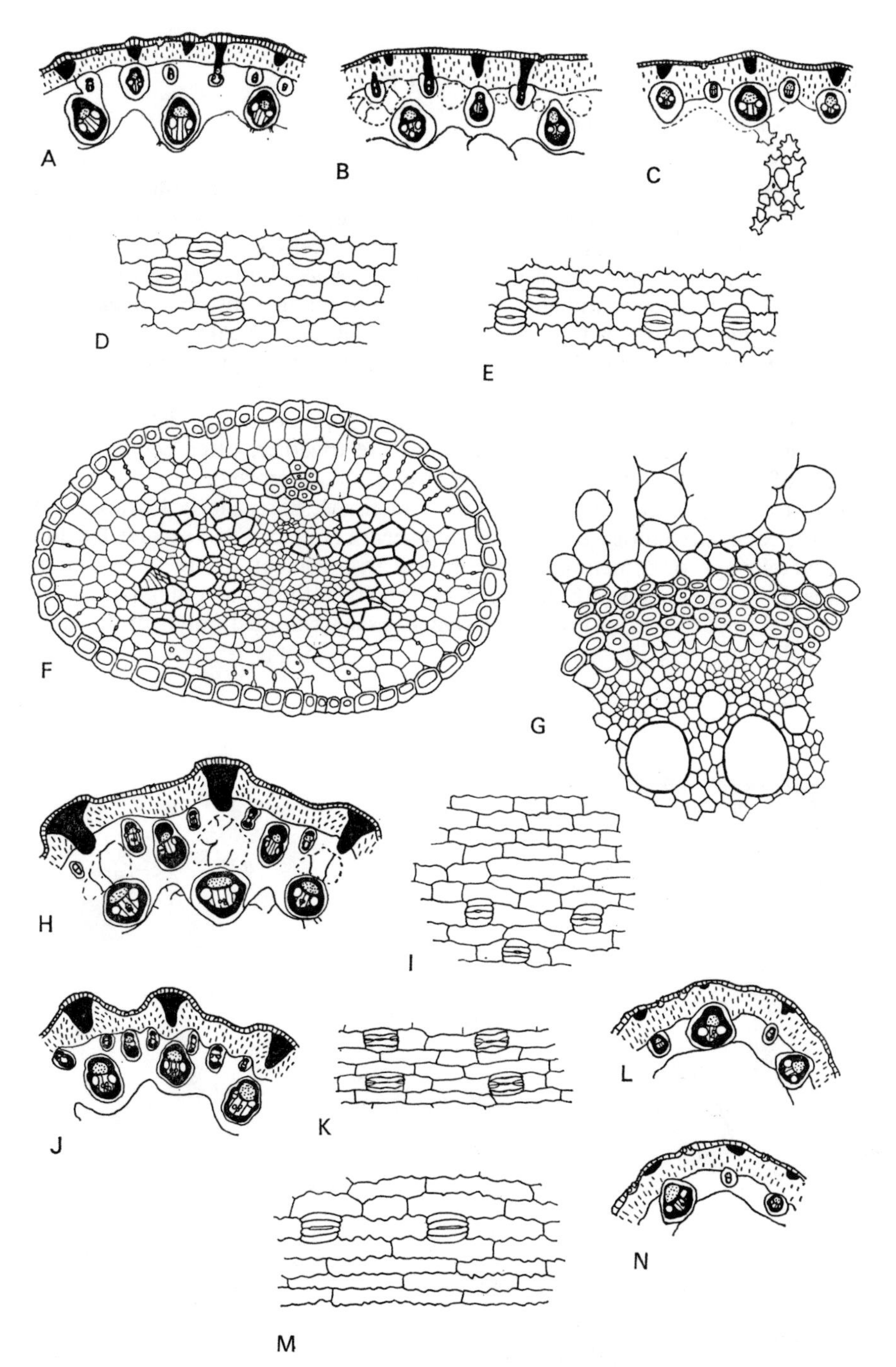

FIG. 4. Juncaceae. A, G, *Juncus conglomeratus*: A, diagram of sector of culm T.S. (×50); G, root T.S., detail of inner part of cortex, endodermis and vascular system (×200). B–F, *J. effusus*: B, diagram of sector of leaf T.S. (×50); C, diagram of sector of culm T.S. (×50); D, surface detail of culm epidermis (×200); E, surface detail of leaf epidermis (×200); F, leaf apex T.S. showing one sclerenchyma strand and arc of vascular tissue (×200). H–K, *J. inflexus*: H, diagram of sector of culm T.S. (×50); I, surface detail of culm epidermis (×200); J, diagram of sector of leaf T.S. (×50); K, surface detail of leaf epidermis (×200). L–N, *J. filiformis*: L, diagram of sector of culm T.S. (×50); M, surface detail of culm epidermis (×200); N, diagram of sector of leaf T.S. (×50).

4. *Junci thalassii* Fr. Buchen.

Leaf T.S.

Outline circular or oval, with slight ridges. **Cuticle** v. thick. **Stomata** (Fig. 5. B, E) sunken (e.g. *J. acutus*, Fig. 5. C) or superficial (e.g. *J. maritimus*). **Chlorenchyma.** 6–7 layers of elongated, palisade cells with slightly swollen-pointed ends. Cells not all perpendicular to surface, but tending to radiate from scl. girders. Procumbent protective cells lining substomatal cavities; cells with thick walls next to cavities and thin walls next to chlorenchyma cells. **Sclerenchyma.** Girders or strands next to epidermis triangular or irregular in outline, mostly broader than those of *J. genuini* and not penetrating through the entire thickness of chlorenchyma in most spp.; some opposite and connecting with bundle sheaths. Strands v. close together, often separated by width of stoma only (Fig. 5. A). Other strands of scl. more or less rounded or oval in outline scattered in central ground tissue, particularly towards leaf centre; each surrounded by parenchyma sheath (Fig. 5. D). **Vascular bundles** in several rings, or sometimes with elliptical arrangement. Sclereids present in phloem. **Bundle sheaths:** both inner sclerenchymatous and outer parenchymatous bundle sheaths present. **Ground tissue:** rounded-polyhedral, wide, parenchymatous cells.

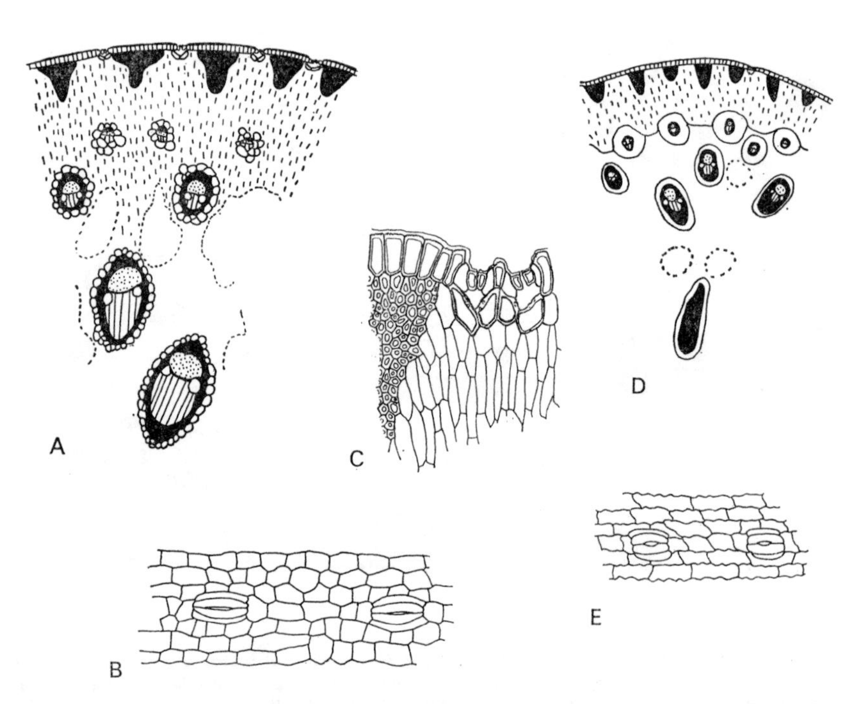

FIG. 5. Juncaceae. A–C, *Juncus acutus*: A, diagram of sector of leaf T.S. (×50); B, surface detail of leaf epidermis (×200); C, detail of leaf T.S. (×200). D–E, *J. maritimus*: D, diagram of sector of leaf T.S. (×50); E, surface detail of leaf epidermis (×200).

Culm T.S.

Outline as for leaf. **Cuticle** v. thick. **Epidermis:** cells over scl. girders and strands taller than remainder in *J. acutus*. **Stomata** and **chlorenchyma** as in leaf. **Sclerenchyma** as in leaf, but also represented by a cylinder 4–5 cells thick embedding most vb's. Individual cells with slightly thickened, lignified walls.

MATERIAL EXAMINED

Juncus acutus L.: (i) Mill, J. S.; nr. Sandwich Sept. 1860 (K). (ii) Wynn-Parry; Bangor 11.10.63 (S). (iii) Sprague and Hutchinson 127; Tenerife 1913 (K). Leaf diameter 3×2·5 mm. Culm diameter 3–4×1·7–2·5 mm.

J. maritimus Lam.: Last, G.V.C. 718/1; Heysham 1894 (K). Leaf diameter 2·5 mm. Culm diameter 2–2·5×1·5 mm.

5. *Junci septati* Fr. Buchen.

Leaf T.S.

Outline oval (Fig. 7. J, M) or flattened, frequently with slight ridges (e.g. *J. alpinoarticulatus* (*J. alpinus*), Fig. 7. G, *J. articulatus*, Fig. 6. C) or with pronounced ridges (e.g. *J. striatus*, Fig. 6. H); conspicuously flattened and ensiform in *J. xiphioides* var. *triandra* (Fig. 9. F, H, p. 40). **Epidermis:** cells exclusively abaxial; in surface view often with wavy walls, cells 2–12 times longer than wide, Figs. 6. F, J; 7. E, H, I; 9. G (2–15 times in *J. alpinoarticulatus*); in T.S. usually between 1½ and 2 times higher than wide. **Papillae** present in *J. mutabilis*. **Stomata** superficial in all spp. examined (Figs. 6. A; 7. B, D) except *J. xiphioides* var. *triandra*. Guard cells of most spp. with outer, cuticular lips, these particularly pronounced in *J. scheuchzerioides* and *J. subnodulosus*; small lips frequently present at inner aperture. **Chlorenchyma:** cells of outer 2–3 layers usually palisade-like, with pegs; cells of inner layers more or less isodiametric, frequently lobed.

Vascular bundles in 1 or 2 rings (Fig. 9. E, p. 40); large vb in keel and pair of small marginal vb's present in *J. xiphioides* var. *triandra*; keel bundle not conspicuous in other spp. Phloem separated from xylem by 1–2 layer(s) of narrow, thick-walled cells in *J. acutiflorus*, *J. articulatus* (weak), *J. scheuchzerioides*, *J. subnodulosus*, and *J. xiphioides* var. *triandra*. Flanking mxv's narrow in *J. scheuchzerioides*. **Bundle sheaths:** O.S. present round some vb's in all spp. examined. **Sclerenchyma** restricted to I.S. **Air-canals:** a single central canal in most spp., e.g. Fig. 6. C; several, separated by longitudinal septa in *J. prismatocarpus* (Fig. 7. J, K). **Diaphragms** transverse, seen in most spp. Cells stellate, sub-stellate or flattened, and accompanied by large intercellular spaces. Diaphragms frequently associated with transverse vb's; concave in *J. subnodulosus*. Longitudinal septa present in *J. prismatocarpus* (Fig. 7. K) and *J. subnodulosus*.

Culm T.S.

Outline circular or oval (Fig. 6. D), with or without slight ridges; ridges irregular and pronounced in *J. striatus* (Fig. 6. G). **Papillae** present in *J. mutabilis*. **Epidermis:** see Figs. 6. E, I; 7. C; 9. D. **Chlorenchyma:** outer layers consisting of palisade cells (Fig. 6. B), layers of more or less isodiametric cells more numerous than in leaf, sometimes all cells isodiametric, e.g. *J. prismatocarpus* and *J. scheuchzerioides*. **Sclerenchyma** as cylinder joining some or all

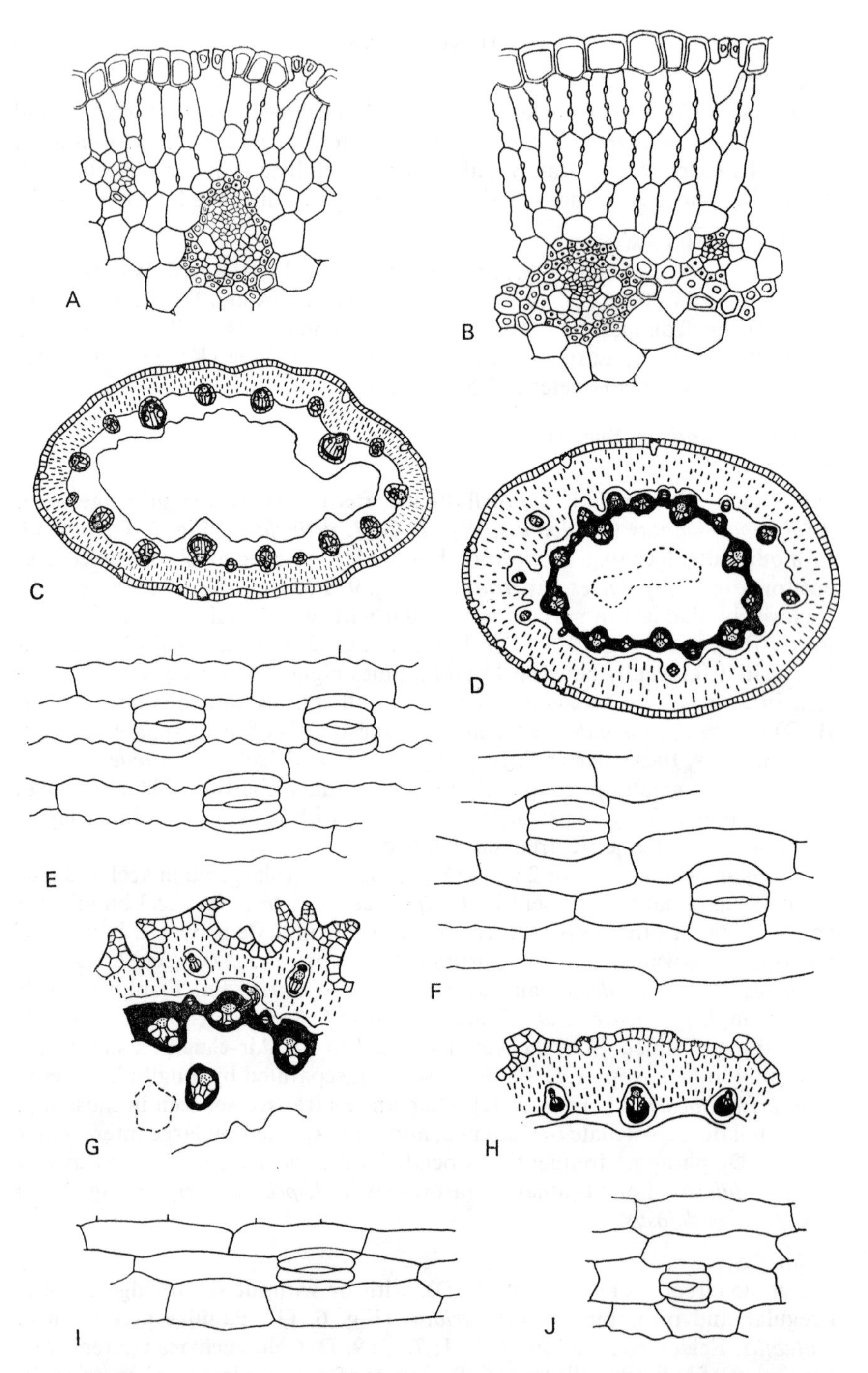

FIG. 6. Juncaceae. A–F, *Juncus articulatus*: A, detail of leaf T.S. (×200); B, detail of culm T.S. (×200); C, diagram of leaf T.S. (×50); D, diagram of culm T.S. (×50); E, surface detail of culm epidermis (×320); F, surface detail of leaf epidermis (×320). G–J, *J. striatus*: G, diagram of sector of culm T.S. (×50); H, diagram of sector of leaf T.S. (×50); I, surface detail of culm epidermis (×200); J, surface detail of leaf epidermis (×200).

inner bundle sheaths in all spp. (Fig. 7. A, F, L) but cylinder weak in *J. subnodulosus* (Fig. 9. C, p. 40). **Vascular bundles** arranged in 2 or more rings (Fig. 6. D). **Air-canals** present in outer cortex of *J. prismatocarpus*. **Central ground tissue** frequently composed of parenchyma, with no central canal (except in *J. articulatus, J. mutabilis* and upper internodes of *J. striatus*).

Rhizome T.S.

J. acutiflorus **and *J. articulatus*.**

Epidermis: cells 1–2 times higher than wide, walls thin (*J. acutiflorus*) or outer walls moderately thickened (*J. articulatus*). **Cortex** differentiated into 3 regions, as in generic description; inner cortex 1–3-layered. **Endodermoid layer** indistinct in *J. articulatus* (material sectioned near to culm base). **Pericycle** distinct only in *J. acutiflorus*, where 2–3-layered; cells narrow, thin-walled.

Root T.S.

Seen in *J. acutiflorus* only. As for generic description. **Inner cortex** 7–8-layered. **Pericycle** 1-layered.

MATERIAL EXAMINED

Juncus acutiflorus Ehrh. ex Hoffm.: (i) Turrill, W. B.; Wimbledon Common 16.8.29 (K). (ii) Cutler, D. F. 3365; Bedgebury, Kent (S). (iii) Kerry, Eire 1951 (S). Leaf diameter 2×1 mm. Culm diameter 2·5×1·8 mm. Rhizome diameter 3·5 mm. Root diameter 1·5 mm.

J. alpinoarticulatus Chaix (*alpinus* Vill.): (i) Townsend, C. C.; Winch Bridge, Yorkshire 19.9.58 (K). (ii) Hort. Kew 1964 (S). Leaf diameter 1·5–2·5× 0·8–1 mm. Culm diameter 1·5–2×1 mm. Rhizome diameter 3 mm.

J. articulatus L.: (i) Metcalfe, C. R.; Milford-on-Sea 1951 (S). (ii) Surrey, July 1951 (S). (iii) Cutler, D. F. 538B; Staines Moor, Middx. 1965 (S). Leaf diameter 1–2×0·8–1 mm. Culm diameter 1–2×0·8–1 mm. Rhizome diameter 2×1–1·2 mm.

J. mutabilis Lam. (*pygmaeus* Rich.): Nicholson, W. E.; The Lizard, Cornwall (K). Leaf diameter 0·6×0·4 mm. Culm diameter 0·8×0·7 mm.

J. prismatocarpus R.Br.: Cheadle, V. I. CA 147; Brisbane, Australia 6.10.59 (S). Leaf diameter 3×1·2 mm. Culm diameter 3×1·5 mm.

J. scheuchzerioides Gaudich.: Vallentin, E.; Shallow Bay Valley, S. America 1911 (K). Leaf diameter 0·7×0·5 mm. Culm diameter 1×0·6 mm.

J. striatus Schousboe: Duval-Jouve, H.; Montpellier, 10/71 (K). Leaf diameter 2×1·5 mm. Culm diameter 1·2×1 mm.

J. subnodulosus Schrank: (i) Wynn-Parry; Bangor, Wales 11.10.63 (S). (ii) Hunt, E.; Chesham Aug. 1869 (K). (iii) Summerhayes, V. S. 1130; N. Somerset 27.8.1930 (K). Leaf diameter 2×1·5 mm. Culm diameter 2×1·5 mm.

J. xiphioides E. Mey. var. *triandra* Engelm.: Fagerland and Mitchell 820; Hawaii Nat. Park (S). Leaf width 5 mm, thickness 1 mm.

6. *Junci alpini* Fr. Buchen.

Leaf T.S.

Outline oval (*J. biglumis*, Fig. 8. A), irregular (*J. triglumis*, Fig. 8. F), flattened or arc-shaped (*J. castaneus*). **Epidermis.** Surface view: abaxial cells

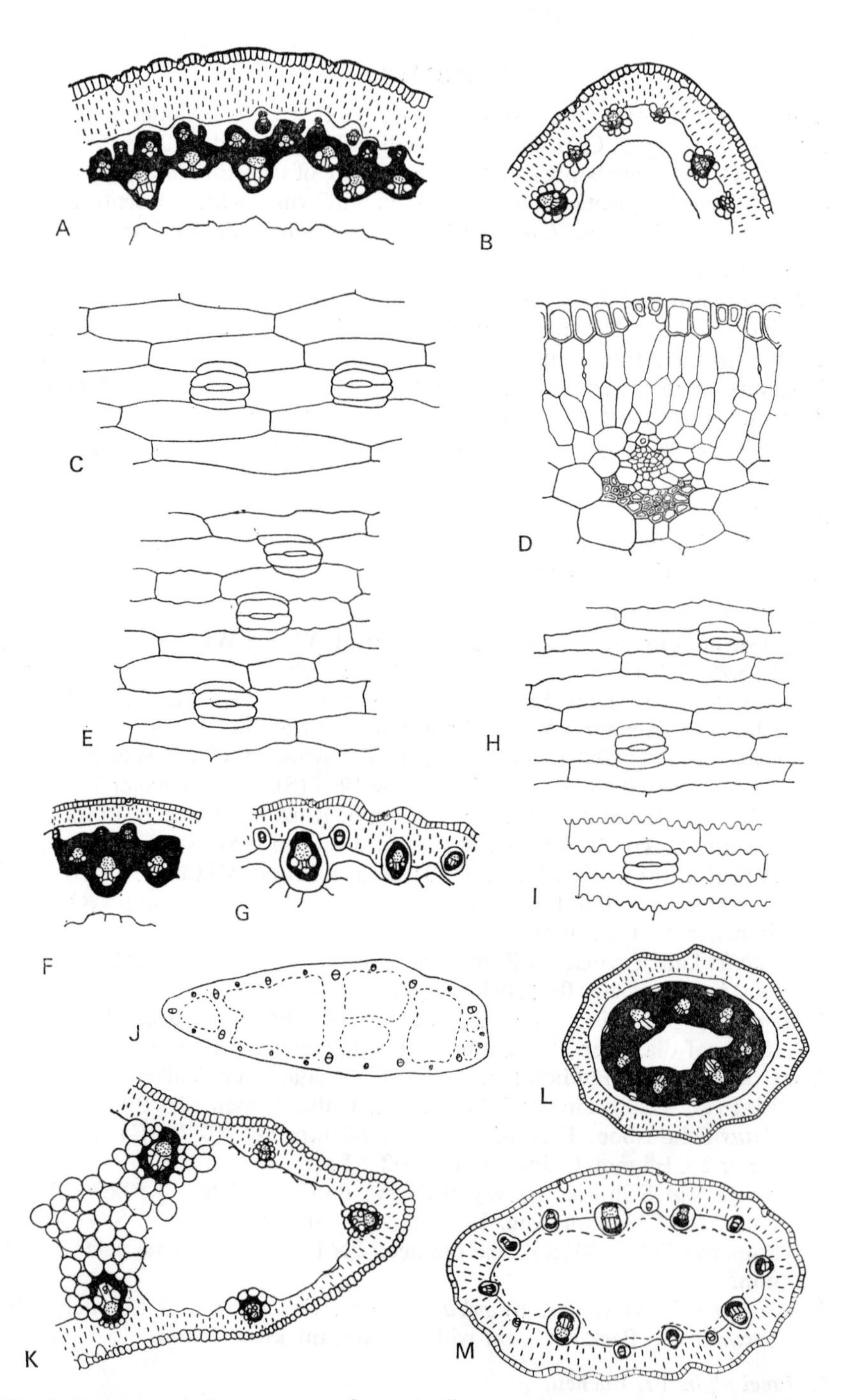

FIG. 7. Juncaceae. A–E, *Juncus acutiflorus*: A, diagram of sector of culm T.S. (×50); B, diagram of sector of leaf T.S. (×50); C, surface detail of culm (×200); D, detail of leaf T.S. (×200); E, surface detail of leaf (×200). F–I, *J. alpinoarticulatus*: F, diagram of sector of culm T.S. (×50); G, diagram of sector of leaf T.S. (×50); H, surface detail of leaf epidermis, low focus (×200); I, as H, high focus (×200). J–K, *J. prismatocarpus*: J, diagram of leaf T.S. (×12); K, diagram of part of leaf T.S. (×200). L–M, *J. scheuchzerioides*: L, diagram of culm T.S. (×50); M, diagram of leaf T.S. (×50).

2–6 times longer than wide except in *J. biglumis* where up to 15 times longer than wide. T.S.: abaxial cells mostly as high as wide, some $1\frac{1}{2}$ times wider than high. Adaxial present in *J. castaneus* only, central adaxial cells about twice as high as abaxial cells, but grading down in size towards leaf margins. **Stomata** superficial; guard cells with prominent lips at inner and outer apertures. In surface view, subsidiary cells frequently aberrant in *J. biglumis* (Fig. 8. D). **Chlorenchyma** completely cylindrical in *J. biglumis* and *J. triglumis*, exclusively abaxial in *J. castaneus*; cells of outer 4–5 layers more or

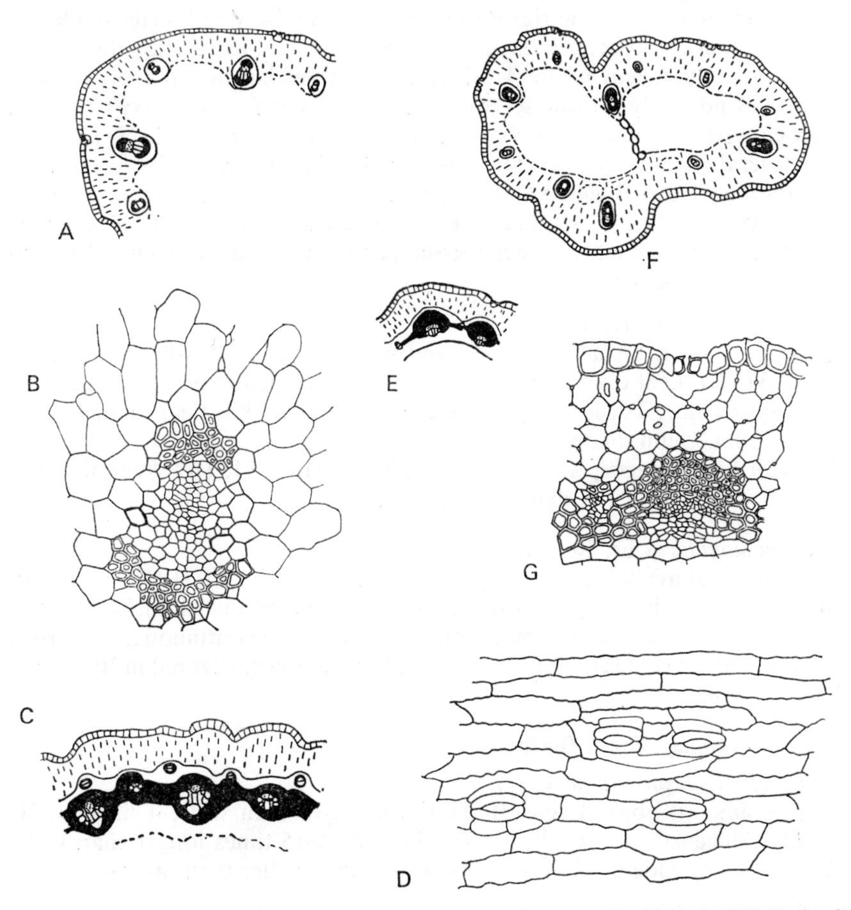

FIG. 8. Juncaceae. A–D, *Juncus biglumis*: A, diagram of sector of leaf T.S. (×50); B, detail of leaf vascular bundle T.S. (×200); C, diagram of sector of culm T.S. (×50); D, surface detail of leaf epidermis, note aberrant subsidiary cells (×200). E–G, *J. triglumis*: E, diagram of sector of culm T.S. (×50); F, diagram of leaf T.S. (×50); G, detail of culm T.S. (×200).

less palisade-like, those of inner 1–2 layers isodiametric. **Vascular bundles** arranged in 1-layered oval ring in *J. biglumis* and *J. triglumis* and in 1-layered arc in *J. castaneus*; flanking mxv's sometimes medium-sized, often narrow (Fig. 8. B). **Bundle sheaths:** O.S. parenchymatous, 1-layered; I.S. represented by weak fibre caps. **Sclerenchyma** restricted to bundle sheaths. **Air-canals:** single central canal in *J. biglumis*, 2 in centre in *J. triglumis*, and many, alternating with vb's in *J. castaneus*. **Diaphragms:** 1, median longitudinal in *J. triglumis*, and several transverse in *J. biglumis*.

Culm T.S.

Outline more or less triangular in *J. biglumis*, circular-oval with slight, domed ridges in *J. castaneus* and *J. triglumis*. **Epidermis** and **stomata** as for leaf of same sp. **Chlorenchyma** 4–6-layered, outer cells somewhat palisade-like, inner more isodiametric. **Parenchyma sheath** 1-layered, cells similar to those of inner chlorenchyma layer. **Sclerenchyma**: cylinder weakly developed, except in *J. castaneus* where 4–5-layered. **Vascular bundles** in 1 or 2 rings; small vb's normally outside scl. cylinder (Fig. 8. C); flanking mxv's narrow, 1 on either flank in *J. biglumis*, 1 or 2 arranged tangentially in *J. triglumis* and several in *J. castaneus*. Mx vessels with oblique, scalariform-reticulate perforation plates. **Bundle sheaths** sclerenchymatous; fibres in few layers at poles except in *J. triglumis* (Fig. 8. E, G), where in 7–8 layers at phloem pole of larger bundles. **Central ground tissue** parenchymatous, breaking down to form central air-cavity.

MATERIAL EXAMINED

Juncus biglumis L.: Mackechnie, R. and Wallace, E. C.; Mid-Perth, Scotland 1937 (K). Leaf diameter 4×0·1 mm. Culm diameter 0·9×0·7 mm.

J. castaneus Sm.: Turner, D.; Ben Lawers, Scotland (K). Leaf diameter 2×0·4 mm. Culm diameter 1·2 mm.

J. triglumis L.: Gregor, A. G.; Ben Laoigh, Scotland 1915 (K). Leaf diameter 0·8×0·5 mm. Culm diameter 0·5 mm.

7. *Junci singulares* Fr. Buchen.

No material available for study. According to Buchenau (1890) similar to *graminifolii*, but laterally compressed leaves arranged distichously on stem. Without transverse diaphragms; central ground tissue continuous; numerous longitudinal air-canals present between vb's in periphery (figured in Buchenau 1906, p. 238).

8. *Junci graminifolii* Fr. Buchen.

Leaf surface (*J. planifolius* only)

Epidermis: (i) abaxial cells 10–14 times longer than wide, 4-sided, walls slightly thickened, not wavy; (ii) adaxial cells 3–15 times longer than wide, about twice as wide as abaxial; cells at margins smaller than others.

Leaf T.S.

Outline: in *J. planifolius* V-shaped (Fig. 9. B), in *J. falcatus* trough-shaped, margins drawn out; in *J. planifolius* of more or less even thickness, but rapidly tapering to margins, one half of lamina wider than other; in *J. capitatus* more or less semi-circular, with slight adaxial groove (becoming flatter towards leaf tip). **Epidermis:** adaxial cells about twice size of abaxial cells except in *J. falcatus* where all cells similar, with thin or slightly thickened walls, outer walls thickest; cells 1½–2 times higher than wide in *J. planifolius* and about as high as wide in *J. capitatus* and *J. falcatus*. **Stomata** superficial, abaxial; guard cells with slight lip at outer aperture. **Chlorenchyma:** palisade or sub-palisade cells in outer 2–3 layers, inner cells about as high as wide. **Vascular bundles:** *J. capitatus* with 3 (5 nearer leaf base) in specimen examined, central bundle largest, none with conspicuous mxv's. *J. planifolius* with 21 in specimen examined, large and small usually alternating, 13 to one side of fold, 8 to other, median fold in lamina not opposite a bundle. *J. falcatus* with 11 vb's, no conspicuous median vb. **Bundle sheaths:** O.S. parenchymatous, 1-layered more or less continuous; I.S. fibres in 1–2 layers in *J. planifolius*, 1–4 layers in *J. falcatus*, and as 2-layered cap at phloem pole and 3–4-layered cap at xylem pole of central vb only in *J. capitatus*. **Sclerenchyma** confined to bundle sheaths. **Air-canals** absent from *J. capitatus*; developed by breakdown of parenchyma between vb's in *J. planifolius*. **Ground tissue:** parenchyma, cells small next to abaxial epidermis and larger towards centre of leaf.

Culm surface

Epidermis: cells 4–6 times longer than wide in *J. capitatus*, 10–14 in *J. planifolius*; transverse end walls raised; walls thin, wavy.

Culm T.S.

Outline oval (*J. falcatus*), oval, with ridges (*J. planifolius*, Fig. 9. A), or more or less polygonal (*J. capitatus*). **Epidermis.** Cells mostly as high as wide; those next to stomata sometimes shorter than others. Outer wall moderately thickened in *J. planifolius* and slightly thickened in *J. capitatus* and *J. falcatus*; other walls thin. As seen in L.S., end walls markedly inclined, one cell overlapping the next at one end in *J. capitatus*. **Chlorenchyma** composed of cells in 2–5 layers; cells slightly higher than wide, with pegs or lobes; air-spaces between cells v. wide as seen in L.S. **Parenchyma sheath** 1-layered opposite larger vb's, often 2–3-layered between them. **Sclerenchyma** restricted to bundle sheaths in *J. planifolius*, as 2–4-layered cylinder between larger vb's in *J. capitatus* and *J. falcatus*. Parallel-sided ridges, 8–9 fibres high, with rounded ends, opposite larger vb's in *J. capitatus*. **Vascular bundles:** large alternating with small; small situated outside scl. cylinder in *J. capitatus* and *J. falcatus*; large in *J. planifolius* and *J. falcatus* with 1 wide mxv per flank; 1–2 narrow to medium-sized mxv's per flank in *J. capitatus*; phloem pole abutting directly on to xylem. **Bundle sheaths:** O.S. parenchymatous, complete in *J. planifolius* and as caps to smaller vb's in *J. capitatus*; I.S. sclerenchymatous, as caps to smallest vb's in *J. planifolius*, completely encircling larger vb's; fibres of sheath to vb's in *J. capitatus* and *J. falcatus* easily distinguishable from those of cylinder by their narrower diameter and thicker walls. **Central ground tissue** composed of wide, thin-walled parenchyma cells. **Tannin** present as dark, granular material in some isolated cells of parenchyma sheath of *J. capitatus*.

MATERIAL EXAMINED

Juncus capitatus Weig: Gewgrays; The Lizard, Cornwall (K). Leaf width 0·4 mm, thickness 0·2 mm. Culm diameter 0·4×0·3 mm.

J. falcatus E. Mey.: Johnson, L. A. S., and Constable, E. F. 19285; Kosciusko District, Australia 22.1.59 (K). Leaf width 5·0 mm, thickness 0·1 mm. Culm diameter 0·9×0·6 mm.

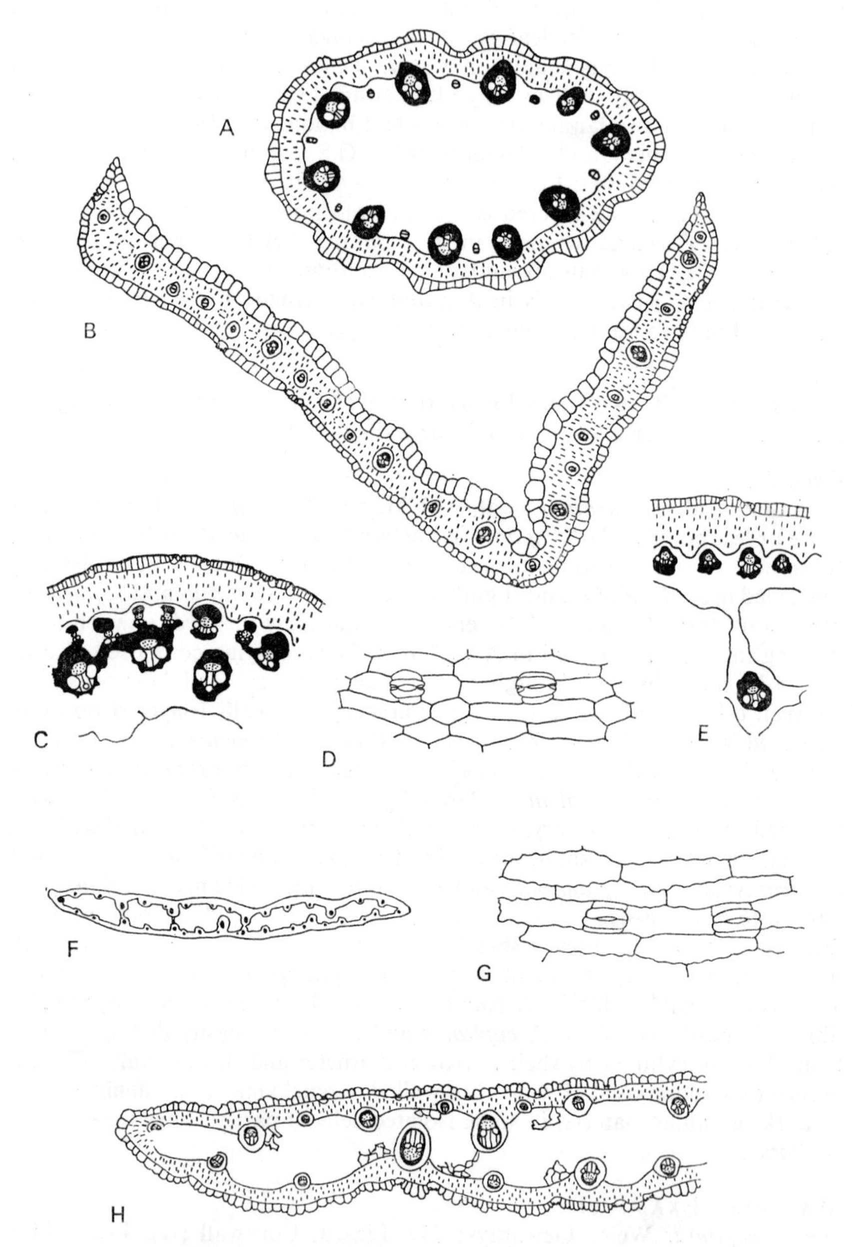

FIG. 9. Juncaceae. A, B, *Juncus planifolius*: A, diagram of culm T.S. (×50); B, diagram of leaf T.S. (×35). C–E, *J. subnodulosus*: C, diagram of sector of culm T.S. (×50); D, surface detail of culm epidermis (×200); E, diagram of sector of leaf T.S. (×50). F–H, *J. xiphioides* var. *triandra*: F, diagram of leaf T.S. (×8); G, surface detail of leaf epidermis (×200); H, diagram of part of leaf T.S. (×35).

J. planifolius R.Br.: (i) Fagerland and Mitchell 546; Hawaii Nat. Park (S). (ii) Cheadle, V. I. CA322; Sydney, Australia 26.10.59 (S). Leaf width $3{\cdot}5 \times 0{\cdot}1$ mm. Culm diameter $1 \times 0{\cdot}5$ mm.

TAXONOMIC NOTES AND INFORMATION FROM THE LITERATURE

Since the division of *Juncus* into 8 subgenera by Buchenau (1875, 1890), several other systems of classification have been suggested. Kuntze (1903) proposed a series of sections based on various genera, subgenera, and sections of earlier authors. Vierhapper (1930) also classified *Juncus* species in sections. He based them on the subgenera and names given by Buchenau, misleadingly attributing the sectional rank to Buchenau. The sections proposed by Kuntze take priority over those of Vierhapper.

Krechetovich and Goncharov (1935) divided the genus into subgenera but in doing so they disregarded Buchenau's subgenera and raised the sections proposed by Kuntze to subgeneric rank. They also proposed some new subgenera. Weimarck (1946) described 7 sections in *Juncus* from Cape Province, South Africa, basing them in part on those recognized by Vierhapper and in part on those described by Krechetovich and Goncharov, and including 2 new sections.

The limits of the subgenera and sections in the various publications sometimes correspond with those of Buchenau's subgenera and are sometimes quite different from them. The way in which anatomical evidence supports or fails to support the various treatments of the genus is discussed in the following paragraphs.

1. *Junci subulati*

Buchenau (1906) figured a culm of *J. subulatus* Forssk. in T.S. showing stellate cells in the pith. Haslinger (1914) stated that there is no scl. cylinder in the culm and this is confirmed by Buchenau's figure. However, a scl. cylinder was found in material examined for the present work. Satake (1933), writing about Japanese Juncaceae, noted that *J. subulatus* is correctly separated from *J. promineus* and *J. falcatus* (*graminifolii*), basing this opinion on anatomical characters of the peduncle. Peisl (1957) considered the *subulati* to be closely related to the *genuini*, but distinguishable from them by their sunken stomata. However, this is not a reliable distinction, since Johnson (1963) records that *J. australis* has sunken stomata and Edgar (1964) observed the same for *J. sarophorus*; both of these are in the *genuini*. The *genuini* lack a scl. cylinder in the culm, however, and this may provide the anatomical distinction between the groups *genuini* and *subulati*.

2. *Junci poiophylli*

Raunkiaer (1895) stated that the leaves of *J. tenageia* Ehrh. are flat in T.S. Buchenau (1890) illustrated the leaf T.S. of *J. squarrosus* and *J. chamissonis* (probably *J. imbricatus* var. *chamissonis*). Each of these has marginal scl. strands, *J. chamissonis* has several abaxial scl. girders. Buchenau (1906) illustrated the leaf T.S. of *J. trifidus* L. which lacks marginal scl. strands, but has adaxial scl. girders. He also illustrated the leaf T.S. of *J. bufonius* L., *J. capillaceus* Lam., *J. plebejus* R. Br., and *J. secundus* Beauv. All of these have scl. strands in the leaf margins and all have abaxial girders extending either from the median vb alone or from several vb's. *J. bufonius* seen by the present author lacks girders. The leaf of *J. capillaceus* in T.S. is more or less semi-circular at the base, but becomes more rounded distally; it has pronounced ridges on the abaxial surface, opposite each of which is a scl. girder from a vb sheath. All the vb's have girders. The culm T.S. of this species (also figured) is oval, and also has several pronounced ridges. The scl. cylinder has a girder or strand opposite each vb but there are more girders than strands. The vb's are arranged in 1 ring, and all are embedded in the scl. cylinder. Illustrations were also given of the root in T.S. of *J. bufonius* L., *J. homalocaulis* F. v. Muell. (= *J. plebejus* R. Br.) and of *J. squarrosus* L., in Buchenau's 1890 monograph.

Castillon (1926) gave illustrations of leaf and culm T.S. of *J. venturianus* Castillon, the anatomy of which is similar to that of many other species of *poiophylli*, that is with marginal scl. strands, and abaxial scl. girders extending from all vb's to the epidermis. The vb's are not grouped in threes. There is a conspicuous, double hypodermis adaxially. The culm is more or less rect-angular in outline; it lacks scl. girders; the vb's are arranged in two rings.

The *poiophylli* include species showing a variety of characters of leaf anatomy. Species can be separated on the basis of presence or absence of marginal scl. strands and sometimes by presence or absence of abaxial scl. girders, although this last character appears to be variable.

Peisl (1957) noted that this group, together with the *genuini* and *subulati*, includes those species which have bracts on the inflorescence axis, but that the species of *poiophylli* have dorsiventrally flattened leaves in which the cells of the adaxial epidermis are larger than those of the abaxial epidermis. He con-sidered the anatomical organization of the *poiophylli* to be v. similar to that of the *genuini*, particularly in species such as *J. setaceus* Rost. where the adaxial leaf surface is v. reduced. He mentioned *J. squarrosus* L., *J. imbricatus* Laharpe, and *J. plebejus* R. Br. as tending to have a reduced adaxial surface. He emphasized the probable close affinities of these groups in an account of seedling development and form.

Guédès (1967) illustrated the leaf of *J. squarrosus* in T.S., cut at different levels, to show the relation of the true margins to the wings of tissue formed at the leaf base.

The *poiophylli* resemble the *alpini* and *graminifolii* in having both adaxial and abaxial leaf surfaces and differ from the other subgenera which have only an abaxial epidermis. The vestigial adaxial epidermis of species of *genuini* may indicate the affinity of that subgenus with *poiophylli*.

3. *J. genuini*

It has been disputed for many years whether species of this group have leaves, or if leaves are replaced by sterile stems. Irmisch (1855) and Adamson (1925) argue that the *genuini* do have leaves. Buchenau (1890) noted a slight groove on the adaxial side of the leaf of *J. jacquinii* L. Peisl (1957) recorded that in the fresh state, the leaf of *J. jacquinii* has a narrow, dark longitudinal stripe on the adaxial side, which consists exclusively of large, colourless 'bulli-form' cells. He also observed a large median vb representing the midrib, joined to the abaxial epidermis by a scl. girder and that 2 adaxial, sub-epidermal scl. strands mark the borders of the adaxial surface.

The arc of bundles near the tip of the organ subtending the inflorescence could be seen in T.S. in most species of the *genuini* examined, e.g. *J. effusus*, Fig. 4. F, p. 31. If this organ were 'stem', a complete ring of bundles would be expected in this position. An arc of bundles is also to be seen at the tips of the so-called 'sterile stems' of some species described in Edgar's (1964) paper. It seems to the present author that the organ subtending the inflorescence, and extending above it in the *genuini*, should for these reasons be regarded as a foliar bract, and that many of the so-called 'sterile stems' of species of the group are, in fact, leaves. Guédès (1967) also provides evidence to support this point of view.

Buchenau (1906) illustrated the stem of *J. uruguensis* Griesb. in T.S. Although unsure of the taxonomic position of this plant, he thought it might have affinities with members of the *poiophylli.* It is unusual anatomically among the *genuini*, since it has numerous rectangular girders extending from the scl. sheath to the epidermis. It is, however, similar to *J. kleinii* Barros (Barros 1962), which is also placed in the *genuini.* Since these two South American species do show anatomical similarities to some members of the *poiophylli* it seems that their present taxonomic position may be incorrect.

Buchenau (1906) also illustrated in T. S. culms of *J. beringensis* Buch., *J. patens* E. Mey., *J. pallidus* R. Br., *J. drummondii* E. Mey., and *J. glaucus* Ehrh. (= *J. inflexus* L.) and the leaf tip (surface view) of *J. arcticus* Willd. In *J. beringensis* the scl. strands next to the epidermis usually extend across the chlorenchyma broadening when they reach the sclerenchymatous bundle sheaths of most of the vb's in the outermost ring of vb's. *J. patens* has similarly arranged girders, but they are opposite to and reach all vb's, tapering towards them. The vb's are arranged in 1 ring. *J. pallidus* has triangular strands opposite to most of the large vb's, but they do not completely traverse the chlorenchyma. The vb's are arranged in 2–3 rings and have well-developed scl. bundle caps, but only a few fibre layers on the flanks. The epidermal cells opposite the scl. strands are taller than the others. *J. glaucus* (*inflexus*) is illustrated with 1 ring of vb's. The scl. strands are triangular–bell-shaped, widening rapidly next to the epidermis; they are opposite to the larger vb's, but whilst completely traversing the chlorenchyma they are separated from the vb's by oval areas of wide, thin-walled cells which break down to form canals. The vb's are each surrounded by a cylinder of parenchyma cells with slightly thickened walls. In the specimen examined at Kew the vb's are in 2–3 rings, and air-canals do not occur opposite every scl. strand. *J. drummondii* lacks scl. strands in the cortex; the vb's are arranged in 1 ring. In a previous paper (1890) Buchenau stated that the *genuini* with smooth stems (i.e. *genuini laeves*) have no subepidermal sclerenchyma.

The position, outline, and degree of penetration of the scl. strands are regarded as of taxonomic significance in the *genuini.* For example, Edgar (1964) studied the New Zealand species, which all belong to Buchenau's *J. genuini valleculati*, and have strands next to the epidermis. She observed that pointed-ended strands or girders occur in *J. effusus*, *J. pallidus*, *J. procerus*, *J. gregiflorus*, and *J. conglomeratus*; in these species they are opposite most vb's and do not always completely penetrate the chlorenchyma. In *J. pauciflorus* and *J. digestus* Edgar found strands and girders with rounded ends; these are opposite all but the smallest vb's. Edgar records that in species with ridged stems from New Zealand, the round-ended scl. girders mostly traverse the entire chlorenchyma (*J. australis*, *J. amabilis*, *J. subsecundus*) but only some of them penetrate the chlorenchyma completely in *J. sarophorus*, *J. filicaulis*, and *J. usitatus*. The present writer saw small, irregular strands as well as the larger ones that extend right through the chlorenchyma in the British material of *J. inflexus*. Small strands were not noted by Edgar in New Zealand specimens.

Stellate cells are characteristic components of transverse diaphragms in many species of the *genuini*. Buchenau (1890) recorded them for the following species: **J. effusus* L., *J. leersii* Marsson, *J. glaucus* Ehrh., *J. mexicanus* Willd. (= *J. balticus* Willd. var. *mexicanus* (Willd.) O. Ktze.), **J. lesueurii* Bolander, *J. andicola* Hook., **J. pallidus* R. Br., *J. procerus* E. Mey., **J. radula* Buchen., *J. smithii* Engelm., **J. patens* E. Mey., and *J. pauciflorus* R. Br. Stellate cells in the species marked * are figured in Buchenau's monograph of 1906.

Healy (1953) in a field key to New Zealand species used the following diagnostic characters: *J. vaginatus* R. Br. no pith; *J. luxurians* Colenso (?) interrupted, loose pith; *J. polyanthemos* Fr. Buchen. pith interrupted or continuous, sometimes with both types in a single stem; *J. pauciflorus* R. Br. pith interrupted, or occasionally continuous, cavities often v. small and sometimes difficult to distinguish; *J. effusus* pith continuous, watery-white to almost colourless, sometimes appearing to be almost honey-combed under a lens; *J. pallidus* pith more cobwebby and denser white than in *J. effusus*. Edgar (1964) stated that air-spaces in pith vary from specimen to specimen. This should be taken into account when interpreting Healy's data.

Johnson (1963) used the position of stomata relative to the culm surface as a character to distinguish between *J. usitatus* (superficial) and *J. sarophorus* (sunken). In the same way *J. australis* (sunken stomata) can be distinguished from *J. amabilis* (superficial stomata).

Peisl (1957) gave a comprehensive, illustrated account of the anatomy of *J. effusus* L.

Barros (1957*b*) described *J. ramboi* Barros from South America. The culm (illustrated by him) is very similar anatomically to that of *J. effusus*, but lacks the central cavity.

4. *Junci thalassii*

Raunkiaer (1895–9) placed *J. maritimus* in a category on its own among the species he examined because the vb's are scattered throughout the culm centre. This character was noted by Peisl (1957) as one distinguishing the *thalassii* from the *genuini*. The pith is continuous, a character used by Buchenau (1890) in defining the sub-group. Peisl (1957) stated that the seedlings are v. similar to those of *J. effusus*. He also drew attention to the protective cells lining the substomatal cavity, which are similar to those in certain species of Restionaceae (see p. 120).

Sclereids are present in the phloem of the vb's. This is unusual among the Juncaceae, but quite common in supposedly related families, e.g. Thurniaceae (see p. 81).

The central scl. strands with their associated sheaths of parenchyma are v. unusual. Their nature is uncertain, but it is possible that they may represent vestigial bundle sheaths, either as tails to blind-ending bundles or just as strands in their own right.

The curving of the palisade chlorenchyma cells towards the outer scl. strands is also characteristic of this group (Peisl 1957). The mechanical significance of this feature was discussed by Schwendener (1874).

5. *Junci septati*

The seedlings of *J. articulatus* are v. similar to those of *J. effusus*, according to Peisl (1957).

He observed that the adaxial groove in *J. articulatus* is more distinct than in *J. subnodulosus*, and is visible for 1–2 cm above the sheath. Towards the middle of the leaf the groove gradually becomes rounded and the underlying tissue no longer consists of colourless parenchyma but is replaced by assimilatory tissue and vb's. Submerged leaves of *J. bulbosus* and *J. heterophyllus* are fine and hair-like. The leaves in *J. articulatus*, *J. anceps*, and *J. validus* are laterally flattened, and yet others, e.g. *J. xiphioides* and *J. ensifolius*, have almost bilateral, ensiform leaves.

Laux (1887) stated that the anatomy of *J. lampocarpus* (= *J. articulatus*) is v. similar to that of *J. acutiflorus*. Duval-Jouve (1872) recorded that *J. lampocarpus* (= *J. articulatus*) has air-canals between the outer vb's of the culm, and strong scl. bundle sheaths. The stomata in *J. fontanesii* are slightly sunken. He stated that in the culm of *J. anceps* the chlorenchyma is separated from the scl. by several layers of parenchyma, and that most vb's are embedded in the scl. Křísa (1963) found that all vb's are enclosed in scl. in *J. alpinus* (= *J. alpinoarticulatus*).

Peisl (1957) followed Weimarck's (1946) treatment of *J. subnodulosus* and included it in the section *subnodulosi*. Weimarck regarded this section as a link between *thalassii* and *septati*; its members are said to be morphologically similar to *thalassii*, but to possess septa. For convenience of description the members of Weimarck's *subnodulosi* are dealt with in *septati sensu* Buchenau in this work, but the presence of the weak scl. cylinder in the culm of *J. subnodulosus* may be used as evidence supporting Weimarck's treatment of this section.

Buchenau (1890) illustrated leaf sections of *J. striatus* and *J. acutiflorus*.

Duval-Jouve's (1872) account of the rhizome vb's in *J. lampocarpus* (= *J. articulatus*) is inaccurate. He referred to some of the mx vessels as protoxylem canals, and described the phloem as 'little fibres'. In the same paper he compared the rhizome anatomy of *J. acutiflorus*, *J. striatus*, and *J. anceps* and stated that they differ only in details and proportions of the various parts. In *J. acutiflorus* the widths of medulla and cortex are similar, the lacunae are large and there are many rings of vb's. In *J. anceps* the medulla is 6 times wider than the cortex, the lacunae are small and the vb's are in 1 ring.

Buchenau (1906) illustrated a culm of *J. prismatocarpus* R. Br. var. *genuinis* Fr. Buchen. in T.S. This has pronounced wings of tissue on 2 sides with 3–4 vb's in either wing. He also illustrated the culm of *J. validus* Coville in T.S.

The outline is oval; vb's are arranged in 1 ring; there is no scl. cylinder. The transverse septa are composed of narrow tracts of tissue extending from opposite the vb's to a central transverse strand. The lamina, also illustrated by Buchenau, lacks the marked longitudinal septa of the culm. The lamina of *J. columbianus* Coville in T.S. is also figured. It is more or less terete, but with a concave face. There is no scl. apart from that in the vb sheaths; the vb's are arranged in 1 (2) rings, the smallest vb's being embedded in the chlorenchyma. Buchenau figured the leaf section of *J. striatus* Schousboe; it is similar to that in his 1890 publication.

Barros (1952) included *J. diemii* Barros in this group, but its leaf anatomy seems to indicate that it would be better placed with the *poiophylli* or *alpini*.

Vierhapper (1930) gave sectional status to the *septati*, excluding from the new group the 5 species making up his new *subnodulosi*.

Guédès (1967) illustrated the leaf of *J. lampocarpus* (= *J. articulatus*) in surface view and cut transversely at different levels. He demonstrated the relation of the ligule to the leaf base, and the extent of the adaxial leaf surface.

6. *Junci alpini*

Satake (1933) divided Japanese *Juncus* species into 3 groups on the basis of distribution of mechanical tissue in the peduncle. *J. triglumis* and *J. maximowiczii* were placed together in the group having scl. confined to the bundle sheaths, but *J. castaneus* was placed in an intermediate position, between a group exhibiting a ring of mechanical tissue in which all vb's are embedded, and a group with a mechanical ring embedding only some of the vb's. Nevertheless, he considered that the 3 species are properly assigned to the group *alpini*. The culm material of *J. triglumis* examined by the writer has a poorly developed scl. cylinder connecting the vb's. This being so it seems possible that Satake's divisions may not be entirely valid.

Buchenau (1906) illustrated a T.S. leaf of *J. maximowiczii* Fr. Buchen., *luzuliformis* Franch. var. *potaninii* Fr. Buchen., and *J. allioides* Franch. (probably = *J. concinuus* D. Don.). Each of these has scl. in the bundle sheaths only. *J. maximowiczii* has a channelled leaf, with 2 large median and 2 small lateral air-canals between the vb's. The leaf of *J. luzuliformis* is flattened, with an arc of vb's and 2 longitudinal air-canals. *J. allioides* has a grooved, compressed cylindrical leaf, with a ring of vb's and a central air-canal. Buchenau also figured T.S. leaves of *J. sphenostemon* Fr. Buchen., *J. przewalskii* Fr. Buchen., and *J. sphacelatus* Decne. The first two have adaxial and abaxial channels, the last has an adaxial channel only. The vb's are arranged in an arc; scl. is restricted to the bundle sheaths. The median vb in *J. sphenostemon* has girders of chlorenchyma extending to it from both surfaces. The epidermal cells in the adaxial groove of this species are larger than the remainder. The specimen of *J. castaneus* (leaf section) illustrated by Buchenau (1890) matches the material seen by the author but exhibits marginal scl. strands not seen in the Turner specimen.

Spinner (1913) described and illustrated culm and leaf sections of *J. leucanthus* Royle, a sp. which fits well in the *alpini*.

This group may not be entirely natural in composition. The leaves of some members have an arc of vascular bundles and the cells of the adaxial epidermis are distinguishable from those of the abaxial epidermis; other members have a ring of vascular bundles and appear to lack an adaxial epidermis.

7. *Junci singulares*

The single species belonging to this subgenus was included in the *graminifolii* (see below) by Vierhapper (1930) and Adamson (1925). Weimarck (1946) thought it is distinctive, and gave this taxon the status of a section as opposed to the subgeneric rank attributed to it by Buchenau (1875). The leaves are laterally compressed and therefore unlike the dorsiventrally compressed leaves of the *graminifolii.*

8. *Junci graminifolii*

Weimarck (1946) included only the perennial species of the subgenus recognized by Buchenau (1875) and of the section recognized by Vierhapper (1930). He placed the annual species in section *Juncinella* erected by V. Krechetovich and Goncharov (1935).

Of the species examined by the author, *J. falcatus* and *J. planifolius* represent Weimarck's section *graminifolii*, and *J. capitatus* represents *Juncinella* V. Krecz. and Gontsch. The two groups differ in many ways. For example *J. capitatus* has unusual cells in the culm epidermis (q.v.) and has narrow, channelled leaves whereas the culm epidermis of *J. falcatus* and *J. planifolius* is similar to that of most other *Juncus* species, and the leaves are wider and flattened. Weimarck's treatment seems to be correct in this instance for the species seen by the author.

Haslinger (1914) described *J. capensis* Thunb. The leaf is flattened; the abaxial epidermal cells are larger than in other species, variable in size and similar to, but rather taller than, the adaxial cells. The stomata are present in strips associated with smaller epidermal cells. The strips of small cells are separated from one another by about 4 files of larger cells. The guard cells of the stomata have outer lips. The leaf margins are produced into wings composed only of abaxial and adaxial epidermal cells. None of the bundle sheaths extends to the epidermis. Scl. strands are absent from leaf margins. Air-canals are present between the vb's. The rhizome has air-canals in the middle cortex; the vb's in the stele are peripheral, the centre being free from them. *J. capensis* is a perennial species, in the section *graminifolii* Vierhapper. Its anatomy matches that of *J. planifolius*, which also belongs in the same group. Buchenau (1906) illustrated the culm of *J. obtusatus* Engelm. in T.S. The species belongs to the perennial group, and is, therefore, also included by Weimarck in section *graminifolii.* The vb's are in 2 rings, both of which are enclosed in the scl. cylinder. In this respect the species is unlike *J. planifolius*, which also belongs to the same group. The presence or absence of a sclerenchyma cylinder appears to be insignificant in the taxonomy of this subgenus, or, if significant, it cuts across the sections proposed by Weimarck.

***Juncus*: General conclusions**

Most subgenera of *Juncus* are anatomically distinct; indeed, anatomical data were applied to the making of the early classifications of the genus (Buchenau 1875). The most readily observed anatomical differences between the members of the subgenera can be seen in the transverse sections of the leaf blade.

The members of only 2 groups, *poiophylli* and *graminifolii*, regularly exhibit a distinctive adaxial epidermis in which the cells are conspicuously larger than those of the abaxial surface. Although an adaxial epidermis can be distinguished from an abaxial epidermis in many species of the *alpini* this subgenus includes species with leaves which, for the most of their length, lack a definable adaxial epidermis.

There are several groups in which the leaf blade is circular or oval in transverse section. Of these the *thalassii* stands out because, besides having the peripheral vascular bundles which are common to all the species with cylindrical leaves, its members have vascular bundles scattered throughout the central ground tissue.

Members of the *genuini*, all of which have cylindrical leaves, are unlike the single species in the *subulati* because none of them has papillate epidermal cells overarching sunken stomata, such as are found in *J. subulatus*. (Some species of the *genuini* do have sunken stomata.)

The *septati* are to be found amongst the groups that have species with leaves that are circular or oval in transverse section, but the transverse septa, which are the special characteristic of the *septati*, afford a ready means of distinguishing them from species of most other subgenera. Transverse septa also occur in some members of the *alpini*.

LUZULA DC.

No. of spp. 80; examined 12.

Genus Description

Leaf surface

Epidermis. (i) Abaxial cells 4-sided, walls slightly thickened, wavy on surface (Fig. 10. D); 3 sizes of cells present: (*a*) cells next to stomatal bands, mostly 1–3 times longer than wide; (*b*) cells between stomatal bands, not over scl., about 7–14 times longer than wide; (*c*) cells over scl., slightly narrower than remainder, 4–8–(20) times longer than wide. (ii) Adaxial cells 4-sided, walls slightly thickened, wavy; cells of 2 sizes (*a*) those not over scl., 1–4 times longer than wide and twice as wide as abaxial cells; (*b*) cells over scl. 4–8–(14) times longer than wide. **Stomata:** guard cells short and wide, with rounded ends, subsidiary cells often shorter than guard cells with rounded, obtuse, or acute ends (Fig. 10. D). Present in abaxial surface only, frequently arranged in longitudinal bands. **Papillae:** marginal cells of most species with papillate extension of distal end of outer wall, e.g. *L. forsteri* (Fig. 10. I). **Hairs:** see leaf T.S. below. **Cuticular marks:** granular, or with longitudinal flecks. **Tannin** absent.

Leaf T.S.

Outline: dorsiventrally flattened, often in the form of a shallow arc (Figs. 10. A, F, K; 11. E); leaf margins rounded, triangular or square (Figs. 10. B, E, I, J; 11. A, B, F). **Hairs** multicellular, arising from small, low multicellular mounds on the leaf margins; shaft of hair 3–4–(5) cells wide and 6–*c*. 36 cells (0·1–3 cm) long, tapering to either end. Individual cells much elongated, 12–20–100+ times longer than wide, shortest near hair bases; walls slightly to moderately thickened. Ends of cells overlapping; many hairs with 'storeyed'

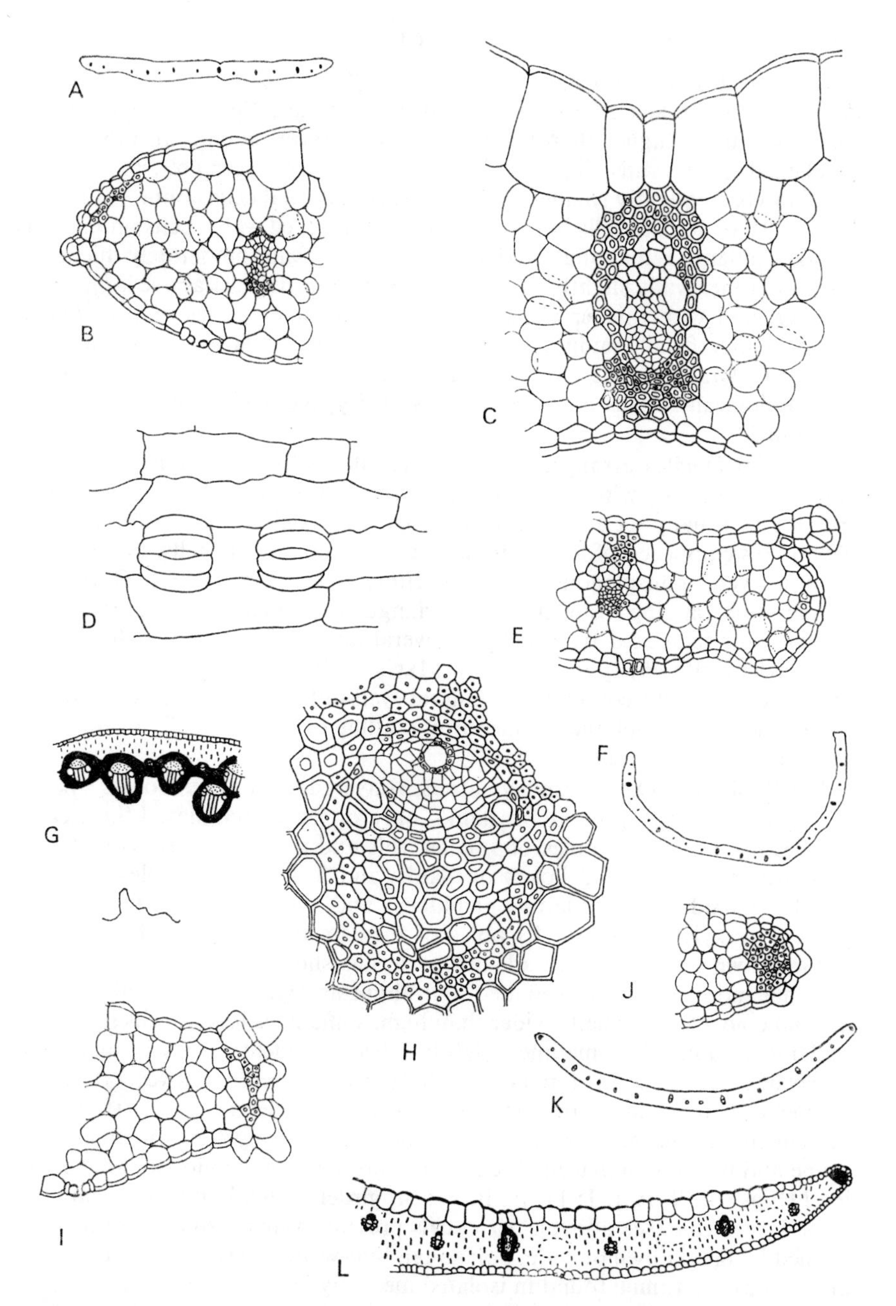

FIG. 10. Juncaceae. A–D, *Luzula campestris*: A, diagram of leaf T.S. (×12); B, detail of leaf margin T.S. (×200); C, detail of midrib vascular bundle of leaf T.S. (×320); D, surface detail of leaf abaxial epidermis (×370). E–H, *L. lutea*: E, detail of leaf margin T.S. (×200); F, diagram of leaf T.S. (×12); G, diagram of sector of culm T.S. (×50); H, vascular bundle from culm in T.S. (×320). I, *L. forsteri*, detail of leaf margin T.S. (×200). J–L, *L. echinata*: J, detail of leaf margin T.S. (×200); K, diagram of leaf T.S. (×12); L, diagram of part of leaf T.S. (×50).

arrangement of cells, the ends of groups of cells frequently occurring more or less at the same level. Cells air-filled in mature hair. **Epidermis.** (i) Adaxial cells 1½–2 times higher than wide, outer wall convex, slightly to moderately thickened, other walls slightly thickened; cells over larger vb's, particularly median vb, frequently smaller than others. (ii) Abaxial cells about ¼–½ height and width of adaxial cells but with similarly thickened walls; cells on margins of leaf similar to remaining abaxial cells, but outer walls moderately thickened to thick, other walls slightly to moderately thickened; cells above larger vb's smaller, and with thicker walls than others in same species (e.g. *L. lutea*). **Stomata** superficial, confined to abaxial epidermis (appearing adaxial at margins in *L. forsteri* owing to extension of abaxial surface round margins on to morphologically adaxial surface in this sp.). **Hypodermis** absent from all spp. examined.

Vascular bundles arranged in 1 row, usually with large or medium-sized bundles alternating with small; median, or vb nearest midrib, normally large. Small vb's usually set at equal distance from abaxial and adaxial epidermis. Small vb's composed of few, narrow tracheids and phloem cells. Largest vb's with 1-several conspicuous mxv's on either flank (Fig. 10. C). Outline of phloem pole ranging from circular to triangular; phloem abutting directly on to xylem or separated from it by 1–several layers of parenchymatous cells. Protoxylem with cavity in some spp. Mxv's medium-sized to large, or small, frequently laterally compressed. Lateral wall-pitting of vessels scalariform; perforation plates oblique, scalariform. Outline of vb's normally oval, sometimes laterally compressed.

Bundle sheaths. O.S. parenchymatous, 1-layered, continuous round smaller bundles, frequently present on flanks only of larger bundles. I.S. sclerenchymatous, as caps at poles of smaller vb's and complete sheaths to larger vb's, where fibres in 1–3 layers on flanks and 3–4 layers at poles. I.S. frequently extending to epidermis at phloem and xylem (Fig. 10. C), or only phloem or xylem (Fig. 10. L) poles of median and largest of lateral vb's. **Chlorenchyma.** Cells next to adaxial epidermis shortly palisade-like in some spp. but more frequently rounded and about as high as wide; cells next to abaxial epidermis normally wider than high, sometimes lobed; cells in middle of lamina rounded, sometimes slightly lobed, frequently loosely arranged. Areas of colourless, polygonal cells with thin walls present between vb's, and in line with them in many spp. (see air-canals). **Sclerenchyma:** (i) strands present in leaf margins of most spp., these 1–2 or many fibres in thickness; shape and position of marginal scl. strands probably of diagnostic value at sp. level (Figs. 10. B, E, I, J; 11. A, B, F); (ii) girders extending from sheaths of some larger vb's (see above). **Air-canals** present in many spp. between vb's, formed by breakdown of wide, polygonal, thin-walled cells; particularly large in *L. sylvatica*. **Tannin** found in isolated mesophyll cells in some spp.

Culm surface

Epidermis: cells 4-sided, walls slightly to moderately thickened, wavy at surface, cells over vb's sometimes longer and narrower than remainder, those next to stomata sometimes shorter than remainder; most frequent size ranges 4–12, 7–8, and 6–10 times longer than wide. **Hairs** absent. **Stomata**

similar to those in leaf; guard cells mostly more or less parallel-sided, with rounded ends, subsidiary cells with rounded, acute, or obtuse ends. **Cuticular marks** slightly granular; longitudinal striations observed in *L. abyssinica.*

Culm T.S.

Outline circular-oval, frequently with shallow grooves. **Cuticle** frequently thin, occasionally slightly to moderately thickened. **Epidermis:** cells usually as high as wide, sometimes wider than high, particularly those situated in grooves; outer walls normally convex and moderately thickened or thick, others slightly thickened, not wavy; ridges present above anticlinal walls in some spp. **Stomata** superficial, normally similar to those of leaf of same sp. **Chlorenchyma** 2–6-layered, cells rounded, wider than high or as high as wide, frequently lobed, with thin walls; longer than high in L.S. Cells often loosely packed in regions between ridges from scl. cylinder. **Sclerenchyma** present as 2–3- or 4–8-layered cylinder linking larger vb's in most spp. (Figs. 10. G; 11. C). Cylinder in some spp. apparently composed of thick-walled parenchymatous cells.

Vascular bundles in 1–2 rings; if 2, then outer composed of small vb's adjacent to but outside scl. cylinder. Larger vb's variable, with 1 or up to 3–4 mxv's on either flank; these mxv's v. small in some spp. (Fig. 10. H), otherwise medium-sized to wide (Fig. 11. D). Lateral wall-pitting of vessels scalariform, scalariform-reticulate, or alternate; perforation plates simple in most spp., scalariform in some, always oblique. Phloem pole either abutting directly on xylem, or separated from it by several layers of narrow, parenchymatous cells. Small sclereids, scattered or in groups to adaxial side of pole (Figs. 10. H; 11. D), present in phloem of some spp. Protoxylem cavity present in some spp., e.g. *L. glabrata* (Fig. 11. D).

Bundle sheaths. O.S. probably represented by 1-layered cylinder of parenchymatous cells to outside of and next to scl. cylinder and around the smallest vb's; indistinct, single-layered parenchymatous sheaths present around all vb's in spp. without scl. cylinder. I.S. sclerenchymatous, present as 2–5-celled cap at phloem pole of smaller vb's and as complete sheaths to larger vb's; fibres in 2–4 layers on flanks and at xylem pole, and 5–7 layers at phloem pole, except in certain spp., e.g. *L. abyssinica*, where 9–10-layered. Fibres of sheaths narrow, thick-walled, frequently readily distinguishable from wider cells of sclerenchyma cylinder. **Central ground tissue** parenchymatous, cells of outer layers with slightly to moderately thickened walls, those of inner layers wider, with thin walls, sometimes breaking down to form central cavity. **Air-canals** present in culm centre of some spp. and also in chlorenchyma; also present between ridges from scl. cylinder in certain spp., particularly towards base of culm.

Rhizome T.S.

Seen in *L. lutea, L. luzuloides,* and *L. sylvatica.*

Epidermis: cells more or less square. **Cortex:** 9–12 layers of more or less isodiametric to rounded parenchyma cells with scattered leaf traces and adventitious roots. **Endodermoid layer** composed of 1–3 layers of cells with U-shaped wall thickenings. **Vascular bundles** mostly amphivasal, in 1–2–several rings close to endodermoid layer; bundles of outermost ring exhibiting much

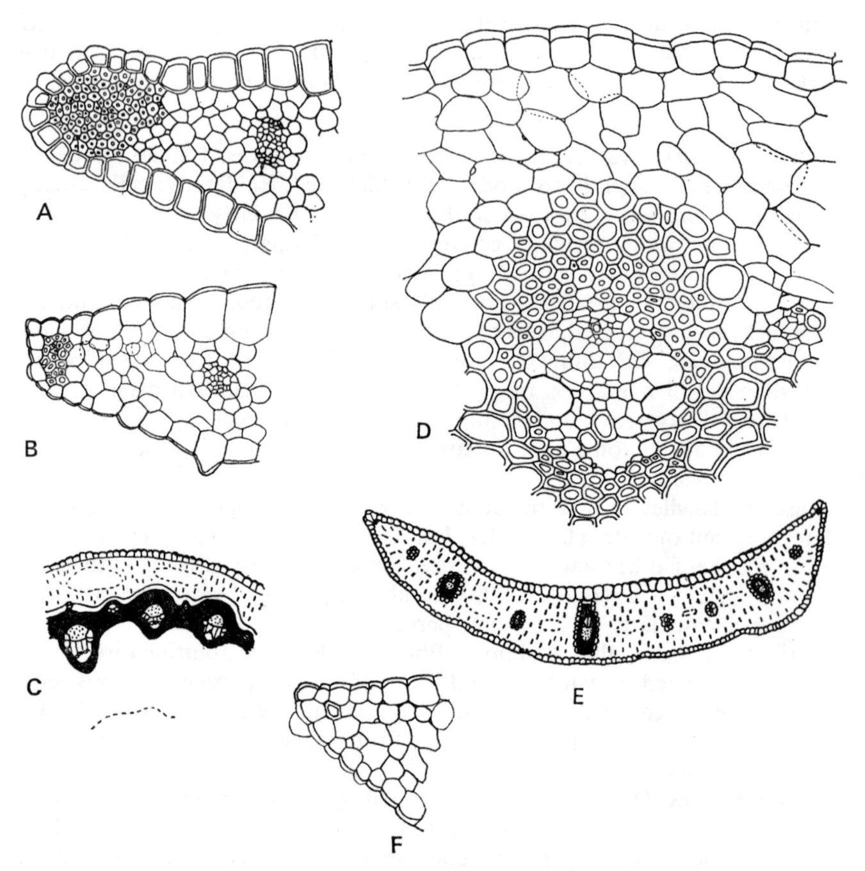

FIG. 11. Juncaceae. A, *Luzula oldfieldii*, detail of leaf margin T.S. (×200). B, *L. pilosa*, detail of leaf margin T.S. (×200). C, D, *L. glabrata*: C, diagram of sector of culm T.S. (×50); D, detail of culm T.S. (×320). E, F, *L. arcuata*: E, diagram of leaf T.S. (×50); F, detail of leaf margin T.S. (×200).

fusion. Most vb's in *L. sylvatica* although amphivasal, exhibiting a well defined px. pole. All vb's crowded in *L. luzuloides*. **Sclerenchyma** present as 2–4-layered sheaths to vb's, and as 1–2 layers of cells to inside of vb ring in *L. luzuloides*. **Central ground tissue** angular, more or less hexagonal, parenchyma cells with some intercellular spaces tannin(?)-filled. Starch present in many cells.

Root T.S.

Seen in *L. campestris*, *L. luzuloides*, and *L. sylvatica*.

Root-hairs arising from thin-walled cells similar to those of epidermis in *L. campestris*. **Epidermis:** cells large, square, thin-walled. **Hypodermis** present in *L. luzuloides*, walls of cells slightly thicker than those of cortical cells. **Cortex:** (i) outer part consisting of 1 layer of thin-walled cells; (ii) middle and (iii) inner parts, 5–10-layered in material examined, cells v. thick-walled in *L. campestris*, thin-walled in *L. luzuloides*, reducing in size towards endodermis, and becoming progressively more flattened. **Endodermis:** 1 layer of cells with v. thick inner and anticlinal walls and thin outer walls. **Pericycle:** 1 layer of narrow thin-walled cells. **Vascular system.** Several wide and medium-sized mxv's arranged in a ring. A v. fine root of *L. sylvatica* contained only 1, central mxv. **Central ground tissue** consisting of narrow thin-walled parenchyma cells.

Material Examined

Luzula abyssinica F. Parl.: Henderson, H. V. H150; Uganda Feb. 1933 (K).
L. arcuata Sw.: Moller H. J.; Asole Lapp Mark Vilhemina Marsfjallen 20.7.1926 (K).
L. campestris (L.) DC.: (i) Melville, R. 1367; Bendigo, Victoria, Australia 1952 (K). (ii) Hort. Kew 1965 (S). (iii) Cutler, D. F.; Bedgebury, Kent 1965 (S). var. *bulbosa* Fr. Buchen.; Drummond; S.W. Australia (K).
L. echinata (Small) F. J. Hermann: Godfrey, A. K. 49021; N. Carolina (K).
L. forsteri (Sm.) DC.: Barcelona 1920 (K).
L. glabrata (Hoppe) Desv.: Hoppe; Sausbergia, Australia (K).
L. lutea (All.) DC.: Hort. Kew 1950 (S).
L. luzuloides (Lam.) Dandy and Wilmott: Hort. Kew 1964 (S).
L. multiflora (Retz.) Lej.: Cutler, D. F.; Hoddesdon, Herts 1963 (S).
L. oldfieldii Hook.: Hort. Kew (Mt. Wellington, N. Zealand) 1964 (S).
L. pilosa (L.) Willd.: Cutler, D. F.; Dorking, Surrey 1966 (S).
L. sylvatica (Huds.) Gaud.: Cutler, D. F.; Lynton, N. Devon 1964 (S).

Species Descriptions

L. abyssinica F. Parl.

Leaf: width 6–8 mm, thickness < 0·1 mm.

Leaf surface

Epidermis: (i) abaxial cells 2–14 times longer than wide; (ii) adaxial cells 2–4 times longer than wide. **Papillae** low, over some abaxial cells. **Cuticle** with longitudinal flecks.

Leaf T.S.

Outline curved strap-shaped, of even thickness, tapering rapidly to bluntly rounded margins. **Epidermis:** abaxial cells about ¼ height of adaxial cells. **Stomata:** subsidiary cells 1½ times height of guard cells; guard cells with pronounced outer lip. **Bundle sheaths:** scl. girders extending to both surfaces from larger vb's. **Sclerenchyma:** marginal strands semicircular in outline. **Air-canals** present.

Culm surface

Epidermis: cells 10–11 times longer than wide.

Culm T.S. Diameter 1·5×0·75 mm.

Mechanical cylinder: 2–3 layers of thick-walled parenchyma cells. **Vascular bundles** with several mxv's per flank, phloem pole rounded, ensheathed. **Bundle sheaths** 1–3-layered at xylem pole and on flanks, pronounced at phloem pole, where 9–10-layered, outline v. rounded at phloem pole. **Air-canals** in chlorenchyma only.

***L. arcuata* Sw.**

Leaf: width 1 mm, thickness 0·1 mm.

Leaf surface

Epidermis: (i) abaxial, cells 4–8+ times longer than wide between stomata, 2–4(–6) times longer than wide next to stomata; (ii) adaxial cells 3–14 times longer than wide. **Stomata:** in bands of 4–6 rows; high proportion of malformed subsidiary cells present, these divided transversely, or with lateral outgrowths.

Leaf T.S.

Outline: flat strap-shaped, tapering abruptly to slightly upturned, acute margins; margins often 2 cells thick (Fig. 11. E, F; Pl. I. B). **Bundle sheaths:** only extensions of median bundle sheath reaching leaf surface. **Sclerenchyma:** marginal strands of 1–3 fibres. **Air-canals** incipient in chlorenchyma.

Culm surface

Epidermis: cells 4–10 times longer than wide.

Culm T.S. Diameter 0·3 mm.

Epidermis: cells with slight ridges over anticlinal walls; outer wall thinnest at centre of each cell. **Vascular bundles:** largest with 1 oval mx vessel per flank; phloem abutting on xylem; px. cavity present. **Air-canal** central.

***L. campestris* (L.) DC.**

Leaf: width 3 mm, thickness 0·2 mm.

Leaf surface

Epidermis: abaxial cells 4–10 times longer than wide in files not including stomata, 1–4 times longer than wide in files including stomata (Fig. 10. D).

Leaf T.S.

Outline strap-shaped, of even thickness, with rounded or angular margins (Fig. 10. A, B). **Vascular bundles:** largest with 1 or 2 mxv's per flank (Fig. 10. C); phloem abutting directly on xylem. **Sclerenchyma:** adaxial and abaxial girders to median vb, adaxial girders to some other large vb's.

Culm surface

Epidermis: cells mostly 7–8 times longer than wide.

Culm T.S. Diameter 0·5 mm.

Sclerenchyma: cylinder weak, 1–2-layered. **Vascular bundles:** flanking mxv's medium-sized to narrow; px flanked and surrounded at pole by 1 layer of thin-walled, parenchymatous cells. Phloem pole tangentially oval, ensheathed. **Air-canal** absent.

***L. echinata* (Small) F. J. Hermann**

Leaf: width 4 mm, thickness 0·1 mm.

Leaf surface

Epidermis: (i) abaxial cells 2–10 times longer than wide, shorter where associated with stomata; (ii) adaxial cells 1–2 times longer than wide. **Stomata** in bands 2–4 cells wide.

Leaf T.S.

Outline shallow U-shaped, of even thickness except for slight constriction at largest (more or less median) vb; margins rounded, curving slightly to adaxial side (Fig. 10. J, K, L). **Vascular bundles.** None v. large, largest with 1 medium-sized to wide mxv per flank; phloem abutting on xylem; px cavity present. Smaller vb's nearer to abaxial than adaxial epidermis. **Bundle sheaths:** I.S. sclerenchymatous. **Sclerenchyma:** abaxial and adaxial girders from larger vb's; incomplete girders from medium-sized vb's; marginal strands semi-circular. **Air-canals** present between some vb's.

Culm surface

Epidermis: cells of 2 sizes (i) between veins 6–10 times longer than wide, and 1½ times wider than those over veins; (ii) over veins 10–11 times longer than wide.

Culm T.S. Diameter 0·8 mm.

Mechanical cylinder composed of *c.* 2 layers of thick-walled parenchymatous cells. **Vascular bundles:** largest with 1 mxv per flank; phloem pole small, ensheathed; px cavity present. **Air-canal** absent.

***L. forsteri* (Sm.) DC.**

Leaf: width 3 mm, thickness 0·1 mm.

Leaf surface

Epidermis: abaxial cells 2–8+ times longer than wide. **Cuticular marks:** longitudinal striations.

Leaf T.S.

Outline U-shaped, of even thickness, margins rounded to slightly squared (Fig. 10. I). **Epidermis:** (i) abaxial cells extending well round leaf margins, most cells of margin with single, short papilla; (ii) adaxial cells slightly smaller over large vb's than over remainder of surface. **Vascular bundles:** larger with 1–2 oval mxv's per flank; phloem abutting directly on xylem; outline of vb's narrow and tall. **Sclerenchyma:** adaxial and abaxial girders to median bundle, and adaxial girders to some other large bundles; fibres in 2–3-layered arcs in leaf margins. **Air-canals** not well developed. **Tannin** present in some larger mesophyll cells.

Culm surface

Epidermis: cells 4–8+ times longer than wide.

Culm T.S. Diameter 1·1×0·8 mm.

Epidermis: cells at base of shallow grooves about twice as wide as high, others about as high as wide. **Stomata** sparse, situated mainly in grooves. **Sclerenchyma:** cylinder absent. **Vascular bundles:** large each with 1–2 rounded-angular, medium-sized mxv's per flank. **Bundle sheaths:** O.S. indistinct, consisting of 1 layer of parenchyma; I.S. sclerenchymatous, fibres narrow, thick-walled, present in 2–3 layers at xylem pole and on flanks and in 6–7 layers at phloem pole.

***L. glabrata* (Hoppe) Desv.**

Leaf: width 6–8 mm, thickness 0·2 mm.

Leaf surface

Epidermis: abaxial cells not seen; adaxial cells as long as to 2–3 times longer than wide, those over veins longer and narrower than remainder.

Leaf T.S.

Outline U-shaped, of even thickness, tapering rapidly to margins. **Epidermis:** adaxial cells over larger vb's smaller than remainder. **Vascular bundles:** largest narrow, flanking mxv's oval and compressed, 1–several per side; phloem ensheathed; px cavity present. **Sclerenchyma.** In leaf margins, about 4 cells high, 8+ cells wide, reaching both abaxial and adaxial surfaces. Girders reaching both surfaces from all but smallest vb's. **Air-canals** present between vb's.

Culm surface

Epidermis: cells 4–12 times longer than wide, shorter next to stomata, longest over veins.

Culm T.S. Diameter $2{\cdot}5 \times 1{\cdot}5$ mm.

Mechanical cylinder composed of 7–8 layers of thick-walled parenchyma cells (Fig. 11. C). **Vascular bundles:** largest with 1–2(–3) medium-sized mxv's per flank; phloem abutting on xylem, containing scattered or grouped sclereids; wide px cavity present (Fig. 11. D).

***L. lutea* (All.) DC.**

Leaf: width 6–7 mm, thickness 0·1 mm.

Leaf surface

Epidermis: (i) abaxial cells $1\frac{1}{2}$–7 times longer than wide; (ii) adaxial cells mostly between $1\frac{1}{2}$ and 3 times longer than wide, but those over veins narrower and 3–4 times longer than wide.

Leaf T.S.

Outline shallow U-shaped, of even thickness (Fig. 10. F). **Epidermis:** adaxial cells smallest over vb's. **Vascular bundles:** large bundles mostly with 1, sometimes 2 mxv's per flank; phloem pole large, radially oval, abutting directly on xylem. **Sclerenchyma.** Girders or partial girders extending to abaxial and adaxial surfaces from larger vb's. Marginal strands absent, or composed of few fibres (Fig. 10. E). **Air-canals** absent.

Culm surface not seen.

Culm T.S. Diameter 1·2 mm.

Sclerenchyma: 6–7-layered cylinder of thick-walled fibres (Fig. 10. G). **Vascular bundles:** large with 1–2 narrow mxv's per flank and 4–5 central rows of mx elements and px elements; with 1–2 rows of narrow, parenchymatous cells separating xylem from I.S. on flanks; phloem separated from xylem by several layers of thin-walled cells; several thick-walled sclereids surrounding a canal present in phloem pole (Fig. 10. H).

***L. luzuloides* (Lam.) Dandy and Wilmott**

Leaf: width 6 mm, thickness $< 0{\cdot}1$ mm.

Leaf surface

Epidermis: (i) abaxial cells in stomatal files 2–8 times longer than wide, those over scl. up to about 20 times longer than wide, about $\frac{2}{3}$ width of others; (ii) adaxial cells between scl. mostly 2–6 times longer than wide, and 4 times wider than widest abaxial cells; those over scl. 6–10 times longer than wide, and about $\frac{1}{2}$ width of others. **Cuticular marks:** longitudinal striations.

Leaf T.S.

Outline strap-shaped, slightly arched, of more or less even thickness, tapering slightly to almost square margins. **Epidermis:** (i) abaxial cells about $\frac{1}{4}$ height of adaxial; (ii) adaxial cells over larger vb's and near margins similar in size to abaxial cells, remainder 2–4 times larger. **Stomata:** guard cells with pronounced outer and slight inner lips; subsidiary cells as high as epidermal cells. **Vascular bundles:** 31 present; largest with 1–2 medium-sized mxv's per flank; phloem ensheathed by sclerenchyma; px cavity present in some vb's. **Bundle sheaths:** O.S. parenchymatous, surrounding smaller bundles, restricted to flanks of larger bundles. **Sclerenchyma:** adaxial or abaxial girders or partial girders to larger vb's; marginal strands 1–2 layers wide, next to epidermis, in broken arc. **Air-canals** present between most vb's. **Tannin** present in scattered cells in mesophyll.

Culm T.S. Diameter 2·0×1·2 mm.

Sclerenchyma cylinder 2–3-layered. **Vascular bundles** close together; 1–2 medium-sized mxv's per flank; phloem pole ensheathed by scl., tangentially oval; px pole elongated, radially oval, frequently with air-cavities. **Bundle sheaths:** O.S. indistinct; I.S. fibres strongest at phloem pole, forming 2–(3–4)-layered arc, extending as far as flanking mxv; thinner walled fibres (?) in 1–2 layers extending around the flanks, and at xylem pole. **Air-canals** small in outer chlorenchyma; central cavity formed by breakdown of thin-walled cells.

Rhizome T.S.

Mainly as in genus description; unusual because of presence of an inner ring of scl. next to vb's.

Root T.S.

See genus description on p. 52.

***L. multiflora* (Retz.) Lej.**

Leaf T.S. only. Width 3–4 mm, thickness 0·1–0·2 mm.

Epidermis: (i) abaxial cells opposite large vb's smaller than remainder; (ii) adaxial cells *c.* 1½ times higher than wide, *c.* 5 times height of abaxial cells. **Vascular bundles:** principal vb's with 1–2 oval, medium-sized flanking mxv's; phloem ensheathed by sclerenchyma. **Bundle sheaths:** O.S. parenchymatous, present on flanks only of larger bundles, completely encircling smaller bundles. **Sclerenchyma:** girders to both surfaces from principal bundles; marginal strands present. **Air-canals** present between vb's **Tannin** present in isolated cells of mesophyll.

***L. oldfieldii* Hook.**

Leaf T.S. Width 4–5 mm, thickness 0·1 mm.

Outline semi-circular, of even thickness, tapering rapidly to rounded margins (Fig. 11. A). **Epidermis:** (i) abaxial cells *c.* $\frac{1}{4}$ height of adaxial cells; of uniform height except opposite largest vb's, where slightly shorter than remainder; (ii) adaxial cells opposite large vb's often only $\frac{1}{2}$ size of those between vb's. **Vascular bundles:** larger and medium-sized vb's with 1–2(–3) rounded angular mxv's per flank. Phloem separated from xylem by several layers of narrow parenchymatous cells. Protoxylem cavity present. **Chlorenchyma:** outermost 2 layers of cells polygonal, with slightly thickened walls.

Sclerenchyma: partial girders to larger vb's; well-developed strands, circular in outline, in either margin. **Air-canals** present between most vb's.

Culm T.S. Diameter 2·0×1·5 mm.

Sclerenchyma: cylinder not developed, cells between vb's parenchymatous, with slightly thickened walls. **Vascular bundles:** large bundles with 1–2 medium-sized mxv's per flank, phloem abutting directly on xylem; protoxylem often with air-cavity, px pole enclosed in arc of or completely encircled by 1–2 layers of thin-walled, narrow, parenchymatous cells. **Bundle sheaths:** O.S. indistinct; I.S. sclerenchymatous, 1–2-layered on flanks and at xylem pole, 4–6-layered at phloem pole. **Air-canal** present at culm centre.

***L. pilosa* (L.) Willd.**

Leaf: width 4–8 mm, thickness < 0·1 mm.

Leaf surface

Epidermis: (i) abaxial cells 3–10 times longer than wide, those next to stomata mostly between 2 and 4 times longer than wide; (ii) adaxial cells as long as to 1½–2 times longer than wide.

Leaf T.S.

Outline strap-shaped, of even thickness, tapering rapidly to obtuse or square margins (Fig. 11. B). **Epidermis**: (i) abaxial about ¼–⅕ height of adaxial cells; (ii) adaxial cells over vb's slightly smaller than remainder. **Vascular bundles:** larger vb's with 1 narrow to medium-sized mxv per flank; phloem pole abutting directly on xylem; no px canals observed. **Sclerenchyma:** abaxial and adaxial girders from larger vb's frequently incomplete; marginal strands well-developed, rectangular (dorsiventral axis longest) or crescentiform. **Air-canals** present between vb's. **Tannin** cells scattered in mesophyll.

Culm T.S. Diameter 1·0 mm.

Outline: circular or slightly polygonal. **Sclerenchyma:** cylinder absent. **Vascular bundles:** principal vb's with 1 medium-sized mxv per flank; phloem pole rounded, either abutting directly on xylem, or separated from it by 1–2 layers of narrow, thin-walled parenchyma cells; protoxylem with parenchyma sheath on flanks and at pole, with air-cavity. **Bundle-sheaths:** O.S. indistinct; I.S. sclerenchymatous, 2–3-layered on flanks and at xylem pole, 6–7-layered at phloem pole. **Air-canals** present in chlorenchyma and also in culm centre.

***L. sylvatica* (Huds.) Gaud.**

Leaf only: width 8 mm, thickness 0·2 mm.

Leaf surface

Epidermis: (i) abaxial cells 1 to 7–8 times longer than wide, those over veins longest; (ii) adaxial cells mostly 1½–4 times longer than wide and 2–3 times wider than abaxial cells.

Leaf T.S.

Outline a shallow arc, of slightly varying thickness; margins 2–3-angled, made irregular by large hair bases. **Epidermis:** (i) abaxial *c.* ¼ height of adaxial cells and about as high as wide, those opposite principal vb's larger than others; (ii) adaxial cells mostly twice as high as wide, smaller opposite principal vb's. **Vascular bundles** (28 present) dorsiventrally elongated in outline, largest with 1 wide, oval mxv per flank; px pole triangular; phloem pole large, semicircular, separated from xylem by several layers of parenchymatous cells. **Sclerenchyma** present as adaxial partial girders to large and medium-sized vb's and abaxial partial girders to principal vb's only. Marginal strands arc-shaped, (1–)2–3 fibres thick and 16–18 fibres high. **Air-canals** large, rectangular, between vb's.

Taxonomic Notes and Information from the Literature

The genus was divided into subgenera by Grisebach (1845), who used characters of gross morphology on which to base his divisions. These subgenera are not anatomically distinct. Ebinger (1963) proposed a new subgenus *Marlenia* for the annual species *L. elegans* Lowe. His description of this subgenus includes anatomical data. *L. elegans* is said to be distinguishable from other species by having straight walls to the leaf epidermal cells, as seen in surface view, to lack strands of sclerenchyma in the leaf margins, and to have evenly thickened endodermal cells in the root.

Some other *Luzula* species have been found by the present writer to have marginal sclerenchyma strands made up of few cells only. This applies to *L. arcuata* but in other species such as *L. lutea* there are no marginal strands at all. So although the lack of marginal strands in *L. elegans* can be used in conjunction with the other characters for the purpose of identification, it is not such a distinctive character as Ebinger supposed. It is possible that further studies may show that the outline and position of marginal sclerenchyma strands are useful characters in distinguishing between species, since the variations in the size and shape of these strands are considerable. The frequency and extent of penetration of sclerenchyma girders may prove to be similarly useful.

Much has been written about the roots of *Luzula* species (e.g. Petersen 1873–4, Laux 1887, Freidenfelt 1904) and in particular it is recorded that they have a 'solid' cortex, in which there are no large air-canals. This contrasts with the cortex of *Juncus* species which is frequently composed in part of radiating plates of cells separated by air-spaces. The distinction is not wholly reliable, however, since the present writer has found some *Juncus* species that lack cortical air-canals in their roots.

Freidenfelt (1904) gave details of the cross section of roots of *L. campestris*, *L. lutea*, *L. nivea*, *L. pilosa*, and *L. spicata*. He found more variation in the cortex than that seen by the present author. Freidenfelt stated that the 4(–6)-layered outer cortex may be composed of cells with all of the walls thickened, or with only the inner walls thickened. He also found that some roots have a central mxv in addition to those in the outer ring.

The presence of bulbils on certain species was discussed by Buchenau (1891). *L. nodosa* E.M. was described with a short, little-branched, horizontal or obliquely ascending rhizome with swollen internodes, which gives the impression of a string of pearls. *L. campestris* (L.) DC. var. *bulbosa* Fr. Buchen. is recorded as having small true bulbils; these have been observed by the present writer in material collected from the Swan River region, South-west Australia.

Buchenau (1906) gave illustrations of leaf sections of *L. crenulata*, *L. parviflora*, and *L. pilosa*. His 1906 work also includes an account of the leaf tip of *L. glabrata*, *L. pilosa*, and *L. purpurea*, with illustrations. The bifid apex of the leaf of *L. purpurea* is worthy of note. Hairs in *L. sylvatica* were also figured.

The closed leaf sheath of *Luzula* distinguishes the genus from *Juncus*. *Prionium* also has a closed leaf sheath (see p. 65).

Luzula seems, on anatomical grounds, to be most closely related to the genus *Juncus* and the *poiophylli* among the subgenera of *Juncus*. The presence of hairs and the general lack of wide mxv's in the leaf vb's are both characters of difference, but the general histology of culm and leaf is quite similar.

MARSIPPOSPERMUM Desv.

No. of spp. 3; examined 1.

Genus and Species Description

Leaf surface

Epidermis: cells 2–10 times longer than wide, walls moderately thickened to thick, wavy (Fig. 14. C, p. 70). **Stomata:** subsidiary cells frequently shorter than guard cells, with rounded ends.

Leaf T.S.

Outline circular, with shallow adaxial and narrower, shallow abaxial grooves (Fig. 14. D, p. 70 ; Pl. II. A). **Cuticle** thick, but grooved above the anticlinal cell walls (cuticle v. thin at these grooves). **Epidermis.** Abaxial cells mostly about twice as high as wide; adaxial cells slightly larger and sometimes taller, confined to strip about 15 cells wide, in shallow groove (Fig. 14. B, p. 70). Outer walls of all cells v. thick, making up $\frac{1}{3}$–$\frac{1}{2}$ of height of cells and produced into slight ridge over anticlinal wall. Anticlinal walls with thickening tapered from outer ends to moderately thickened inner walls. **Stomata:** superficial, subsidiary cells slightly taller than epidermal cells and raising guard cells slightly above surface; outer half of each subsidiary cell compressed, inner half bulbous; cuticular lips present at inner and outer apertures, the latter well developed and directed outwards. **Chlorenchyma** 4–5-layered, cells with broad short pegs. Cells below stomata with larger pegs than remainder. Cells of outer layers palisade-like, 4–5 times higher than wide, those of inner layers shorter and wider. Ring of chlorenchyma interrupted adaxially by 14–15 files of parenchymatous ground tissue cells extending from leaf centre to epidermis; also interrupted abaxially by 4–5 files of thin or thick-walled cells extending from scl. sheath of median vb to epidermis (Pl. II. A).

Vascular bundles arranged in 1 ring, keel bundle largest (Fig. 14. A; Pl. II. A), smaller vb's alternating with larger, bundles flanking adaxial groove small (Fig. 14. B). Most vb's with 1–2(–3+) narrow to medium-sized mxv's on either flank, rather low down on xylem pole; bulk of xylem composed of narrow tracheids; phloem pole about $\frac{1}{4}$ height of vb, ensheathed by narrow, thick-walled cells; collapsed or tannin-filled cells present in some phloem poles. **Bundle sheaths:** O.S. not distinguishable from ground tissue. I.S. sclerenchymatous, represented by crescentiform caps at poles and extending a short way along bundle flanks. Phloem cap about 8–10 cells thick at centre, xylem cap 2–4(–6) cells thick at centre. **Ground tissue** composed of rounded, parenchymatous cells in 1–2 layers between vb's and chlorenchyma, and extending in band to adaxial groove. Central cells breaking down to form cavity.

Culm surface

As leaf (Fig. 14. F, G, p. 70).

Culm T.S.

Outline circular. **Epidermis:** cells and **stomata** as in leaf. **Chlorenchyma** consisting of palisade cells with short, wide pegs; walls slightly thickened; 2-layered opposite vb's of outer ring, 2–6-layered opposite vb's of inner ring and up to 10-layered where not opposite any vb's. **Vascular bundles** in 2 main rings: medium-sized and small vb's alternating, in outer ring and more or less embedded in chlorenchyma; those of inner ring, large and flanked on either side by air-canals (Fig. 14. E, p. 70; Pl. II. B). Principal and some lesser vb's with 1–4 medium-sized, angular mxv's on either flank. These mxv's with oblique, scalariform-recticulate perforation plates and scalariform lateral wall-pitting. Phloem as in leaf vb's. Outline of outer, smaller vb's rounded, that of medium-sized and larger vb's oval, the longest being 3–4 times taller than broad. Protoxylem canals present in larger vb's. More or less transverse vb's present at irregular intervals. **Bundle sheaths.** O.S. 1-layered, parenchymatous, composed of cells with slightly thickened walls. I.S. reduced to scl. caps; phloem cap 6–7-layered at widest point in smallest vb's and 15–16-layered in largest vb's. Xylem cap consisting of 2–6 fibres in smallest vb's and of a 2–4-layered U in largest vb's.

Sclerenchyma represented by I.S. only, except near to inflorescence where forming a cylinder 10–12 layers thick, embedding all vb's. **Central ground tissue:** areas between larger vb's composed of wide, thin-walled cells, these breaking down to form canals; rounded-angular cells with slightly thickened walls present in 6–7 layers to inner side of vb's, with 1–2-layered extension from this tissue extending along flanks of larger vb's; central area, occupying *c.* $\frac{1}{3}$ diameter of culm, composed of wide, thin-walled cells breaking down to form central cavity. **Silica** and **crystals** absent.

MATERIAL EXAMINED

Marsippospermum grandiflorum (L.f.) Hk.f.: (i) Capt. King 34; Port Famin, Tierra del Fuego (K). (ii) Ball 1882; Porto Bueno, S.E. Patagonia (K). Culm and leaf diameter 2 mm.

TAXONOMIC NOTES

Buchenau (1890) believed *Marsippospermum* to be most closely related to species of *Juncus* belonging to the subgenus *genuini*. This appears to be reasonable on anatomical grounds. The structure of leaf and stem is unlike that of *Rostkovia*, with which *Marsippospermum* has been placed by some earlier authors. Although anatomically similar to the *Junci genuini*, *Marsippospermum* is sufficiently different to warrant its separate generic status.

OXYCHLOË Phil.

(including *Andesia* but excluding *O. clandestina*; see pp. 64–5)

No. of spp. 4?; examined 3.

GENUS DESCRIPTION

Leaf surface

Epidermis: cells 4-sided, between 2 and 9 times longer than wide; walls thick, wavy. **Stomata** not seen.

Leaf base T.S.

Outline sheathing, tapering from centre to either margin, with median, abaxial groove. **Epidermis:** (i) abaxial cells slightly higher than wide, outer walls slightly thickened, others thin; (ii) adaxial cells thin-walled, slightly wider than high. **Stomata:** none seen. **Vascular bundles** in 1 row, medium-sized and small vb's mostly alternating; vessels and tracheids all narrow. **Bundle sheaths.** O.S. parenchymatous, incomplete; I.S. sclerenchymatous, represented by caps at poles. Some caps at phloem poles extending as parallel-sided girders to epidermis. **Chlorenchyma:** none recognizable. **Ground tissue** consisting of 2–3 layers of rounded cells next to adaxial epidermis, 1 abaxial layer and of 2–4 layers of parenchyma cells to either side of each vb. **Air-canals** present between parenchyma girders.

Lamina T.S. (Fig. 12. A, C)

Outline oval, with slight ridges opposite some scl. girders (and hence, opposite vb's). **Epidermis:** cells wholly abaxial; 4-sided, mostly about as high as wide; outer walls thick; anticlinal walls with tapering thickening, being thickest at their outer ends; inner walls slightly thickened. **Stomata** superficial; subsidiary cells thin-walled, as tall as or taller than epidermal cells and about twice as tall as guard cells; guard cells with lips at inner and outer apertures, lumina broadly triangular. **Chlorenchyma** 4–8-layered, cells of outer 4–5 layers interdigitating, each $1\frac{1}{2}$–2 times higher than wide, cells of inner layers more or less hexagonal; pegs present, intercellular spaces narrow. Layers interrupted by scl. girders. In tangential view, cells arranged in more or less transverse plates, some cells of adjacent plates touching by means of pegs, giving a net-like appearance.

Vascular bundles 12–19 in 1 ring in specimens examined; of 3 distinct sizes, largest each with group of 2–4 medium-sized angular vessels on either flank, smallest with only narrow tracheary elements. Perforation plates of medium-sized mxv's oblique, scalariform; lateral wall-pitting also scalariform. Protoxylem canals absent. Phloem pole circular, situated in arms of xylem arc and abutting directly on xylem. Keel bundle recognizable as one of larger bundles; large and smaller bundles not regularly alternating. **Bundle sheaths.** O.S. parenchymatous, cells rounded, with slightly thickened walls, completely ensheathing smallest bundles, but present only at flanks of largest bundles, where difficult to distinguish from mxv's in T.S.; I.S. sclerenchymatous, not or poorly developed round smallest bundles, present as caps to larger vb's. Fibres in 2–4 layers at xylem pole, frequently extending into girders at phloem pole. Girders of larger vb's, more or less parallel-sided or widest next to epidermis, 2–5 cells wide and *c.* 20 cells high. Girders to medium-sized bundles incomplete, more or less triangular with apex directed outwards. **Sclerenchyma** restricted to bundle sheaths and girders just described. **Ground tissue:** wide, polyhedral cells with thin walls; air-spaces present at angles and between adjacent wall-pairs. **Air-canals** and **tannin** absent. **Crystals** solitary, present in some cells of chlorenchyma and epidermis; irregular in outline.

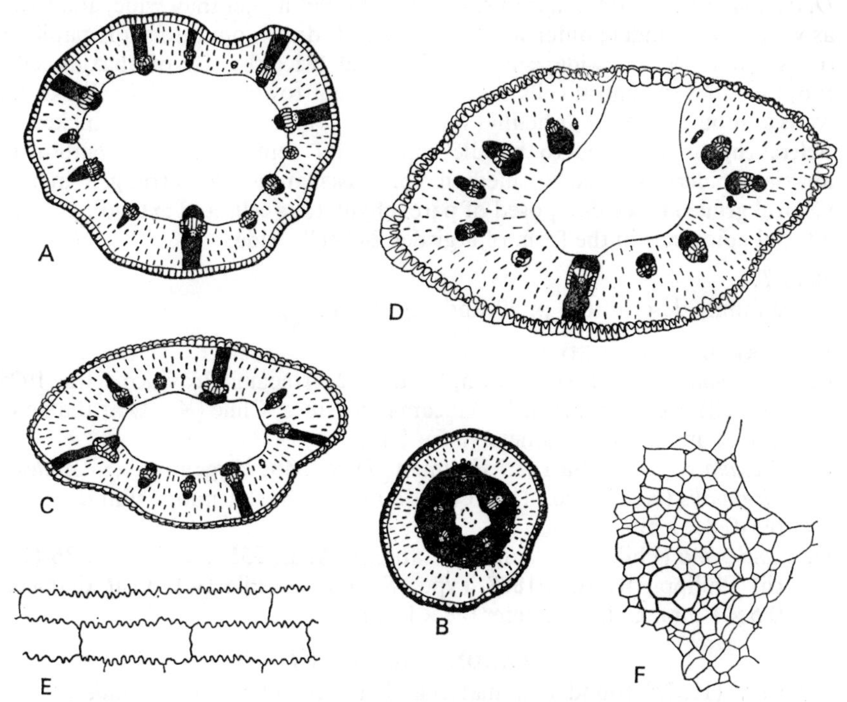

FIG. 12. Juncaceae. A, B, *Oxychloë bisexualis*: A, diagram of leaf T.S. (×50); B, diagram of culm T.S. (×50). C, *O. andina*, diagram of leaf T.S. (×50). D–F, *O. clandestina*: D, diagram of leaf T.S. (×50); E, surface detail of leaf epidermis (×200); F, detail of root T.S., showing inner part of cortex, endodermis and vascular tissue (×200).

Culm surface: as in leaf.

Culm T.S. (Fig. 12. B)

Outline sub-circular. **Epidermis:** cells as in leaf. **Stomata:** none seen. **Chlorenchyma** consisting of 3–4 layers of more or less isodiametric cells with intercellular spaces at angles. **Parenchyma sheath** discontinuous, 1-layered, cells rounded with slightly thickened walls; present to inner side of chlorenchyma. **Sclerenchyma** as 6–8-layered cylinder with all vb's embedded in it. **Vascular bundles** in 1–2 obscurely defined rings; smallest vb's outermost, with small phloem pole and few narrow tracheids; largest with 1–2 medium-sized-narrow, thin-walled, angular mxv's on either flank; px without air-cavity; phloem abutting directly on xylem. **Bundle sheaths** not distinct from scl. cylinder. **Central ground tissue** parenchymatous, cells breaking down at culm centre. **Silica and tannin:** none seen.

O. clandestina differs from *O. andina* and *O. bisexualis* in the following respects.

Leaf surface: see Fig. 12. E.

Leaf T.S. (Fig. 12. D)

Outline oval, often with adaxial groove. **Epidermis.** (i) Abaxial cells as in *O. andina* except at false margins, where 2–3 times higher than wide, and twice as wide as remainder; outer walls of some cells drawn out into short papillae. (ii) Adaxial 8–9 cells wide, central cells about twice the size of abaxial cells; those to sides of central cells reducing in height towards abaxial cells. Outer walls and outer ends of anticlinal walls v. thick, inner walls slightly thickened. **Sclerenchyma** restricted to bundle caps and 1 central girder extending to abaxial epidermis opposite median vb. **Vascular bundles** arranged in arc. **Central ground tissue** composed of parenchymatous cells and extending to the adaxial epidermis in the form of a girder 2–4 cells wide.

Root T.S.

See family description, p. 20 and Fig. 12. F.

MATERIAL EXAMINED

Oxychloë andina Phil.: (i) Johnson, I. M. 6052; Atacama Province, Chile 1926 (K). (ii) Werdermann 453; Atacama Province, Chile (K). Leaf diameter 0·8 × 1 mm. Culm diameter 0·5 × 1 mm.

O. bisexulais (Kze.) Barros: Hauman Type 346; Argentina (= *Andesia bisexualis* (Kze.) Kze.) (K). Leaf diameter 0·8 × 1·1 mm. Culm diameter 0·5–0·8 × 1 mm.

O. clandestina (Phil.) Hauman: (i) Cabrera, A. L. 3557; Chile 19.1.36 (K). (ii) Philippi, R. A. 2/1888 Type; Santiago, Chile (K). Leaf diameter 0·5 × 1 mm. Culm diameter 0·8 × 1 mm.

TAXONOMIC NOTES

Barros (1957*a*) found new material that links *Oxychloë* to *Andesia*, and cited *Andesia* as a synonym for *Oxychloë*.

The leaf of *O. clandestina* (= *Patosia clandestina*) often has a groove right up the blade as far as the mucro. It is not cylindrical as figured by Hauman (1915). It has only 1 scl. girder, as stated by Buchenau (1906) and Barros (1953*b*), not several, as described by other authors. Examination of type material by the present writer confirms the observations made by Barros (1953*b*) and shows clearly that other authors after Buchenau have not described the correct species under the name *O. clandestina*. They could possibly have mistaken small specimens of *O. andina* or *Andesia* for *O. clandestina*.

Castillon (1926) followed Hauman's treatment of *O. clandestina*, but found it necessary to revive the genus *Patosia* and named a new species, *P. tucumanensis*. He gave a photograph of the leaf of this species in T.S., which shows a very similar structure to that found by Buchenau (1906) for *Patosia clandestina*. The main points of difference are slight, namely the lack of a median, abaxial sclerenchyma girder and papillate marginal cells in Castillon's material.

Since the leaves of *O. clandestina* and *P. tucumanensis* are so similar to one another and different anatomically from those of other members of *Oxychloë* seen by the present writer, there seems to be good reason for reseparating *Patosia*, as suggested by Barros (1953*b*). The taxonomy of the genus *Oxychloë* is obviously in need of revision.

Guédès (1967) described the leaf anatomy of *O. andina* and *O. castellanosii*, mentioning in particular the median ligule present in both species. Guédès did not disclose the source of his material, but his description of the anatomy of *O. andina* corresponds closely with the structure found in material described by the present author.

PRIONIUM E. Mey.

Genus and Species Description

Leaf. Leaves *c.* 1 m long, 10–20 cm wide at base, tapering at first rapidly and then gradually to apex; maximum thickness 0·7 mm. Base completely encircling rhizome.

Leaf surface (Fig. 13. E)

Epidermis. Cells of both surfaces alike, but consisting of 2 types: (i) cells in bands between stomata, 2–3 times wider than long, 4–6-sided; (ii) cells in same bands as stomata, irregular in outline, often larger than those of type (i), particularly at ends of stomata. Walls thin or slightly thickened, not wavy. Cells of both types arranged in regular, longitudinal files (Fig. 13. E). **Stomata** paracytic (sometimes appearing tetracytic, but not really so); subsidiary cells as long as guard cells, normally with rounded ends; walls of guard cells slightly thickened, ends rounded. Cells of leaf margin and midrib forming small teeth or prickles (see figs. in *Bot. Mag.* 5722). **Cuticular marks** and **tannin:** none seen.

Leaf T.S. (Fig. 13. A–D, F; Pl. III. A)

Outline V-shaped, triangular near apex (Fig. 13. F); wings tapering to margins and to either side of midrib; midrib thick, pointed abaxially and domed adaxially (Fig. 13. C). One wing of lamina wider than other. **Prickles** on margins and midrib, multicellular, cells thick-walled. **Epidermis:** abaxial and adaxial cells alike, except at midrib, where cells on adaxial surface much larger than others; most cells small, with thin or slightly thickened walls; as wide as to twice as wide as high; anticlinal walls not wavy. **Stomata** situated in slight longitudinal furrows, superficial, present on both surfaces. Guard cells with pronounced cuticular lips at inner and outer apertures, walls evenly and slightly to moderately thickened; subsidiary cells about 1½ times higher than guard cells, extending inwards beyond them. **Hypodermis:** adaxial 1-layered, abaxial 1–2-layered, cells thin-walled, translucent, hexagonal, as wide as, but twice as high as epidermal cells. **Vascular bundles** *c.* 80 in specimen examined, arranged in 1 row, equidistant from each leaf surface. Transverse connections between longitudinal vb's frequent and fairly regular. Larger and medium-sized vb's alternating with smaller vb's. Outline of bundles oval (Fig. 13. B; Pl. III. A); phloem pole occupying about 2/5 height of vb, semicircular in outline, with numerous, small sclerified cells similar in diameter to sieve-tubes and particularly numerous to abaxial side of phloem. Phloem pole in large bundles separated from xylem by 1 or 2 layers of narrow cells with slightly thickened walls. Xylem pole of larger bundles with 1 or 2 wide, angular to oval mxv's on either flank. Perforation plates slightly oblique, scalariform-reticulate. Lateral wall pitting also scalariform.

Bundle sheaths: O.S. parenchymatous, cells wide, 1–2-layered, completely encircling most bundles but absent from flanks of some of largest; produced into girders 2–4 cells wide at either pole; I.S. sclerenchymatous, 1–2(–3–4)-layered, well-developed round large bundles only, confined to group of 2–3 fibres at phloem pole of small bundles; fibres at phloem poles with thickest walls. **Chlorenchyma** arranged in blocks between vb's and their accompanying girders of parenchyma, cells of outer layers palisade-like, those of inner region isodiametric, all with pegs or lobes; each block of chlorenchyma *c.* 8–10 cells wide at outer ends, but reduced to *c.* 3–4 cells towards the middle. The broad portion, particularly the abaxial, frequently with a central region of wide, thin-walled, parenchymatous cells lacking chloroplasts. At certain places individual chlorenchyma strands clearly exhibiting a figure-of-eight outline (Fig. 13. D; Pl. III. A). **Sclerenchyma:** fibres in strands of 1–25 cells; strands scattered through parenchyma girders and also in inner hypodermis. **Air-canals** formed by breakdown of regions of thin-walled cells in chlorenchyma. Canals septate, diaphragms chlorenchymatous, frequently occurring at same level in both canals of any one chlorenchyma strand. **Tannin** present in occasional cells throughout parenchyma and mesophyll.

Culm T.S. 1 cm × 5 mm.

Outline tetragonous. **Cuticle** slightly thickened. **Epidermis:** cells as high as wide, outer wall convex, moderately thickened, other walls slightly thickened. **Stomata** superficial; subsidiary cells slightly higher than guard cells; guard cells with pronounced thin lips at outer and inner apertures. **Hypodermis** represented by 2 layers of wide, thin-walled cells, slightly larger than those of epidermis; interspersed with small, longitudinal strands of fibres. **Chlorenchyma:** 7–8-layered, consisting of more or less isodiametric cells with intercellular spaces at angles. **Sclerenchyma** present as bundle sheaths and hypodermal strands; strands usually 2–3 cells wide and composed of between 1 and 10 fibres. **Vascular bundles** numerous, scattered throughout culm centre; smallest outermost, largest near centre. Large bundles with 1–4 medium-sized or wide mxv's on either flank. Phloem pole separated from xylem by 1–2 layers of narrow cells with slightly thickened walls. Narrow sclereids scattered throughout phloem and aggregated into a strand either at centre of pole, or adjacent to scl. sheath. **Bundle sheaths:** O.S. parenchymatous, distinguishable as 1 layer of cells around smallest, outer bundles only. I.S. sclerenchymatous, fibres in 3–4 layers at poles and on flanks; fewer layers present round smaller bundles. **Central ground tissue** composed of medium-sized, more or less isodiametric, thin-walled cells. **Tannin** scattered in isolated cells throughout ground tissue.

Rhizome morphology

Upright, massive, 5–6 cm in diameter for most of its length, tapering rapidly to apex; about 10 cm long. Clothed by persistent leaf bases, each older leaf surrounding rhizome for about $\frac{9}{10}$ circumference, younger leaves completely encircling. Sticky, resinous substance present between leaves at base. Numerous adventitious roots present. Apical meristem at base of conical depression.

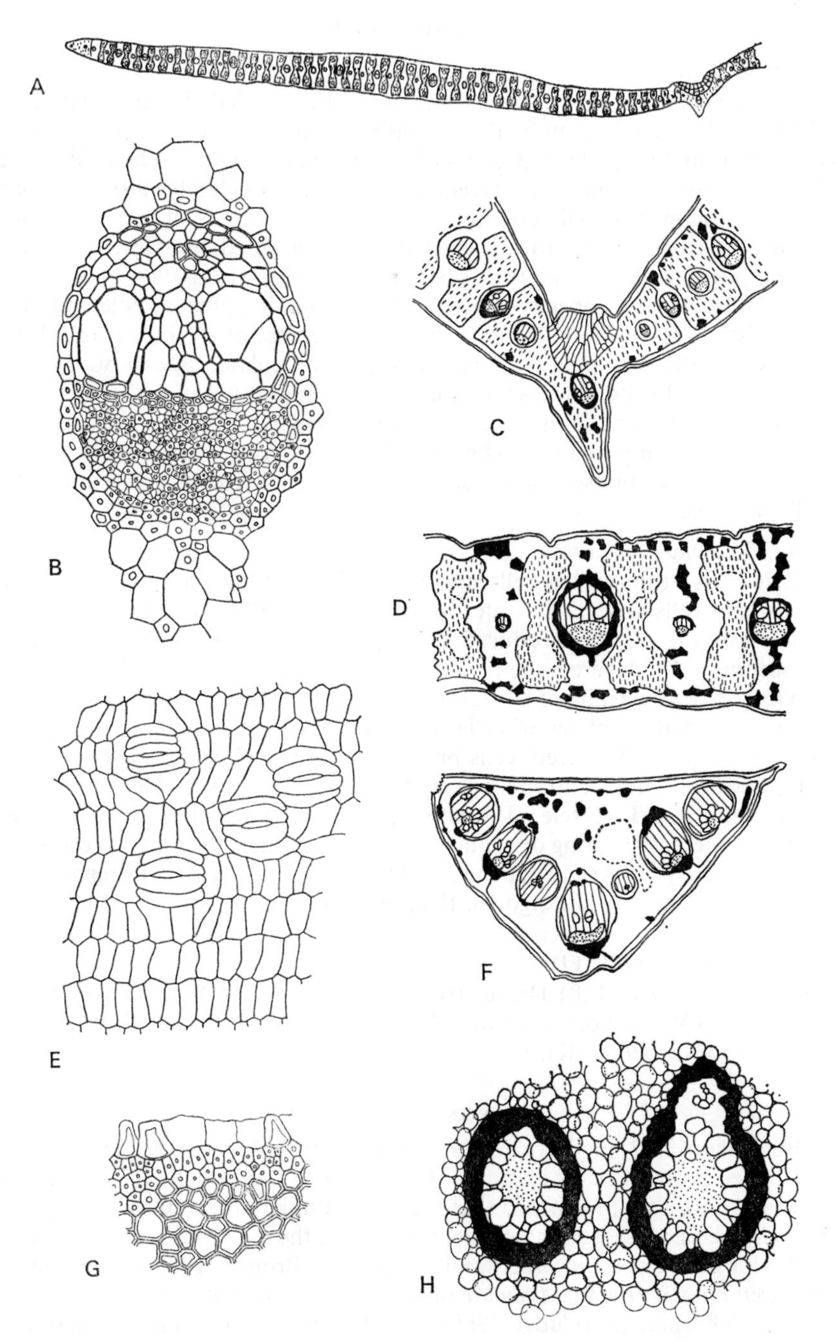

FIG. 13. Juncaceae. A–H, *Prionium serratum*: A, diagram of half of leaf T.S. (×8); B, vascular bundle of leaf in T.S. (×200); C, diagram of leaf midrib T.S. (×50); D, diagram of part of leaf blade T.S. (×50); E, surface detail of abaxial leaf epidermis (×200); F, diagram of leaf T.S. near apex (×50); G, detail of outer part of root T.S. showing thick-walled cells in epidermis (×200); H, diagram of vascular bundles from T.S. of upright 'rhizome' (×50).

Rhizome T.S.

Outline circular. **Epidermis:** cells small, thin-walled. **Stomata** not seen. **Cortex** 7–8-layered, composed of rounded, thin-walled cells separated from central ground tissue by 2–3 layers of hexagonal cells with slightly thickened, lignified walls. **Endodermoid layer** not distinct. **Vascular bundles** in 3 main zones as follows: (i) collateral leaf traces each with an arc of mxv's enclosing rounded phloem pole; (ii) intermediate zone of fused bundles; (iii) inner, major portion of rhizome with scattered, amphivasal bundles, each having 1 ring of wide, angular mxv's surrounding a rounded phloem pole and often with distinct px pole outside mxv ring, facing centre of rhizome (Fig. 13. H). Vessel perforation plates slightly oblique, reticulate; lateral wall-pitting scalariform. **Bundle sheaths:** 1–3 layers of narrow fibres with moderately thickened walls and pointed ends. **Sclerenchyma** present as subepidermal strands, each composed of 2–6 fibres with moderately thickened walls. **Central ground tissue** composed of wide, round, thin-walled cells and large intercellular spaces.

Root T.S. Diameter 3 mm.

Epidermis: thin-walled cells and root-hairs making up most of surface; occasional cells with thick outer and anticlinal walls and thin inner walls; outer wall of such cells produced into a point, dome, or mucro (Fig. 13. G). **Exodermis:** 1–2-layered, cells narrow, hexagonal, thick-walled. **Cortex.** (i) Outer part consisting of 8+ layers of hexagonal cells. (ii) Middle part composed of radiating plates of cells separated from one another by air-spaces. (iii) Inner part 2–3-layered, cells procumbent, oval, thin-walled. **Endodermis** 1-layered; cells similar to those of inner cortex, but with thickened inner and anticlinal walls. **Pericycle:** 1–2 layers of small, thin-walled cells. **Vascular system:** xylem with 1 ring of solitary or paired angular, oval mxv's; each mxv surrounded by 1 layer of narrow, flattened tracheids. **Central ground tissue** composed of narrow, hexagonal, thick-walled cells.

Material Examined

Prionium serratum (L.f.) Drège: (i) Strey, R. G. 6228; Natal, Uvongo, *c.* 8 miles S.W. of Port Shepstone 19.12.65 (S). (ii) Codd, L. E. 9704; Port Shepstone 6.9.56 (K).

Taxonomic Notes

Buchenau (1893) made an extensive study of this plant and has given a well-illustrated account of its anatomy. He described the upright rhizome as the 'inflorescence axis'. He noted that the plant, unlike others in the family, has leathery perianth segments. He mentioned that the scl. strands not associated with vb's are not rare in other families, quoting Bromeliaceae as an example. It is well known now that they are present in numerous other families, including the Thurniaceae (Cutler 1965) and Rapateaceae (Carlquist 1966), and even within the Juncaceae, e.g. in *J. maritimus* and *J. acutus.*

Buchenau considered the longitudinal air-canals of the leaf to be characteristic of water plants and he erroneously stated that they had not been found previously in Juncaceae. As we have seen, however, air-canals between the vb's or in the centre of the leaf of some cylindrical-leaved species are common in the Juncaceae.

Tannin was recorded in isolated cells of root and rhizome by Buchenau. He also mentioned (1890) a 'wax' covering to the epidermis (cuticle?).

Haslinger (1914) published a shortened version of Buchenau's account, and did not reach any fresh conclusions regarding the affinities of *Prionium*.

Zimmermann and Tomlinson (1968) studied the course of vb's in the rhizome.

Prionium appears to be only distantly related to other genera in the Juncaceae. Apart from the anatomical distinctiveness, there are morphological differences. The closed leaf sheath is a feature shared with *Luzula*, but the floral parts are like those of *Juncus*.

The present author holds the view that *Prionium* should properly be removed from the Juncaceae. This opinion, which is based on anatomical and morphological observations (e.g. the unusual leaf chlorenchyma with air-canals, the distribution of sclerenchyma in leaf and culm), needs testing with additional evidence from cytology, palynology, etc.

ROSTKOVIA Desv.

No. of spp. 2; examined 1.

Genus and Species Description

Leaf surface

Epidermis. Cells 4-sided, walls v. wavy, slightly to moderately thickened. Adaxial cells 2–3 times wider than abaxial. Cells between stomata (over fibres) as long as or up to 3 times longer than wide; cells surrounding stomata mostly about as long as wide, some up to 3 times longer than wide. **Stomata** as in culm, restricted to 2 bands, each up to about 15 cells wide, one on either flank of abaxial surface. **Cuticular marks:** fine granules in longitudinal files.

Leaf T.S (Fig. 14. H, J). Width 0·9–1·1 mm, thickness 0·6–0·7 mm.

Outline sub-hemispherical with slight, median, abaxial groove; adaxial surface slightly concave, edges rounded. **Cuticle** slightly thickened. **Epidermis.** (i) Abaxial cells mostly slightly wider than high and $\frac{1}{2}$–$\frac{2}{3}$ width of largest adaxial cells. Outer walls of all cells v. thick, irregular, convex and slightly raised over anticlinal walls; inner and anticlinal walls slightly to moderately thickened. (ii) Adaxial cells confined to concave portion of leaf surface, central cells largest, about as high as wide, cells reducing in size to either side, the marginals being similar to abaxial cells. Epidermal cells more flattened in sections near to leaf tip. **Stomata** confined to bands on leaf flanks; superficial; guard cells with pronounced lips at outer and inner apertures; subsidiary cells about twice the height of guard cells, narrow next to guard cell and wider to inside. **Hypodermis** present except opposite stomatal bands; composed of 4–6 layers of narrow, thick-walled fibres. Fibres absent from areas opposite adaxial epidermis in some leaves. **Chlorenchyma** composed of 4–5 layers of low palisade cells with numerous, short pegs; occurring to inside of hypodermis; cells of inner layers grading into rounded cells surrounding air spaces and vb's. Cells opposite stomatal bands taller than remainder, in

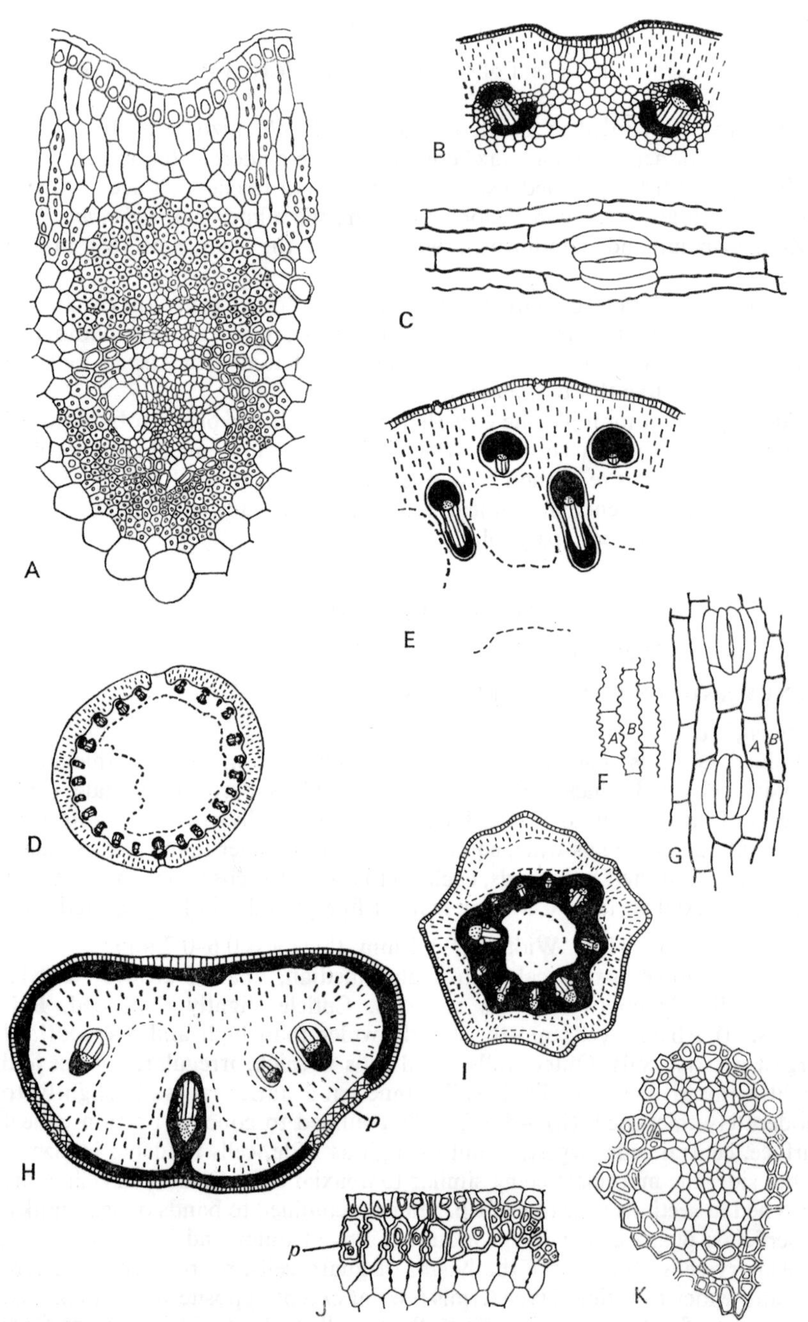

FIG. 14. Juncaceae. A–G, *Marsippospermum grandiflorum*: A, T.S. vascular bundle opposite abaxial groove of leaf (×200); B, diagram of adaxial groove of leaf T.S. (×50); C, surface detail of leaf epidermis (×200); D, diagram of leaf T.S. (×12); E, diagram of sector of culm T.S. (×50); F, detail of epidermal cells of culm, surface view, high focus (×200); G, as F, low focus (×200); cells marked A and B correspond in Figs. F and G. H–K, *Rostkovia magellanica*: H, diagram of leaf T.S. (×50); I, diagram of culm T.S. (×50); J, detail of leaf T.S. showing protective cells (p) beneath the stomatal region of the epidermis (×200); K, culm vascular bundle in T.S. (×320).

1 layer, with slightly to moderately thickened walls and few, large pegs (Fig. 14. J); similar to protective cells of Restionaceae (see p. 120).

Vascular bundles in arc of (3)–5 bundles, median largest. Frequently only median bundle exhibiting conspicuous mxv's, and these often narrow and numerous. Xylem pole composed mainly of medium-sized tracheids and narrow xylem parenchyma cells. Phloem pole semicircular or more or less triangular in outline, abutting directly on xylem; often with several scattered, sclerified cells. **Bundle sheaths:** O.S. parenchymatous, cells medium-sized, in 1(–2) layer(s), frequently encircling lateral vb's, discontinuous round phloem pole of median vb; I.S. sclerenchymatous, those of smaller vb's frequently reduced to caps, the caps at the phloem poles being largest; sheaths of large and median bundles complete and joined abaxially (at phloem pole) to hypodermis. **Air-canals** present between vb's, formed by breakdown of large, thin-walled translucent cells. **Tannin** and **crystals:** none seen.

Culm surface

Epidermis: cells 4-sided, 2–6 times longer than wide, walls wavy, slightly thickened. **Stomata:** subsidiary cells longer than guard cells, guard cells with rounded ends and narrow apertures.

Culm T.S. (Fig. 14, I, K). Diameter 0·5–0·8 mm.

Outline circular, with several unevenly spaced ridges. **Epidermis:** cells about as high as wide; outer walls thin, convex, with ridges over anticlinal walls, other walls moderately thickened. **Stomata** superficial, as in leaf. **Hypodermis** 1-layered, cells with slightly thickened walls, otherwise similar to chlorenchyma. **Chlorenchyma:** cells irregularly lobed, mostly wider than high, several times longer than wide as seen in L.S.; cells in 6–8 layers, extending to scl. cylinder, with large air-spaces between ridges of scl. cylinder. **Sclerenchyma:** cylinder with ridges opposite all vb's, ridges largest opposite largest vb's; cylinder 10–12 cells thick between vb's, walls of cells moderately thickened, lumina wide. **Vascular bundles** in 1 ring, larger alternating with smaller; 2-several conspicuous, medium-sized to narrow mxv's on flanks of largest bundles only. Phloem poles rounded, many small cells next to xylem (Fig. 14. K). **Bundle sheaths** sclerenchymatous, continuous with scl. cylinder. **Central ground tissue** parenchymatous, breaking down to form central cavity. **Air-canals** present in chlorenchyma and culm centre.

MATERIAL EXAMINED

Rostkovia magellanica (Lam.) Hk. f.: (i) Vallentin, E.; W. Falkland Is. Sept.–Oct. 1910 (K). (ii) Vallentin, E.; Shallow Bay, Jan. 1911 (K). (iii) Smith, J.; S. Georgia M 1014, 28.11.57 (K). (iv) Smith, J.; King Edward's Pt., S. Georgia M 1015, 28.11.57 (K). (v) Campbell Is. 1640, An. 1840 (K).

TAXONOMIC NOTES

The unusual leaf of *Rostkovia*, with stomata in 2 bands and a well-developed hypodermis, is unlike that of other genera in the family. This applies even to species of *Juncus*, in the subgroup *poiophylli*, to which *Rostkovia* shows its strongest morphological resemblance. However, the anatomical variation is well within the limits generally accepted for family likeness in other families, for example, the Restionaceae.

VOLADERIA R. Benoist

No. of spp. 1.

GENUS DESCRIPTION

(Adapted from original description by Benoist 1937.)

Leaf T.S. only

Outline channelled. **Cuticle** thick. **Epidermis:** cells with thick outer walls, inner and anticlinal walls thin. **Stomata** superficial. **Chlorenchyma** not described. **Vascular bundles:** 5, in arc, all with xylem poles facing adaxial surface; median bundle largest, flanking bundles smallest. **Bundle sheaths:** O.S. not described; I.S. sclerenchymatous. **Sclerenchyma:** abaxial girders joined to I.S. of each vb; also 3 rectangular adaxial girders, one opposite median and either flanking vb; several rows of parenchymatous cells of ground tissue joining median adaxial girder to median vb. **Ground tissue** composed of thin-walled, parenchymatous, compact cells. **Air-canals** absent.

TAXONOMIC NOTE

The status of *Voladeria* is uncertain since it was originally described from male material only. The gross morphology is similar to that of *Oxychloë* (incl. *Andesia*) and *Distichia*, but it has 3 stamens only. An illustration of the leaf in T.S. is given by Benoist (1937).

JUNCACEAE.
LISTS OF GENERA AND SPECIES IN WHICH CERTAIN DIAGNOSTIC CHARACTERS OCCUR

Species with more or less dorsiventrally flattened or channelled leaves which are without marginal sclerenchyma strands

Juncus bulbosus
*J. capensis**
J. capitatus
J. castaneus (small, in Buchenau 1890; absent from Turner material)
J. falcatus
J. luzuliformis var. *potnanii**
*J. maximowiczii**
J. planifolius
*J. sphacelatus**
*J. sphenostemon**
J. trifidus
*Luzula elegans**
L. lutea (sometimes a few fibres present)
Prionium serratum

Species with papillae on leaf (L) or culm (C)

Juncus mutabilis (L, C)
J. subulatus (L, C)
Luzula abyssinica (L)

Species in which largest leaf vb's are without conspicuous mx vessels

Distichia muscoides
D. tolimensis
Juncus capitatus
J. falcatus
J. scheuchzerioides

* In literature.

Species with no adaxial leaf epidermis for most of their length

Distichia muscoides
D. tolimensis
Juncus acutiflorus
J. acutus
J. alpinoarticulatus
J. amabilis†
J. articulatus
J. australis†
J. biglumis
J. conglomeratus
J. effusus
J. filicaulis†
J. filiformis
J. gregiflorus†
J. inflexus
J. maritimus
J. mutabilis
J. pallidus†
J. pauciflorus†
J. prismatocarpus
J. procerus†
J. sarophorus†
J. scheuchzerioides
J. striatus
J. subnodulosus
J. subsecundus†
J. subulatus
J. triglumis
J. usitatus†
J. xiphioides var. _triandra_
Oxychloë andina
O. bisexualis

Species with reduced or very reduced adaxial leaf epidermis

Juncus bulbosus
_J. imbricatus_*
_J. jacquinii_*
_J. plebejus_*
_J. setaceus_*
J. squarrosus
Marsippospermum grandiflorum

Species with abaxial sclerenchyma girder(s) in leaf
All _Junci genuini_ and _thalassii_ seen

Juncus bufonius (some specimens)
_J. capillaceus_*
_J. imbricatus_*
_J. plebejus_*
_J. secundus_*
_J. tenuis_*
_J. venturianus_*
Luzula abyssinica
L. arcuata
L. campestris
L. echinata
L. forsteri
L. glabra
L. lutea
L. luzuloides
L. multiflora
L. pilosa
L. sylvatica
Marsippospermum grandiflorum
Oxychloë andina
O. bisexualis
O. clandestina (partial)
_Voladeria_*

Species with adaxial sclerenchyma girder(s) in leaf

_Juncus imbricatus_*
_J. leptocaulis_*
_J. marginatus_*
Luzula abyssinica
L. arcuata
L. campestris
L. echinata
L. forsteri
L. glabra
L. lutea
L. luzuloides
L. multiflora
L. pilosa
L. sylvatica
_Voladeria_* (partial)

Species with air-canals present between leaf vb's

Distichia muscoides (leaf sheath)
D. tolimensis (leaf sheath)
Juncus bufonius
J. bulbosus
J. castaneus
J. compressus
J. gerardii
_J. plebejus_*
_J. secundus_*
_J. singularis_*
J. tenuis
J. trifidus
Luzula abyssinica
(_L. arcuata_)
(_L. campestris_)
L. echinata
(_L. forsteri_)
L. glabrata
L. luzuloides
L. multiflora
L. oldfieldii
_L. parviflora_*
L. pilosa
L. sylvatica
Rostkovia magellanica

(   ) = poorly developed.

† 'Stems' of these spp. are described by Edgar 1964; it is highly probable that the leaf and stem anatomy is identical in these spp., except at the apex; see p. 42.
* In literature.

Species with sunken stomata in leaf

Juncus acutus
*J. australis**
*J. fontanesii** (slightly)
J. planifolius
*J. sarophorus**
J. subulatus
J. xiphioides var. *triandra*

Species with subepidermal sclerenchyma strands or girders in culm

Juncus acutus
*J. amabilis**
*J. australis**
*J. beringensis**
J. conglomeratus
*J. digestus**
J. effusus
*J. filicaulis**
J. filiformis
*J. gregiflorus**
J. inflexus
J. maritimus
*J. pallidus**
*J. patens**
*J. pauciflorus**
*J. procerus**
*J. ramboi**
*J. sarophorus**
*J. subsecundus**
*J. usitatus**
Prionium serratum

Species with mechanical cylinder in the culm

Distichia muscoides
D. tolimensis
Juncus acutiflorus
J. alpinoarticulatus
J. articulatus
J. biglumis
J. bufonius
J. bulbosus
*J. capillaceus**
J. capitatus
J. castaneus
J. compressus
J. falcatus
J. gerardii
J. mutabilis
*J. obtusatus**
J. obtusiflorus (weak)
J. prismatocarpus
J. scheuchzerioides
J. squarrosus
J. striatus
J. subnodulosus (weak)
J. subulatus
J. tenuis
J. trifidus (weak)
J. triglumis
J. uruguensis
J. xiphioides var. *triandra*
Luzula abyssinica
L. arcuata
L. campestris
*L. confusa**
L. echinata
L. glabra
L. lutea
L. luzuloides
*L. spicata**
Marsippospermum grandiflorum (near apex)
Oxychloë andina
O. bisexualis
Rostkovia magellanica

* In literature.

Species in which largest culm vb's are without conspicuous mx vessels

Juncus biglumis
J. castaneus
J. triglumis

Bibliography for Juncaceae

Abramski, T. (1911) *Beiträge zur Kenntnis der Juncaceen*. Diss. Breslau. pp. 53 (see *Just's bot. Jber*. **41** (2), Sec. 20, No. 45 (1913).)

Adamson, R. S. (1925) On the leaf structure of *Juncus*. *Ann. Bot*. **39**, 599–612.

Andersson, S. (1888) Om de primära kärlsträngarnes utveckling hos monokotyledonerna. *Bih. svenska VetenskAkad. Handl*. **13** (12), 1–23.

Arber, A. (1922) Leaves of the Farinosae. *Bot. Gaz*. **74**, 80–94.

Areschoug, F. W. C. (1878) *Jemförande undersökningar öfver bladets anatomi*. Lund. pp. 193–5.

Arzt, T. (1937) Die Kutikula bei einigen Pteridophyten, Gymnospermen und Monokotyledonen. *Ber. dt. bot. Ges*. **55**, 437–64.

Backer, C. A. (1951) Juncaceae in *Flora Malesiana* ser. 1, **4**, 210.

Balfour, J. H. (1871) *Juncus effusus* with both spirally twisted and spirally curled leaves. *J. Bot., Lond*. **9**, 281.

Barnard, C. (1958) Floral histogenesis in the Monocotyledons. *Aust. J. Bot*. **6**, 285–98.

Barros, M. (1952) Notas sobre Juncáceas. *Darwiniana* **10**, 65–8.

—— (1953*a*) Las Juncáceas de la Argentina, Chile y Uruguay. *Darwiniana* **10**, 279–460.

—— (1953*b*) Notas sobre Juncáceas. Los géneros *Patosia* y *Oxychloe*. *Lilloa* **26**, 343–6.

—— (1957*a*) Nueva contribucion al conocimiento de los pequeños géneros andinos de las Juncáceas. *Lilloa* **28,** 207–8.
—— (1957*b*) Un *Juncus* nuevo del Brasil meridional. *Darwiniana* **11,** 283–5.
—— (1958) Juncaceae in *Angely. Catál Estatist. Gên. Bot. Fanerogam.* 45.
—— (1962) Las Juncáceas del Estado de Santa Catalina. *Sellowia* **14,** 9–45.
BENOIST, R. (1937) Phanérogames nouvelles de l'Amérique méridionale. *Bull. Soc. bot. Fr.* **84,** 632–9.
BENTHAM, G. (1883) Juncaceae in Bentham and Hooker, *Genera plantarum*, Vol. 3, pp. 861–9.
BLAU, J. (1904) *Vergleichend-anatomische Untersuchung der schweizerischen Juncus-Arten.* Diss. Zürich. pp. 82.
BUCHENAU, F. (1871) Kleinere Beiträge zur Naturgeschichte der Juncaceen. *Abh. naturw. Ver. Bremen* **2,** 365–404.
—— (1874) Die Deckung der Blattscheiden bei *Juncus. Abh. naturw. Ver. Bremen* **4,** 135–8.
—— (1875) Monographie der Juncaceen vom Cap. *Abh. naturw. Ver. Bremen* **4,** 393–512.
—— (1886) Über die Randhaare (Wimpern) von *Luzula. Abh. naturw. Ver. Bremen* **9,** 293–9. See also p. 319.
—— (1890) Monographia Juncacearum. *Bot. Jb.* **12,** 1–495.
—— (1891) Über Knollen- und Zwiebelbildung bei den Juncaceen. *Flora, Jena* **74,** 71–83.
—— (1893) Über den Aufbau des Palmiet-Schilfes (*Prionium serratum* Drège) aus dem Caplande. Eine morphologisch-anatomische Studie. *Biblthca bot.* **5** (27), pp. 26.
—— (1906) Juncaceae in *Das Pflanzenreich* IV, 36, pp. 284.
BUCHHOLZ, M. (1921) Über die Wasserleitungsbahnen in den interkalaren Wachstumszonen monokotyler Sprosse. *Flora, Jena* **114,** 119–86.
CARLQUIST, S. (1966) Anatomy of Rapateaceae—roots and stems. *Phytomorphology* **16,** 17–38.
CASTILLON, L. (1926) Contribución al conocimiento de las Juncáceas argentinas. *Mus. Hist. nat., Univ. Tucumán* (7), pp. 1–50.
CATESSON, E. A. M. (1953) Structure, évolution et fonctionnement du point végétatif d'une monocotylédone: *Luzula pedemontana* Boiss. et Reut. (*Joncacées*). *Annls Sci. nat.*, Bot. ser. 11, **14,** 253–91.
CHAUVEAUD, G. (1897*a*) Sur le rôle des tubes criblés. *Revue gén. Bot.* **9,** 427–30.
—— (1897*b*) Recherches sur le mode de formation des tubes criblés dans la racine des monocotylédones. *Annls Sci. nat.*, Bot. ser. 8, **4,** 307–81.
CHEADLE, V. I. (1955) The taxonomic use of specialization of vessels in the metaxylem of Gramineae, Cyperaceae, Juncaceae and Restionaceae. *J. Arnold Arbor.* **36,** 141–57.
CUTLER, D. F. (1965) Vegetative anatomy of Thurniaceae. *Kew Bull.* **19,** 431–41.
DUFOUR, L. (1896) Note sur les relations qui existent entre l'orientation des feuilles et leur structure anatomique. *Bull. Soc. bot. fr.* **33,** 268–76.
DUVAL-JOUVE, J. (1869) Sur quelques tissus de *Juncus* et de Graminées. *Bull. Soc. bot. fr.* **16,** 404–10.
—— (1871) Sur quelques tissus de Joncées, de Cyperacées et de Graminées. *Bull. Soc. bot. fr.* **18,** 231–9.
—— (1872) Des quelques *Juncus* à feuilles cloisonnées. *Revue Sci. nat.* 117–50.
—— (1873) Diaphragmes vasculifères des monocotylédones aquatiques. *Mém. Acad. Montpellier* **8,** 157–76.
EBINGER, J. E. (1963) A new subgenus in *Luzula* (Juncaceae). *Brittonia* **15,** 169–74.
EDGAR, E. (1964) The leafless species of *Juncus* in New Zealand. *N.Z. Jl. Bot.* **2,** 177–204.
ERDTMAN, G. (1952) *Pollen morphology and plant taxonomy. Angiosperms.* Chronica Botanica Co., Waltham, Mass. 539 pp.
—— BERGLUND, B., and PRAGLOWSKI, J. (1961) An introduction to a Scandinavian pollen flora. *Grana palynol.* **2,** 3–92.
FREIDENFELT, T. (1904) Der anatomische Bau der Wurzel in seinem Zusammenhange mit dem Wassergehalt des Bodens. *Biblthca bot.* **12** (61), pp. 118.
GEESTERANUS, R. A. M. (1941) On the development of the stellate form of the pith cells of *Juncus* spp. *Proc. K. Ned. Akad. Wet.* **44,** 489–501, 648–53.
GÉRARD, R. (1881) Recherches sur le passage de la racine à la tige. *Annls Sci. nat.*, Bot. ser. 6, **11,** 279–430.

GRAEBNER, P. (1895) Studien über die norddeutsche Heide. *Bot. Jb.* **20,** 500–654.
—— (1934) Juncaceae in: Kirchner, O. von, Loew, E. and Schröter, C. *Lebensgesch. Blütenpfl. Mitteleur.* **1,** 3, 80–221.
GREISS, E. A. M. (1957) Anatomical identification of some ancient Egyptian plant materials. *Mém. Inst. Égypte* **55,** 165 pp.
GRISEBACH, A. (1845) *Spicilegium florae rumelicae et bithynicae.* II, pp. 404–5.
GUÉDÈS, M. (1967) Stipules médianes et stipules ligulaires chez quelques Liliacées, Joncacées et Cypéracées. *Beitr. Biol. Pfl.* **43,** 59–103.
GULLIVER, G. (1863) On the pith-cells of Juncaceae. *A. Mag. nat. Hist.* ser. 3, **12,** 472.
HÄMET-AHTI, L. (1965) *Luzula piperi* (Cov.) M. E. Jones, an overlooked woodrush in western North America and Eastern Asia. *Aquilo,* ser. Bot. **3,** 11–21.
HASLINGER, H. (1914) Vergleichende Anatomie der Vegetationsorgane der Juncaceen. *Sber. Akad. Wiss. Wien* Math.-naturw. Kl. **123,** 1147–94.
HAUMAN, L. (1915) Note sur les Joncacées des petits genres Andins. *An. Mus. nac. Hist. nat. B. Aires* **27,** 285–306.
HAURI, H. (1916–17) Anatomische Untersuchungen an Polsterpflanzen nebst morphologischen und ökologischen Notizen. *Beih. bot. Zbl.* **33** (1), 275–93.
HEALY, A. J. (1953) The identification and distribution of rushes in New Zealand. *Proc. 6th N.Z. Weed Control Conf.,* pp. 5–16.
HEGNAUER, R. (1963) *Chemotaxonomie der Pflanzen. II Monocotyledoneae.* Birkhäuser Verlag, Basel and Stuttgart. pp. 540.
HERRIOTT, E. M. (1905) On the leaf-structure of some plants from the Southern Islands of New Zealand. *Trans. N.Z. Inst.* **38,** 377–422.
HOLM, T. (1926) The bulbiferous form of *Luzula multiflora. Rhodora* **28,** 133–8.
—— (1929) The application of the term 'rhizome'. *Rhodora* **31,** 6–17.
IRMISCH, T. (1855) Morphologische Mittheilungen über die Verzweigung einiger Monocotylen. *Bot. Ztg.* **13,** 57–63.
JOHNSON, L. A. S. (1963) New species of *Juncus* in Australia and New Zealand. *Contr. N.S.W. Herb.* **3,** 241–4.
KORSMO, E. (1954) *Anatomy of weeds.* Grøndahl and Søns Forlag, Oslo. pp. 413.
KRECHETOVICH, V. I., and GONCHAROV, N. F. (1935) Juncaceae Vent. in Komarov, V. L., *Flora of the U.S.S.R.,* Vol. **3,** pp. 504–76 (400–55 in English translation, 1964).
KŘÍSA, B. (1963) Taxonomische Stellung der Art *Juncus alpinus* Vill. s.l. in der europäischen Flora. *Novit. bot. Hort. Bot. Univ. Carol. Prag.,* pp. 28–35.
KUNTZE, O. (1903) in POST, T. VON, and KUNTZE, O. *Lexicon Generum Phanerogamarum,* p. 303. Stuttgart.
LAURENT, M. (1904) Recherches sur le développement des Joncées. *Annls Sci. nat.,* Bot. ser. 8, **19,** 97–194.
LAUX, W. (1887) Ein Beitrag zur Kenntnis der Leitbündel im Rhizom monokotyler Pflanzen. Diss. Berlin. pp. 49. Also *Verh. bot. Ver. Prov. Brandenb.* **29,** 65–111.
LE BLANC, — (1912) Sur les diaphragmes des canaux aérifères des plantes. *Revue gén. Bot.* **24,** 233–43.
LEWIS, F. T. (1926) An objective demonstration of the shape of cells in masses. *Science* **63,** 607–9.
LINSBAUER, K. (1930) *Die Epidermis* (*Handb. PflAnat.,* 1/2, Vol. 4). Berlin. 283 pp.
LÖV, L. (1925–6) Zur Kenntnis der Entfaltungszellen monocotyler Blätter. *Flora, Jena* **120,** 283–343.
MARTIN, D. J. (1955) Features on plant cuticle. *Trans. Proc. bot. Soc. Edinb.* **36,** 278–88.
MAYR, F. (1914) Hydropoten an Wasser- und Sumpfpflanzen. Diss. Erlangen. pp. 98. Also *Beih. bot. Zbl.* **32** (1), 278–371 (1914–15).
MEYER, F. J. (1962) *Das trophische Parenchym. A. Assimilationsgewebe* (*Handb. PflAnat.* **4,** 7A). 2nd edn. Borntraeger, Berlin. 188 pp.
NORDENSKIÖLD, H. (1961) Modes of species differentiation in the genus *Luzula. Recent Adv. Bot.* **2,** 1469–73.
PEISL, P. (1957) Die Binsenform. *Ber. schweiz. Bot. Ges.* **67,** 99–213.
PETERS, T. (1927) Über die Bedeutung der inversen Leitbündel für die Phyllodien-Theorie. *Planta* **3,** 90–9.

PETERSEN, O. G. (1873–4) Bemaerkninger om den anatomiske bygning af rod og rodstok hos nogle Monocotyledoner. *Bot. Tidsskr.* ser. 2, **3,** 210–11.

PORSCH, O. (1905) *Der Spaltöffnungsapparat im Lichte der Phylogenie.* Gustav Fischer, Jena. pp. 196.

PRIESTLEY, J. H., and HINCHLIFF, M. (1922) The physiological anatomy of the vascular plants characteristic of peat. *Naturalist, Hull,* 263–8.

RAUNKIAER, C. (1895–9) *De danske blomsterplanters naturhistorie,* Vol. 1, pp. 383–417. Kjøbenhavn.

ROTHERT, W. (1885) *Vergleichend-anatomischen Untersuchungen über die Differenzen im primären Bau der Stengel und Rhizome krautiger Phanerogamen.* Diss. Dorpat.

SATAKE, Y. (1933) Systematic and anatomical studies on some Japanese plants, II (Juncaceae). *J. Fac. Sci. Tokyo Univ.* **4,** 131–223.

SCHWENDENER, S. (1874) *Das mechanische Princip im anatomischen Bau der Monokotylen mit vergleichenden Ausblicken auf die übrigen Pflanzenklassen.* Leipzig. 179 pp.

SHIELDS, L. M. (1952) Architecture of herbaceous plants. *Scient. Mon., N.Y.* **74,** 269–72.

SIEDLER, P. (1887–92) Über den radialen Saftstrom in den Wurzeln. *Beitr. Biol. Pfl.* **5,** 407–42.

SIFTON, H. B. (1945) Air-space tissue in plants. *Bot. Rev.* **11,** 108–43.

SNOGERUP, S. (1960) Studies in the genus *Juncus* II. Observation on *Juncus articulatus* L. x *bulbosus* L. *Bot. Notiser* **113,** 246–56.

SNOW, L. M. (1914) Contributions to the knowledge of the diaphragms of water plants. I. *Scirpus validus. Bot. Gaz.* **58,** 495–517.

—— (1920) —— II. Effect of certain factors upon development of air chambers and diaphragms. *Bot. Gaz.* **69,** 297–317.

SOUÈGES, R. (1933) Recherches sur l'embryogénie des Joncacées. *Bull. Soc. bot. Fr.* **80,** 51–69.

SPINNER, H. (1913) Étude anatomique de quelques phanérogames rapportées de l'Himalaya par le Dr. Jacot-Guillarmod. *Bull. Soc. neuchâtel. Sci. nat.* **39,** 3–19.

STAUDERMANN, W. (1924) Die Haare der Monokotylen. *Bot. Arch.* **8,** 105–84.

STEBBINS, G. L., and KHUSH, G. S. (1961) Variation in the organization of the stomatal complex in the leaf epidermis of monocotyledons and its bearing on their phylogeny. *Am. J. Bot.* **48,** 51–9.

TOBLER, J. (1957) *Die mechanischen Elemente und das mechanische System* (*Handb. PflAnat.* **4,** 6). Borntraeger, Berlin.

VAN TIEGHEM, P. (1887*a*) Sur les poils radicaux géminées. *Annls Sci. nat.*, Bot. ser. 7, **6,** 127–8.

—— (1887*b*) Structure de la racine et disposition des radicelles dans les Ériocaulées, Joncées, Mayacées et Xyridées. *J. Bot., Paris* **1,** 305–15.

—— (1891) Sur les tubes criblés extralibériens et les vaisseaux extraligneux. *J. Bot., Paris* **5,** 117–28.

VIERHAPPER, F. (1929) Kerner von Marilaun, A. *Das Pflanzenleben der Donauländer.* Innsbruck Univ. Verlag.

—— (1930) Juncaceae in Engler and Prantl, *Die natürlichen Pflanzenfamilien,* 2nd edn, Vol. **15A,** pp. 192–224.

VORONIN, N. S. (1920) Nekotorye osobennosti korneobrazovaniya u *Juncus.* (Some specific characteristics of root-formation in *Juncus.*) *Bot. J. U.S.S.R.* **45,** 1359–60.

WEBERBAUER, A. (1905) Anatomische und biologische Studien über die Vegetation der Hochanden Perus. Vorläufige Mitteilung. *Bot. Jb.* **37,** 60–93.

WEIMARCK, H. (1946) Studies in Juncaceae, with special reference to the species in Ethiopia and the Cape. *Svensk bot. Tidskr.* **40,** 141–78.

WRIGHT, F. R. E. (1948) Notes on growth form of tall rush-like plants in Braunton Burrows. *NWest Nat.* **23,** 15–19.

ZIEGENSPECK, H. (1939) Die Micellierung der Turgeszenzmechanismen. I: Die Spaltöffnungen (mit phylogenetischen Ausblicken). *Bot. Arch.* **39,** 268–309, 332–72.

ZIMMERMANN, M. H., and TOMLINSON, P. B. (1968) Vascular construction and development in the aerial stem of *Prionium* (Juncaceae). *Am. J. Bot.* **55,** 1100–9.

THURNIACEAE

(Figs. 15–19; Pl. III. B)

INTRODUCTION

THE family consists of 3 species belonging to the single genus *Thurnia*. It is endemic to Guyana and certain parts of the Amazon valley. All 3 species are sedge-like herbs, bearing elongated, leathery leaves with sheathing bases.

Formerly the genus *Thurnia* was included in the Rapateaceae (Baillon 1894) or in the Juncaceae as a *genus anomalum* (Bentham 1883). It was later given the status of a distinct family, the Thurniaceae, within the Farinosae (Desde 1907) and placed beside the Rapateaceae and Bromeliaceae in the Bromelineae. Hutchinson (1959) has more recently placed the family in the Juncales, together with Juncaceae, Restionaceae, and Centrolepidaceae.

Two species, *Thurnia jenmanii* and *T. sphaerocephala*, were available for study.

GENUS AND SPECIES DESCRIPTION

Leaf surface

Epidermis. Cells of costal zones as long as to 2–3 times longer than wide, those of intercostal zones either similar to costal cells, or up to 3 times wider than long; walls thin in *T. jenmanii* (Fig. 15. B, E), thick in *T. sphaerocephala* (Fig. 17. C, F), slightly wavy. Occasional enlarged, tannin-containing cells present in *T. sphaerocephala*. Cells over median vascular bundle bulliform in *T. jenmanii* (Fig. 15. I) and wider than other cells in *T. sphaerocephala*. **Stomata** confined to intercostal zones, more frequent on abaxial than adaxial surface; paracytic to rarely tetracytic. **Hypodermis:** parenchymatous hypodermal cells protruding between epidermal cells in *T. jenmanii* in distal $\frac{3}{5}$ of leaf; such cells reaching leaf surface; occasional fibres also superficial. See Figs. 15. D, F, G; 16. A–D.

Leaf T.S. (Figs. 15. A; 17. A; Pl. III. B)

Outline: dorsiventrally flattened, V-shaped in *T. sphaerocephala*, flat in *T. jenmanii*. **Epidermis:** cells wider than high to slightly higher than wide; those of adaxial surface the shorter in *T. jenmanii*, cells of both surfaces of similar height in *T. sphaerocephala*; walls thick in *T. sphaerocephala*, thin in *T. jenmanii*. **Stomata** superficial. **Hypodermis** developed beneath both surfaces as large parenchymatous cells and 1–2-layered, longitudinal fibre strands; in *T. jenmanii*, adaxial fibre strands 2–(6)–8 cells wide and 1–2 cells deep; abaxial fibre strands composed of 2–(4)–8 cells; some parenchymatous cells of abaxial side of leaf reaching surface (see above and Fig. 15. C, D, G). In *T. sphaerocephala* scl. fibres occurring in a more or less continuous single or double layer, interrupted at intervals by files of parenchymatous cells (Fig. 17. D, E).

Vascular bundles of main system lying parallel to long axis of leaf in plates of parenchymatous tissue connected to abaxial and adaxial hypodermis, separated from one another by air-canals. Large vb's each with 1 or 2 (3)

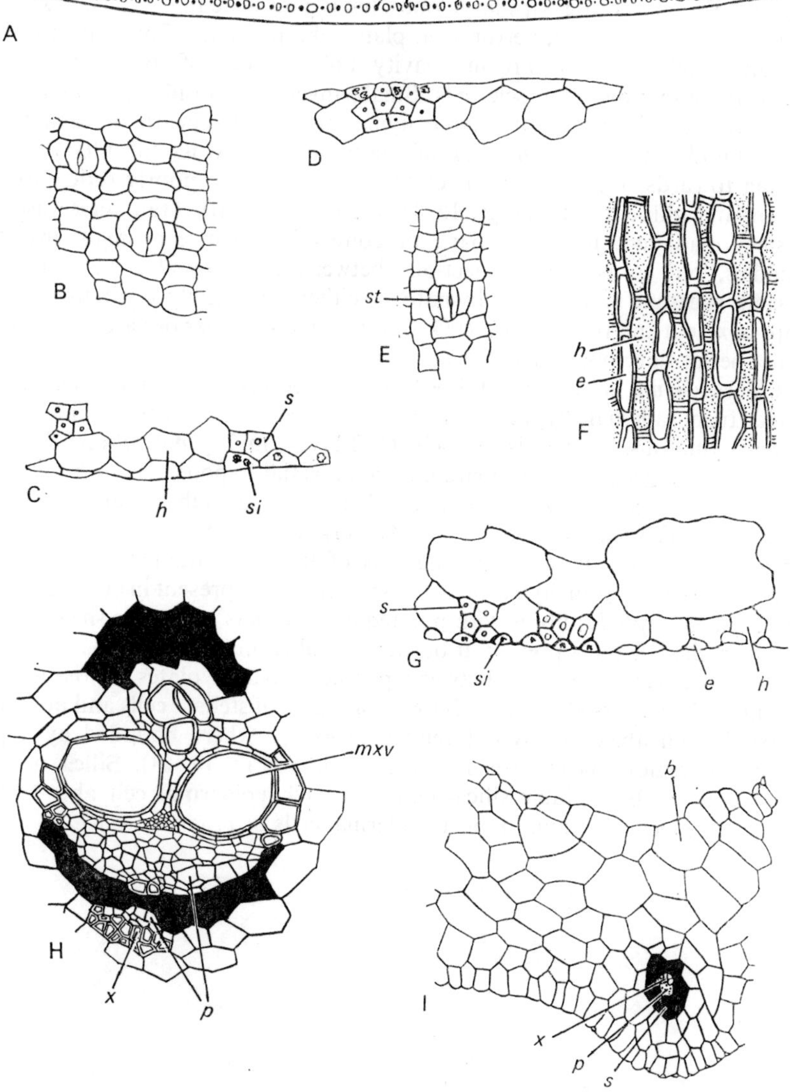

FIG. 15. Thurniaceae. A–I, Leaf anatomy of *Thurnia jenmanii*: A, diagram of leaf T.S. showing vascular bundles and air-canals (×5); B, surface view of abaxial epidermis from basal 1/5 of leaf (×130); C, as B, in T.S., showing hypodermal fibre strands and epidermal silica-bodies (×315); D, adaxial epidermis T.S. from basal 1/5 of leaf (×315); E, as D, surface view (×130); F, surface view of abaxial epidermis from half-way up leaf showing files of epidermal cells separated by extrusive outer walls of hypodermal cells (stippled) (×315); G, as F, in T.S., showing hypodermal fibre strands, extrusive hypodermal cells and epidermal cells with silica-bodies (×315); H, larger vascular bundle in T.S., with inverted subsidiary bundle (×180); I, midrib T.S., showing small vascular bundle with sclerenchyma sheath and adaxial epidermal bulliform cells (×130).
a = air-canal; b = bulliform cells; e = epidermal cell; h = hypodermal cell; mxv = metaxylem vessel; p = phloem; s = sclerenchyma; si = silica-body; st = stoma; x = xylem.

conspicuous angular mx vessels occurring on either flank (Figs. 15. H; 17. B). Small vb's, alternating with large ones, having no wide mx vessels. Mx vessel wall-pitting scalariform, perforation plates oblique, scalariform. Protoxylem usually breaking down to form a cavity. Phloem poles of vb's in *T. sphaerocephala* containing small sclereids. Small, inverted 'subsidiary' vb's present next to abaxial surface of sclerenchymatous bundle sheath of each large, main bundle, with *phloem poles* of main bundle and its subsidiary strand facing towards one another. Inverted vb with well or poorly differentiated xylem and phloem poles (Figs. 15. H, 17. B; Pl. III. B). Transverse vascular system made up, for most part, of connections between small, inverted bundles, but transverse connections between main bundles also present. **Bundle sheaths:** I.S. sclerenchymatous, either completely encircling, or as caps at xylem and phloem poles; O.S. consisting of 1 layer of parenchymatous cells; better developed in *T. sphaerocephala.*

Chlorenchyma composed of 1–9 layers of spongy, thin-walled cells, separated from adaxial hypodermis by several layers of large, thin-walled, parenchymatous, 'water tissue' cells (1–2 layers in *T. jenmanii* and 3 layers in *T. sphaerocephala*), also separated from abaxial hypodermis by 5 layers of similar cells in *T. sphaerocephala.* Thick-walled substomatal sclereids present in *T. sphaerocephala* (Fig. 17. G). **Sclerenchyma** represented by fibre strands lying parallel to long axis of leaf and situated in epidermis, hypodermis and parenchymatous plates; strands also present in chlorenchyma of *T. sphaerocephala*; fibres with pointed ends. Fibres of bundle sheaths with square ends. Sclereids present at borders of substomatal cavity in *T. sphaerocephala.* **Air-canals** present between parenchymatous plates of tissue supporting vb's; traversed at intervals by diaphragms of stellate cells and bordered adaxially and abaxially by chlorenchyma except where bounded by hypodermis in regions near base and apex of leaf (Fig. 17. H). **Silica** present as several small, nodular silica-bodies in each epidermal cell above fibre strands in *T. jenmanii* and in most epidermal cells of *T. sphaerocephala.*

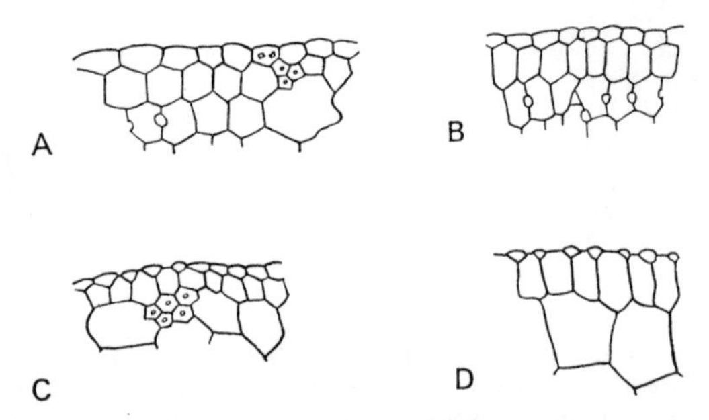

FIG. 16. Thurniaceae. *Thurnia jenmanii*: leaf adaxial epidermis, transverse sections taken at different distances from the base. Total length of leaf 150 cm. Distances from leaf base: A, 23 cm; B, 52 cm; C, 60 cm; D, 70 cm. For explanation, see text, p. 78.

Culm surface

Epidermis: cells more or less rectangular in outline, many 2–3 times longer than wide, long axes parallel with that of culm; walls thin (*T. jenmanii*, Fig.

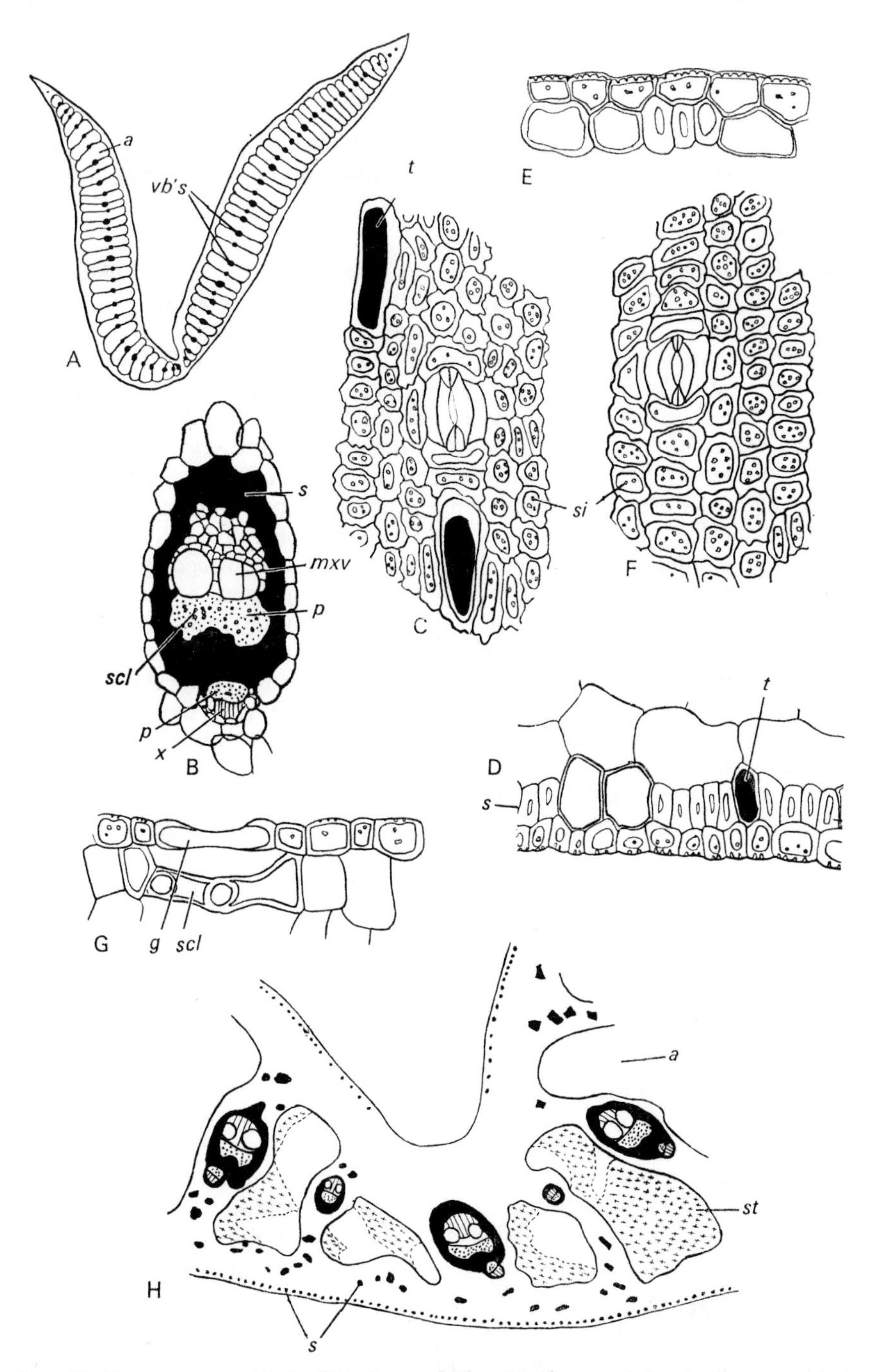

FIG. 17. Thurniaceae. A–H, Leaf anatomy of *Thurnia sphaerocephala*: A, diagram of leaf T.S. (×2·5); B, T.S. larger vascular bundle with subsidiary inverted bundle; note sclereids in phloem (×130); C, surface view of abaxial epidermis, showing thick-walled epidermal cells containing small silica-bodies, and large, tannin-filled cells (×315); D, as C, in T.S.; note pitting of outer wall of epidermal cells, and fibres arranged in a more or less continuous hypodermal layer (×315); E, adaxial epidermis T.S. (×315); F, as E, surface view (×315); G, as C, in L.S., showing guard cell of stoma and thick-walled substomatal sclereids (×315); H, diagram of median fold of leaf in T.S., showing distribution of stellate diaphragm cells (×30).

a = air-canals; g = guard cell; mxv = metaxylem vessel; p = phloem; s = sclerenchyma; scl = sclereid; si = silica-body; st = stellate cells; t = tannin-cells; x = xylem.

18. D) or thick (*T. sphaerocephala*, Fig. 19. C). **Stomata** paracytic, some in *T. sphaerocephala* tetracytic; restricted to regions between subepidermal scl. strands.

Culm T.S. (Figs. 18. A; 19. A)

Outline 3-angled (*T. sphaerocephala*), or 4-angled (*T. jenmanii*). **Epidermis:** cells as high as wide, or up to twice as wide as high. **Stomata** superficial; guard cell walls most heavily thickened at mouth of aperture (Fig. 18. E). **Hypodermis:** in *T. jenmanii* 1–2 layers of large, thin-walled cells and strands of fibres in groups of 1–(6)–20+; in *T. sphaerocephala* fibres in 1 layer, in groups of 2–6 cells separated by files of thin-walled cells. **Cortex:** spongy parenchyma confined to regions next to stomata in *T. sphaerocephala* (Fig. 19. B) but in several layers next to hypodermis in *T. jenmanii*; other cortical cells large, parenchymatous, grading into ground tissue of central cylinder. **Vascular bundles** of 2 main types. (i) Small, peripheral, collateral strands (Fig. 18. B) having phloem situated between arms of arc formed by xylem; mx vessels small, with scalariform wall-pittings and oblique, scalariform perforation plates. (ii) Large vb's in *T. jenmanii* (Fig. 18. C) with 1–3 or more conspicuous mx vessels on either side of xylem arc, the arms of which partly enclose the phloem; protoxylem and tracheids occurring in central part of arc. Vb's of *T. sphaerocephala* exhibiting a range of variation: (*a*) bundles in outer ring with 1 conspicuous mx vessel on either side and 1 central radial row of medium-sized mx vessels terminating at px pole; (*b*) remaining bundles larger, scattered throughout central ground tissue, each with several conspicuous mx vessels on either flank and 1–3 rows of medium-sized mx vessels terminating at px poles. Phloem either U-shaped, with arms of U enclosing an invagination from scl. sheath (Fig. 19. D), or containing a central scl. strand (Fig. 19. E). Subsidiary mx poles frequent on either side of the sclerenchymatous invaginations, or on side away from main xylem pole of a scl. strand embedded wholly in phloem. **Bundle sheaths** sclerenchymatous, frequently widest at phloem pole, appearing toothed when coming into contact with and conforming to shape of large cells of adjacent parenchymatous ground tissue. **Sclerenchyma** represented by hypodermal strands and by the bundle sheaths mentioned above; and in *T. sphaerocephala* also represented by additional strands of fibres, 1–(8)–20 cells in T.S., situated throughout cortex and ground tissue. Sclereids present round border of substomatal cavity in both spp. **Central ground tissue** consisting of wide, thin-walled, polygonal, parenchymatous cells. **Silica** in *T. jenmanii* as (i) sand-like particles in certain parenchyma cells and (ii) silica-bodies resembling those of leaf, in epidermal cells above hypodermal fibre strands. **Tannin** cells scattered.

Root T.S. (Figs. 19. F, G)

Root-hairs thin-walled, scattered; basal cells small. **Cortex** in 3 main zones: (i) outer zone, consisting of 2 layers of small, thick-walled cells grading into 8–10 layers of thin-walled parenchymatous cells of increasing size towards centre of root, the parenchymatous cells being separated from one another by progressively larger intercellular spaces; (ii) middle zone, composed of radiating plates of thin-walled cells, the plates being up to 20 cells deep and 2 cells wide and separated from one another by air-spaces; (iii) inner zone,

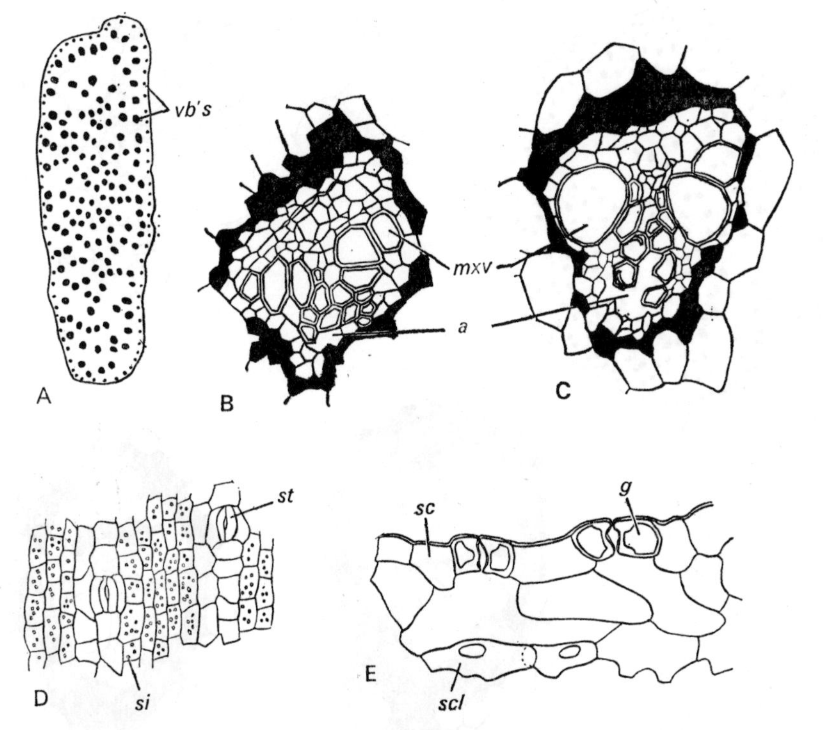

FIG. 18. Thurniaceae. A–E, Culm anatomy of *Thurnia jenmanii*: A, culm T.S. (×5); B, vascular bundle from periphery, T.S. (×180); C, vascular bundle from medulla, T.S. (×180); D, surface view of epidermis; note files of silica-containing cells associated with hypodermal fibres (×130); E, detail of outer part of culm T.S. showing 2 stomata sharing a subsidiary cell (×315).
a = air-canal; g = guard cell; mxv = metaxylem vessel; sc = subsidiary cell; scl = sclereid; si = silica-body; st = stoma; vb's = vascular bundles.

made up of 3–6 layers of large, thin-walled, loosely packed cells. **Endodermis** 1–2-layered; cells with uniformly thick walls and square in T.S.; no passage cells observed. **Pericycle** 1-layered, consisting of small, thin-walled cells. **Vascular system:** phloem strands alternating with px poles, strands in *T. sphaerocephala* enclosed by U-shaped groups of scl. Mx vessels forming 1 ring; wall-pitting scalariform, perforation plates scalariform, oblique/transverse. **Central ground tissue** composed of small, thick-walled cells.

Rhizome T.S.

(*T. sphaerocephala* only)

Exhibiting 3 main regions: (i) outer region, with entrant leaf bases and small, collateral vb's; (ii) intermediate region, with larger vb's fused together; (iii) inner region, with larger, variously orientated vb's. Mx vessel elements short; lateral wall pitting scalariform, perforation plates scalariform, oblique to slightly oblique. **Bundle sheaths** composed mainly of thick-walled, elongated fibres with pointed ends, but fibres next to phloem poles shorter and wider, with thinner walls. **Tannin** scattered in ground tissue.

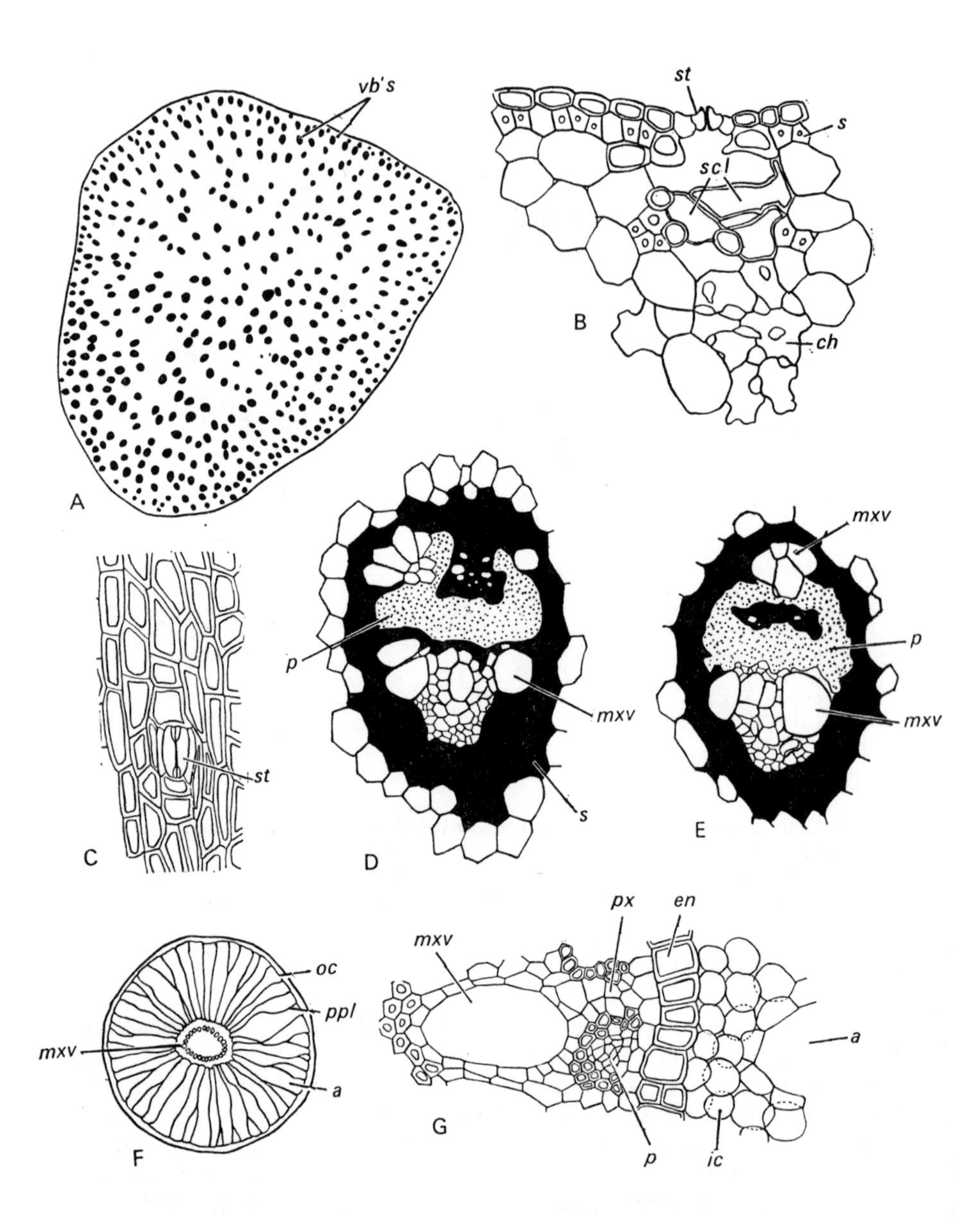

FIG. 19. Thurniaceae. A–G, Culm and root anatomy of *Thurnia sphaerocephala*: A, culm T.S. (×5); B, detail of A, showing hypodermal fibres, substomatal sclereids and spongy chlorenchyma (×150); C, surface view of culm epidermis (×120); D, diagram of T.S. of a medullary vascular bundle showing subsidiary metaxylem poles and invagination from sclerenchyma sheath into phloem (×225); E, as D, but with strand of sclerenchyma in phloem (×225); F, root T.S. (×5); G, detail of root T.S. showing thick-walled endodermis, conspicuous, thin-walled metaxylem vessel, and phloem in horse-shoe of sclerenchyma (×180).

a = air-canal; ch = chlorenchyma; en = endodermis; ic = inner cortex; mxv = metaxylem vessel; oc = outer cortex; p = phloem; ppl. = parenchyma plate; px = protoxylem; s = sclerenchyma; scl = sclereid; st = stoma.

Material Examined

Thurnia jenmanii Hook. f.: Sandwith, N.Y. 1444 (S). Leaf width 17 mm; max. thickness 0·7 mm. Culm diameter 8×2·5 mm. Root diameter 2·8 mm.
T. sphaerocephala (Rudge) Hook. f.: (i) Maguire and Wurdack 41471 (K). (ii) Maguire and Fanshawe 23540 (K). Leaf width 28 mm; max. thickness 2·5 mm. Culm diameter *c.* 14 mm. Root diameter 4·5 mm.

Taxonomic Notes

The main anatomical basis for the separation of *Thurnia* from other supposedly related families is provided by the vascular anatomy of the leaf. Inverted bundles with their distinctive arrangement were first described for *Thurnia* by Cutler (1963) and, as stated then, they are at present unknown in the leaves of any other plants. Solereder and Meyer (1929) observed these bundles but erroneously described them as strands of phloem or procambial tissue. They were unable to comment on their orientation, because they did not recognize them as complete vascular bundles. Inverted bundles occur in equitant leaves of some *Iris* spp. and in various other monocotyledonous families (Metcalfe 1963), but the xylem poles of these bundles face towards one another (Arber 1925). No member of the Juncaceae has leaf bundles in 2 rows; bundles may be concentric or elliptic in arrangement but their xylem poles face towards one another (e.g. Buchenau 1890; this work, pp. 33, 40). Metcalfe (1963) has described a system of large and small double-tiered vascular bundles for the leaves of *Lagenocarpus* (Cyperaceae) but shows the smaller bundle of each pair to have its xylem pole facing towards the phloem pole of the larger bundle.

Silica-bodies are found in the cells adjacent to fibre strands, a common occurrence in certain families of monocotyledons, for example Cyperaceae, Gramineae, Restionaceae, and Rapateaceae. The silica-bodies, being spheroidal and nodular, are unlike any found in the Gramineae (Metcalfe 1960), but are similar to those found in the Rapateaceae (Carlquist 1966), Restionaceae (Solereder and Meyer 1929, Cutler 1965) and some members of the Cyperaceae (Metcalfe 1963). Silica-bodies have not been seen in Juncaceae (pp. 6, 8 of this work).

The unusual way in which cells of the hypodermis of the leaf of *T. jenmanii* penetrate to the surface is of interest. This mode of development is not known to have been described in any other monocotyledon and provides one more character which points to an isolated taxonomic position for Thurniaceae.

It is also interesting to note that while stomata are mainly paracytic some are occasionally tetracytic. In this respect Thurniaceae are not unlike Cyperaceae where tetracytic stomata are known to occur sporadically in a few genera amongst the many in which stomata are paracytic.

The culm has a scattered distribution of vascular bundles, a character shared by many other monocotyledons and represented in the Cyperaceae and Gramineae. The distribution is unlike that in members of the Juncaceae (except *Prionium*), Restionaceae, or Centrolepidaceae. The vascular bundles themselves in *T. sphaerocephala* are unusual (p. 82). It is possible that they may represent regions of bundle fusion, a situation which could be clarified by a study of serial sections.

As stated by Cutler (1965) the two species of the Thurniaceae studied share several distinctive characters which serve to hold them together taxonomically and equally to separate them from families that have been suggested as close relatives. It would seem advisable, in view of the anatomical evidence, to exclude the Thurniaceae from the Juncaceae, *sensu* Hutchinson. There is no anatomical support for placing them near to any particular one of the other suggested relatives.

Bibliography for Thurniaceae

Adamson, R. S. (1925) On the leaf structure of *Juncus*. *Ann. Bot.* **39,** 599–612.

Arber, A. (1925) *Monocotyledons*. Cambridge Botanical Handbooks, p. 232.

Baillon, H. E. (1894) *Hist. pl., Genève* **13,** 238–40, 244.

Bentham, G. (1883) Juncaceae in Bentham and Hooker, *Genera plantarum* Vol. **3,** pp. 861–9.

Buchenau, F. (1890) Monographia Juncacearum. *Bot. Jb.* **12,** 1–495.

Carlquist, S. (1966) Anatomy of Rapateaceae—roots and stems. *Phytomorphology* **16,** 17–38.

Cutler, D. F. (1963) Inverted vascular bundles in the leaf of the Thurniaceae. *Nature, Lond.* **198,** 1111–12.

—— (1965) Vegetative anatomy of Thurniaceae. *Kew Bull.* **19,** 431–41.

Desde, —— (1907) Thurniaceae in Engler, *Syllabus Pflanzenfamilien*, 5th edn, p. 94.

Hutchinson, J. (1959) *The families of flowering plants* II. *Monocotyledons*, 2nd edn, pp. 698–9. Clarendon Press, Oxford.

Metcalfe, C. R. (1960) *Anatomy of the monocotyledons* I. *Gramineae*, p. lx. Clarendon Press, Oxford.

—— (1963) Comparative anatomy as a modern botanical discipline. *Adv. bot. Res.* **1,** 101–47.

Solereder, H., and Meyer, F. J. (1929) *Systematische Anatomie der Monokotyledonen*, Heft 4, pp. 70–3.

CENTROLEPIDACEAE

(Figs. 20–4)

INTRODUCTION

MEMBERS of the Centrolepidaceae are small, tufted perennial or annual herbs, similar in habit to the smaller representatives of the grasses, sedges, or rushes; some are moss-like in appearance. They are mostly 1·5–5(–7·5) cm tall, but one (*Centrolepis cambodiana*) reaches 25 cm. The leaves are linear or thread-like and crowded or closely imbricate. They often have an acute hyaline apex, and those of many spp. are ligulate. The inflorescence, subtended by 1–3 glume-like bracts, is raised on a short culm. The rhizome, present in perennial species, is short and upright. It is frequently morphologically indistinguishable from the culm base.

The family is distributed mainly in Australia and New Zealand but members also occur in New Guinea, Borneo, Philippines, Cambodia, and South America. There are 6 genera and about 35–40 species. The taxonomic limits of the genera are very confused in the literature. Dr. R. Melville, The Herbarium, Kew, has very kindly advised on the correct nomenclature of the species described here.

MATERIALS AND METHODS

About half the material examined was freshly fixed; the remainder came from reconstituted herbarium specimens. Some spp. are very small and it was necessary to embed them in wax prior to sectioning. The remainder of the material was prepared and sectioned according to the methods described on p. 12. Hairs were examined in the freshly fixed spirit material, or in the boiled and fixed herbarium material, without clearing or staining.

FAMILY DESCRIPTION

LEAF

Abaxial surface

Epidermis. Cells usually 4-sided but 6-sided in some spp., from 1–(15)–30+ times longer than wide; marginal cells frequently narrower and more elongated than others, present in leaves with flattened or arc-shaped section. Walls of cells wavy in few spp., normally moderately thickened. (In most spp. of all genera except *Gaimardia* the end walls of epidermal cells are oblique with respect to the leaf surface. One end of each cell shortly overlaps the next cell in the same file. These end walls are deeply concave in certain spp., with a spur or lobe on either side of the concavity. The spurs are directed towards the base of the leaf; see Fig. 20. B, F.)

Hairs multicellular, with slightly or moderately thickened or thin walls, present in certain species of *Aphelia*, *Centrolepis*, *Gaimardia*, and *Pseudalepyrum*. Hair-types divisible into 2 main classes: A, unbranched filaments; B, branched filaments. Class A consisting of 5 categories. (i)–(iv) With tapering,

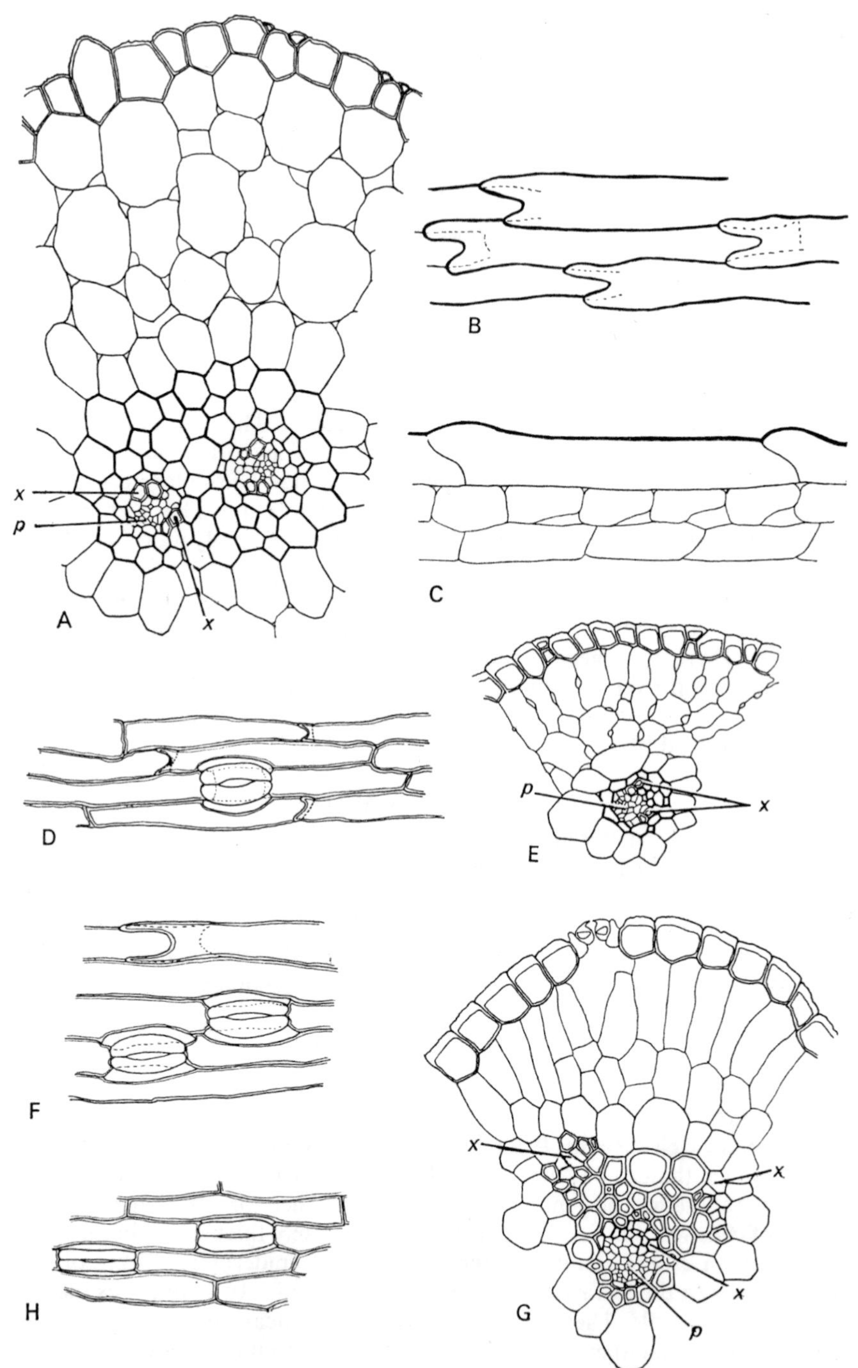

FIG. 20. Centrolepidaceae. A–C, E, *Pseudalepyrum monogynum*: A, detail of culm T.S. (×200); B, leaf surface showing overlapping end walls of epidermal cells (×200); C, leaf, outer part, L.S. showing overlap of end walls of epidermal cells and arrangement of chlorenchyma cells (×200); E, detail of leaf T.S. (×200). D, *Centrolepis cambodiana*, abaxial leaf surface (×320). F, *C. exserta*, abaxial leaf surface showing concave, overlapping transverse wall of epidermal cell (×320). G, H, *C. polygyna*: G, detail of culm T.S., showing group of 3 vascular bundles (×320); H, detail of leaf surface (×320).
p = phloem; x = xylem.

pointed apical cell and 2–7 cells in filament. (i) Basal cell inflated, wider and frequently shorter than cells of filament (Fig. 21. B–E). (ii) Basal cell short, next cell inflated (Fig. 21. F, G). (iii) Basal cell with expanded foot, tapering distally, where only slightly wider than filament cells (Fig. 21. H–K). (iv) Basal cell shorter than but otherwise similar to filament cells (Fig. 21. A). (v) Basal cell as in type (iii) but apical cell with reflexed hook (Fig. 21. L, M). Class B, consisting of one category only. (i) Basal cell with enlarged foot, filament single or double, branching at 1 or several levels, with 2 or many branches, producing a woolly, tangled mat. Branches of hairs arising as lateral outgrowths of cells in a manner abnormal for angiosperms (Fig. 21. N–Q). Hairs also noted by Solereder and Meyer (1929) in *Centrolepis drummondii* (of type iii), *C. strigosa*, and *Aphelia cyperoides* (thick-walled and 2–5-celled, the last sp. having a crooked or recurved terminal cell, i.e. of type (v)).

Papillae dome-shaped, present in certain spp. at proximal end of each cell, above or near to inclined transverse cell wall. **Stomata** paracytic in all spp. seen except *Trithuria filamentosa*, where anomocytic; guard cell pair oval in outline, slightly constricted centrally in many spp. (e.g. Fig. 20. H, *Centrolepis polygyna*); subsidiary cells frequently with acute ends (e.g. Fig. 20. F) and often longer than guard cells. **Cuticular marks:** faint, parallel, longitudinal striations present in many spp. **Inclusions:** rectangular bodies present in some epidermal cells of *Gaimardia australis*; granular material found in some epidermal cells of one sample of *G. setacea.*

Adaxial surface

As abaxial surface in most spp. **Epidermis:** cells wider than those of abaxial surface in some spp. of *Gaimardia.* **Stomata** absent from certain spp. with distinctive adaxial surface.

LEAF T.S.

Outline circular, oval, dorsiventrally flattened, or oblong truncate. **Cuticle** frequently with several small ridges to each epidermal cell. **Epidermis.** Cells always in 1 layer, but frequently appearing 2-layered by overlap of ends of cells (see surface description). (i) Abaxial cells usually about as high as wide; walls uniformly thickened, or outer wall v. much thicker than others; anticlinal walls not wavy. (ii) Adaxial resembling abaxial cells in most spp. but sometimes larger.

Hairs and **papillae:** see leaf surface. **Stomata** superficial; guard cells frequently with small lip at outer aperture: subsidiary cells often taller than guard cells and extending further into leaf. Stomata absent from adaxial surface in some spp. and not seen in either epidermis in certain other spp. **Hypodermis:** complete in *Trithuria* only, as 1 layer of cells twice as wide as those of epidermis (Fig. 22. J). **Vascular bundles.** Majority of spp. with 1, central vb only; most other spp. with 3, comprising 1 large central and 2 smaller, lateral vb's. Xylem pole composed of narrow vessels and tracheids or, in some spp., possibly of tracheids only. Lateral wall-pitting of metaxylem vessels scalariform or scalariform-reticulate, perforation plates (where seen) oblique, scalariform. Phloem pole small, outline oval or semicircular in most spp., conspicuously overarching and partially enclosing xylem in *Centrolepis aristata.*

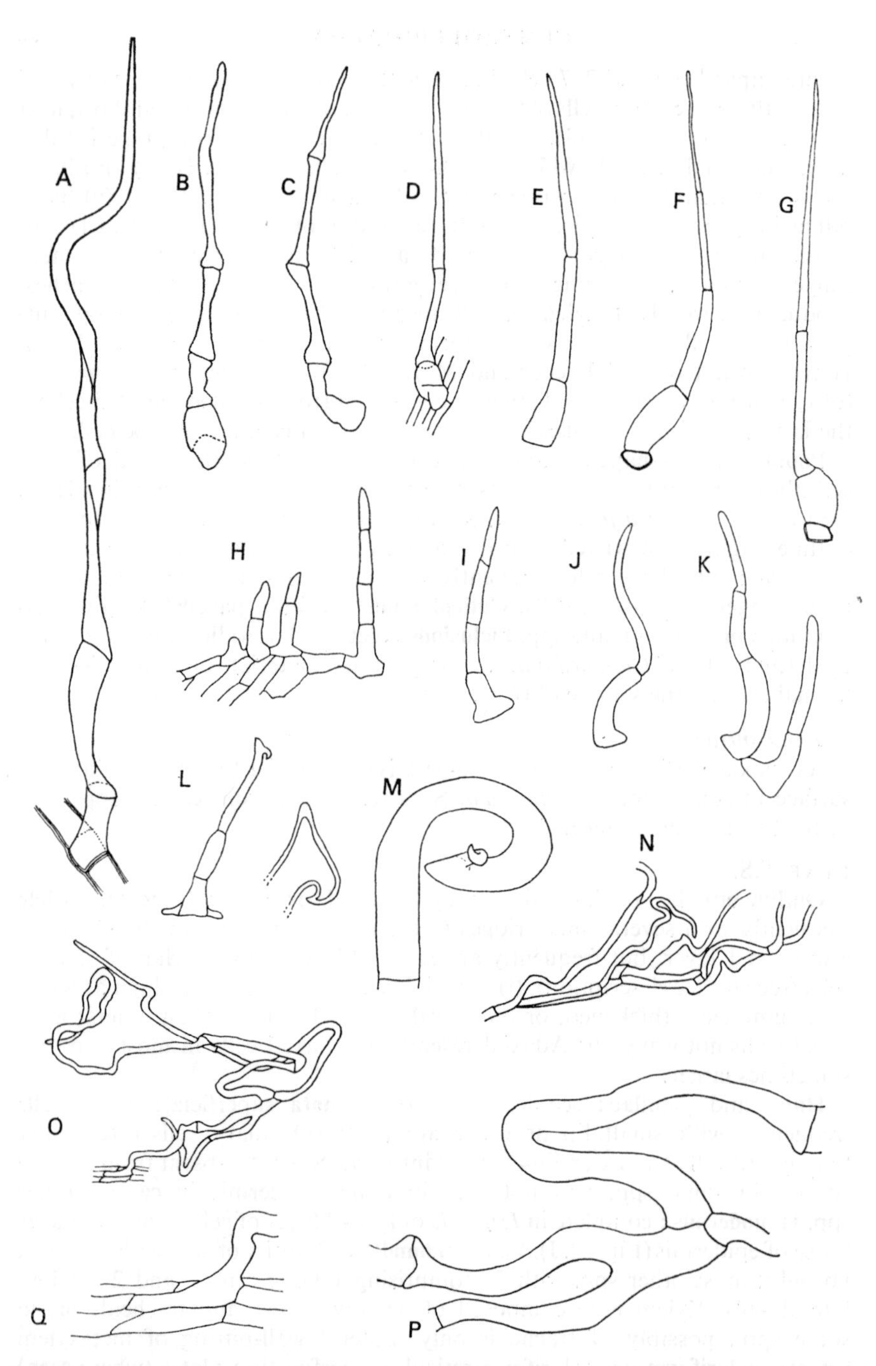

FIG. 21. Centrolepidaceae. A–Q, Hairs from leaf, or ligule margins: A, *Centrolepis exserta* (×200); B, C, *C. fascicularis* (×50); D, E, *Pseudalepyrum ciliatum* (×50); F, G, *Pseudalepyrum monogynum* (×50); H, I, *Aphelia pumilio* (×50); in H, part of ligule margin also shown; J, K. *C. drummondii* (×50); L, M, *Aphelia cyperoides*: L, whole hair (×50) and detail of apex (×200); M, apical cell showing variation of hook-type (×200); N–Q, *Gaimardia fitzgeraldi*: N, part of hair (×50); O, entire hair (×50); P, detail of branched cell from O (×200); Q, basal cell of hair O (×200).

Bundle sheaths. I.S. sclerenchymatous, 1–2- or, occasionally, 3–4-layered; fibres narrow, walls thin or moderately thickened, occasionally more heavily thickened; U-thickening present in some cells of *C. polygyna*, or cells prosenchymatous. O.S., when distinct from mesophyll, usually 1-, sometimes 2-layered; parenchymatous, cells normally wide; with U-shaped wall thickenings in *Gaimardia australis* (Fig. 23. G). **Chlorenchyma** composed of 1–2 or 3–6 layers of cells; layers most numerous to either side of vb's in leaves having semicircular outline in T.S. Cells either all more or less isodiametric or somewhat lobed, or those of outer layer(s) palisade-like. Peg-cells of restionaceous type (see p. 119) absent, but in certain spp. some cells of outer layers with occasional pegs (e.g. *Gaimardia fitzgeraldi*, Fig. 23. K). Cells often loosely packed, and separated from one another by wide intercellular spaces. Protective cells (see p. 120) absent. **Sclerenchyma** confined to inner bundle sheaths. **Air-canals** present to either side of or between vb's in few spp. (e.g. *G. fitzgeraldi*, Fig. 23. J). **Tannin** v. rare, noted in *Aphelia cyperoides*. **Inclusions:** silica-like, granular material observed in certain spp. (see also leaf surface). **Crystals:** none observed.

CULM SURFACE

Epidermis. Cells usually 4-, occasionally 6-sided, mostly 3–12, occasionally 13–20 times longer than wide. Walls slightly or moderately thickened, not wavy in majority of spp. Ends of cells of many spp. overlapping, as in leaf (see p. 87). **Hairs** seen[1] in *C. cambodiana* and *C. exserta*. **Papillae** low, present in many spp., one near to end of each cell, over or close to inclined transverse wall. **Stomata** similar to those of leaf, guard cells v. narrow in certain spp. and guard cell pair outline lacking central constriction in most spp. (see Stomata, leaf surface, p. 87). **Cuticular marks:** longitudinal, more or less parallel, faint striations present in many spp. **Silica** and **tannin:** none seen.

CULM T.S.

Outline circular, oval, or polygonal. **Cuticle:** several ridges present over each epidermal cell in most spp. **Epidermis.** Cells in 1 layer, but appearing 2-layered in places in certain spp. owing to overlap of ends of cells (as in leaf), see Fig. 20. A, *P. monogynum*. Cells mostly about as high as wide, occasionally up to 3 times higher than wide if sectioned through a papilla. Outer walls moderately thick to thick, other walls slightly to moderately thickened; anticlinal walls not wavy. **Papillae** present in many spp., particularly at region of cell overlap (see culm surface, p. 89), frequently up to twice as high as wide. **Stomata** superficial; guard cells with lip at outer aperture; subsidiary cells usually extending further into culm than guard cells.

Chlorenchyma. Cells in 1–3 or 4–5–(7) layers, all rounded-polygonal in T.S. or palisade-like and occasionally with few, short pegs; cells of inner layers sometimes irregularly lobed. Cells in certain spp. L-shaped as seen in L.S. with base of L next to epidermis. Cells in other spp. 3–5+ times longer than wide. Protective cells absent. **Parenchyma sheath** present in few spp. only, 1–2 layers of thin-walled colourless cells to inner side of chlorenchyma and next to

[1] Hairs reported for culms of some spp. by Hieronymus (1873), where apparently similar to those of leaves.

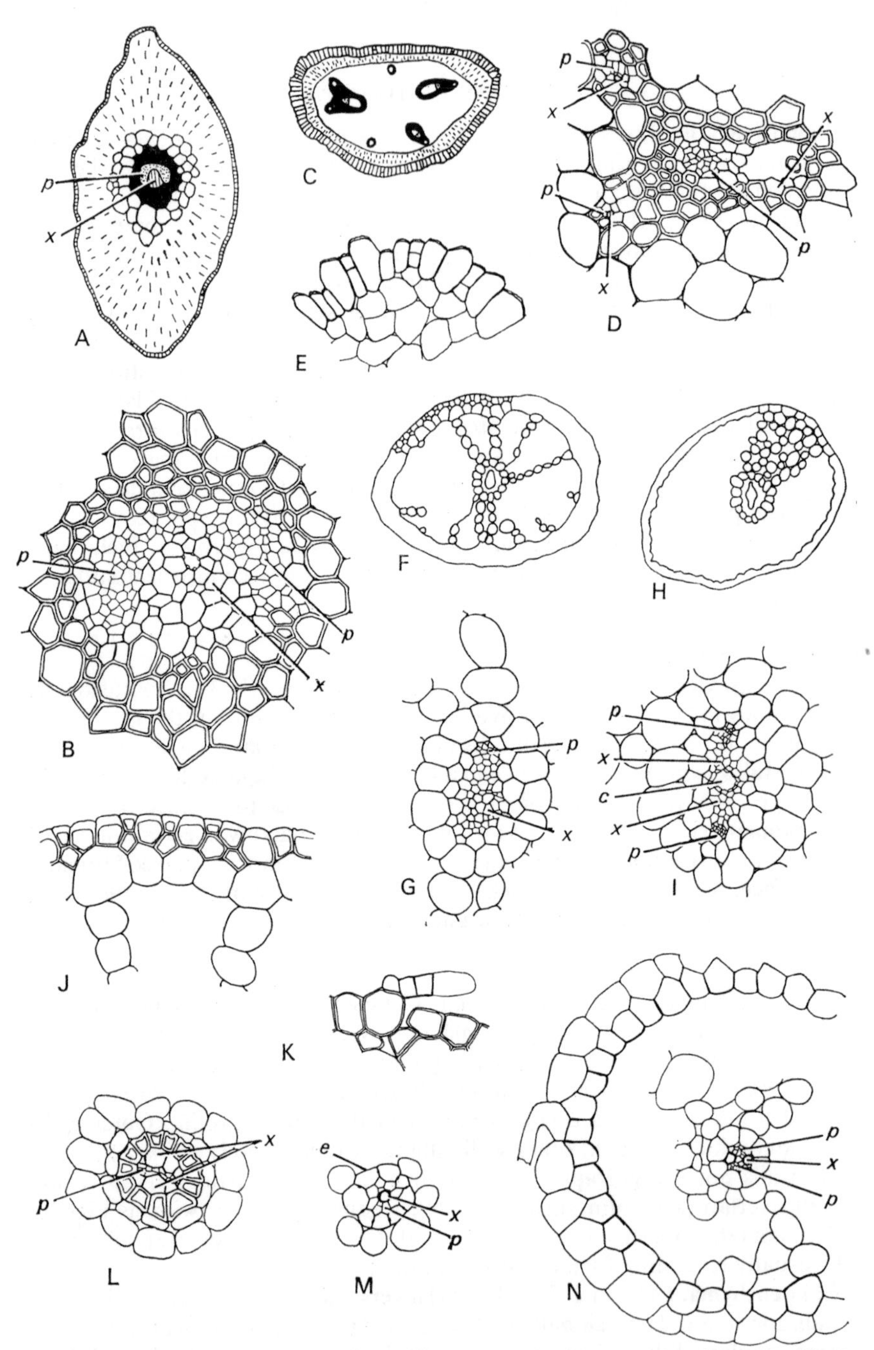

FIG. 22. Centrolepidaceae. A, B, *Centrolepis aristata*: A, diagram of leaf T.S. (×50); B, detail of vascular bundle of leaf in T.S. (×320). C–E, *Aphelia cyperoides*: C, diagram of culm T.S. (×50); D, detail of vascular bundle group of culm in T.S. (×320); E, detail of culm epidermis and chlorenchyma T.S. (×200). F–K, M, *Trithuria filamentosa*: F, diagram of leaf T.S. (×50); G, detail of leaf vascular bundle in T.S. (×200); H, diagram of culm T.S. (partly drawn) (×50); I, vascular bundle pair with central canal from culm T.S. (×200); J, detail of leaf margin T.S. (×200); K, rhizome, part of epidermis T.S. showing hair (×200); M, centre of root T.S., note Casparian strips on endodermal cells and solitary central mx vessel (×320). L, *C. drummondii*, centre of root T.S., note phloem between 2 xylem poles (×320). N, *Pseudalepyrum ciliatum*, part of root T.S.; note root-hair arising to one side of cell, exodermis, and 2 xylem poles of root (×200).
c = central canal; e = endodermis; p = phloem; x = xylem.

sclerenchyma cylinder. Cells of this type elongated in L.S., with square ends. (Note that this parenchyma sheath is not the same as the outer bundle sheaths and is similar to the parenchyma sheath of Restionaceae, p. 121.) **Sclerenchyma.** Present either as (i) a 1–several-layered cylinder surrounding all vb's; (ii) confined to I.S. of vb's, or (iii) in 1–3 layers surrounding small groups of vb's. Fibres medium-sized to narrow, with moderately thick or thick walls. **Vascular bundles** of 2 distinct sizes in most spp. (i) Small outer vb's, with few, narrow tracheids and phloem cells. (ii) Larger, inner vb's, with several narrow or medium-sized mx vessels on either flank, these having scalariform or reticulate lateral wall-pitting and oblique, scalariform, or scalariform-reticulate perforation plates. Protoxylem vessels often breaking down to form cavity. Vb's varying in number, 1 or 2 present in several spp. and up to 15 in others. Large and small bundles occurring either in equal numbers, or about twice as many small as large vb's present.

Bundle sheaths. If distinguishable, O.S. represented by 1 layer of rounded parenchymatous cells; I.S. sclerenchymatous, fibres medium-sized or narrow, with slightly thickened to thick walls; in certain spp. some cells, particularly those next to phloem, with U-shaped wall thickening. Fibres of sheath often indistinguishable from those of scl. cylinder. I.S. of adjacent vb's frequently fused. **Central ground tissue** parenchymatous, cells often with thickened (lignified) walls; sclerenchymatous in certain spp., e.g. those with 2 central vb's. **Inclusions:** silica-like material present in few spp. as irregular bodies or granular material. **Tannin** present in few spp. **Crystals** absent.

RHIZOME T.S.

Epidermis: cells frequently wider than high, occasionally higher than wide (*C. fascicularis*); walls all thin or slightly thickened. **Hairs** multicellular, short filamentous, seen in *Trithuria filamentosa* only (Fig. 22. K). **Exodermis** present in few spp. as 1 layer of thin-walled cells similar in size and shape to those of epidermis. **Cortex** not normally divided into distinct zones, but cells of outermost layers frequently more compact and smaller than remainder. Cells isodiametric and lobed or polygonal, thin-walled, in 1–2 or 3–15 layers. **Endodermoid sheath:** cells in 1–2 or 3–4 layers; outermost cells frequently with U-shaped wall thickenings, others with even wall thickening.

Vascular bundles amphivasal (some collateral in *G. australis*), arranged in ring next to endodermoid sheath; ring divided into 4 sectors by ground tissue in some spp. Bundles seldom exhibiting individual identity. Vessels wider than those of culm, lateral wall-pitting and perforation plates mostly reticulate, rarely scalariform-reticulate. Xylem v. irregular in *T. filamentosa*, with small groups of vessels dispersed in parenchyma. **Central ground tissue** parenchymatous, cells with slightly thickened or thick walls.

ROOT T.S.

Epidermis: cells thin-walled, as high as wide or up to 3 times wider than high. **Root-hairs** about $\frac{1}{4}$–$\frac{1}{3}$ width of cell from which they arise; originating from one side of cell in all material examined. **Exodermis** present in most spp., 1-layered, cells similar in size and shape to those of epidermis, with slightly thickened walls. **Cortex.** 1–6+ layers, not divided into conspicuous zones

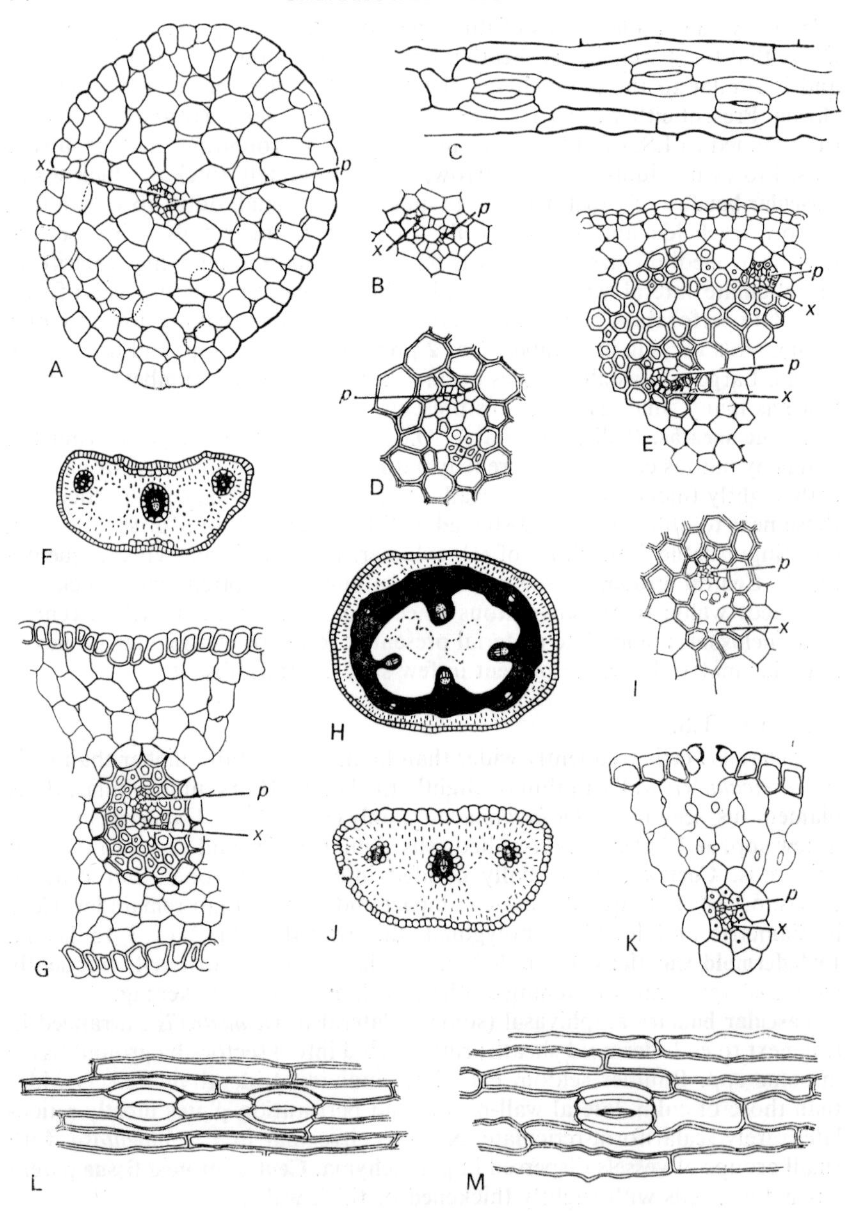

FIG. 23. Centrolepidaceae. A–C, *Pseudalepyrum ciliatum*: A, leaf T.S. (×200); B, leaf vascular bundle in T.S. (×320); C, abaxial leaf surface (×200). D, E, *Gaimardia setacea*, vascular bundles from culm T.S.: D (×320); E (×200). F, G, L, *G. australis*: F, diagram of leaf T.S. (×50); G, median vascular bundle of leaf in T.S. (×200); L, abaxial leaf surface (×200). H–K, M, *G. fitzgeraldi*; H, diagram of culm T.S. (×50); I, one of larger culm vascular bundles in T.S. (×200); J, diagram of leaf T.S. (×50); K, detail of leaf T.S. showing stoma and vascular bundle (×200); M, abaxial leaf epidermis (×200).
p = phloem; x = xylem.

but in certain spp. some cells arranged in radiating plates. Most cells more or less rounded, with intercellular spaces. **Endodermis** usually 1-layered (rarely more in certain spp.). Walls all evenly thickened, or inner and anticlinal walls more heavily thickened than outer walls. Casparian strips well developed in *T. filamentosa* (Fig. 22. M). **Pericycle** absent, or indistinguishable from phloem. **Vascular tissue:** xylem composed of narrow mx vessels in 1, 2, or 3 poles. Pole central when solitary, but situated next to endodermis when 2 or 3 present. Arcs either isolated or meeting in centre of root. Phloem present between xylem poles (Fig. 22. L, M, N). **Central ground tissue** absent. **Crystals, silica,** and **tannin** absent.

Additional Notes from the Literature

Vessels are reported to be present in all vegetative parts of the plant (Solereder and Meyer 1929), an observation confirmed for all material examined except the leaf of *Trithuria filamentosa*. Solereder and Meyer also recorded that the pericycle of the roots is broached by the xylem arcs. The roots of species examined for this work did not appear to have a pericycle. Cells which might possibly be regarded as pericycle appear to be derived from, or part of, the phloem. Hieronymus (1873) stated that for *Centrolepis tenuior* cambiform phloem cells are present where pericycle cells would normally be expected to occur, and he noticed that the xylem arcs reach the endodermis at each of the poles. Van Tieghem's (1887) observations confirmed those of Hieronymus and other authors regarding the xylem poles. He examined material of *Centrolepis fascicularis*, *C. muscoides*, *Aphelia cyperoides*, *Alepyrum monogynum* (now *Pseudalepyrum monogynum*), and *Gaimardia australis*. He stated that lateral roots arise from between the xylem poles.

The observation made by Solereder and Meyer that epidermal cells all have straight walls is true for many species, but some species have wavy walls to these cells.

Malmanche (1919) considered the culm anatomy of Centrolepidaceae to be similar to but easily distinguishable from that of the Eriocaulaceae. He noticed that the culm anatomy of Restionaceae is very different, but thought that the roots are similar in Centrolepidaceae and Restionaceae. However, all members of the Restionaceae examined to date have been found to possess a well-developed pericycle. Of all the genera, Malmanche considered *Gaimardia* to be most similar to the Restionaceae.

Hamann (1963) reported that the stomata of *Centrolepis aristata* develop as in the Gramineae, and the subsidiary cells are not derived from the guard-cell-mother-cells, as stated by Solereder and Meyer (1929).

Taxonomic Notes

The taxonomic position of Centrolepidaceae with respect to other families is still being debated. Originally the family was included in the Restionaceae, but Desvaux defined it as a separate family in 1828 and it has been recognized as distinct since that date. It was for some time included in the Enantioblastae with Restionaceae and part of the Eriocaulaceae, e.g. Wettstein (1935), and was placed in the Farinosae of Engler (1936). Bentham and Hooker (1883) considered it to belong in the Glumaceae with Gramineae and Cyperaceae. Malmanche (1919) thought it to be closely related to Eriocaulaceae, Restionaceae, Xyridaceae, Philydraceae, and Mayacaceae. Hutchinson (1959) included it with Juncaceae and Thurniaceae in the Juncales. The unusual inflorescence structure and the reduced nature of the plants have both contributed to the taxonomic confusion.

The most recent extensive work on the family has been carried out by Hamann (1961, 1962*a*, *b*, 1963).

In his 1961 paper Hamann compared the families of Farinosae with one another and worked out similarity quotients using a wide range of characters for his comparisons. On this basis he suggested that Gramineae are the closest relatives to Centrolepidaceae, and Restionaceae are a more distant second. He mentioned that there is a connection whose significance must not be overlooked between both Cyperaceae and Pandanaceae and Centrolepidaceae. It must be commented, however, that Hamann used the range of characters that he could glean from the literature in constructing his comparison tables. This may not prove to be the most satisfactory of lists and the results could lead to some misinterpretation. At the end of his survey of Centrolepidaceae he noted that Centrolepidaceae might not be a homogeneous group, since *Centrolepis* and *Gaimardia* show considerable differences.

In July 1962 Hamann published a paper on the embryology of Centrolepidaceae. He reported that the Centrolepidaceae are very similar in their embryology to Gramineae, and they share an unusual character with Restionaceae. He felt that in their floral morphology, the Centrolepidaceae represent a cul-de-sac. He criticized Hutchinson, who considers Juncaceae, Thurniaceae, Cyperaceae, Centrolepidaceae, and Gramineae to be derived from one group, and stated that in his view the evidence points to the existence of two groups, one Juncaceae, Thurniaceae, and Cyperaceae, the other Gramineae, Restionaceae, and Centrolepidaceae. He found no proof for a liliaceous ancestor for these families, as did Hutchinson, but considered that Gramineae and Restionaceae have much closer links with the Commelinales.

Hamann, in his paper of October 1962, restated his opinion that Centrolepidaceae are most closely related to Gramineae and Restionaceae. It is of interest to note that the ligule, absent from Restionaceae and Juncaceae, is represented in members of the Centrolepidaceae and Gramineae, and some Cyperaceae.

The taxonomic value of anatomical characters possessed by members of the Centrolepidaceae is difficult to assess. Reduced gross morphology is linked with a simple form of anatomical organization. The plants exhibit little by way of anatomical data that can be used in comparison with other families. There are, nevertheless, some important points that can be raised. First, the possession by some species of the unusual type of epidermal cell arrangement whereby one cell overlaps the end of the next in a file (see p. 87) might indicate that the family has been separated from its nearest relatives for a very long time, because this character is not present in any other monocotyledons so far examined. Second, the apparent lack of pericycle in the root is another distinctive character. Third, the root-hairs arise from one side of root epidermal cells—another unusual character. This probable long isolation could mean that we are unable to recognize the families that are closely related to

Centrolepidaceae, because their present-day representatives have had so much time to evolve along different paths.

Centrolepidaceae like Juncaceae lack silica-bodies; they differ from Gramineae and Cyperaceae in which characteristic silica-bodies are present. The stomatal type of most species (paracytic) of Centrolepidaceae (except *Trithuria*) is similar to that of Restionaceae, Juncaceae, Thurniaceae, Gramineae, Cyperaceae, and most families that have previously been associated with Centrolepidaceae in the taxonomic literature.

The vascular bundles very rarely have wide metaxylem vessels; they are most similar to the smaller vascular bundles of other families. The last-formed vessel elements are elongated, narrow, and have oblique, scalariform or scalariform-reticulate perforation plates. They are, therefore, of the type described by Cheadle (1942, 1943*a*, *b*, 1944) as primitive. Whilst the vessel elements themselves are of the primitive type, their distribution throughout the whole plant in the majority of species examined (except *Trithuria filamentosa*) is advanced according to Cheadle's (l.c.) and Bierhorst and Zamora's (1965) theories. However, the stage of evolution reached by the vessels in the modern representatives of the Centrolepidaceae could either be similar to that of its ancestors, or it might be more advanced. Consequently the stage reached by vessel development cannot be taken as evidence, on its own, of phylogenetic relationships of the Centrolepidaceae with other, present-day taxa.

Palynological studies (Bortenschlager *et al.* 1966) have shown that Centrolepidaceae contain two groups; pollen of the genera *Hydatella* and *Trithuria* differs from that of the rest of the family. The presence of anomocytic stomata in *Trithuria*, and the relationship of *Hydatella* and *Trithuria* to other members of the family are discussed on p. 113.

The relationships of individual genera one to another are discussed after the account of each genus.

APHELIA R. Br. (including *Brizula* Hieron.)

No. of spp. 5, examined 5.

Genus and Species Descriptions

Leaf surface

Epidermis: cells of both surfaces similar, 4- or 6-sided, 4–10 or 14+ times longer than wide; ends of cells overlapping, end walls frequently deeply concave at leaf surface (see p. 87). **Hairs** in *A. cyperoides* of type (v), see p. 89, with hooked apical cell (Fig. 21. L, M); those in *A. pumilio* of type (iii), consisting of a basal cell with a foot, and filament of 2–3 cells (Fig. 21. H, I). **Papillae** short, 1 each to most epidermal cells, and situated near to overlapping ends. **Stomata:** guard cells narrowest at centre in *A. pumilio*, not narrowing in other spp.; subsidiary cells longer than guard cells, usually with acute, sometimes with rounded ends.

Leaf T.S.

Outline more or less oval, or flattened oval, about twice as wide as thick in *A. cyperoides*, *A. gracilis*, and *A. pumilio*, and about 4 times wider than thick in *A. brizula*. **Epidermis:** cells of both surfaces similar, mostly 1–2 times higher than wide, but frequently wider than high in *A. brizula*; appearing 2-layered owing to overlap of cell ends. **Papillae** frequent, short. **Stomata:** guard cells with outer lip, this pronounced in *A. cyperoides*. **Vascular bundles:** 1, central, in all spp. except *A. cyperoides*, where 3, in a row. Mx vessels narrow, with scalariform lateral wall-thickenings and oblique, scalariform perforation plates. **Bundle sheaths:** I.S., in most spp. composed of narrow, thick-walled cells, those in *A. cyperoides* wide, thick-walled. O.S. usually composed of 1 layer of wider, parenchymatous cells but not distinguishable in *A. cyperoides*. **Mesophyll** cells in 1–3 layers, slightly higher than wide to more or less isodiametric, with or without lobes, interspersed by wide intercellular spaces. **Inclusions:** irregularly shaped bodies of silica-like material present in some epidermal cells of *A. cyperoides*. **Tannin** present in some cells of *A. cyperoides*.

Culm surface seen in *A. cyperoides*, *A. gracilis*, and *A. pumilio*.

Epidermis: cells 3–4 times longer than wide in *A. cyperoides* and between 8–20 times longer than wide in other spp.; ends of cells with pronounced overlap. **Papillae:** 1 present on each of most cells near to end-wall overlap; well developed in *A. cyperoides*. **Stomata:** guard cells narrow, subsidiary cells longer than guard cells, with acute ends.

Culm T.S. (Fig. 22. C)

Outline circular, oval, or sub-triangular. **Epidermis:** cells of irregular size and proportions, mostly about twice as high as wide, some reaching *c*. 3 times higher than wide in *A. cyperoides* (Fig. 22. E), but wider than high in regions of cell overlap. **Stomata** as in leaf. **Chlorenchyma** 2–5-layered, cells loosely packed, variable in size in *A. cyperoides*, outermost cells in *A. gracilis* similar to those of epidermis, others *c*. 3 times larger. As seen in L.S., cells of outer layer in *A. pumilio* inverted L-shaped. **Sclerenchyma** present only in vb sheaths. **Vascular bundles:** in *A. brizula* 3 inner and 5 outer; *A. cyperoides* 3 inner, 6 outer (Fig. 22. C, D); *A. drummondii* 2 inner, 3 outer; *A. gracilis* 2 inner, 2 outer (all embedded in scl. matrix at culm centre); *A. pumilio* 3 inner, 6 outer. **Bundle sheaths** sclerenchymatous; fibres with moderately thickened walls in all spp. except *A. gracilis* and *A. pumilio*, where thin. **Central ground tissue** present in *A. cyperoides* and *A. pumilio* only, parenchymatous. **Tannin** present in some cells of *A. cyperoides*.

Rhizome T.S. seen in *A. drummondii*, *A. gracilis*, and *A. pumilio*

Epidermis: cells thin-walled, wider than high. **Exodermis** present in *A. drummondii* as 1 layer of cells similar to those of epidermis. **Cortex** 5–9+ layers; cells polygonal or rounded, with short lobes or wide pegs. **Endodermoid sheath** composed of 1–2 layers of polygonal cells with slightly to moderately thickened walls. **Vascular bundles** amphivasal or irregular, arranged in ring next to endodermoid sheath. **Central ground tissue:** wide parenchymatous cells with thin or slightly thickened walls.

Root T.S. seen in *A. cyperoides* and *A. drummondii*.

Epidermis and **root-hairs** as for family description. **Exodermis** present in *A. drummondii* as cells about twice as high as wide, arranged in 1 layer. **Cortex:** 3–5 layers; differentiated into radiating files of narrow cells alternating with larger cells in *A. drummondii*. **Endodermis** 1-layered, cells with thickened inner and anticlinal walls. **Vascular tissue:** xylem triarch in *A. cyperoides*, diarch in *A. drummondii*, phloem between arms of xylem and either parenchyma or phloem cells between xylem poles in *A. drummondii*.

Material Examined

Aphelia brizula F. v. Muell.: Cheadle, V. I. CA 333; Perth, W. Australia 5.11.1959 (as *Brizula muelleri* Hieron.) (S). Leaf blade width $< 0{\cdot}1$ mm, thickness $\leqslant 0{\cdot}1$ mm. Culm diameter 0·1 mm.

A. cyperoides R. Br.: Cheadle, V. I. CA 59; Perth, W. Australia 23.9.1959 (S). Leaf blade width 0·5 mm, thickness 0·2 mm. Culm diameter $0{\cdot}5 \times 0{\cdot}2$ mm. Root diameter 0·1 mm.

A. drummondii Benth.: Cheadle, V. I. CA 337; Perth, W. Australia 11.11.1959 (as *Brizula drummondii* (Benth.) Hieron.) (S). Leaf blade width and thickness $< 0{\cdot}1$ mm. Culm diameter 0·2 mm. Rhizome diameter 1 mm. Root diameter *c.* 0·05 mm.

A. gracilis Sond.: Curtis, W. M.; Bridport, Tasmania, Nov. 1952 (K). Leaf blade width $< 0{\cdot}1$ mm, thickness $\leqslant 0{\cdot}1$ mm. Culm diameter $< 0{\cdot}1$ mm. Rhizome diameter 0·8 mm.

A. pumilio F. v. Muell. ex Sond.: Melville, R. 106.54; Quail Is., Victoria, Australia (as *Brizula pumilio* (F. v. Muell. ex Sond.) Hieron.) (K). Leaf blade width 1 mm, thickness 0·8 mm. Culm diameter 0·8 mm. Rhizome diameter 1·5 mm.

Additional Notes from the Literature

Malmanche (1919) described the stem anatomy of *A. cyperoides*. He found 5 small, outer vb's and 3 large, inner bundles. He regarded the 3 large bundles as constituting separate steles. The description corresponds in all other respects with the material examined for this work.

Solereder and Meyer (1929) noted that the terminal hair cell in *A. cyperoides* is hooked. They also recorded only 1 central leaf vb for this sp. and *Brizula muelleri* (= *Aphelia brizula*).

Taxonomic Notes

The genus *Aphelia*, including *Brizula*, seems to form a group that is anatomically uniform and probably natural. The only exception is *A. cyperoides*, which differs from other species in a number of characters. The apices to the hairs are hooked, not tapered to acute points, the leaf may have 3 vascular bundles, not 1. The average length of the epidermal cells is shorter, and the range of size of cells smaller in *A. cyperoides* than in the other species. When considering a family of reduced plants that exhibit few characters of taxonomic significance, it is very easy to give undue importance to small structural differences. However, these differences taken in combination, may have taxonomic significance, and the total number of characters by which *A. cyperoides* differs from other species of the genus *Aphelia* may indicate that it should be removed from the genus. There are several genera which share many of the features of *A. cyperoides*, but there is none in which hooked hairs are recorded.

CENTROLEPIS Labill.

No. of spp. *c*. 24; examined 8.

GENUS DESCRIPTION

Leaf surface

Epidermis: both surfaces similar, but abaxial cells sometimes the longer; cells usually 4-sided, mostly between 7 and 10 times but some 1–6 or 11–18 times longer than wide; walls moderately thickened, not wavy; end walls inclined and cell-ends overlapping in all spp. examined except *C. polygyna*. **Hairs:** see individual species; represented by types i–iv (p. 89). **Papillae** slight in *C. drummondii* and *C. fascicularis*, well developed in *C. exserta*. **Stomata** equally frequent on both surfaces, or more frequent abaxially; guard cells not constricted centrally except in *C. polygyna*; subsidiary cells with acute ends. **Cuticular marks:** longitudinal striations.

Leaf T.S.

Outline more or less circular or oval, sometimes with flattened or grooved adaxial face. Leaf of *C. aristata* elliptical (Fig. 22. A). **Epidermis.** Cells of both surfaces similar, appearing 2-layered in places in all spp. except *C. polygyna*; cells mostly from as high as to 1½ times higher than wide. Cells of unequal heights in *C. fascicularis*, mostly wider than high in *C. philippinensis*, where marginal cells smaller and with thicker walls than others. Outer walls thick or moderately thickened, other walls usually slightly thickened. **Stomata:** guard cells with outer lip. **Vascular bundle** 1, central; phloem pole small, except in *C. aristata* where overarching xylem (Fig. 22. B) and *C. polygyna* where equal to about 2/3 cross-sectional area of vb. All mx vessels narrow. **Bundle sheaths.** I.S. sclerenchymatous, fibres narrow to wide, walls thin to heavily thickened, usually lignified except in *C. humillima* where those at phloem pole cellulosic; U-shaped wall thickenings present in 1 specimen of *C. polygyna*; fibres in 1 or 2–(3) layers. O.S., 1 (2) layer(s) of wide, parenchymatous cells. **Mesophyll:** cells in 1–4–(5) layers, most layers in wings to either side of vb, cells of outer layers frequently palisade-like, occasionally with several pegs, inner cells (and sometimes outer cells also) more or less isodiametric or lobed and loosely arranged. **Air-canals** sometimes forming in old leaves of *C. fascicularis*, 1 to either side of vb. **Inclusions:** some cells containing granular particles in *C. drummondii* and fine 'sand'-like particles also present in certain cells of *C. polygyna*.

Culm surface

Epidermis: cells similar to those of leaf of corresponding sp. but cells more often longer. **Hairs** in 2 spp. **Papillae** slight, in *C. aristata* and *C. fascicularis*. **Stomata** as in leaf of same sp. **Cuticular marks:** longitudinal striations usually prominent.

Culm T.S.

Outline oval or elliptical to circular or polygonal. **Epidermis:** cells mostly about as high as wide; appearing 2-layered in most spp. except *C. polygyna*. **Stomata:** guard cells with outer lip. **Chlorenchyma:** 1–3 or 4–5 layers, all cells irregularly circular or lobed, or outer cells palisade-like, sometimes with few pegs. (Palisade-like cells inclined as seen in L.S., and irregularly circular cells elongated longitudinally.) **Parenchyma sheath** frequently distinguishable as 1 cell layer between chlorenchyma and scl. cylinder. **Sclerenchyma** as cylinder in all spp. except *C. humillima.* Fibres medium-sized or narrow, walls ranging from slightly thickened to thick. Mechanical tissue in *C. fascicularis* grading into central ground tissue; many cells with intercellular spaces. **Vascular bundles** ranging in number from 1 to 15, frequently with about twice as many peripheral, smaller bundles as larger, more central bundles. See individual spp. descriptions. Xylem of larger vb's frequently with medium-sized, flanking mx vessels. **Bundle sheaths:** I.S. sclerenchymatous, merging into scl. cylinder in most spp.; O.S. composed of parenchymatous cells in 1 layer in spp. with 1 or 3 vb's. **Central ground tissue** when present composed of polygonal parenchymatous cells with moderately thickened (lignified) walls. **Inclusions:** silica-like granular material in certain epidermal cells of *C. humillima;* see also *C. exserta.*

Rhizome T.S. seen in *C. drummondii, C. fascicularis, C. humillima, C. polygyna.*

Outline irregular but more or less circular. **Epidermis:** cells upright, thin-walled. **Cortex** several-layered, cells thin-walled, rounded or polygonal. **Endodermoid sheath** 1–3-layered, cells with moderately thickened walls. **Vascular bundles** in 1 peripheral ring, amphivasal; vessels polygonal, medium-sized (*C. fascicularis*), with short elements; lateral wall-pitting and perforation plates reticulate. **Central ground tissue** composed of parenchymatous cells with thickened walls.

Root T.S.

Epidermis: cells thin-walled, mostly wider than high. **Root-hairs** arising to one side of epidermal cells. **Exodermis** present as distinct layer of cells in several spp., walls of cells frequently slightly thickened, sometimes thin. **Cortex** 1–4-layered, either homogeneous (most spp.), or with radiating files of small, rounded cells in matrix of wider cells (*C. aristata, C. polygyna*). **Endodermis:** 1–(2) layer(s) of cells with evenly thickened walls (outer wall thinnest in *C. drummondii*). **Vascular tissue:** xylem with 2 poles, except in *C. fascicularis* where 3. *C. aristata* and *C. drummondii* (Fig. 22. L) unusual in having phloem and narrow parenchymatous cells between the 2 xylem poles.

Material Examined

Centrolepis aristata (R. Br.) Roem. *et* Schult.: Cheadle, V. I. CA 58; Perth, W. Australia 1959 (S).

C. cambodiana Hance: (i) Poilane, E. 2863g; Plateau des Boloven, Laos. 26. 4.1939 (K). (ii) Sleumer 4781; Loie, Phutrading, Thailand 11.9.1963 (K).

C. drummondii (Nees) Walp.: Cheadle, V. I. CA 336; Perth, W. Australia 1960 (S).

C. exserta Roem. *et* Schult.: Specht, R. L. 492; Arnhem Land Aboriginal Reserve, Australia 4.6.1948 (S).

C. fascicularis Labill.: (i) Van Steenis 8577 (K). (ii) Cheadle, V. I. CA 292; Coff's Harbour, N.S.W., Australia 1959 (S).

C. humillima F. v. Muell.: Cheadle, V. I. CA 64; Perth, W. Australia 1959 (S).

C. philippinensis Merr.: type specimen; N. Borneo (K).

C. polygyna (R. Br.) Hieron.: (i) Adelaide Botanic Garden, Hort. Kew, 685.63 (S). (ii) Cheadle, V. I. CA 65; Perth, W. Australia 1959 (S).

SPECIES DESCRIPTIONS

***C. aristata* (R. Br.) Roem. *et* Schult.**

Leaf surface not seen.

Leaf T.S. (Fig. 22. A). Leaf blade width 1 mm; thickness 0·5 mm.

Outline elliptical. **Stomata:** guard cells with small outer lip. **Vascular bundle** 1, central; phloem overarching xylem, becoming widest on either flank of bundle and narrowest on abaxial face (Fig. 22. B). Crushed px cells in centre of xylem pole. This single vb continuing into leaf base; in basal region, phloem pole strikingly dumb-bell-shaped. **Mesophyll** 2–3-layered, outer cells palisade-like, inner cells irregularly lobed.

Culm surface

Papillae low, dome-shaped, present one to a cell on certain epidermal cells. **Stomata** v. narrow.

Culm T.S. Culm diameter 1 × 0·5 mm.

Outline elliptical. **Chlorenchyma:** outer cells palisade-like. **Parenchyma sheath** probably represented by inner chlorenchyma layer. **Vascular bundles** comprising 8 small, outer vb's with xylem almost encircling phloem pole together with 4 large and 1 small inner vb's. Vessels; lateral wall-pitting scalariform to alternate.

Root T.S. Root diameter < 0·01 mm.

Exodermis: 1 layer of cells with slightly thickened inner and anticlinal walls. **Cortex:** radiating plates of rounded cells, 3–4 deep, dividing groups of wide cells into irregular sectors. **Endodermis:** cells v. thick-walled. **Vascular tissue** composed of 2 xylem poles separated by several files of phloem cells.

***C. cambodiana* Hance**

Leaf surface

Epidermis. Cells mostly between 4 and 8 times longer than wide. End walls overlapping, oblique, sometimes curved (Fig. 20. D). Elongated pits present on inner walls, conspicuous in 2863g. **Stomata:** see figure. **Hairs** as in culm. **Cuticular marks** as pronounced longitudinal striations.

Leaf T.S. Leaf blade width 0·5 mm; thickness 0·3 mm.

Outline circular-polygonal, with slightly grooved adaxial face. **Epidermis:** cell walls more heavily thickened than in most spp. **Vascular bundle:** vessels and tracheids overarched by crescent-shaped phloem pole. **Bundle sheaths:** I.S. composed of 1–3 layers of v. thick-walled fibres. **Mesophyll:** cells palisade-like, about twice as high as wide, mostly arranged in 2 or 3 layers.

Culm surface

Epidermis: walls of cells moderately thickened; cells 2–12+ times longer than wide; ends of cells slightly overlapping. **Hairs** each composed of a v. enlarged basal cell with moderately thickened walls, the remainder of the hair uniseriate, consisting of usually 2–3 (sometimes more) elongated cells with slightly thickened walls; cells widest at cross walls, terminal cell ending in a fine point (type i).

Culm T.S. Culm diameter 1·0 mm.

Outline more or less circular. **Epidermis** mostly appearing 1-layered, but occasionally becoming 2-layered by overlap of cells. **Stomata:** guard cells raised above general level of culm surface by subsidiary cells. **Chlorenchyma:** cells in 2–3 layers, rounded in T.S., elongated as seen in L.S.; cells of innermost layer probably representing parenchyma sheath. **Sclerenchyma:** fibres in 2–5- or 6-layered cylinder; walls thick. **Vascular bundles:** 9 peripheral vb's embedded in scl. cylinder and 6 inner, larger vb's scattered in outer part of ground tissue; inner bundles with conspicuous px cavities. **Bundle sheaths:** I.S. distinguishable as narrow cells with moderately thickened walls; cells in 1–2 layers at xylem pole and in 4–5 layers at phloem pole, joining on to inner side of scl. cylinder. **Central ground tissue:** outer 3–4 layers of cells wide, with moderately thickened walls, other cells with thin or slightly thickened walls, walls decreasing in thickness in central cells, and these broken down to form central canal.

C. *drummondii* (Nees) Walp.

Leaf surface not seen

Leaf T.S. Leaf blade width $< 0{\cdot}4$ mm; thickness $\leqslant 0{\cdot}1$ mm.

Outline oval. **Hairs:** see Figs. 21. J, K. **Mesophyll:** cells rounded or slightly elongated, in 2–4 layers. **Inclusions:** granular, silica-like particles present in some cells.

Culm T.S. Culm diameter $< 0{\cdot}1$ mm.

Outline circular. **Chlorenchyma:** cells in 1–2 layers, rounded. **Parenchyma sheath** 1-layered. **Sclerenchyma:** cylinder of narrow, thick-walled cells in 3–4 layers. **Vascular bundles** consisting of 4 small vb's, immediately to outside of scl. cylinder, together with 3 large inner vb's.

Rhizome: see genus description; diameter 0·6 mm.

Root T.S. Root diameter 0·1 mm.

Cortex 1-layered. **Vascular tissue:** (Fig. 22. L); as in *C. aristata*.

C. *exserta* Roem. *et* Schult.

Leaf surface

Epidermis: cells (3)–12–(15+) times longer than wide; narrowest at either end; ends of cells overlapping, end walls v. concave (Fig. 20. F). **Papillae** pronounced; present on many cells. **Hairs** uniseriate, see Fig. 21. A.

Leaf T.S. Leaf blade width 0·4 mm; thickness *c.* 0·1 mm.

Outline semicircular with concave adaxial surface. **Mesophyll:** cells more or less isodiametric, in 2–3 layers. **Vascular bundle:** few mx elements; px broken down. **Bundle sheaths:** I.S., 1 layer of narrow cells with slightly thickened walls; O.S., 1 layer of wide, parenchymatous cells.

Culm surface

Similar to that of leaf. **Inclusions:** non-crystalline, small particles with wedge-shaped outline and rounded corners present in some cells.

Culm T.S. Culm diameter 0·8 mm.

Chlorenchyma composed of 2–3 layers of palisade-like cells with abundant intercellular spaces. **Parenchyma sheath** well defined, cells in 1 layer, wide, thin-walled. **Sclerenchyma :** cylinder composed of 1–3 layers of narrow, thick-walled cells. **Vascular bundles** consisting of 6 small, peripheral vb's embedded in outer part of scl. cylinder, and of 3 larger, inner vb's, each with large px cavity. **Bundle sheaths** represented by I.S. only; sclerenchymatous, made up of narrow cells in 1–2 layers at phloem pole, joining on to scl. cylinder, and 1 layer of wider cells around xylem pole, indistinguishable from those of the ground tissue. **Central ground tissue** composed of polygonal parenchymatous cells with slightly thickened walls. **Inclusions** represented in certain cells of ground tissue by non-crystalline bodies apparently enclosed in gelatinous sheaths, each body composed of two cones joined at their bases.

C. *fascicularis* Labill.

Leaf surface

Epidermis : transverse walls deeply concave. **Hairs :** cells widest at and near transverse walls (Fig. 21. B, C). **Papillae** low, 1 on most cells, situated near to region of cell overlap.

Leaf T.S. Leaf blade width 0·5 mm; thickness 0·4 mm.

Outline oval, with flattened adaxial surface. **Stomata :** subsidiary cells tall and narrow; guard cells with pronounced outer lips. **Mesophyll :** 1 layer of palisade-like cells with few pegs present next to adaxial epidermis and at leaf margins, remaining 1–2–(3) layers of cells more or less isodiametric.

Culm surface similar to leaf surface, but hairs not seen.

Culm T.S. Culm diameter 0·25 mm.

Outline circular–polygonal. **Chlorenchyma** 4–(5)-layered, cells of outer layer palisade-like, others slightly higher than wide. **Parenchyma sheath** distinct, cells as high as wide, with rounded corners and slightly thickened walls. **Vascular bundles.** 6 small, outer and 3 larger, inner (8577) or 7 small, outer and 4 larger, inner (CA 292). Metaxylem vessels with scalariform or reticulate lateral wall-pitting and oblique, scalariform or reticulate perforation plates.

Rhizome: see genus description; diameter 0·7 mm.

Root T.S. Root diameter $< 0{\cdot}1$ mm.

Cortex 2–(3)-layered. **Vascular tissue :** 3 xylem poles radiating from centre, phloem between arms of xylem.

C. *humillima* F. v. Muell.

Leaf surface

Stomata : subsidiary cells acute at one end and rounded or perpendicular at other end.

Leaf T.S. Leaf blade width 0·3 mm; thickness $< 0{\cdot}1$ mm.

Outline oval, with flattened adaxial surface. **Bundle sheaths :** I.S. cells at phloem pole with cellulosic wall thickening.

Culm surface

Stomata : subsidiary cells acute at both ends.

Culm T.S. Culm diameter 0·25 mm.

Outline circular–oval. **Chlorenchyma** composed of wide, rounded cells; abundant intercellular spaces. **Sclerenchyma** confined to I.S. **Vascular bundles** 1, central; all vessels narrow, lateral wall-pitting spiral-annular, or sometimes intermediate between spiral and scalariform. **Inclusions:** granular, silica-like material in certain epidermal cells.

Rhizome T.S. See genus description; diameter 0·7 mm.

Root T.S. See genus description; diameter < 0·1 mm.

***C. philippinensis* Merr.**

Leaf surface

Epidermis: cells with deeply concave transverse walls. Cells on leaf margins narrower than remainder. **Hairs:** none seen. **Stomata:** those on abaxial surface with acute ends to subsidiary cells; those on adaxial surface with 1 acute and 1 perpendicular end to each subsidiary cell. **Tannin** present in some cells.

Leaf T.S. Leaf blade width 0·5 mm; thickness 0·2 mm.

Outline oval, with concave adaxial surface. **Mesophyll:** cells more or less isodiametric, 4–5-layered to either side of vb and 1–2-layered above and below it.

Culm surface not seen.

Culm T.S. Culm diameter 0·5 mm.

Outline oval. **Stomata** not seen. **Chlorenchyma:** cells of outer 1–(2) and inner 2–3 layers thin-walled, middle 2–3 layers composed of v. thin-walled cells, these partly collapsed in material examined. **Vascular bundles:** 2 small vb's embedded in outer layers of scl. cylinder, 4 larger vb's in ring to inside of scl. cylinder.

Root T.S. Root diameter < 0·1 mm.

Endodermis: cell walls all evenly and heavily thickened. **Vascular tissue:** 2 xylem poles.

***C. polygyna* (R. Br.) Hieron.**

Leaf surface

Stomata long and narrow in CA 65, guard cells narrowing slightly at centre in 685.63 (see Fig. 20. H). **Hairs:** not seen.

Leaf T.S. Leaf blade width < 0·1 mm; thickness $\ll$ 0·1 mm.

Outline oval, with slightly flattened adaxial face. **Mesophyll:** palisade-like cells in 1–2 layers in 685.63, more or less isodiametric cells in 2–4 layers in CA 65.

Culm surface

Epidermis: cells mostly 6-sided, end walls with little or no overlap. **Stomata:** guard cells narrowest at centre in 685.63.

Culm T.S. Culm diameter 0·4 mm.

Chlorenchyma: cells of outer layer palisade-like, with few pegs, those of inner 2–3 layers about as high as wide in T.S. (elongated longitudinally). **Sclerenchyma** in 685.63, represented by a 4–6-layered cylinder; in CA 65, by I.S. only. **Vascular bundles.** In 685.63, consisting of 6 small, outer vb's (3 of them composed of 2–3 xylem elements only, but others with several phloem cells as well) together with 3 larger, inner vb's. In CA 65, 2 small, outer vb's apparently represented by several xylem elements only, attached by their inner bundle sheaths to that of the central vb. One large, central vb only present in this specimen (Fig. 20. G). **Bundle sheaths:** I.S. sclerenchymatous, indistinguishable from scl. cylinder in 685.63; O.S. parenchymatous, enclosing all 3 vb's in CA 65, and probably equivalent to parenchyma sheath of 685.63. **Central ground tissue** composed of parenchymatous cells with slightly lignified walls in 685.63, absent from CA 65.

Rhizome and *root*: see genus description; respective diameters 0·5–1·5 and 0·1 mm.

Additional Notes from the Literature

Arber (1922) described and illustrated leaf sections of *C. aristata*, and recorded the presence of 1, central vb. She noted that Goebel (1913) regarded the leaf of this sp. as being of the extremely reduced ensiform type, with a second vb above the median vb resulting from a fusion of the 2 laterals. Arber has seen this only in bracts. The unusual outline of the vb is recorded and illustrated in this work, pp. 92, 102, Fig. 22. A, B. It appears to be a single bundle, but the idea that it may be a fusion product cannot be ruled out on the basis of the limited information to hand.

Malmanche (1919) described the culm anatomy of *C. aristata* Roem. *et* Schult., *C. fascicularis* Labill., and *C. tenuior* Roem. *et* Schult. The descriptions of *C. aristata* and *C. fascicularis* correspond closely with those for material described in this work. The range in number of vb's is further extended for *C. fascicularis*, where 5 small, outer and 5 large, inner bundles are recorded. The culm of *C. tenuior* is reported to have the following structure. Culm Surface. **Epidermis:** cells irregular. **Stomata** rare. **Hairs** absent. Culm T.S. **Outline** more or less circular, with 6 barely distinguishable faces. **Chlorenchyma:** cells all similar, with slightly thickened walls. **Sclerenchyma:** cylinder. **Vascular bundles:** 6 small, to outside of scl. each opposite 1 face of stem; 3 large, embedded in scl. cylinder. **Central ground tissue:** polygonal, parenchymatous cells with sclerified walls.

Solereder and Meyer (1929) reported the presence of 2–5-celled hairs in *C. strigosa* Roem. *et* Schult.; the walls of the cells are thickened. This sp. has a scl. cylinder in the culm, with 5 large, inner vb's and 11 small, outer vb's. The same authors also report that the root of *C. strigosa* has an exodermis; the xylem is triarch, each pole is adjacent to the endodermis; the phloem poles meet centrally and extend to the pericycle.

C. drummondii, also examined by Solereder and Meyer, was reported to have 2 large, inner and 4 small, outer vb's in the culm.

C. tenuior leaf has filamentous hairs according to Hieronymus (1873). The cells of the chlorenchyma (mesophyll) are elongated axially. The inflorescence axis below the leaf insertion shows no distinction between the cells of the cortex and ground tissue repectively, all of which are large. In the inflorescence axis above the leaf insertion there is a cylinder of prosenchyma which separates cortical and ground tissue cells. The outer chlorenchyma cells are spongy, rounded, about twice as long as high or wide. In the axial direction these cells are arranged slightly obliquely. Cells of innermost layer are 10+ times longer than wide, with small intercellular spaces. No stomata were found on the inflorescence axis above the insertion of the leaves.

The I.S. to the stem vb's of *C. tenuior* are 4–5-layered (Hieronymus).

The leaf vascular strands gradually lose phloem towards the apex, and terminate in solitary vessels (Hieronymus).

Taxonomic Notes

Centrolepis shares the character of overlapping epidermal cells with all other genera except *Gaimardia*.

It seems unlikely that the number of vascular bundles in the culm T.S. can generally be used to distinguish between individual species. *C. philippinensis* is the only species in the genus examined which has fewer small, outer bundles than large, inner ones. It is of interest that the small outer bundles are frequently about twice as numerous as the large inner bundles in most species.

When hairs occur they could provide a guide to the identity of a species, since a range of forms has been noted.

GAIMARDIA Gaudich.

No. of spp. *c.* 3; examined 3.

Genus and Species Descriptions

Leaf surface

Epidermis. Cells on both surfaces similar; mostly 4-sided, ranging from as long as to 15 times longer than wide; mostly 10–12 times longer than wide in *G. australis* (Fig. 23. L) and 2½–15 times longer than wide in *G. fitzgeraldi* (Fig. 23. M). Ends of some cells in *G. australis* slightly inflated, but no evidence of overlapping of ends seen in any spp. Walls moderately thickened, slightly wavy. Marginal cells often narrower than remainder. Pits in lateral (anticlinal) walls pronounced in *G. setacea*[1]. **Hairs:** branched filaments in *G. fitzgeraldi* (Figs. 21. N–Q) (see family description, p. 89.) **Papillae:** none seen. **Stomata:** subsidiary cells frequently slightly longer than guard cells, usually with acute or rounded ends (Fig. 23. L, M). **Inclusions:** bodies or silica-like inclusions of irregular size but rectangular outline observed in certain epidermal cells of *G. australis*. (These inclusions did not turn pink when treated with carbolic acid solution, as many silica bodies do but are not apparently crystalline.)

Leaf T.S.

Outline circular (*G. setacea*) or more or less oval, with flattened (*G. fitzgeraldi*, Fig. 23. J) or grooved adaxial surface (*G. australis*, Fig. 23. F), and flattened abaxial surface. **Epidermis:** adaxial larger than abaxial cells, mostly about as high as wide, cells on margins often smaller than others; outer walls moderately thick to thick, other walls thin to moderately thickened. **Stomata:** guard cells with thin but pronounced outer lip. **Hypodermis** in *G. australis*, 1 layer of cells with slightly thickened walls present opposite adaxial trough, and adaxial face; also 1 or 2 isolated cells in margins. **Vascular bundles:** 1 in *G. setacea*, 3 in other spp. Mx vessels narrow in all except central vb of *G.*

[1] Leaf terminating in hyaline point, cells much elongated, narrow, separating into several filaments at apex.

australis, where medium-sized (Fig. 23. G, K). **Bundle sheaths:** I.S. sclerenchymatous, fibres thick-walled, in 1–3 layers; O.S. cells in 1 layer, parenchymatous, in *G. australis* with U-shaped wall thickenings (Fig. 23. G). **Mesophyll:** outer cells palisade-like, occasionally with pegs (Fig. 23. K), inner cells more or less isodiametric. **Sclerenchyma** confined to bundle sheaths. **Air-canals** present in spp. with 3 vb's, between vb's. **Inclusions** present in epidermal cells of *G. australis* (see leaf surface).

Culm surface not seen.

Culm T.S. seen in *G. fitzgeraldi* and *G. setacea.*

Outline sub-circular (Fig. 23. H). **Epidermis:** cells about as high as wide; outer walls moderately thickened or thick, others thin. **Chlorenchyma** mainly 2–3-layered, 4-layered in places; cells rounded-polygonal, innermost cells with wide intercellular spaces. **Sclerenchyma:** cylinder, fibres in 1–4+ layers. **Vascular bundles:** 4 small, outer vb's and 3 or 4 larger, inner vb's, all enclosed in scl. cylinder and none exhibiting wide mx vessels (Fig. 23. D, E, I). **Bundle sheaths** sclerenchymatous, fibres indistinguishable from those of scl. cylinder in which vb's are embedded. **Central ground tissue** composed of parenchymatous cells with slightly thickened walls.

Rhizome T.S.

Outline more or less circular. **Epidermis:** cells of variable size, walls thin or slightly thickened. **Exodermis:** cells in 1 layer, slightly larger than epidermal cells, sometimes with thickened inner and anticlinal walls. **Cortex:** cells in 2–4 layers, wide, thin-walled (with thickened inner and anticlinal walls in *G. australis*). **Endodermoid sheath** several-layered, all cells or only those of outermost layer with U-shaped wall thickenings. **Vascular bundles** in peripheral ring, most amphivasal, few collateral in *G. australis*, distributed in 4 blocks in *G. fitzgeraldi* and *G. setacea.* **Central ground tissue** composed of thick-walled cells.

Root T.S.

Epidermis: cells thin-walled, wider than high. **Root-hairs:** each about $\frac{1}{4}$–$\frac{1}{3}$ width of epidermal cell; arising to one side of cell. **Exodermis** 1-layered, cells similar in size to those of epidermis, with evenly thickened walls (*G. fitzgeraldi*) or inner and anticlinal thicker than outer walls. **Cortex** consisting of 2–3 layers of rounded cells, with thin or slightly thickened walls. Larger cells arranged in radiating plates situated between smaller cells in *G. fitzgeraldi.* **Endodermis:** in *G. fitzgeraldi* and *G. setacea* 1-layered, cells with slightly or moderately thickened inner and anticlinal walls; in *G. australis* 3–4-layered, cells narrow, with v. thick inner and anticlinal walls. **Vascular tissue:** 3 xylem poles in *G. australis*, 2 in other spp.

MATERIAL EXAMINED

Gaimardia australis Gaudich.: Savatier 281; S. Argentina (K). Leaf blade width 0·8 mm, thickness 0·3 mm. Root diameter < 0·1 mm.

G. fitzgeraldi F. v. Muell. and Rodw.: Curtis, W. M., 1948; Nat. Park, Tasmania (S). Leaf blade width 0·8 mm, thickness 0·3 mm. Culm diameter 0·4 mm. Root diameter < 0·1 mm.

G. setacea Hook. f.: (i) Pulle, A. 1035; W. New Guinea (K). (ii) Cheadle, V. I. CA 412; Hobart, Tasmania 10.2.1960 (S). (iii) G. 4153; Lincoln, New Zealand (S). Leaf blade width < 0·1 mm. Culm diameter 0·5 mm. Root diameter ⩽ 0·1 mm.

Additional Notes from the Literature

Arber (1922) illustrated leaf and leaf base in T.S. of *G. australis.*

Malmanche (1919) gave an account of the culm anatomy of *G. australis.* His description matches that for the material examined for this present work, except that he records the presence of only 2 large inner vb's. He stresses the fact that there are no free, cortical vb's, and mentions that the outermost cells of the sclerenchyma cylinder have horseshoe-shaped wall thickenings. He considers *Gaimardia* to be distinct from the rest of the family, unlike the Restionaceae since it has a small number of vascular bundles and has a different type of chlorenchyma, and more like the Eriocaulaceae than other genera.

Vessel elements in root of *G. australis* are said to be spiral by Hieronymus (1873). His account of the root anatomy of this sp. corresponds exactly with that described here.

Taxonomic Notes

Gaimardia occupies an isolated position in the family anatomically since it is the only genus in which the ends of the epidermal cells do not overlap. It is also unusual since some species have leaves with 3 vascular bundles; one central bundle is the general rule, except in *Aphelia cyperoides.* The branched hairs of *G. fitzgeraldi* are without parallel in the rest of the family, and the way in which individual cells in the hair branch is remarkable in a higher plant.

HYDATELLA Cheesem.

No. of spp. 2?; examined 1.

Leaf material only was examined; it is anatomically similar to *Trithuria filamentosa* (see p. 113 for discussion).

Material Examined

Hydatella inconspicua (Cheesem.) Cheesem.: Carse, H., 1796; Lake Orgatu, N. Auckland, New Zealand Jan. 1913 (K).

PSEUDALEPYRUM Dandy

No. of spp. 4; examined 2.

Genus and Species Descriptions

Leaf surface

Epidermis. Cells of both surfaces similar, but those of adaxial surface larger than abaxial cells in *P. ciliatum.* Cells of occasional files narrower than most. Cells mostly above 10 and up to 25+ times longer than wide, those in files with stomata normally shorter; cells more or less 4-sided, with marked overlap of ends. Transverse wall at overlap deeply concave, with 2 lobes (see family description, p. 87, and Fig. 20. B, C), cells frequently widest at region of overlap. Walls slightly or moderately thickened. **Hairs** either with small basal cell, inflated second cell and 2–3-celled uniseriate filament (*P. monogynum*, Fig. 21. F, G) or consisting of an inflated basal cell and normally 2 other elongated cells (*P. ciliatum*, Fig. 21. D, E). **Stomata.** Subsidiary cells frequently longer than guard cells, with acute ends (Fig. 23. C). Stomata in more or less well-defined longitudinal files. No stomata observed in adaxial surface of *P. ciliatum*.

Leaf T.S.

Outline subcircular (*P. ciliatum*, Fig. 23. A), or with flattened or slightly concave adaxial face; somewhat wider than thick. **Epidermis:** cells of both surfaces similar, but abaxial frequently shorter than adaxial cells; outer walls moderately thickened (*P. ciliatum*, Fig. 23. A) or thick (*P. monogynum*, Fig. 20. E). **Hairs:** see 'leaf surface'. **Stomata** with lip at outer aperture. **Vascular bundle** 1, central, with narrow mx vessels (Figs. 20. E; 23. B). **Bundle sheaths:** I.S. sclerenchymatous, fibres with slightly thickened or thick walls, in 1, 2, or rarely 3 layers; O.S. parenchymatous, cells wider than fibres of I.S., in 1 layer. **Mesophyll:** 2–3–(5) layers of cells, those of outer layer(s) palisade-like, infrequently with lobes, those of inner layers more nearly isodiametric. **Air-canals:** none well developed, but 1 sometimes present to either side of vb in basal material of *P. ciliatum*. **Tannin** present in some epidermal cells of *P. monogynum*.

Culm surface

Epidermis: cells similar to those of leaf. **Hairs:** none seen. **Papillae:** 1 on most cells, near to overlapping end. **Stomata:** subsidiary cells frequently extending proximally beyond end of guard cells.

Culm T.S.

Outline circular, with 6 rounded ridges in *P. ciliatum*, circular or oval in *P. monogynum*. **Epidermis** as in leaf, but cells of *P. monogynum* with only slightly or moderately thickened walls (Fig. 20. A). **Stomata** as in leaf; infrequent in *P. monogynum*. **Chlorenchyma** 3–5-layered, cells appearing more or less isodiametric, some with lobes. **Parenchyma sheath** 2-layered, seen in *P. ciliatum* to inside of chlorenchyma. **Sclerenchyma:** 2–4-layered cylinder enclosing larger vb's; smaller vb's (*P. ciliatum* only) adjacent to outer side of scl. **Vascular bundles:** central pair of large vb's (Fig. 20. A); 2 small vb's also present in *P. ciliatum*; all mx vessels narrow or medium-sized. **Bundle sheaths:** I.S. sclerenchymatous, more or less distinct from scl. between vb's. O.S. indistinct, or (*P. ciliatum*) probably represented by parenchyma sheath. **Central ground tissue** absent. **Inclusions:** particles with appearance of silica? embedded in some cell walls (*P. monogynum*). **Tannin** present in some cells (*P. monogynum*).

Rhizome T.S.

Outline more or less circular or irregular. **Epidermis:** cells thin-walled, slightly wider than high. **Cortex.** In *P. monogynum* outer part consisting of 2–3 layers of rounded, compact cells; middle part of 1–2 layers of wider, thin-walled cells; cells of innermost layer narrow, thin-walled. In *P. ciliatum* outer part composed of 2 layers of polygonal cells; middle part of 10 layers of rounded cells with short lobes (these cells containing starch grains); inner part consisting of 2–3 layers of polygonal cells. **Endodermoid sheath** cells in

1–2 layers, parenchymatous in *P. ciliatum*, with U-shaped wall thickenings in *P. monogynum*. **Vascular bundles** distinct or fused laterally, some collateral but mostly more or less amphivasal. **Central ground tissue:** polygonal cells with moderately thick or thick walls; conspicuous, simple pits in cell walls of *P. ciliatum*. (*Note*: The material of both spp. has more the appearance of basal culm than normal rhizome.)

Root T.S. seen in *P. ciliatum* (Fig. 22. N).

Epidermis: cells about as high as wide, thin-walled, occasionally 2-layered. **Root-hairs** arising to one side of epidermal cells. **Exodermis** 1-layered; cells with slightly thickened inner and anticlinal walls. **Cortex:** outer part consisting of 1 interrupted layer of small, rounded cells; middle part frequently composed of fragmented cells (in best preserved material small rounded cells present in *c.* 5 radiating plates, the area between the plates being occupied by larger, thin-walled cells); inner part composed of a single layer of cells *c.* $\frac{1}{3}$ height of others. **Endodermis** composed of 8–12 cells; inner and anticlinal walls moderately thickened. **Vascular tissue:** 2 mx vessels; phloem present to either side of mx vessels, extending to endodermis.

Material Examined

Pseudalepyrum ciliatum (Hook. f.) Dandy: (i) Auckland Islands Expedn. Jan. 1963; grown at Lincoln, N.Z. as G. 4172 (as *Gaimardia ciliata* Hook. f.) (S). Leaf blade width 0·4 mm, thickness $<$ 0·1 mm; (ii) Table Top Tararuas; grown at Lincoln, N.Z. as G. 3894 (S). Leaf blade width 0·5 mm, thickness *c.* 0·1 mm. Rhizome diameter 1·5 mm. Root diameter 0·6 mm; (iii) Oliver, W. R. B.; Upper Bealey Valley 16.1.1928 (K). Leaf blade width 0·2 mm, thickness 0·1 mm. Culm diameter 0·5 mm.

P. monogynum (Hook. f.) Dandy: (i) Dobson, L.; Tasmania, 7.2.1966 (living material) (as *Centrolepis monogyna* Benth.) (S). Leaf blade width 0·4 mm, thickness 0·1 mm. Culm diameter 0·8 mm. Rhizome 0·8 mm.; (ii) Curtis, W. M.; Mt. Field Nat. Park, Tasmania 12.3.1951 (S). Leaf blade width 0·4 mm, thickness 0·2 mm. Culm diameter 0·5 $\times$ 0·8 mm.

Taxonomic Notes

Pseudalepyrum, proposed by Dandy (1932), was based on *Alepyrum* Hieron. (1873), not *Alepyrum* R. Br. (1810). The species *A. polygonum* R. Br., *A. pumilio* R. Br., and *A. muticum* R. Br., together with *A. muscoides* Hook. f. and *A. muelleri* Hook. f. were put into *Centrolepis* Labill. by Hieronymus. Dandy argued that *Alepyrum pallidum* Hook. f., cannot be legitimately retained as the type for the monotypic genus *Alepyrum* as was done by Hieronymus (1873).

The specimens examined under *Pseudalepyrum* for this present work were housed in the Kew herbarium under their names accepted prior to Dandy's revision, i.e. *P. ciliatum* (Hook. f.) Dandy as *Gaimardia ciliata* Hook. f. and *P. monogynum* (Hook. f.) Dandy as *Centrolepis monogyna* Benth.

Anatomical evidence is of little help in either supporting or refuting the combination of *G. ciliata* and *C. monogyna* in *Pseudalepyrum*. Both have 1 central vascular bundle in the leaf, a feature more prevalent in *Centrolepis* than *Gaimardia*, and both have overlapping leaf epidermal cells—a character not recorded for *Gaimardia*. This suggests that *G. ciliata* was probably misplaced in *Gaimardia*. Additional material of a wider range of *Pseudalepyrum* species needs to be examined anatomically before more conclusive statements can be made relating to the taxonomy of the genus.

TRITHURIA Hook. f. (*Juncella* F. v. Muell.)

No. of spp. 3?, examined 2.

Genus and Species Descriptions

Leaf surface

Epidermis: both surfaces similar; cells 2–35 but mostly between 12 and 30 times longer than wide. Ends of some cells overlapping. **Stomata** seen only in *T. submersa*; anomocytic (Fig. 24. A).

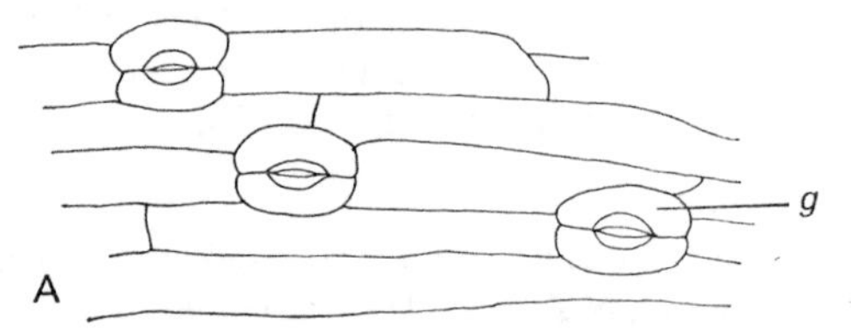

Fig. 24. Centrolepidaceae. A, *Trithuria submersa*, leaf epidermis (×320). g = guard cell.

Leaf T.S.

Outline subcircular to more or less oval, adaxial surface slightly more flattened than abaxial (Fig. 22. F). **Epidermis:** cells 4–6-sided, becoming locally 2-layered by overlap (Fig. 22. J). **Stomata:** none seen. **Hypodermis:** cells in 1 layer, with thin walls, mostly twice as wide as epidermal cells, some larger than others. **Vascular bundle** 1, central; all xylem elements narrow (Fig. 22. G). **Bundle sheaths:** I.S., 1 layer of narrow, thin-walled cells, probably parenchymatous; O.S., 1 layer of wide, thin-walled parenchymatous cells. Cells surrounding distal tip of vb in *T. submersa* all with spiral or reticulate wall thickenings. **Mesophyll:** plates of rounded cells radiating from I.S. to hypodermis; wide air-cavities present between cells.

Culm surface

Epidermis: cells similar to those of leaf, but mostly between 10 and 20 times longer than wide; ends rarely overlapping. **Stomata:** none seen.

Culm T.S.

Outline more or less subcircular (Fig. 22. H). **Epidermis:** cells rarely in 2 layers. **Stomata:** none seen. **Chlorenchyma:** cells in 5–7 layers, rounded, thin-walled, loosely packed; large intercellular spaces present. **Vascular bundles** 2, small, central; px poles next to circular arrangement of small cells lining central duct; nature of duct not determined (see Fig. 22. I). **Bundle sheath:** all cells thin-walled, parenchymatous, those of inner layer(s) smaller than outermost.

Rhizome T.S.

Epidermis: cells about as high as wide, 2-layered in places. **Hairs** short, multicellular, filamentous, composed of single basal cell, 2 stalk cells each about half width of basal cell and slightly wider than high, and 1 terminal cell about 3 times higher than wide and slightly wider than stalk cells (Fig. 22. K). **Cortex:** outer 2 layers composed of polygonal cells, remaining cells rounded, with pegs and separated by large intercellular spaces. **Endodermoid sheath:** cells in 1 layer, with slightly thickened walls. **Vascular bundles** v. irregular; fragments of xylem dispersed in parenchymatous matrix. Wall thickening of most vessels spiral-annular.

Root T.S.

Exodermis: cells similar to those of epidermis, in 1 layer, all walls evenly and slightly thickened. **Cortex:** (i) outer part with cells in 1 layer, similar to those of exodermis; (ii) middle part composed of radiating plates of rounded cells; (iii) inner part 1-layered, cells rounded, thin-walled. **Endodermis:** cells with conspicuous Casparian strips (Fig. 22. M). **Vascular tissue:** 1 central mx vessel surrounded by phloem.

MATERIAL EXAMINED

Trithuria filamentosa Rodw.: (i) Gibbs, L. S. 6674; Central Plateau, Tasmania Jan. 1915 (K). (ii) Curtis, W. M.; Mt. Field Nat. Park, Tasmania 12.3.51 (S). Leaf blade width 0·8 mm, thickness 0·5 mm. Culm diameter 0·5 mm.

T. submersa Hook. f.: (i) Gibbs, L. S. 6809; Lake Hardy, Tasmania 1914–15; leaf only (K). (ii) as *Juncella tasmanica* F. v. Muell., Tasmania, on type sheet, bottom left-hand specimen; leaf only (K). Leaf blade width 0·8 mm, thickness 0·5 mm.

ADDITIONAL NOTES FROM THE LITERATURE

Solereder and Meyer (1929) reported that chlorenchyma cells in the leaf of *Juncella tasmanica* F. v. Muell. (= *Trithuria submersa*) are elongated axially. The culm of this sp. was described as having 3 vb's arranged round a small pith.

TAXONOMIC NOTES

Edgar (1966) reported that male material recently collected shows that *Hydatella* should be regarded as a genus in its own right, and distinct from *Trithuria*. The pollen of *Hydatella inconspicua* and *Trithuria macranthera* was examined by Bortenschlager *et al.* (1966). They stated that the two genera have different pollen grains and that the grains are not similar to those of the rest of the Centrolepidaceae. They indicated that a new family may be needed to accommodate these genera.

The anatomy exhibited by *Hydatella* and *Trithuria* is the most reduced in the family Centrolepidaceae. The leaf anatomy of the two genera is very similar. These annual plants are partly or totally submerged for a large part of the year, and the aquatic habit is reflected by the large proportion of air-cavities present in both culm and leaf, by the lack of mechanical tissue, and by the weakly developed xylem.

The anomocytic stomata seen in the leaf of *T. submersa* are unique in the family (Cutler 1968). Their occurrence was also noted by Solereder and Meyer (1929), where the plant examined was called *Juncella tasmanica* F. v. Muell.

Ecological adaptations may mask or emphasize characters of taxonomic significance. It is, therefore, difficult to say whether or not the anatomical distinctions between these two genera on the one hand, and the others in the family on the other, are taxonomically important. It is possible to support the view that the genera *Hydatella* and *Trithuria* should be moved from the family, in particular because of the presence of anomocytic stomata in *T. submersa*, but evidence should also be sought from other sources.

There is no evidence from leaf anatomy which suggests that *Hydatella* and *Trithuria* should be distinct genera, but again, the similarity could be due to similar environmental adaption in the two. Unfortunately, no stomata could be found on any of the leaves of *Hydatella* and the useful information they would have provided is lacking.

On grounds of gross morphology, *Trithuria* and *Hydatella* are distinguishable from all other genera in the Centrolepidaceae because they have bilocular as opposed to unilocular anthers.

Bibliography for Centrolepidaceae

Arber, A. (1922) Leaves of the Farinosae. *Bot. Gaz.* **74**, 80–94.

Bentham, G. and Hooker, J. D. (1883) *Genera Plantarum*, 3, 1025–7.

Bierhorst, D. W. and Zamora, P. M. (1965) Primary xylem elements and element associations of angiosperms. *Am. J. Bot.* **52**, 657–710.

Bortenschlager, S., Erdtman, G., and Praglowski, J. (1966) Pollenmorphologische Notizen über einige Blütenpflanzen incertae sedis. *Bot. Notiser* **119**, 160–8.

Chanda, S. (1966) On the pollen morphology of the Centrolepidaceae, Restionaceae and Flagellariaceae, with special reference to taxonomy. *Grana palynol.* **6**, 355–415.

Cheadle, V. I. (1942) The occurrence and types of vessels in the various organs of the plant in the Monocotyledoneae. *Am. J. Bot.* **29**, 441–50.

—— (1943*a*) The origin and certain trends of specialization of the vessel in the Monocotyledoneae. *Am. J. Bot.* **30**, 11–17.

—— (1943*b*) Vessel specialization in the late metaxylem of the various organs in the Monocotyledoneae. Ibid. 484–90.

—— (1944) Specialization of vessels within the xylem of each organ in the Monocotyledoneae. *Am. J. Bot.* **31**, 81–92.

Cutler, D. F. (1968) Anatomy and taxonomy of certain monocotyledonous families. *Proc. Linn. Soc. Lond.* **179**, 261–7.

Dandy, J. E. (1932) Some new names in the Monocotyledones II. *J. Bot., Lond.* **70**, 328–32.

Desvaux, M. (1828) Observations sur quelques familles des plantes monocotylédones d'après les manuscrits de feu le Baron Palisot de Beauvois. *Ann. Sci. nat.*, Bot., ser. 1, **13**, 37–52.

Edgar, E. (1966) The male flowers of *Hydatella inconspicua* (Cheesem.) Cheesem. (Centrolepidaceae). *N.Z. J. Bot.* **4**, 153–8.

Engler, A. (1936) *Syllabus der Pflanzenfamilien*, 11th edn. Berlin.

Erdtman, G. (1952) *Pollen morphology and plant taxonomy*. Chronica Botanica Co., Waltham, Mass.

Goebel, K. (1913) *Organographie der Pflanzen*. I. *Allgemeine Organographie*, 2nd edn, p. 285. Gustav Fischer, Jena.

Hamann, U. (1960) Die Chromosomenzahl von *Centrolepis strigosa* (Centrolepidaceae). *Naturwissenschaften* **47**, 360.

—— (1961) Merkmalsbestand und Verwandtschaftsbeziehungen der Farinosae. Ein Beitrag zum System der Monokotyledonen. *Willdenowia* **2**, 639–768.

—— (1962*a*) Weiteres über Merkmalsbestand und Verwandtschaftsbeziehungen der Farinosae. *Willdenowia* **3**, 169–207.

—— (1926*b*) Beitrag zur Embryologie der Centrolepidaceae mit Bemerkungen über den Bau der Blüten und Blütenstände und die systematische Stellung der Familie. *Ber dt. bot. Ges.* **75**, 153–71.

—— (1963) Über die Entwicklung und den Bau des Spaltöffnungsapparats der Centrolepidaceae. *Bot. Jb.* **82**, 316–20.

HAURI, H. (1917) Anatomische Untersuchungen an Polsterpflanzen nebst morphologischen und ökologischen Notizen. *Beih. bot. Zbl.* **33** (1), 275–93.

HEGNAUER, R. (1963) *Chemotaxonomie der Pflanzen.* II. *Monocotyledoneae.* Birkhäuser Verlag, Basel and Stuttgart.

HIERONYMUS, G. (1873) Beiträge zur Kenntniss der Centrolepidaceen. *Abhandl. naturf. Ges. Halle* **12** (3 and 4), 108 pp.

HOU, D. (1957) Centrolepidaceae in *Flora Malesiana* ser. 1, **5**, 421–8.

HUTCHINSON, J. (1959) *The families of flowering plants II. Monocotyledons*, 2nd edn, pp. 699–700. Clarendon Press, Oxford.

MALMANCHE, L.-A. (1919) *Contribution à l'étude anatomique des Eriocaulonacées et des familles voisines: Restiacées, Centrolépidacées, Xyridacées, Philydracées, Mayacacées.* Thesis, Paris. pp. 165.

SOLEREDER, H., and MEYER, F. J. (1929) *Systematische Anatomie der Monokotyledonen.* Heft 4, pp. 30–3.

STEBBINS, G. L., and KHUSH, G. S. (1961) Variation in the organization of the stomatal complex in the leaf epidermis of monocotyledons and its bearing on their phylogeny. *Am. J. Bot.* **48**, 51–9.

VAN TIEGHEM, P. (1887) Structure de la racine et disposition des radicelles dans les Centrolépidées, Eriocaulées, Joncées, Mayacées et Xyridées. *J. Bot., Paris* **1**, 305–15.

WETTSTEIN, R. v. (1935) *Handbuch der systematischen Botanik*, 4th edn. Leipzig and Vienna.

RESTIONACEAE

(Figs. 25–40; Pl. IV. B; V–VIII)

Introduction

THE Restionaceae is a family of perennial, herbaceous plants bearing a close morphological resemblance to members of Juncaceae and Cyperaceae (see p. 1).

The distribution is almost entirely in the southern hemisphere, the main concentration being in south and south-western Australia and coastal to subcoastal regions of Cape Province, South Africa. Species are also found in Tasmania, New Zealand and the Chatham Islands, Madagascar, the Malay peninsula, and Chile and Patagonia in South America. One species is found north of the equator, in South Vietnam, and another occurs in Malawi.

In South Africa the family is represented by 12 genera, all found between the latitudes 31S and 35S, in Cape Province. They are most prolific in the south-west corner of the province, the area defined by Tulbagh in the north, Riversdale in the south-east and the coast. This is the wettest region, receiving up to 60 in of rain per annum locally in a normal winter. The complete range of the family is defined by two arms radiating from this region, one to the NNW, ending at the coast in Little Namaqualand, and the other to the ENE, following the lowland strip up the coast as far as Pondoland. The density becomes very low, only about 1–5 species in a Division, towards the end of these arms. The normal rainfall in these regions is from 10 to 20 in per annum. Species have been recorded at up to and over 6000 ft. In the Basutoland Highlands they may attain greater altitudes.

The Australian representatives of the family occur both in the east and the west. Species are also recorded in Queensland. The distribution of species is discontinuous, very few growing in both east and west; *Leptocarpus tenax* R. Br. is a striking exception.

Gardner (1941–2) gave an account of the vegetation and climate of Western Australia in which he described the habitats in which the Restionaceae grow. For the majority of species, the sandy, seasonably wet regions provide suitable conditions.

Diels (1904) stated that some of the Australian species grow in drier regions than the most drought-resistant species of South Africa.

In general, the habitats favoured by the Restionaceae are fairly moist, with winter rains; they are subjected to a period of drying normally corresponding to the dormant phase of the plants' life cycle.

No species has been found to occur both in South Africa and Australia, although at present several genera are thought to be represented in both countries. The anatomical evidence indicates, however, that there are probably no genera occurring in both countries.

Family Description

Leaf

Leaf blades are rarely present, but are described in T.S. when suitable material was available; descriptions given here frequently refer to the sheathing leaf base, which is more readily available. The base of the leaf overlaps and encircles the culm by $1\frac{1}{3}$–$1\frac{1}{2}$ times at its region of fusion with the culm.

LEAF SURFACE

Hairs uncommon; if present, unicellular or multicellular. **Papillae** more frequent than hairs, often occurring in spp. with papillate culms. Individual papillae short, rounded, usually 1 to a cell. **Epidermis.** Leaf blade, abaxial and adaxial surfaces sometimes similar. Leaf base material usually with dissimilar surfaces: (i) abaxial cells exhibiting same range of variation as found in culm, often similar to those of culm of same sp., but sometimes shorter in axial direction; (ii) adaxial cells 4-, 5- or 6-sided, frequently v. elongated, sometimes hexagonal and 2–3 times wider than long, walls straight or wavy, thin to thick, usually thinner than those of abaxial cells. **Stomata** paracytic, arranged with long axis parallel to that of leaf; exhibiting same range of variation as found in those of culm; frequently present on both surfaces of leaf blade but normally present only on abaxial surface of sheathing leaf base. Often similar to culm stomata of same sp. **Silica**: spheroidal-nodular bodies present in occasional epidermal cells of one sp. of *Lepyrodia.* **Crystals** absent. **Tannin** frequently present, particularly in epidermal cells of sheathing leaf bases. **Cuticular marks** frequently granular, occasionally with longitudinal striations.

LEAF T.S.

Flattened dorsiventrally, leaf blade usually with parallel faces and rounded margins. Leaf bases usually with parallel faces in central region, but tapering gradually to margins. Sections described here all taken from the thicker part of leaf, and not at the margins. Leaf base T.S. sometimes ridged opposite to vb's on abaxial side. Veins running longitudinally in both blade and sheathing base, thus cut transversely in T.S.

Hairs and **papillae**: see leaf surface. **Cuticle** normally thin, sometimes slightly thickened. **Epidermis.** (i) Abaxial cells showing the same range of outline and wall thickness as in the culm, cells frequently similar to those from culm of same sp. (ii) Adaxial cells in leaf blade similar to those of abaxial surface; cells in sheathing base frequently unlike those of abaxial surface, normally ranging from 2–4–(5) times wider than high, 4- or 5-sided and thin- to thick-walled, usually about $\frac{1}{4}$–$\frac{1}{3}$ of height of abaxial cells. **Stomata** present in both surfaces of leaf blade; normally absent from adaxial surface of leaf sheath; superficial or sunken, showing same range of variation in guard cells and subsidiary cells as occurs in culm stomata. **Chlorenchyma** composed of palisade-like or more or less isodiametric cells, frequently with pegs (see culm chlorenchyma on p. 119); cells occupying all space between vascular bundle sheaths and epidermis in most leaf blades, but present in 1 or 2–(3) layers next to abaxial epidermis in sheathing leaf bases; layers interrupted by bundle sheaths in leaf bases of some spp. Protective cells (see p. 120) sometimes present in spp. having them in culm chlorenchyma.

Vascular bundles: (*a*) in blade, usually 3, but sometimes up to 12–15, arranged in 1 row, all with phloem poles facing abaxial surface; (*b*) in sheathing base, 14–30–(40), in 1 row, small and medium-sized (or large) frequently alternating, orientated as in blade. Bundles usually with small phloem and xylem poles. Tracheids of xylem all narrow or medium-sized in smaller bundles, sometimes wide on flanks in larger bundles; vessels not seen. Phloem composed of narrow sieve-tubes and companion cells. **Bundle sheaths:** I.S. sclerenchymatous, fibres usually narrow or medium-sized, rarely wide, usually thick- or v. thick-walled, arranged in 1–3–(4+) layers, sometimes continuous on flanks with scl. of ground tissue; O.S. parenchymatous, present as complete, usually 1-layered sheath round leaf blade bundles. O.S. in leaf sheaths often reduced to a 1-layered cap at phloem pole, or in 1 continuous layer between bundles to abaxial side of ground tissue scl.

Sclerenchyma either confined to bundle sheaths or sometimes also present as 1–3 layers of ground tissue between bundles in leaf base material of certain spp. **Ground tissue** chlorenchymatous in leaf blades, either all parenchymatous or partly sclerenchymatous (see above) in sheathing leaf bases; when partly sclerenchymatous, scl. often separated from adaxial epidermis by 1 or 2 layers of parenchyma cells. Parenchyma cells usually hexagonal in outline (3–6 times longer than wide, with transverse end walls as seen in L.S.), or lobed and aerenchymatous; occupying all space not filled by other tissues except where air-cavities present. **Air-cavities** rare, present in occasional *Restio* group 3 spp., formed by breakdown of parenchyma cells between vb's. **Silica** present as (i) granular amorphous deposits in various tissues of occasional species; (ii) spheroidal-nodular bodies in cells of O.S. or in stegmata in outer layer of I.S. at phloem pole in some spp.; also present in some epidermal cells in certain *Lepyrodia* spp. **Tannin** frequent, particularly in epidermal cells of sheathing leaf bases.

CULM SURFACE

Hairs present in some spp. in several genera; unicellular or multicellular, ranging from simple or branched filaments to flattened, fan-shaped hairs closely applied to epidermis. **Papillae** short, rounded, frequent in few spp. **Epidermis.** Cells arranged in longitudinal files, exhibiting a wide range of outline; 4-, 5-, or 6-sided, all of similar size or some longer than others; ranging from as long as wide to 7–9 times longer than wide, occasionally wider than long; walls straight or wavy, or wavy at surface only and straight at lower focus, thin to v. thick. Epidermal cells with rounded or oval gaps between transverse walls in some genera (e.g. *Chondropetalum*), with cells of inner epidermal layer visible through such gaps. **Stomata** paracytic; long axis of stomata parallel to that of culm; guard cells and subsidiary cells showing a wide range of form. **Silica** present as spheroidal-nodular bodies or as granular material in spp. of *Harperia*, *Lepyrodia*, and *Thamnochortus bachmannii*. **Crystals** absent. **Tannin** present in epidermal cells of many spp. **Cuticular marks** frequently granular, sometimes with longitudinal striations; wavy lines sometimes present above anticlinal walls of epidermal cells; short lines rarely present above, and at right angles to, anticlinal walls of epidermal cells.

CULM T.S.

The tissues are arranged in the following sequence: epidermis, chlorenchyma, parenchymatous sheath, sclerenchymatous sheath (enclosing peripheral vb's); medullary vb's, all, some or none embedded in sclerenchyma sheath; if free, then scattered throughout entire ground tissue. Inner part of ground tissue, if free from vb's, normally v. thin-walled. (See Fig. 25 general plan.)

CULM T.S.: DETAILS OF TISSUES

Outline circular, oval, or kidney-shaped, 4-sided or polygonal, or circular or oval with low, dome-shaped or flat-topped emergences, or circular and grooved. Solid or hollow at internodes, solid at nodes. **Cuticle** thin to v. thick; clear, or with granular inclusions; smooth, or granular, or with small ridges; usually continuous over guard cells and often produced into cuticular lips at outer and inner apertures of guard cells (Fig. 25. G–J); thin cuticular covering also present as lining to substomatal tube formed by protective cells. **Hairs** and **papillae**: see Culm surface. **Epidermis.** Cells normally 4-sided; usually present in 1, sometimes in 2, or occasionally in 3–(4) layers; cells wider than high, as high as wide, or up to 8–9+ times higher than wide; all of the same height or some taller than others. Taller cells present (i) next to stomata, raising them above culm surface, or extending into chlorenchyma and lining substomatal tube (Fig. 30. G, p.170), or (ii) between stomata, or (iii) opposite to pillar cells (p. 121). Occasional cells shorter than others in some spp. When present in 2 layers, cells of outer layer sometimes with gaps between them, outer walls of some cells of inner layer then becoming superficial (Fig. 28. A, p. 152); in one sp. (*Elegia neesii*) cells of outer layer reduced to small areas, the outer surface of culm made up largely of outer walls of cells of inner epidermal layer. Cells of grossly unequal sizes in *Dielsia* and *Restio leptocarpoides*. Cells overarching stomata in some spp. notably *Sporadanthus*. Outer walls slightly thickened to v. thick; straight, convex, or concave, curving inwards at margins or irregular. Anticlinal walls straight or wavy, thin to thick, evenly thickened or with tapering thickening, when thick at outer end, thickening tapering gradually to inner end, or thick for part of length and thickening tapering rapidly, inner part of wall less heavily thickened (Fig. 39. A, p. 309). Inner walls usually slightly or moderately thickened, occasionally thick.

Stomata superficial or sunken, raised on to mounds or ridges or sunken in grooves. Subsidiary cells normally thin-walled, sometimes with slightly or moderately thickened walls; guard cells sometimes with lips at either inner or outer apertures or both, or with ridge on inner walls (Fig. 25. J); lumina variable, ranging from narrow-lenticular to broad, triangular.

Chlorenchyma normally composed of 1, occasionally of 2–5 layers of thin-walled palisade-like cells; layers uninterrupted, or divided into sectors by other tissues. Palisade-like cells of variable height, anticlinal walls straight, or with peg-like processes (when termed **'peg-cells'**), characteristic of many spp. of the family (Cutler 1964), each peg-cell having (1–) 6–12(–20+) tubular peg-like outgrowths or projections from each of the 2–6 anticlinal walls, the pegs being arranged in longitudinal files; pegs of adjacent cells opposite and apposed; some pairs with conspicuous pits in their mutual end walls. Pegs varying in length from half to about twice as long as wide. Cells with pegs normally arranged in close lattice as seen in T.L.S., each cell normally joined to its neighbours by (4–)6 files of pegs (Fig. 25. B, C). Cells without

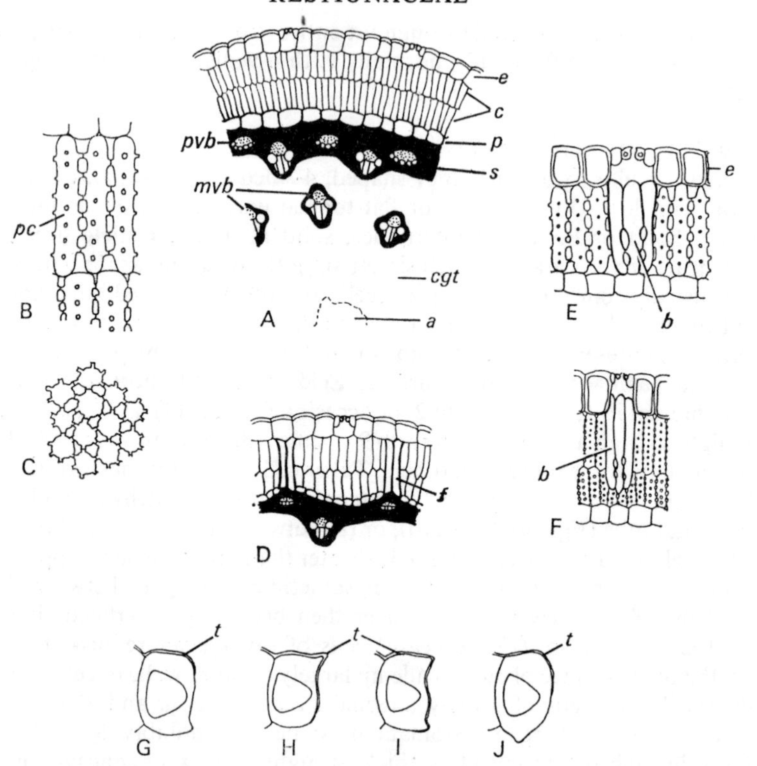

FIG. 25. Restionaceae. Diagrams of culm histology and cell types. A, sector of culm T.S. to show typical arrangement of tissues. B, peg-cells from chlorenchyma (p. 119) as seen in culm T.S. and C, as seen in culm T.L.S. D, pillar cells from part of culm T.S. (p. 121). E, F, protective cells (p. 120); those in E extending from epidermis to parenchyma sheath, those in F extending from part way up substomatal tube formed by epidermal cells, into inner chlorenchyma layer. G–J, guard cells of stomata seen in T.S.: G, with lip at inner aperture; H, with lip at outer aperture; I, with lips at both outer and inner apertures; J, with ridge on inner wall.
a = central air-cavity; b = protective cell; c = chlorenchyma; cgt = central ground tissue (composed of parenchyma cells); e = epidermal cells; f = pillar cells; mvb = medullary vascular bundles; p = parenchyma sheath; pc = peg-cells; pvb = peripheral vascular bundles; s = sclerenchyma sheath or cylinder; t = cuticle (this may extend in a readily definable layer part way or entirely round the guard cell).

pegs or with v. few pegs normally arranged in transverse plates, in chequer-board pattern, or in longitudinal plates as seen in T.L.S.

Protective cells (Fig. 25. E, F): present in many spp., especially those from South Africa.

(These cells consist of modified palisade- or peg-cells, and normally have their walls slightly to moderately thickened (lignified), or the thickening may be restricted to the walls which border on a substomatal cavity. The cells surround a substomatal cavity and form a tube, the walls of which are 1 cell thick. The tube is 1–3 layers of cells deep and extends from the inner wall of the epidermal cells [or part way up the substomatal cavity lined by epidermal cells] right to the parenchyma sheath, or only part way into the chlorenchyma. The tube is frequently closed at its base because the inner ends of the protective cells are curved towards one another and meet. The tube communicates with the chlorenchyma through apertures between the anticlinal walls of the protective cells. The apertures, which may be oval, circular, lenticular, or very elongated, are sometimes present at the outer or inner ends of the tube. Alternatively they may occur where layers of protective cells join one another, or they may be present at intervals throughout the whole length of the tube. Occasionally the cells that form the tube are in contact with one another only at their inner and outer ends.)

Parenchyma sheath composed of 1–several layers of cells, bordering on the chlorenchyma to the outside and on the sclerenchyma sheath to the inside. (i) Cells all 4–6-sided, with rounded corners, as high as wide, or up to 2(–3) times wider than high or higher than wide. (Cells rectangular, normally 1½–4 times or up to 7 or 8 times longer than wide with simple, rounded pits as seen in L.S.). Cells all of more or less uniform size, or some larger than others, walls usually thin or slightly thickened, sometimes moderately thickened, thickening cellulose (becoming more heavily thickened in basal culm material). (ii) In certain genera, cells of 2 different types: (*a*) as those in type (i); (*b*) **'pillar cells'** (Fig. 25. D) (Cutler 1964), elongated, palisade-like cells, normally 3–4 but sometimes up to 10 times higher than wide, usually with moderately thickened walls (lignified), infrequently with slightly thickened or thick walls. Pillar cells radiating, singly or in groups of 2 or 3, from ridges or girders from sclerenchyma sheath to epidermis, dividing chlorenchyma into longitudinal canals. Most pillar cells nearly parallel-sided, some with slightly expanded ends. Parenchyma sheath including **stegmata** in some spp. Stegmata consisting of short cells with U-shaped thickenings (lignin) on inner and anticlinal walls and with thin outer walls, each cell containing a single spheroidal-nodular silica-body.

Sclerenchyma: sheath developed in all genera; present to inner side of parenchyma sheath, and bordering directly on to it. Outline of sheath frequently following that of culm surface, with low, dome-shaped ridges opposite to small peripheral vb's. Girders or ridges present in some spp., opposite to or alternating with peripheral vb's; girders wedge-shaped or more or less rectangular, extending from scl. sheath into chlorenchyma, sometimes reaching to epidermis, e.g. *Dielsia*, *Restio leptocarpoides*, or joined to epidermis by pillar cells (see above), e.g. some *Leptocarpus* and *Willdenowia* spp.; T-shaped, reaching epidermis in *Anthochortus*. Sheath varying in thickness in different spp. (or different specimens of same sp.), ranging from 2 to 3 layers in some to *c.* 25 layers in others, but normally 6–8-layered. Fibres of girders and outer layers of sheath frequently narrow, with thick or v. thick walls and narrow lumina; fibres of inner layers often wider, with thick or moderately thickened walls and wider lumina; fibres to outer side of peripheral vb's in some spp. with moderately thickened walls. Inner boundary of scl. sheath distinct or indistinct from ground tissue. Outline of fibres usually hexagonal, sometimes slightly rounded and with intercellular spaces between them filled with extra-cellular substances. Fibres elongated, often many times longer than wide, frequently with square (transverse) end walls, or end walls oblique, or fibres pointed as seen in L.S.; wall-pitting simple, pits oblique, slit-like. Sheath enclosing all peripheral vb's in all spp. except *Hypolaena graminifolia*. Some, all, or no medullary vb's embedded in sheath; medullary vb's sometimes attached to sheath by fibres at phloem pole.

Vascular bundles collateral; of 2 distinct types. (i) Peripheral: confined to outer ring, enclosed in scl. sheath, more or less evenly spaced; each with small, sometimes poorly developed xylem pole consisting of a 1- or several-layered arc or horse-shoe of tracheids (narrow vessels noted in v. few spp.); all tracheids of similar size (narrow to wide), or central ones widest, or flanking ones wider than the rest. Phloem pole small, composed of sieve-tubes and companion cells, rounded, or oval, situated in concavity of arc formed by xylem, facing outer surface of culm; phloem occasionally separated from xylem by 1 layer of narrow cells with slightly or moderately thickened walls. (ii) Medullary: present to inner side of ring formed by peripheral vb's; all, some, or none embedded in scl. sheath; if free, then scattered throughout central ground tissue or confined to outer region. Outermost medullary bundles smallest, innermost largest, with intermediate sizes between them. Outline of bundles circular or oval. Bundles most frequently with 1 conspicuous mxv on either flank; mxv's narrow, medium-sized or wide, rounded, angular or many-sided, walls slightly thickened to thick, occasionally v. thick; narrowest mxv's usually found in outermost bundles.

Vessel lateral wall-pitting usually scalariform, sometimes alternate, occasionally reticulate; pits simple or bordered (usually bordered in thick-walled vessels). Perforation plates oblique or transverse, simple, scalariform (with few or many bars), scalariform-fenestrate, or reticulate. Flanking mxv's sometimes close together, characteristically widely separated by several rows of narrow cells (xylem parenchyma) in some spp. Protoxylem present in most larger, inner bundles and absent from smaller, outer bundles; vessels with annular or spiral wall thickenings; lysigenous cavities rarely present at px pole. Phloem poles rounded, tangentially or radially oval, or over-arching flanking mxv's; abutting directly on xylem in certain spp., or, more usually, separated from xylem by 1 layer of narrow cells; narrow cells usually with thin walls next to phloem and other walls slightly or moderately thickened. Phloem composed of sieve-tubes and companion cells; some sieve-tubes wide in certain spp., all narrow in others; sieve-tube elements short or elongated, sieve-plates oblique in narrower elements, nearly transverse in wider ones; narrow sclereids present in phloem of some spp.

Bundle sheaths sclerenchymatous, variable in thickness, but normally with more fibre-layers at poles than on flanks; fibres at phloem pole normally with thickest walls, often narrow, arranged in 1, 2, or 3–(4) layers; those on flanks narrow or medium-sized, with slightly or moderately thickened or, occasionally, thick walls, usually arranged in 1, occasionally in 2 layers; fibres at xylem pole often narrow, sometimes medium-sized, usually with moderately thickened or thick walls, arranged in 1 or 2–(3) layers. Thickness of walls variable; normally thickest in specimens with numerous layers of v. thick-walled fibres in scl. sheath. Fibres of bundle sheaths distinguishable from those of scl. sheath by their higher birefringence when viewed in T.S. between crossed polars.

Central ground tissue parenchymatous, cells usually hexagonal in T.S. (with transverse end walls, usually from 4 to 8 times longer than wide as seen in L.S., with round, simple pits in thin-walled cells; pits often elliptic, oblique in cells next to scl. sheath). Cells of outer layers, nearest to scl. sheath and often surrounding medullary vb's, frequently with moderately thickened (partly lignified) walls (cells of outermost layers sometimes with v. thick (lignified) walls). Inner cells wider than outer, sometimes with slightly thickened (cellulose) walls. Central cells free from vb's in most spp., thin-walled, often breaking down to form central cavity. Central cells cut off from other cells by single ring (sheath) of fibres in some *Leptocarpus* and 1 *Restio* spp. Some spp., notably of *Thamnochortus*, with scattered areas of thin-walled cells amongst those with moderately thickened walls in outer layers.

Crystals rhombic, present in *Lyginia barbata* only, in cells of central ground tissue. **Silica** occurring in either of 2 forms. (i) Spheroidal-nodular silica-bodies present in certain spp. sometimes in unspecialized cells or in stegmata forming part of the parenchyma sheath or of the outer layer of scl. sheath. Bodies of this kind also noted in special epidermal cells in some *Lepyrodia* spp.; also occasional in some pillar cells, protective cells or cells of inner chlorenchyma layer but then usually with irregular outlines. (ii) Granular siliceous material (silica sand) present in certain spp., most frequently occurring in occasional cells of central ground tissue, parenchyma sheath, or infrequently in pillar cells, protective cells or chlorenchyma cells; usually filling lumen. **Tannin:** dark-brown, refractive, hard material, of uncertain chemical composition, frequent in some epidermal cells, occasional in cells of parenchyma sheath or central ground tissue, and also noted in phloem and xylem in pathological material.

RHIZOME T.S.

Sections usually taken in region between buds or culm bases.

Hairs present in certain spp.; usually as short multicellular, uniseriate filaments. **Epidermis:** cells 1–2–3 times wider than high, normally 4-, sometimes 6-sided; walls usually all of similar thickness, ranging from thin to thick. **Stomata** infrequent; superficial; showing the same range of variation as in culm. **Hypodermis** occasionally distinct, present as 1–5 layers of cells, but normally either absent or indistinguishable from cells of outer cortex. **Cortex** variable in composition. Cells either all parenchymatous, or mainly parenchymatous but accompanied by strands or several complete layers of sclereids. Cortex occasionally composed almost entirely of sclereids; inner layers often aerenchymatous.

Endodermoid[1] **sheath** often distinct as 1–2–(3) layers of cells with all walls thick to v. thick, or with thin outer walls. **Vascular bundles** all amphivasal, or outer, smaller bundles collateral. Vessels sometimes narrow, more often medium-sized or wide, angular or sometimes round or oval; with scalariform wall pitting and simple, more or less transverse perforation plates. Phloem pole rounded, composed of sieve-tubes and companion cells. Bundles scattered throughout all of central ground tissue, or absent from small, central region. **Bundle sheaths** sclerenchymatous, occasionally distinct from ground

[1] *Sensu* Tomlinson, 1964: there is no evidence that this is a true endodermis.

tissue; 1–4-layered, fibres with moderately thickened or thick walls. **Central ground tissue** composed of 2 sorts of cell. (*a*) Cells next to bundle sheaths, sclerenchymatous, with moderately thickened or thick walls. (*b*) Parenchymatous, often aerenchymatous, thin-walled, lobed, more or less isodiametric cells; cells situated in strands scattered amongst thickened cells, and often also at centre of rhizome.

Silica present either in the form of granular material in some parenchyma cells in occasional spp., or as spheroidal-nodular silica-bodies in some cells of cortex or ground tissue. **Crystals:** none seen. **Tannin** present in some cells of parenchyma of cortex, ground tissue, or epidermis in certain spp.

ROOT T.S.

Root-hairs developing from cells similar in size to those making up remainder of epidermis; basal part of hair normally 2–3 times wider than shaft. **Epidermis:** cells 4-(6-) sided, as wide as high to 2–3 times wider than high, normally with slightly thickened walls. **Hypodermis** 1–3-layered, parenchymatous, composed of cells similar to those of epidermis; present in few spp. **Exodermis** rarely present, composed of 1–3 layers of cells with thickened inner and anticlinal and thin outer walls.

Cortex either wholly parenchymatous (cells of outer layers occasionally with slightly thickened walls), or divided into the following 3 regions: (i) outer part composed of several layers of parenchymatous, 4- or 6-sided cells with slightly or moderately thickened walls; (ii) middle part consisting of radiating parenchymatous plates normally 1 cell wide and up to 20 cells high, with air-spaces between them; (iii) inner part composed of 1–3 layers of parenchymatous cells, individual cells normally narrower than those in plates. **Endodermis** frequently 1-, occasionally 2-layered; cells usually with thick or v. thick inner and anticlinal walls and thin outer walls; passage cells present in some spp., wall-pitting simple, often branched. **Pericycle** normally 1- or 2-layered, but sometimes up to 10-layered, cells thin or moderately thick-walled (occasionally thick-walled).

Vascular tissue: region immediately inside pericycle composed of alternating strands of px and phloem surrounded by cells of ground tissue. Mxv's present (i) in 1 ring, to inner side of px and phloem, or (ii) in 2 rings, or (iii) in 1 outer ring and scattered in central ground tissue. Individual mxv's circular, oval, or angular, usually with scalariform lateral wall-pitting and simple, transverse perforation plates. Unusual phloem distribution noted in some spp.; strands of phloem associated with central mxv's, or scattered in central ground tissue. **Central ground tissue** parenchymatous, all cells with slightly to moderately thickened or thick walls, or, in some spp., central cells thin-walled. **Silica:** none seen, but see p. 302. **Crystals:** none seen. **Tannin** present in occasional ground tissue cells or vessels in some spp.; possibly pathological.

INFORMATION FROM LITERATURE

The selection of characters and their taxonomic importance outlined above agree with the findings of previous authors to some extent. Gilg (1890) did not find stomatal position in relation to the culm surface as seen in T.S. to be of any importance. He mistakenly described the silica-bodies as probably being mucilage, and therefore failed to realize their taxonomic importance. In his classification, however, he gave weight to the importance of chlorenchyma characters, as well as to the structure of the sclerenchyma sheath, the presence or absence of pillar cells or protective cells, and the number of layers of epidermal cells. Malmanche (1919) was aware of the taxonomic value of the chlorenchyma types and indicated that epidermal characters were of taxonomic significance at the species level. Gilg (1890) and Meyer (1929) likewise attached taxonomic importance to the presence or absence of free vascular bundles in the central ground tissue. As a result of the present investigation this has been found to be incorrect. For example, *Calorophus minor* described by the present writer has both conditions, and several other species of which more than one sample has been examined show that the diameter of the culm is more likely to be the influencing factor. Wider culms tend to have free vascular bundles whereas narrow culms (or sections taken from terminal internodes) do not. No pattern of organization of medullary vascular bundles could be recognized.

The study of the nodal organization in *Leptocarpus tenax* shows that vascular bundles do not hold the same spatial relationships to one another at successive levels in the same internode.

Meyer recognized that the position of stomata relative to the culm surface was of taxonomic significance, but only at the species level. Gilg-Benedict (1930) made taxonomic use of the same anatomical characters as did her husband, E. Gilg, in 1890.

Before giving an account of individual genera, brief critical comments must be made on the general anatomical findings of Gilg (1890) and Meyer (1929). Gilg described the general anatomy of the culm of the family by starting from the centre and working outwards and the same sequence has been adopted here for describing the tissues.

Cells of the ground tissue were stated by Gilg to be distinguishable from those of the sclerenchyma sheath because 'round pits are present in the walls of the parenchyma cells, and oblique slit-pits are found in the fibre walls'. The present author has found that there may be a transition region between the two types, in which pits are elliptic and oblique (see family description, p. 123).

Gilg stated of vascular bundles that some small (i.e. peripheral) vascular bundles in certain *Thamnochortus* species are amphivasal. The present author has been unable to confirm this report. Cells next to the outside of the phloem pole of such bundles are always found to be sclerenchymatous; they do, however, compare in size and wall thickness with the tracheids of the vascular bundles, and could easily be confused with them. Gilg noticed vessels in the small (peripheral) vascular bundles in *T. platypteris* Kunth. However, it is difficult to differentiate between vessels and tracheids in these bundles with certainty. Almost all the material examined by the present author has shown that the xylem of the peripheral bundles is made up entirely of tracheids; a middle lamella could be discerned in the end walls. *Thamnochortus* is possibly an exception; more detailed studies on macerated material are needed before this point can be clarified.

Gilg also stated that he found small (peripheral) bundles in direct contact with the parenchyma sheath in some species, and even extending into the chlorenchyma in *Lyginia barbata*, and he went on to elaborate on the physiological significance of such a condition. The present author failed to find a single instance of such an occurrence during the whole of this study, and it can only be suggested that Gilg may have misinterpreted what he saw in rather thick, hand-cut sections. Gilg's comment that he saw no collenchyma is endorsed.

In describing the mechanical ring, Gilg stated that the fibres have pointed ends. Relatively few were found to have pointed ends as a result of the present work; most fibres are more or less square-ended in the Restionaceae.

The cells of the parenchyma sheath are described by Gilg as normally being cubic and sometimes 2–3 times longer than wide. The present work has shown, however, that they are often up to 6 or 7 times longer than wide in many species. Gilg failed to recognize the silica-bodies often found in parenchyma sheath cells.

Gilg's account of the structure of peg-cells is of interest. He regarded these cells as having canals encircling them (radial canals) and as often accompanied by longitudinal canals also. This is the opposite of the 'peg' concept put forward by Cutler (1964). Gilg recognized that species with pillar cells have a different type and arrangement of chlorenchyma cells from those in species without pillar cells. He stated, however, that there are no canals around such cells, and that the cells are arranged in transverse plates. It has been established by the present author that a few pegs may be present on chlorenchyma cells in certain species with pillar cells, and that although the palisade cells may be arranged in transverse plates in some species, they are in a chequerboard pattern as seen in T.L.S. in others. Protective cells were first described by Pfitzer as developing from ordinary palisade cells. Gilg discovered that they are covered by cuticle on the faces lining the substomatal cavity. He stated that protective cells are only absent when the passage of air can be slowed down by other devices. This does not appear to be true for a number of species. For example in *Calorophus minor* and the Australian *Restio* species belonging to group 1 in this work no special air-speed reducing devices could be discerned. The '*Leptocarpus*' from the Chatham Islands (later shown to be *Lepyrodia traversii* = *Sporadanthus traversii* by Gilg-Benedict) was stated to have protective cells which also act as supporting cells. This now appears to be true of many species where protective cells extend from the epidermis to the parenchyma sheath.

Gilg claimed that the epidermal cells in surface view are quadratic, sometimes 2–3 times longer than wide but never wider than long. Had he examined further material he would have found cells over 5 times longer than wide, e.g. in *Calorophus gracillimus*, and many instances of cells which are wider than long, particularly those at either end of stomata, as in many *Chondropetalum* spp. He noticed wavy anticlinal walls, as seen in T.S., and mentioned that many are thick at the outer ends and thinner at the inner ends, e.g. in *Willdenowia* species. He found that when species lacked mechanical support from sclerenchyma girders or pillar cells, they usually had large, strong epidermal cells. This has been confirmed by the present author, but like many statements about biological subjects, it must be made with some reservations.

For example, *Coleocarya gracilis* provides an example of a sp. in which there is no conspicuous wall thickening or increase in size of epidermal cells, despite the fact that there are no pillar or protective cells, or sclerenchyma ribs.

Whilst noting that *Elegia, Dovea* (= *Chondropetalum*), and *Lamprocaulos* (= *Elegia*) have a 2-layered epidermis, Gilg went on to say that the inner layer was probably 'water conducting', as opposed to Tschirch's view that it was concerned with reducing the intensity of light falling on the chlorenchyma. Comment here is of little value since these hypotheses have not been checked by experiment. Gilg did not notice the gaps between the end walls of certain epidermal cells which are to be seen in surface view in these genera. The observation that tannin is present in nearly all Restionaceae is not accurate; many species have been examined in which no tannin could be found. The taxonomic significance of this fact will not be fully apparent until a greater range of specimens of individual species is examined. It is not known whether or not the deposition of tannin in the Restionaceae is genetically controlled, or even if there is seasonal variation in the amounts present.

Hair-form was surveyed by Gilg, and is dealt with under individual species in the section beginning on p. 131.

Stomata were reported to have 2 subsidiary cells in all species examined by Gilg and no deviation from this rule has been found as a result of the present work. He reported that most of the 21 stomatal positions given by Tschirch (1880) could be illustrated by members of the Restionaceae. The epidermal cells surrounding the stomata in *Restio nitens* Nees were correctly interpreted as extending into the chlorenchyma to form a substomatal tube. This has been recorded for *Lepidobolus* species, amongst others, by the present author. Gilg's idea that such cells must be regarded as 'subsidiary cells' does not fit in with the general concept of the nature of subsidiary cells in this family. To be consistent he would also have to call cells which surround raised stomata (e.g. in many South African *Restio* species) subsidiary cells and this is obviously incorrect.

The sections in which are described leaf, root, and rhizome are very short in Gilg's work.

Meyer's account of the family (1929) duplicated much of Gilg's work. Many of his new findings relate to individual species examined by him, and are here discussed in the section that follows below. He provided, for rapid reference, interesting lists of species, with diagnostic characters by which they can be recognized. These lists included one dealing with species in which particular sorts of perforation plates could be found, another of examples of species with various types of chlorenchyma, a third dealing with species that have stegmata, and a fourth species with sunken or superficial stomata. The lists of species with and without some free medullary vascular bundles respectively have been discussed on p. 125.

Meyer was not certain whether concentric vascular bundles were present in the outer bundle ring as recorded by Gilg; he reserved his judgement on this point. His doubts are seen to be justified by the present author, who has found only collateral bundles in this position.

Meyer applied the concept of the dimorphic epidermis to the Restionaceae, noting species with larger cells opposite to pillar cells, and the mounds formed by larger cells in such species as *R. cincinnatus*, and species of *Hypolaena* in Africa.

He saw the two different sizes of epidermal cell in *R. fastigiatus*, *R. leptocarpoides* and also in *Lyginia barbata*. Meyer saw what he interpreted as 'bordered pits' in the anticlinal walls of epidermal cells of certain species as seen in surface view. These could not be seen in any of the samples examined by the present author. It is possible that Meyer saw this phenomenon in species that have wavy anticlinal walls. The surface sinuations in these walls are sometimes in the opposite direction to those that are visible at a lower focal plane. If the two views had been seen as if they were in the same plane the misinterpretation could have resulted.

Meyer was the first to record the presence of silica-bodies in the family. He noted them in epidermal cells of some *Lepyrodia* species and also in the parenchyma sheath of many species of the family. Although he attributed no taxonomic importance to their distribution it now seems possible that the presence or absence of silica-bodies, or their position in the culm, may be of diagnostic value.

The absence of raphides was recorded. This observation together with the statement that crystals are present only in *Lyginia barbata* have been confirmed.

Meyer's description of the leaf, root, and rhizome are as brief as Gilg's.

The work of Gilg-Benedict (1930), which also contains anatomical data, is discussed in conjunction with individual genera.

Taxonomic Notes on Anatomy of Family as a Whole

Culm material has provided most of the characters of taxonomic value, mainly because leaves are poorly developed in most species. This makes comparison with members of leaf-bearing families more difficult. Material of root and rhizome was not readily available.

As a result of the study of the family as a whole, the following culm characters are considered to be of diagnostic importance in the Restionaceae. (Note: s = of significance at the specific level, g = of significance at the generic level: if either symbol is in brackets, it is of limited application.)

Culm surface

Hairs	g, s
Papillae	s
Epidermal cell outline	(g), s
Stomata	s
Silica	g
Tannin	(s)
Cuticular marks	(s)

Culm T.S.

Outline	s
Hairs	g, s
Papillae	s
Epidermal cells (number of layers)	g, s

Stomata	(g), s
Chlorenchyma	g, s
Protective cells	g
Pillar cells	g
Parenchyma sheath	(g)
Sclerenchyma sheath outline	g, (s)
Peripheral vascular bundles	(g)
Medullary vascular bundles	(g), (s)
Bundle sheaths	(s)?
Central ground tissue	(g), (s)
Silica	g, (s)

Artificial Key to Genera and some Species of Restionaceae based on Anatomical Characters of the Culm and Geographical Distribution

1 Plants from South Africa 2
Plants not from South Africa 11
2 With protective cells and well-developed, more or less parallel-sided, sclerenchyma ridges from sclerenchyma sheath extending into chlorenchyma; pillar cells often present—*Hypodiscus* spp. group 1, *Willdenowia* spp. group 1
Protective cells but no pillar cells present, sclerenchyma girders sometimes present; sclerenchyma ridges, if present, not extending into chlorenchyma 3
3 With several areas of thin-walled cells and often with tannin-filled cells scattered throughout central ground tissue; with or without hairs; epidermal cells normally with thickened, wavy anticlinal walls as seen in T.S.—*Thamnochortus*
Without areas of thin-walled cells (except some spp. of *Cannomois* and *Restio* group 3) and usually without tannin-filled cells in ground tissue; without hairs; epidermal cells normally with straight anticlinal walls as seen in T.S. 4
4 Without silica-bodies in parenchyma sheath or outer layer of sclerenchyma sheath 5
With silica-bodies in parenchyma sheath or outer layer of sclerenchyma sheath 6
5 Epidermis 2- or more-layered—*Elegia*, *Chondropetalum* spp.
Epidermis 1-layered—*Chondropetalum* spp., *Leptocarpus* spp., some *Restio* spp. from groups 3 and 4, *Staberoha.*
6 With T-shaped sclerenchyma girders—*Anthochortus*
Without T-shaped girders 7
7 Culms square in T.S.—*Restio quadratus*, *R. tetragonus*
Culms not square in T.S. 8
8 With rectangular sclerenchyma girders—*Phyllocomos*
Without rectangular girders 9
9 Stomata raised on mounds—*Restio* spp. group 4, *Leptocarpus* spp., *Mastersiella* spp.
Stomata not raised on mounds 10
10 Stomata sunken, or if superficial then plants with sclerenchyma sheath produced into low ridges interrupting parenchyma sheath between peripheral vb's—*Cannomois*, *Hypodiscus* spp. group 2
Stomata superficial, parenchyma sheath entire—*Leptocarpus* spp., *Mastersiella* spp., *Hypolaena* spp. group 1, some *Restio* spp. group 3, *Willdenowia* spp. group 2
11 Plants from Australia, Tasmania, or New Zealand 12
Plants not from Australia, Tasmania, or New Zealand 33
12 Epidermal cells grossly dimorphic 13
Epidermal cells not grossly dimorphic 14
13 Large and small cells alternating in longitudinal files—*Restio leptocarpoides*
Large cells in files at angles of culm only—*Dielsia*
14 Sclerenchyma girders present—*Restio applanatus*, *R. tremulus*
Sclerenchyma girders absent; ridges from sclerenchyma sheath, if present, sometimes extending into chlorenchyma 15

15 Stomata arranged in longitudinal grooves—*Restio fastigiatus*
Stomata not arranged in grooves 16
16 Protective cells present 17
Protective cells absent 20
17 Stomata, as seen in R.L.S., inclined at angle of 45° to culm surface—*Lyginia*
Stomata in same plane as culm surface as seen in R.L.S. 18
18 Chlorenchyma with few or no pegs, cells arranged in longitudinal plates—*Lepyrodia* spp. group 1, *Sporadanthus*
Chlorenchyma with pegs, cells not arranged in longitudinal plates 19
19 Silica-bodies present in some epidermal cells—*Lepyrodia* spp. group 2
No silica-bodies present in epidermal cells—*Calorophus elongatus*
20 Pillar cells present 21
Pillar cells absent 27
21 Culm compressed—*Restio complanatus, R. sphacelatus*
Culm more or less terete 22
22 Hairs present 23
Hairs not seen 26
23 Hairs filamentous, not closely applied to culm surface—*Leptocarpus spathaceus, Loxocarya pubescens, Restio megalotheca*
Hairs multicellular, peltate or subsessile—*Hypolaena fasciculata, H. fastigiata*
Hairs closely applied to culm surface 24
24 Central hair-cells thin-walled, bordering hair-cells thick-walled—*Meeboldina*
Hairs composed of thick-walled cells 25
25 Epidermal cells surrounding stomata inclined inwards—*Leptocarpus tenax, Restio laxus*
Epidermal cells surrounding stomata not inclined inwards—*Chaetanthus, Hypolaena exsulca, Leptocarpus anomalus, L. aristatus, L. brownii, L. canus, L. coangustatus, L. erianthus, L. simplex.*
26 With inwardly elongated epidermal cells next to stomata—*Restio nitens*
Without inwardly elongated epidermal cells—*Leptocarpus ramosus, Restio* spp. remaining from group 7.
27 Epidermal cells next to stomata inwardly elongated 28
Epidermal cells surrounding stomata not extending inwards into chlorenchyma 31
28 Hairs present 29
Hairs absent 30
29 Hairs with short, multicellular stalk and branches formed from files of elongated cells, joints angled (Fig. 34. F, p. 224)—*Loxocarya* spp. excluding *L. pubescens*
Branch cells without angular joints—*Lepidobolus, Restio confertospicatus*
30 Elongated epidermal cells often extending to parenchyma sheath—*Harperia*
Elongated epidermal cells extending only a short way into chlorenchyma, never as far as parenchyma sheath—*Coleocarya, Onychosepalum*
31 Stomata superficial—*Calorophus* spp. excluding *C. elongatus, Restio* spp. from group 1, excluding *R. australis* and *R. dimorphus*
Stomata sunken 32
32 Chlorenchyma regularly 3-layered—*Hopkinsia*
Chlorenchyma 1- or rarely 2-layered—*Restio australis* and *R. dimorphus*
33 Plants from S. America—*Leptocarpus chilensis*
Plants from Malaysia or South Vietnam—*Leptocarpus disjunctus*
Plants from Madagascar—*Restio madagascariensis*

Genus Descriptions

ANTHOCHORTUS Nees

No. of spp. 1; examined 1.

Genus and Species Description

Leaf surface not seen.

Leaf T.S.

Outline: enclosing culm in $1\frac{1}{2}$ turns, central region thick, with deep, widely spaced longitudinal grooves in abaxial surface; leaf *c.* $\frac{1}{2}$ as thick at grooves as between them (Fig. 26. C). **Epidermis.** (i) Abaxial cells of 2 main sizes; (*a*) between grooves more or less square to slightly higher than wide; anticlinal walls not wavy; (*b*) at base of grooves cells *c.* $\frac{1}{2}$–$\frac{1}{3}$ as wide and high as type (*a*), all walls thin. (ii) Adaxial cells 4–5-sided, *c.* $\frac{1}{3}$ as wide and $\frac{1}{5}$ as high as those of abaxial surface; walls slightly thickened. **Stomata** present in abaxial surface only, among cells of type (*a*), superficial, as in culm. **Hypodermis** absent. **Chlorenchyma** confined to thicker parts of central part of leaf; composed of 2–3 layers of cells situated below abaxial epidermis; cells of outer layer lobed, as wide as or $1\frac{1}{2}$ times wider than high, walls slightly thickened (as in culm); cells of inner layers palisade-like, ranging from 2 to 7 times as wide as high, walls thin.

Vascular bundles of 2 main sizes. (i) Larger vb's in grooves; tracheids all narrow, arranged in an arc, or flanking tracheids medium-sized; phloem partially enclosed by arc. (ii) Smaller vb's opposite chlorenchyma in wider parts of leaf; all tracheids narrow. **Bundle sheaths:** O.S., if present, not distinguishable from ground tissue; I.S. sclerenchymatous, fibres 1-layered around smaller bundles, thin-walled; fibres in 1–2 layers at phloem poles of larger bundles, with thick walls; those on flanks and at xylem pole present in 1 layer, with thin walls. **Sclerenchyma** present only as bundle sheaths. **Ground tissue** parenchymatous, in 3–4 layers occupying all space between vb's, chlorenchyma and adaxial epidermis. **Air-cavities** absent. **Silica:** spheroidal-nodular bodies present in some cells situated between larger vb's and abaxial epidermis. **Tannin:** none seen.

Culm surface (Fig. 26. B)

Hairs and **papillae:** none seen. **Epidermis:** cells 1–3 times longer than wide; all walls wavy. **Stomata** *c.* $\frac{2}{3}$ length of epidermal cells; subsidiary cells with perpendicular or acute ends. **Silica:** bodies absent. **Tannin** in some cells. **Cuticular marks** granular.

Culm T.S. (Fig. 26. A, D)

Outline sub-circular, slightly flattened on one side and tending to be polygonal. **Epidermis:** cells square to $1\frac{1}{2}$ times higher than wide. Anticlinal walls straight. **Stomata** superficial; guard cells with lips at inner and outer apertures. **Hypodermis** absent. **Chlorenchyma** divided into 5 longitudinal chambers. Cells of 2 types. (i) Lobed cells more or less isodiametric to slightly higher than wide, walls slightly thickened (staining light pink with safranin); in 1 outer layer; with large, intercellular air-spaces between them. (ii) Palisade cells mostly *c.* 4–5 times higher than wide, walls thin (staining blue with haematoxylin); present in 1–2 inner layers; air-spaces v. infrequent. **Parenchyma**

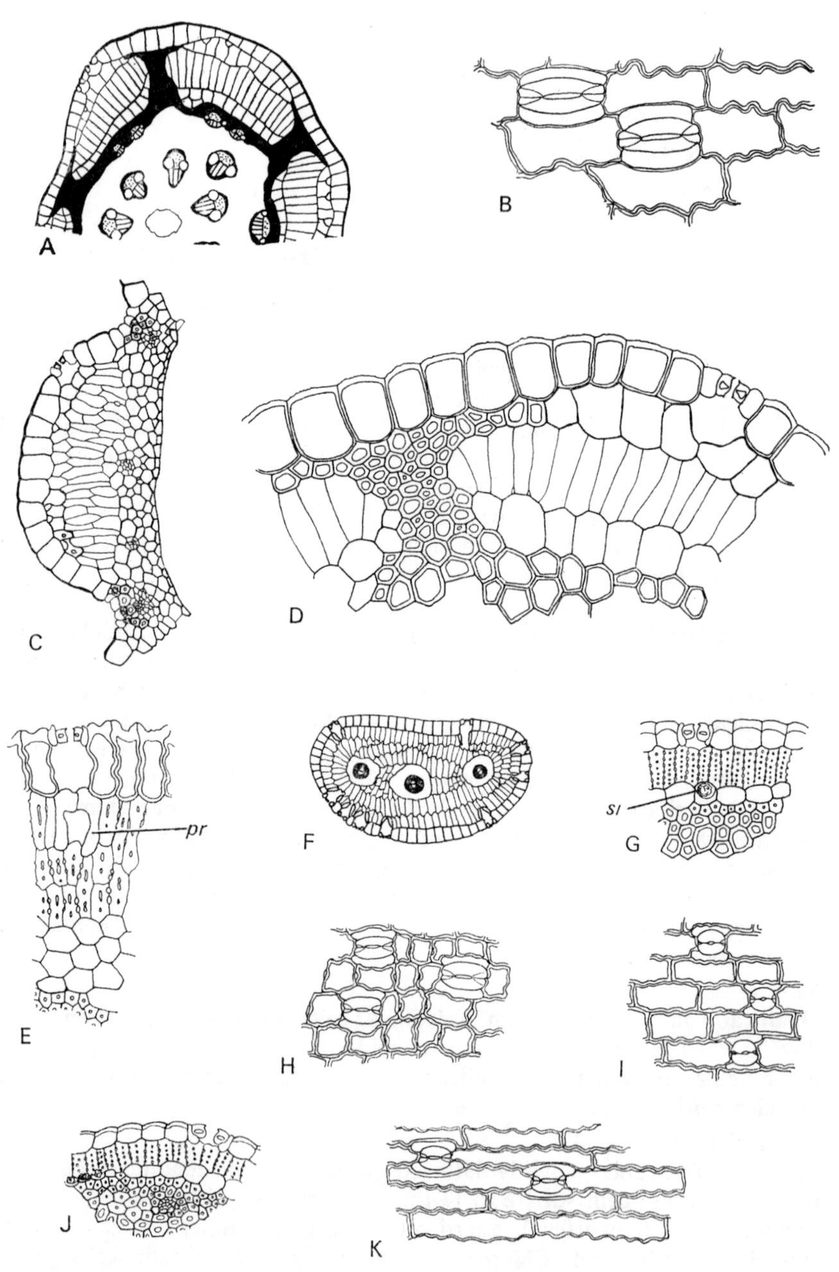

Fig. 26. Restionaceae. A–D, *Anthochortus ecklonii*: A, diagram of culm T.S. (× 50); B, culm epidermis (× 185); C, part of leaf T.S. (× 50); D, detail of culm T.S. (× 185). E, F, H, *Calorophus elongatus*: E, detail of culm T.S. (× 240); F, leaf T.S. (× 60); H, culm epidermis (× 240). G, I, *C. minor*: G, detail of culm T.S. (× 240); I, culm epidermis (× 240). J, K, *C. gracillimus*: J, detail of culm T.S. (× 240); K, culm epidermis (× 240).
pr = protective cell; si = silica-body.

sheath: cells in 1 (occasionally 2) layer(s), largest next to scl. sheath, smallest next to girders. **Sclerenchyma** sheath poorly developed, cells in 1–3 layers; walls moderately thickened or thick. Sheath produced at regular intervals into strong girders extending outwards to epidermis (5 present in specimen examined). Individual girders T-shaped, the column having concave sides and being from 4 to 7 cells wide, the bar being 1 cell thick and 12–15 cells wide.

Vascular bundles. (i) Peripheral vb's well developed, each having 1 medium-sized, angular tracheid on either flank; phloem abutting directly on xylem. (ii) Medullary vb's radially elongated; each with 1 medium-sized to wide, many-sided mxv on either flank; lateral wall pitting of vessels opposite scalariform; perforation plates oblique, many-barred, scalariform-reticulate. Phloem pole abutting directly on xylem, over-arching and extending between flanking mxv's. **Bundle sheaths:** fibres at bundle poles thick-walled, arranged in 1–2 layers; those on flanks moderately thick-walled, arranged in 1 layer; fibres at xylem pole v. narrow, with slightly thickened walls. **Silica:** spheroidal-nodular bodies present in some cells of parenchyma sheath. **Tannin** in some epidermal cells.

Rhizome and *root* not seen.

MATERIAL EXAMINED

Anthochortus ecklonii Nees; Esterhuysen 10901; S. Africa (K). Culm diameter 1 by 0·75 mm.

TAXONOMIC NOTES

This is an anatomically distinctive, monospecific genus. It has been recognized without change since it was first described.

The T-shaped sclerenchyma girders and v. weak sclerenchyma sheath are diagnostic. It is unusual in having chlorenchyma rather than sclerenchyma opposite the angles of the culm. Another feature not found elsewhere in the family is the layer of cells with slightly thickened walls between the chlorenchyma and the epidermis. This layer appears to be formed from modified chlorenchyma cells and is therefore not a true hypodermis. Gilg (1890) and Gilg-Benedict (1930) both described the anatomy of the genus accurately, and regard *Hypolaena tenuis* Mast. as a synonym, but the fact that no member of the genus *Hypolaena* was found to have sclerenchyma girders of the type found in *Anthochortus* favours the separation of the latter genus.

Despite the distinctiveness of *Anthochortus*, anatomical evidence indicates that it should remain in the family Restionaceae.

CALOROPHUS Labill.

No. of spp. 3; examined 3.

GENUS DESCRIPTION

Leaf surface

Hairs absent. **Epidermis:** cells square to oblong, anticlinal walls wavy. **Stomata:** subsidiary cells with wavy long walls and obtuse ends. **Silica** absent. **Tannin:** none seen. **Cuticular marks:** longitudinal striations present in some spp.

Leaf T.S. (Fig. 26. F)

Blade seen in *C. elongatus* only.

Outline subhemispherical. **Epidermis:** all cells more or less square. **Stomata** superficial, as in culm; subsidiary cells large, wider and deeper than guard cells, each being half as deep as epidermal cells, with thin walls. **Hypodermis** absent. **Chlorenchyma** composed of palisade peg-cells and lobed cells in 2–3 layers, occupying all the space around the bundle sheaths and continuous to the epidermis. Protective cells present, of same type as those in culm. **Vascular bundles:** 3 present; median bundle largest, with 1 angular, medium-sized, mx tracheid on either flank and central group of narrower tracheids; xylem of lateral bundles made up of narrow tracheids; cells of phloem abutting directly on xylem. **Bundle sheaths:** O.S. composed of 2–3 layers of thin-walled parenchymatous cells, I.S. composed of 1 layer of narrow, thick-walled fibres. **Sclerenchyma** represented by bundle sheaths only. **Ground tissue** all chlorenchymatous. **Air-cavities:** none present. **Silica** and **tannin:** none seen.

Leaf sheath

Seen in *C. minor* (G 3616) only.

Encircling the culm, with overlap of $\frac{1}{3}$ turn; 220 μm at thickest part.

Epidermis: (i) abaxial cells more or less square, outer walls moderately thickened, inner and anticlinal walls thin; anticlinal walls straight, outer walls slightly corrugate; (ii) adaxial cells 4- or 5-sided, slightly smaller than abaxial cells, walls slightly thickened. **Stomata** found in abaxial epidermis only, superficial; subsidiary cells narrow at the top, ballooning out below; *c.* $1\frac{1}{2}$ times higher than guard cells, the top being level with that of guard cell. Guard cells with ridge on inner wall. **Chlorenchyma** consisting of 1 layer of peg-cells.

Vascular bundles in 1 row, of 2 sizes, large and small, mostly alternating. (i) Larger bundles, with numerous narrow tracheids at the xylem pole and 1–3 medium-sized, angular tracheids on either flank. (ii) Smaller bundles, with narrow tracheids only. Phloem poles consisting of only 5–6 cells in smaller bundles and up to 25 cells in larger bundles; abutting directly on xylem on abaxial side. **Bundle sheaths:** O.S. parenchymatous, present to outer side of phloem pole only, becoming continuous with parenchyma from adjacent bundles and forming a layer, 1 cell thick, immediately to inside of chlorenchyma; I.S. sclerenchymatous, 1–2 cells wide, fibres narrower and having thicker walls at phloem than xylem pole.

Sclerenchyma as bundle sheaths and also as a layer, 2–4 cells wide, joining bundle sheaths laterally, and situated immediately to inside of parenchyma layer. Cells closest to parenchyma with thickest walls. **Ground tissue:** 1–2 layers of polygonal cells, occurring between scl. and adaxial epidermis. **Air-cavities** absent. **Silica** present as spheroidal-nodular bodies in specially thickened cells of parenchyma sheath; thickenings present on all but outer walls.

Culm surface (Fig. 26. H, I, K)

Hairs absent. **Epidermis:** cells square or rectangular, walls moderately thickened, wavy. **Stomata:** subsidiary cells of 2 main types: (i) outer part of axial, anticlinal walls wavy (Fig. 26. H, I); (ii) centre compressed almost to exclusion of lumen, but ends uncompressed and bulbous (Fig. 26. K). Guard cells frequently shorter than subsidiary cells. **Silica :** none seen. **Tannin :** none seen. **Cuticular marks** granular; longitudinal striations present in one specimen, thickenings present over transverse end-walls in another.

Culm T.S.

Outline circular, or circular but with one side flattened (latter normal condition of side branches). **Epidermis :** cells twice as wide as high to twice as high as wide; outer walls thick, other walls thin or moderately thickened. **Stomata** superficial; subsidiary cells frequently compressed, lumina either wider above and below, or only below guard cells; guard cells frequently with cuticular lips at outer and inner apertures. **Hypodermis** absent. **Chlorenchyma** composed of peg-cells in 1, 2, or 3 layers. Protective cells, when present, unusual, being palisade-like and equal in height to cells of outer chlorenchyma layer. Each protective cell with short arms on outer end (immediately below subsidiary cells), each arm arranged at about 90° to long axis of cell and equal to it in width. Adjacent arms touching and so forming a ring complete at outer end only, the main, palisade axes being well separated from one another; walls moderately thickened. **Parenchyma sheath** 1- or several-layered, continuous around culm. **Sclerenchyma :** well-developed cylinder enclosing peripheral and all or some medullary bundles; outline of cylinder either circular, or with rounded ridges opposite to and including the peripheral vb's.

Vascular bundles. (i) Peripheral vb's each with a small, rounded phloem pole partially enclosed by arc of tracheids. Tracheids medium-sized and narrow; tracheid end walls oblique, scalariform, side walls with scalariform pitting. (ii) Medullary vb's; mxv's wide or medium-sized, angular (*a*) 1 on either bundle flank, widely spaced or close together, or (*b*) single and central (in outermost bundles). Lateral wall pitting alternate- or opposite-scalariform; perforation plates simple, transverse to slightly oblique; phloem abutting directly on xylem, with strands of fibres in *C. gracillimus*. **Bundle sheaths :** fibres 1- or 2-layered at poles and 1-layered on flanks. **Silica :** spheroidal-nodular bodies present in some cells of parenchyma sheath; granular material sometimes present. **Tannin :** none seen.

Rhizome T.S.

Seen only in *C. minor* Hk. f. (G3616). Diameter 2 mm.

Hairs multicellular, thin-walled; (i) unbranched filaments, up to 4 cells long, cells 2–6 times as long as wide; (ii) bifurcating, stalk 2-celled, branches each 2 cells long. **Epidermis :** cells thin-walled, more or less rectangular, twice as wide as high. **Hypodermis** absent. **Cortex :** (i) outer part parenchymatous, cells polygonal, more or less straight-sided, in up to 5 layers; (ii) inner part consisting of up to 7 layers of lobed, more or less isodiametric cells, with large intercellular spaces between them. **Endodermoid layer** composed of 1–2 layers of thin-walled, more or less isodiametric cells, some containing small, spheroidal-nodular silica-bodies.

Vascular bundles of 2 main types; (i) collateral, having an arc of wide, thin-walled, angular mxv's partially enclosing the phloem; (ii) amphivasal, having a circle of wide, thin-walled, angular mxv's enclosing the phloem. **Central ground tissue :** cells of 2 types; (i) angular, walls slightly thickened, surrounding vb's; (ii) lobed, more or less isodiametric, thin-walled, occupying central region. **Silica:** bodies present in endodermoid sheath and central ground tissue. **Tannin:** none seen.

Root T.S.

Seen in *C. minor* Hk. f. (Cheadle CA331 and G3616).

Root-hairs sparse, arising from cells similar to those of remainder of epidermis; hair shaft half as wide as base. **Epidermis:** cells thin-walled, more or less square. **Cortex.** (i) Outer part composed of 1 layer of thin, cellulose-walled cells situated externally to 5–12 layers of angular cells with slightly lignified, thickened walls. (ii) Inner part, consisting of 7–8 layers of square, thin-walled cells diminishing regularly in size towards centre of root; cells arranged in radial files or plates. Small intercellular spaces occurring at angles. **Endodermis** composed of 1 layer of large, thin- to moderately thick-walled, lignified cells, up to twice as high as wide. **Pericycle:** cells v. narrow, in a single layer about $\frac{1}{6}$ height of endodermal cells. **Vascular tissue:** mxv's large, thin-walled, rounded, arranged in a single ring. **Central ground tissue:** cells parenchymatous, polygonal, walls slightly thickened.

Material Examined

Calorophus elongatus Labill.: Hinsby, W. M. C. Stranan; Tasmania (3 specimens) (K).

('*Calostrophus*') *gracillimus* F. v. Muell.: (i) Drummond, Herb. Hook. An. 1867; Swan R., Australia (K). (ii) Andrews, C.; Bayswater, Perth, Australia (K).

C. minor Hk. f.: (i) Moore G3616; Salt marsh, Moana Tua Tua, N. Zealand (S). (ii) Cheadle CA331; nr. Sydney, Australia (S). (iii) Drummond, Herb. Hook. An. 1867; Swan R., Australia (K). (iv) Hubbard, C. E. 4609; Fraser Is., Australia, An. 1930 (K). (v) Cunningham, A.; Austr. Herb., An. 1862, Australia (K). (vi) Port Jackson, 28, Australia (K). (vii) Herb. Hook. An. 1867, 599, Australia (K). (viii) Port Jackson, 49, Australia (K.)

Species Descriptions

***C. elongatus* Labill.** Culm diameter: 0·75–1·5 mm.

Leaf; see genus description, p. 133.

Culm surface (Fig. 26. H)

Epidermis: cells 1–1½ times longer than wide. **Stomata:** subsidiary cells with wavy walls; guard cells shorter than subsidiary cells. **Cuticular marks:** thickenings present over transverse end-walls in one specimen.

Culm T.S. (Fig. 26. E)

Outline circular but slightly flattened. **Epidermis:** cells *c.* 1½–2 times as high as wide. Anticlinal walls slightly to v. wavy; outer walls corrugate (with slight papillae over anticlinal walls in 1 sample). **Stomata:** subsidiary cells not compressed. **Chlorenchyma** 2–3-layered; pegs large, few, mostly on upper and lower faces; protective cells having lobed outer ends (Fig. 26. E). **Parenchyma sheath** 2–3-layered.

***C. gracillimus* F. v. Muell.** Culm diameter 0·6, 0·8 mm.

Culm surface (Fig. 26. K)

Epidermis: cells v. elongated, mostly 4–5 times longer than wide. **Stomata:** subsidiary cells constricted at middle; guard cells much shorter than subsidiary cells.

Culm T.S. (Fig. 26. J.)

Outline circular, slightly flattened on one side. **Epidermis:** cells mostly about as wide as high. **Stomata:** subsidiary cells compressed at outer end, bulbous at inner end; guard cells with outer lips. **Chlorenchyma** cells in 1 layer; protective cells absent. **Parenchyma sheath:** cells in 1 layer opposite peripheral vb's; regions of numerous stegmata between peripheral vb's. **Sclerenchyma:** sheath circular in outline; fibres thick-walled, in 5–6 layers. **Silica:** bodies v. frequent in stegmata of parenchyma sheath. Granular material present sporadically in most tissues.

***C. minor* Hk. f.** Culm diameter: 0·75–2·0 mm.

Culm surface (Fig. 26. I)

Epidermis: cells oblong, 2–3 times longer than wide. **Stomata:** subsidiary cells extending beyond guard cells at either end; compressed centrally; guard cell pair with more or less circular outline. **Cuticular marks** granular, except in G3616, from New Zealand, where longitudinally striate.

Culm T.S. (Fig. 26. G)

Outline circular, with one flattened side. **Epidermis:** cells as high as or slightly higher than wide. **Chlorenchyma:** peg-cells in 1 layer (2 layers in No. 599), cells mainly 4–6 times higher than wide; protective cells absent. **Parenchyma sheath** 1-layered, with occasional stegmata. **Sclerenchyma:** sheath mainly only 3–4-layered, outline circular. **Silica** present as (i) bodies, in stegmata of parenchyma sheath; (ii) granular material, in some cells of ground tissue, and occasionally in some chlorenchyma cells.

Leaf, *rhizome*, and *root*: see genus description.

Taxonomic Notes

A relationship between *Calorophus* and some Australian *Restio* species was thought to be possible by R. Brown (1810). Endlicher (1836) included it in the genus *Restio* and Kunth (1841) used the name *Restio lateriflorus* for *C. elongatus*. Masters (1878) regarded this species as a member of the genus *Hypolaena*, calling it *Hypolaena lateriflora* Mast. Gilg (1890) recognized that it should be distinct from *Hypolaena* on the basis of its anatomy and it was treated as a distinct genus by Gilg-Benedict in 1930.

The culm anatomy of *Calorophus* is of the simplest found in the Restionaceae. This simplicity might be the result of reduction, and so it does not necessarily indicate that the genus is 'primitive'.

There is some anatomical discontinuity between the species. *C. elongatus* Labill. has protective cells, but lacks silica-bodies. *C. gracillimus* F. v. Muell. and *C. minor* Hk. f. lack protective cells but have silica-bodies in stegmata in the parenchyma sheath. The subsidiary cells of the stomata of *C. gracillimus* and *C. minor* are compressed centrally as seen in surface view. Subsidiary cells of this type have not been seen in other species or genera.

The lack of pillar cells separates *Calorophus* from the Australian species of *Hypolaena* (see p. 191), in which genus it was placed by Masters and retained by Bentham and Hooker (1883). Gilg's (1890) re-separation of the genus is seen to be correct.

The simple culm organization may indicate a relationship between *C. minor* and *C. gracillimus* on the one hand and certain species of *Restio* which are here treated as group 1 on the other. The anatomy thus supports the ideas of R. Brown (1810). The similarity between the two could, however, be interpreted as an example of parallel evolution just as reasonably as an indication of true affinity.

C. elongatus does not appear to fit into the genus very well. Investigation of the gross morphology of this species may help to clarify its taxonomic status. Johnson and Evans (1966) also remark on this discontinuity.

CANNOMOIS Beauv. ex Desv.

No. of spp. 7; examined 6.

Genus Description

Leaf surface not seen.

Leaf T.S.

Sheath only, seen in *C. acuminata* and *C. virgata*; overlapping culm by 1½ turns.

Hairs: none present. **Epidermis.** (i) Adaxial cells 5-sided, outer walls flat, anticlinal walls short, inner walls with 2 faces, all walls moderately thickened. (ii) Abaxial cells 4-sided, of unequal heights, 2–3 times wider than those of adaxial surface; individual cells either as high as wide, or 1½–2 times higher than wide; all walls moderately thickened, outer walls convex or low dome-shaped, anticlinal walls not wavy. **Stomata** present in abaxial surface only; sunken; guard cells mounted low down on subsidiary cells, with ridge on inner wall; walls to pore convex. **Chlorenchyma** represented by more or less isodiametric, lobed cells in 2–4 layers; restricted to small areas opposite to stomata and between sheaths of vb's (i.e. in longitudinal channels below abaxial epidermis as seen in surface view). Protective cells absent.

Vascular bundles: (i) small; tracheids few, all narrow; phloem poles composed of 10–20 cells; (ii) v. small bundles consisting of 2–3 tracheids and 3–4 phloem cells alternating with (i). **Bundle sheaths:** O.S. parenchymatous, 1-layered, present on abaxial side and flanks of bundles; I.S. sclerenchymatous, completely encircling bundles, fibres medium-sized and narrow, thick-walled, in 2–3 layers at phloem pole, 3–4 layers on flanks, and 6–8 layers at xylem pole. Outline of I.S. rounded or oval, with bundle positioned eccentrically, near the abaxial surface. **Sclerenchyma** represented by that of inner bundle sheaths and also present in ground tissue at narrowing leaf margins. **Ground tissue** composed of wide-celled parenchyma; arranged in 2–3 layers between vb sheaths and adaxial epidermis, and in 3–4 layers between bundles, abutting on chlorenchyma to abaxial side and on adaxial epidermis to adaxial side. **Air-cavities** absent. **Silica:** (i) spheroidal-nodular bodies, present in some cells of O.S. and some outer cells of I.S.; (ii) granular material, in some cells of ground tissue and in occasional xylem and phloem cells. **Tannin:** none seen.

Culm surface

Hairs and **papillae** absent. **Epidermis:** cells 4-sided and as long as or up to 4 times longer than wide, or 5–6-sided and up to 3 times as long as wide; shortest cells next to stomata; walls slightly or moderately thickened, or thick, straight or wavy. **Stomata:** subsidiary cells and guard cells variable. **Silica:** none seen. **Tannin** in epidermal cells in some spp. **Cuticular marks** granular, with or without longitudinal striations.

Culm T.S. (Pl. VI. A)

Outline circular. **Cuticle** thick or v. thick. **Epidermis:** cells as high as to slightly higher than wide or 2–3 or 5–8 times higher than wide; outer walls thick, frequently convex, other walls moderately thickened, anticlinal walls straight or wavy. **Stomata** superficial in spp. with shorter epidermal cells, sunken in the others; subsidiary cells with outer wall frequently produced into a conspicuous ridge; guard cells with lip at outer and/or inner aperture(s) in some spp.; other spp. with ridge on inner wall; walls to pore frequently convex. **Hypodermis** absent. **Chlorenchyma** composed of 2 uninterrupted layers of peg-cells; protective cells present in all spp. but more strongly developed in some than others. **Parenchyma sheath** usually 1-layered but 2-layered in places in some spp.; cells opposite peripheral vb's larger than the others. **Sclerenchyma:** sheath well developed in all spp., particularly wide in some; outline circular, with slight ribs between peripheral vb's interrupting parenchyma sheath in some spp.

Vascular bundles. (i) Peripheral vb's usually well developed; outline circular or oval; xylem composed of wide, angular tracheids with slightly to moderately thickened walls, arranged in a single-layered arc; phloem pole situated in concavity of arc. (ii) Medullary vb's: outline circular or oval, each with 1 or 2 narrow, medium-sized or wide, rounded-angular, thick-walled mxv's on either flank, these separated from one another by 2–3 files of narrow parenchymatous cells; phloem only occasionally abutting directly on xylem, normally separated from it by a layer of narrow parenchymatous cells with thin or slightly thickened walls. Px poles present in all but outermost smallest bundles. Lateral pitting of flanking mxv's scalariform; perforation plates oblique and scalariform or almost transverse and simple. Many bundles free from culm scl. sheath in all spp. **Bundle sheaths:** fibres thick-walled, either completely encircling, 1-layered on flanks and 2–3-layered at poles, or present as 1–3-layered caps.

Central ground tissue parenchymatous, walls thin or slightly to moderately thickened; small, scattered groups of narrower, thin-walled cells present in some spp. Central cells frequently thin-walled and breaking down to form central cavity. **Silica:** spheroidal-nodular bodies present in some smaller cells of parenchyma sheath and some outer cells of ribs from scl. sheath. **Tannin** in certain epidermal cells of some spp.

Rhizome T.S.

Seen in *C. acuminata* and *C. virgata.*

Epidermis: cells 4-sided, up to $1\frac{1}{2}$–2 times higher than wide; outer walls slightly to moderately thickened, other walls slightly thickened. **Cortex.** (i) Outer part *c.* 10-layered in *C. acuminata*, composed of parenchymatous cells interspersed with tangential bands of sclerenchymatous cells; all cells parenchymatous in *C. virgata*, in *c.* 20 layers. (ii) Inner part composed of 4–8 layers of lobed, more or less isodiametric, thin-walled cells. **Endodermoid layer:** cells with thin outer walls, other walls moderately thickened (*C. virgata*) or thick (*C. acuminata*). **Vascular bundles** amphivasal and collateral, scattered throughout ground tissue; mxv's usually wide, angular but some outer vb's without wide mxv's; lateral wall-pitting scalariform; perforation plates simple, more or less transverse. **Bundle sheaths** indistinguishable from ground tissue. **Ground tissue** parenchymatous, cells either (i) angular, 6-sided, thick-walled (forming bulk of tissue), or (ii) thin-walled, lobed (in scattered strands). **Silica:** occasional granular bodies present in some lobed cells of central ground tissue. **Tannin:** none seen.

Root T.S.

Seen in *C. acuminata* and *C. virgata.*

Epidermis: cells 4–5-sided, slightly wider than high, thin-walled. **Root-hairs** arising from cells similar to others in epidermis, shaft *c.* $\frac{1}{3}$–$\frac{1}{2}$ of width of base. **Cortex.** (i) Outer part composed of 2–4 layers of 5–6-sided parenchyma cells with slightly thickened walls. (ii) Middle part consisting of radiating plates composed of 1 or 2 layers of wide, rounded parenchyma cells with slightly thickened walls; air-spaces present between plates of cells. (iii) Inner part made up of 1–3 layers of parenchymatous cells, individual cells rounded or slightly wider than high, with slightly to moderately thickened walls. **Endodermis** 1-layered, cells with thin outer walls, all other walls v. thick; lumina occluded in *C. acuminata*, v. reduced in *C. virgata.* **Pericycle** 2–3-layered in *C. virgata*, *c.* 7-layered in *C. acuminata*; cells 4–6-sided, mostly as high as wide, with thick walls. **Vascular system.** Phloem strands composed of 2–4 wide sieve-tubes and several companion cells; strands arranged (i) in outer ring, immediately inside pericycle (ii) to inner side of, but close to, mxv's of outer ring (iii) close to mxv's scattered in central region of root, 1–3 strands to each vessel or vessel multiple. Mxv's radially oval-angular, either solitary, or in tangential or radial multiples of 2 or 3, each vessel or multiple ensheathed by single or double layer of compressed tracheids; mxv's arranged in outer ring and also scattered in central ground tissue. **Central ground tissue** parenchymatous, all, except for few central cells in *C. acuminata*, thick-walled. **Tannin:** none seen.

MATERIAL EXAMINED (all South African)

Cannomois acuminata (Thunb.) Pillans: (i) Univ. Cape Town 8735 ♂; South Africa (K). (ii) Acocks, 14706 (in Kew Herb. as *C. simplex* Kunth) (K).

C. congesta Mast.: Burchell; (as *C. scirpoides* Mast.) (M).

C. nitida (Mast.) Pillans: (i) Esterhuysen 8382 (K). (ii) Esterhuysen, 10085 (K).

C. parviflora (Thunb.) Pillans: Compton 2626 (K).

C. scirpoides (Kunth) Mast.: Nat. Herb., Pretoria, 18266; (as *C. drègei* Pillans) (K).

C. virgata (Rottb.) Steud.: (i) Cheadle, V. I. CA742 (S). (ii) Burchell (M). (iii) Burchell (M).

Species Descriptions

***C. acuminata* (Thunb.) Pillans.** Culm diameter 1·5, 2·4 mm.

Culm surface (Fig. 27. F)

Epidermis: cells variable: (i) those in 14706, at ends of stomata as long as wide, partly overhanging subsidiary cells, others 2–3 times longer than wide; walls moderately thickened, not wavy; (ii) those in 8735, at ends of stomata as long as wide, others mostly 3 times, some up to 4 times longer than wide; walls slightly thickened, wavy. **Stomata:** guard cells narrower than in most spp. **Tannin** in some epidermal cells. **Cuticular marks:** longitudinal striations.

Culm T.S. (Fig. 27. C, Pl. VI. A)

Epidermis: cells as high as wide to slightly higher than wide, some up to twice as wide as high in 14706, the shorter cells being next to stomata; anticlinal walls straight. **Stomata** superficial in 8735, slightly sunken in 14706. Subsidiary cells with bulbous inner ends; guard cells with no lips in 8735, with slight lips at inner aperture in 14706. **Chlorenchyma:** cells mostly 4 times higher than wide, those opposite protective cells normally twice as high as wide; protective cells extending in 1 layer from epidermis to about half way into inner chlorenchyma layer, each cell expanded at outer and inner ends, making contact with its neighbours, middle part of walls free, bordering wide, elongated apertures. **Parenchyma sheath:** cells in 1 layer, but interrupted between peripheral vb's by scl. ribs. **Sclerenchyma:** sheath 3–6- or 6–9-layered, outline circular, with short, wide rectangular ribs between peripheral vb's.
Leaf, rhizome, and *root T.S.*: see genus description.

***C. congesta* Mast.** Culm diameter: 1 by 2 mm.

(Sample examined is an inflorescence branch, with a semicircular outline, but otherwise resembles a normal culm.)

Culm surface

Epidermis: cells 4–6-sided, mostly 1–2 times longer than wide; walls moderately thickened, wavy. **Stomata:** subsidiary cells with wavy walls; guard cell pair lemon-shaped. **Tannin** in many epidermal cells. **Cuticular marks:** longitudinal striations.

Culm T.S.

Outline semicircular. **Epidermis:** cells mostly as high as wide, some slightly higher than wide; anticlinal walls slightly wavy or straight. **Stomata** superficial; guard cells with lip at outer aperture and ridge on inner wall. **Chlorenchyma:** cells $3\frac{1}{2}$–4 times higher than wide; protective cells extending from epidermis to about half way into inner chlorenchyma layer, joined to their neighbours at slightly expanded inner and outer ends; apertures between cells wide. **Parenchyma sheath** 1-layered. **Sclerenchyma:** sheath 4–7-layered; cells of outer layers between peripheral vb's narrow, v. thick-walled, with narrow lumina; cells to outside of peripheral vb's and in inner layers of sheath moderately thick-walled, and with wider lumina.

C. nitida **(Mast.) Pillans.** Culm diameter: 1·25 and 2 mm.

Culm surface

Epidermis: cells 4–6-sided, as long as or up to 1½ times longer than wide, walls not wavy. **Stomata** overarched by adjacent epidermal cells, visible through transversely orientated, rectangular or bone-shaped aperture. Subsidiary cells with rounded ends, slightly longer than guard cells. **Tannin** in many epidermal cells. **Cuticular marks:** wide, longitudinal striations.

Culm T.S.

Epidermis: cells 1½–3 times higher than wide, anticlinal walls straight or wavy. **Stomata** sunken, positioned about half-way down anticlinal walls of adjacent epidermal cells, and slightly overarched by their outer ends, particularly at either end of the stoma; subsidiary cells with v. bulbous inner portion; guard cells with slight lips at inner aperture, the outer wall sloping sharply towards subsidiary cells at either end of the stoma, due to increase in length of walls to pore in those regions. **Chlorenchyma:** cells 7–8 times higher than wide, except when opposite protective cells, where about 5 times higher than wide; protective cells extending from just inside substomatal cavity lined by epidermal cells to about half-way into inner chlorenchyma layer; individual cells with sharply pointed inner ends; apertures between cells elongated, with pointed ends, present towards inner end of substomatal cavity. **Parenchyma sheath** 1-layered, cells slightly wider than high or up to 1½ times higher than wide. **Sclerenchyma:** sheath 6–7-layered, fibres thick-walled.

C. parviflora **(Thunb.) Pillans.** Culm diameter 3 mm.

Culm surface (Fig. 27. D)

Epidermis: cells 4–6-sided, as long as to 1½ times longer than wide; walls thick or v. thick, sometimes wavy. **Stomata** close together; visible at much lower plane of focus, through longitudinally orientated, rectangular apertures with thick cuticular rims formed by overarching epidermal cells. Subsidiary cells slightly longer than guard cells, with rounded ends; guard cells narrow. **Tannin:** none seen. **Cuticular marks** finely granular.

Culm T.S. (Fig. 27. B)

Epidermis: cells 5–6 times higher than wide, outer walls flattened-convex, cuticle extending between adjacent cells to depth equivalent to thickness of outer walls. Outer ends of anticlinal walls thick, thickening tapering and walls moderately thickened at inner ends, wavy. **Stomata** deeply sunken, attached to anticlinal walls of surrounding epidermal cells about half-way down their length. Subsidiary cells with outer wall extended into ridge at either end of stoma, following increased height of guard cells in these regions (Fig. 27. B); guard cells with slight lips at inner and outer apertures.

Chlorenchyma: cells of outer layer *c.* 15 times higher than wide, inner cells *c.* 5–10 times higher than wide; protective cells occupying *c.* ½ of volume of outer chlorenchyma layer (because of density of stomata); individual cells moderately thick-walled, extending from *c.* ⅓ of way up substomatal cavity lined by epidermal cells, to over half-way into inner chlorenchyma layer; ends rounded; apertures between protective cells short and narrow, occurring in files in inner half of substomatal tube. **Parenchyma sheath** 1–2-layered, cells

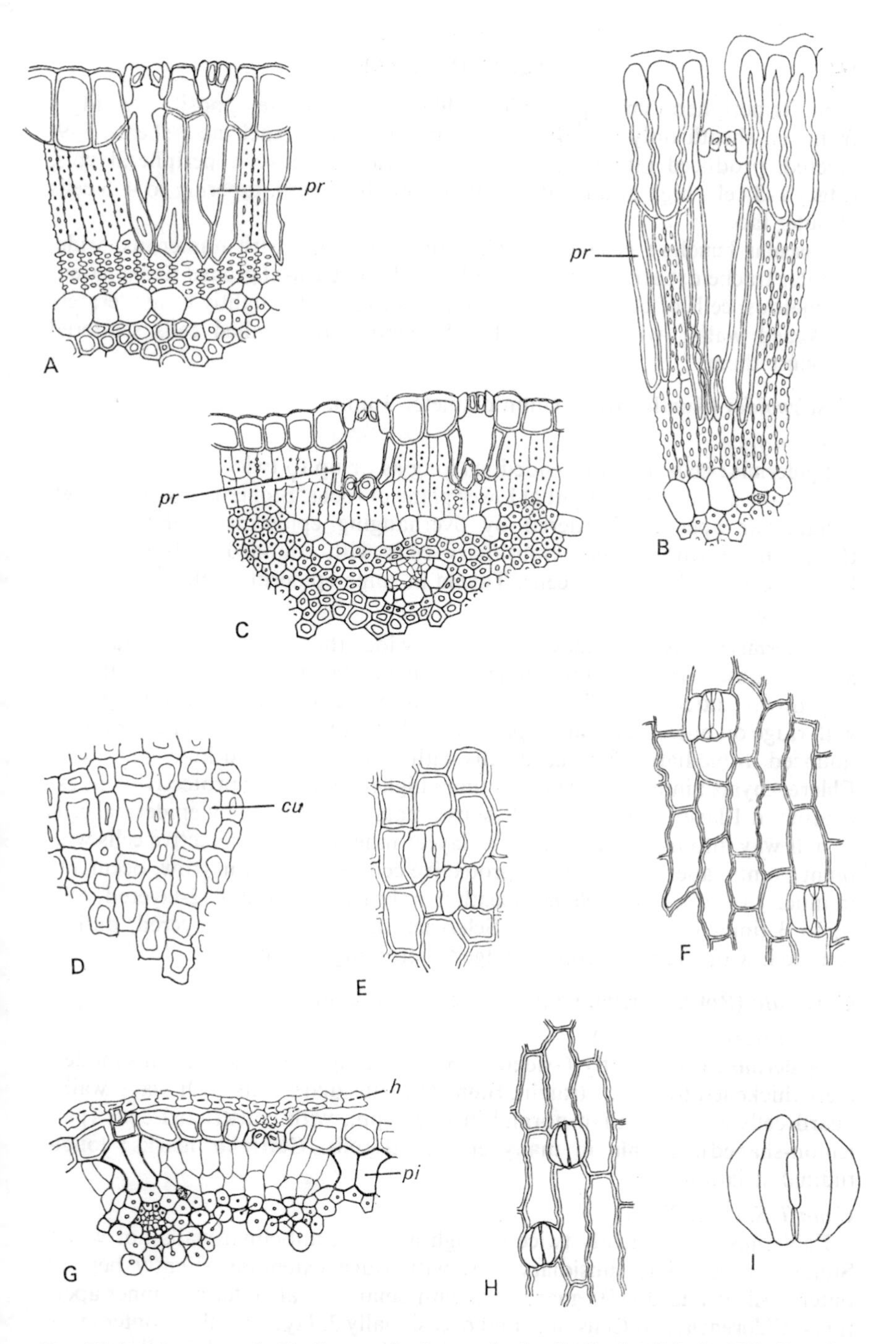

FIG. 27. Restionaceae. A, E, *Cannomois virgata*: A, detail of culm T.S. (×185); E, culm epidermis (×185). B, D, *Cannomois parviflora*: B, detail of culm T.S. (×185); D, culm epidermis (×185). C, F, *Cannomois acuminata*: C, detail of culm T.S. (×185); F, culm epidermis (×185). G–I *Chaetanthus leptocarpoides*: G, detail of culm T.S. (note hair, h) (×150); H, culm epidermis (×150); I, stoma (×450).
cu = cuticle; h = hair; pi = pillar cell; pr = protective cell.

opposite peripheral vb's up to twice as high as wide, those opposite low ridges from scl. sheath only slightly higher than wide. **Sclerenchyma:** sheath 8–9-layered, produced into low, dome-shaped ridges between peripheral vb's, ridges 1–3 cells high and *c.* 10 cells wide next to sheath, tapering to 4–5 cells at outer face.

Vascular bundles: peripheral vb's unusually large, with numerous, wide, angular tracheids[1] arranged in single-layered arc. **Central ground tissue** parenchymatous, cells of outer layers around vb's wide, with slightly to moderately thickened walls; with strands of 2–5 narrow, thin-walled cells scattered amongst them.

***C. scirpoides* (Kunth) Mast.** Culm diameter 3 mm.

Culm surface

Epidermis: cells 4–5- or 6-sided, as long as or up to twice as long as wide, walls wavy at surface, straight at lower plane of focus. **Stomata** visible through rectangular aperture bordered by overhanging epidermal cells; opening thickly lined with cuticle. Subsidiary cells with rounded ends extending beyond each end of guard cells. **Tannin** in some epidermal cells.

Culm T.S.

Epidermis: cells 5–6 times higher than wide; those next to stomata over-arching subsidiary cells; anticlinal walls wavy. **Stomata** sunken to about half-way down anticlinal walls of surrounding epidermal cells; subsidiary cells with ridge on outer wall, this v. pronounced at either end of stomata, less pronounced in median T.S.; guard cells with lips at outer and inner apertures. **Chlorenchyma:** individual cells 7–9 times higher than wide; protective cells extending in 1 layer from just inside substomatal cavity lined by epidermal cells to half way into inner chlorenchyma layer; inner ends of protective cells with blunt points; apertures between protective cells elongated, narrow, occurring in inner ⅓ of tube. **Parenchyma sheath** mainly 1-, occasionally 2-layered, cells 2½ to 3 times higher than wide. **Sclerenchyma:** sheath 6–8-layered, outline rounded, with irregular, convex ridges alternating with peripheral vb's.

***C. virgata* (Rottb.) Steud.** Culm diameter 4·5, 3, and 3 mm.

Culm surface (Fig. 27. E)

Epidermis: cells mostly 6-sided, 1–3 times longer than wide, walls moderately thickened to thick, straight. **Stomata:** subsidiary cells with wavy walls; guard cells narrower than normal in most spp., with constricted ends (pair lemon-shaped). **Tannin** in many epidermal cells. **Cuticular marks:** longitudinal striations.

Culm T.S. (Fig. 27. A)

Epidermis: cells mostly twice as high as wide; anticlinal walls not wavy. **Stomata** superficial; subsidiary cells with ridge extending slightly beyond outer wall of guard cell; guard cells with small lips at outer and inner apertures. **Chlorenchyma.** Cells in 2, and occasionally 3, layers; cells of outer layer *c.* 5–6 times higher than wide, those of inner layer *c.* 4 times higher than wide.

[1] It is possible that some of the wider xylem cells of the peripheral bundles may be vessels in this sp.; a distinct wall could not be seen between the scalariform thickenings of end walls (perforation plates).

Protective cells extending in 1 layer from epidermis to over half way into inner chlorenchyma layer. Cells either expanded and touching one another at outer and inner ends, with a wide, elongated aperture, or with elongated pegs dividing large aperture into several shorter ones. **Parenchyma sheath** 1- or 2-layered, interrupted in places by ribs from scl. sheath. **Sclerenchyma:** sheath 5–7-layered, with short ribs 1–3 cells high and 3–7 cells wide between peripheral vb's, partially or entirely penetrating parenchyma sheath.

Leaf, *rhizome*, and *root T.S.*: see genus description.

Taxonomic Notes

This genus is stated by Pillans to have as synonyms the genera *Mesanthus* Nees and *Cucullifera* Nees, and some of the species are given with synonyms in *Thamnochortus*, *Hypolaena*, and *Restio*.

The species of this genus show only a limited range of anatomical characters. All have protective cells and chlorenchyma composed of peg-cells. Silica-bodies are found in some species. The stomata are never raised on mounds.

The genus can be divided into 2 groups on stomatal position: (*a*) stomata superficial in *C. acuminata* (Thunb.) Pillans (8735), *C. congesta* Mast., and *C. virgata* (Rottb.) Steud; (*b*) stomata sunken in *C. acuminata* (Thunb.) Pillans (14706), *C. nitida* (Mast.) Pillans, *C. parviflora* (Thunb.) Pillans, and *C. scirpoides* (Kunth) Mast.

The parenchyma sheath is interrupted at intervals by low ridges from the scl. sheath in species belonging to group (*a*) as well as in *C. acuminata* (Thunb.) Pillans (14706) of group (*b*). The situation is also very similar in some species of *Hypodiscus*.

C. virgata (Rottb.) Steud. is synonymous with *Mesanthus* species according to Masters and Gilg. It shows no anatomical characters which justify its separation from *Cannomois*. It particularly resembles species of *Cannomois* with superficial stomata and also species which are treated as members of *Willdenowia* group (ii) of this work (see p. 323). This is of interest because *Mesanthus* was included in the synonymy of *Willdenowia* by Endlicher in 1836. *Willdenowia* needs revision. When this is undertaken a possible relationship with certain *Cannomois* species will need to be considered.

C. acuminata (Thunb.) Pillans has amongst its synonyms *Cucullifera dura* Nees, *Cannomois simplex* Kunth, and *Thamnochortus strictus* Kunth, according to Pillans (1928). The species as examined here is represented by specimens with either superficial or slightly sunken stomata and with slight differences in epidermal cell outline. These distinctions, coupled with the difference in outline of the vb's, suggest that the material examined here may be a taxonomic mixture. The specimens show pronounced affinities with the other species in the genus *Cannomois*, and this suggests that it would be unwise to reinstate *Cucullifera* as a genus in its own right solely on anatomical evidence. The specimens all have small ribs from the sclerenchyma sheath which interrupt the parenchyma sheath at intervals. This character is unknown in *Thamnochortus*, but it occurs in some species of *Hypodiscus* indicating that *C. acuminata* might be related to the latter genus. If the specimens on which these studies were based are all correctly named, the evidence suggests that *C. simplex* Kunth may be a species in its own right.

C. parviflora (Thunb.) Pillans and *C. scirpoides* (Kunth) Mast. both have areas of thin-walled cells scattered in the ground tissue, as in *Thamnochortus* species. *C. scirpoides* (Kunth) Mast. is synonymous with *T. scirpoides* Kunth according to Pillans. Both *C. parviflora* and *C. scirpoides* have sunken stomata, a character which is uncommon in *Thamnochortus*, and the anticlinal walls of the epidermal cells, although wavy like those of most *Thamnochortus* species, do not exhibit the tapered thickening which is so characteristic of that genus. Further study of taxonomic evidence from other sources is necessary before the affinities of these two species can be indicated with certainty.

C. nitida (Mast.) Pillans, synonymous with *Hypodiscus nitidus* Mast. according to Pillans, is probably correctly placed in *Cannomois* rather than *Hypodiscus* because the entire type of parenchyma sheath found here is not recorded for *Hypodiscus*.

C. nitida can be distinguished from *C. parviflora* and *C. scirpoides* by (i) the lack of areas of thin-walled tissue in the ground parenchyma (ii) the more or less uniform size of cells of the parenchyma sheath (i.e. not largest opposite the peripheral vb's), and (iii) the small peripheral vb's with a shallow arc rather than a deep horseshoe of tracheids.

CHAETANTHUS R. Br.

No. of spp. 1; examined 1.

Genus and Species Description

Leaf not seen.

Culm surface (Fig. 27. H, I)

Hairs multicellular, flattened, covering culm surface, each attached by single basal cell; form much as in some Australian spp. of *Leptocarpus* (Fig. 27. G). **Epidermis:** cells rectangular, up to 6 times longer than wide, anticlinal walls moderately thickened, wavy. **Stomata:** subsidiary cells flattened dome-shaped, slightly shorter than guard cells. **Cuticular marks:** occasional slight ridges.

Culm T.S. (Fig. 27. G)

Outline circular. **Hairs:** see Culm surface; basal cell of hair slightly sunken into epidermis. **Epidermis:** cells more or less square, outer walls thick, slightly dome-shaped, anticlinal walls moderately thickened, straight, inner walls moderately thickened. **Stomata** slightly sunken, guard cells with lip at outer aperture and ridge on inner wall. **Hypodermis** absent. **Chlorenchyma** composed of palisade cells without pegs; cells present in 2 similar layers and divided into 14 sectors by pillar cells. Palisade cells arranged in transverse plates usually 1 cell thick as seen in L.S.; air-spaces present between adjacent cells and also between plates of cells. **Parenchyma sheath** v. discontinuous, represented by: (i) small, thin-walled cells, many containing silica-bodies, and (ii) pillar cells in longitudinal files, normally 2, but occasionally 1 or 3 cells wide, joining ridges from scl. cylinder to epidermis and dividing chlorenchyma into 14 or more sectors. Pillar cells 5–6 times as high as wide, walls thick, with simple pits; individual cells widest at outer end, tapering away from epidermis but widening again slightly at inner end. **Sclerenchyma:** sheath composed of 2–5 layers of v. thick-walled fibres; individual fibres having rounded outline, intercellular spaces filled with extra-cellular substances; outline of sheath rounded, with ridges occurring opposite the peripheral vb's.

Vascular bundles all embedded in scl. sheath. (i) Peripheral vb's situated in ribs of scl. sheath, each having 1 wide, angular tracheid on either flank, or with several angular, medium-sized tracheids in a single-layered arc; phloem well developed, walls of cells slightly thickened. (ii) Medullary vb's with 1 wide, angular, thin-walled mxv on either flank; vessels separated from one another by *c.* 3 rows of tracheids and narrow vessels; perforation plates simple, oblique, lateral wall-pitting scalariform; numerous narrow vessels and tracheids present at xylem pole; phloem ensheathed by single layer of narrow, slightly thickened cells. **Bundle sheaths:** fibres in 1 layer on flanks and 1–2 layers at xylem and phloem poles. **Central ground tissue** parenchymatous, walls slightly to moderately thickened; cells breaking down to form central cavity. **Silica:** present as (i) spheroidal-nodular bodies in some cells of parenchyma sheath, and (ii) granular material in some pillar cells. **Tannin:** none seen.

Rhizome and *Root* not seen.

MATERIAL EXAMINED

Chaetanthus leptocarpoides R. Br.: Pritzel No. 136; Australia (K).
Culm diameter 1·0 mm.

TAXONOMIC NOTES

This is a monotypic genus, which has been accepted without change since it was first described. Masters (1878) included it among genera of uncertain taxonomic position and quoted Steudel's (1855) *Prionosepalum* as a synonym. Gilg (1890) and Gilg-Benedict (1930) both noticed anatomical similarities between this genus and *Leptocarpus* species from Australia.

Material from Kew Herbarium from a sheet bearing both the names *Prionosepalum* and *Chaetanthus* has been studied. As indicated by Gilg and Gilg-Benedict the hair structure, arrangement of pillar cells, parenchyma sheath, silica-bodies and chlorenchyma suggest affinities with Australian *Leptocarpus* species. *Chaetanthus* also resembles the following species in all characters except for the type of hair: Australian *Hypolaena* species, *Meeboldina denmarkica* Suess., *Loxocarya pubescens* Benth., and species of *Restio* here included as group 7 (see p. 282). The great similarity in the basic anatomical plan of the culm of the genera and species just mentioned suggests that they are probably closely related to one another.

CHONDROPETALUM Rottb.

No. of spp. 18; examined 16.

GENUS DESCRIPTION

Leaf surface not seen.

Leaf T.S.

Examined in 1 sp. only (*C. deustum* Rottb.).

Sheathing base, encircling culm 1½ turns; section parallel-sided, tapering to margins.

Epidermis. (i) Adaxial cells 5-sided, outer wall flat, anticlinal walls short and straight and inner side bounded by 2 straight walls with an included angle of *c.* 110°; all walls slightly thickened. (ii) Abaxial cells more or less square, outer wall thick, inner and anticlinal walls moderately thickened, pits simple, numerous on anticlinal walls. **Stomata** superficial. **Hypodermis** absent. **Chlorenchyma** consisting of lobed cells, about as wide as high, in 2–3 irregular layers immediately below abaxial epidermis. Protective cells present, similar to lobed cells but thick-walled. **Vascular bundles:** phloem abutting directly on xylem; xylem in larger bundles with 1 wide, angular tracheid on either flank and a row of narrower tracheids between them; xylem in smaller bundles represented by a group of narrow tracheids. **Bundle sheaths:** O.S. parenchymatous, as cap to phloem pole only, but extended laterally in a single layer to adjacent bundles; I.S. sclerenchymatous, 1-layered. **Sclerenchyma:** adaxial girders 1–2 layers thick and up to 12 cells wide, joining on scl. of bundle sheath at xylem pole. **Ground tissue** appearing to consist of isolated parenchyma cells. **Air-spaces:** none present. **Tannin:** some mx. tracheids occluded. **Silica:** none present.

Culm surface

Hairs absent. **Papillae** infrequent (*C. macrocarpum*). **Epidermis:** cells either all hexagonal, square, or oblong, or with several different types, i.e. 4-, 5-, 6-, or 7-sided, present in the same epidermis. Walls straight or wavy, moderately thickened or thick. Lumina outline following that of cells, or circular to oval. Gaps between transverse (end) walls of cells frequent; rounded in outline; base of cavity so formed lined by outer wall of cell of inner epidermal layer (see culm T.S.). **Stomata:** subsidiary cells with curved or more or less parallel sides, or with 3 faces; guard cells normally thick-walled, longer than or the same length as subsidiary cells, parallel-sided and with rounded ends, or with slightly constricted ends, outline lemon-shaped. **Silica:** bodies absent; silica-like material present in a few cells of one sp. **Tannin** frequent, in cells of inner epidermis (see culm T.S.). **Cuticular marks** mainly granular; occasional striations.

Culm T.S. (Pl. V. A)

Outline circular or, infrequently, oval. **Cuticle** thick, continuous into depressions in epidermis. **Epidermis:** cells in 1, 2, 3 or, infrequently, 4 layers. Cells of inner layers frequently protruding between those of outer layer (alternating with them as seen in L.S. in most spp.). Cells varying in height; proportionate heights of layers differing in various spp. Wall thickness: outer walls thick to v. thick; other walls of outer layer of cells usually moderately thickened; all walls of inner cell layer(s) thin or slightly to moderately thickened, but when exposed to surface as thick as normal outer wall. Anticlinal walls usually straight. Outer walls sometimes concave. **Stomata** superficial or sunken; exhibiting a range of types of guard and subsidiary cells. **Chlorenchyma** consisting of peg-cells in 2 and sometimes 3 layers; protective cells always present, extending almost to parenchyma sheath in most species, apertures between protective cells rounded or elongated.

Parenchyma sheath of 1 or several layers of cells; cells normally rounded-hexagonal (end-walls transverse as seen in L.S.); walls slightly or moderately thickened. Circular or oval simple pits present. **Sclerenchyma:** sheath normally well developed; outline frequently slightly ridged to accommodate peripheral vb's, and with slight ridges between them. Sheath containing all peripheral vb's and sometimes some outer medullary vb's. Fibres thick-walled, either wide and with a wide lumen or narrow and with a narrow lumen; outline hexagonal, corners sometimes rounded. Fibres normally distinct from cells of central ground tissue.

Vascular bundles. (i) Peripheral vb's: phloem pole rounded or oval, abutting directly on 1- or rarely 2-layered horseshoe of medium-sized to narrow, angular tracheids. Xylem otherwise with 1 wider tracheid on either flank and narrower ones in between. Tracheids with scalariform wall-pitting. (ii) Medullary vb's: phloem pole circular to oval, often enclosed by single layer of narrow cells with thin walls next to phloem but with other walls thickened. Xylem: medium-sized or wide, angular or rounded mxv on either bundle flank, normally separated from one another by 2–3 files of narrow tracheids or xylem parenchyma cells. Vessels: lateral walls thick or slightly to moderately thickened; pitting opposite scalariform; perforation plates slightly to v. oblique, scalariform, reticulate, scalariform-fenestrate, or scalariform-reticulate. Protoxylem pole developed in all, or only larger (more central) bundles.

Bundle sheaths: at either pole normally consisting of 2 layers of narrow, thick-walled fibres with narrow lumina; at bundle flanks composed of 1 layer of wider fibres, with moderately thick to thick walls and wider lumina. Sheath reduced to caps only in some spp. **Central ground tissue** parenchymatous, cells wide, usually hexagonal, walls slightly or moderately thickened. Intercellular spaces sometimes present. Central cavity rare. **Silica:** bodies absent; silica-like material present in some cells of *C. capitatum*. **Tannin** frequent in inner epidermal layers, occasional in central ground tissue.

Rhizome T.S.

Seen in *C. rectum* (Mast.) Pillans, *C. microcarpum* (Kunth) Pillans, and *C. tectorum* (L.f.) Pillans.

Epidermis: cells thin-walled, twice as wide as high. **Hypodermis** 5-layered; outer layer similar to epidermis, cells of other layers with similar dimensions, but walls thickened. **Cortex:** *c.* 7 layers of thin-walled, rounded cells; intercellular spaces large; cells of 2 distinct sizes in *C. tectorum*; cells of innermost layer narrow. **Endodermoid sheath:** cells in 1 layer, inner and anticlinal walls heavily thickened, outer wall thin, lumina partly or entirely occluded.

Vascular bundles mainly collateral, some nearly amphivasal; orientation frequently random; each with 1, 2 or several wide, rounded mxv's and several narrower, angular vessels and tracheids. Vessel lateral wall-pitting scalariform, opposite and alternate, bordered; perforation plates scalariform-fenestrate, v. oblique. Tracheids with scalariform pitting in end walls. Phloem pole ensheathed by a single layer of narrow, thin- or thick-walled cells. **Bundle sheaths:** fibre ends pointed as seen in L.S. **Ground tissue:** cells parenchymatous, walls thick, indistinguishable from cells of bundle sheath when seen in T.S.; strands of thin-walled cells also present in *C. rectum* and *C. tectorum*. **Tannin** and **Silica:** none seen.

Root T.S.

Seen in *C. nudum* Rottb., *C. rectum* (Mast.) Pillans, and *C. tectorum* (L.f.) Pillans.

Root-hairs arising from cells similar to those of epidermis, shaft narrower than cell base. **Epidermis:** cells thin-walled, square to upright. **Hypodermis** absent. **Cortex.** (i) Outer part, 3-layered, outermost cells thin-walled, inner layers thick-walled. (ii) Middle part composed of large, thin-walled cells with large intercellular spaces between them. (iii) Inner part 2–3-layered, cells small, thin-walled. **Endodermis** 1-layered, cells with thin outer wall and heavily thickened anticlinal and inner walls; passage cells not seen. **Pericycle** 2–3-layered, cells angular, thick-walled, slightly wider than high; lumina polygonal, with rounded corners. **Vascular tissue** in 1 ring, vessel lateral wall-pitting scalariform, perforation plates simple, more or less transverse; walls thin or moderately thickened. **Central ground tissue** parenchymatous, cell walls thick, pitting simple. **Tannin:** none seen.

MATERIAL EXAMINED

Chondropetalum aggregatum (Mast.) Pillans: Schlechter 9821 (K).
C. andreaeanum Pillans: Esterhuysen 11523 (K).
C. capitatum (Steud.) Pillans: Drège 2510, An. 1840 (type) (K).
C. chartaceum (Pillans) Pillans: Esterhuysen 24298, An. 1955 (K).
C. deustum Rottb.: Parker, R. N. 4913 (K).
C. ebracteatum (Kunth) Pillans: Burchell 52 Δ (K).
C. hookerianum (Mast.) Pillans: (i) Burchell (2 specimens) (M). (ii) Schlechter 10341 (K).
C. macrocarpum (Kunth) Pillans: Acocks, J. P. 19651 (K).
C. marlothii (Pillans) Pillans: Esterhuysen 24469, An. 1955 (K).
C. microcarpum (Kunth) Pillans: Schlechter 10503, det. Mast. (K).
C. mucronatum (Nees) Pillans: Parker, R. N. 4912 (K).
C. nitidum (Mast.) Pillans: Esterhuysen 14589 (K).
C. nudum Rottb.: Prior, A.; Cape Town, An. 1947 (K).
C. paniculatum (Mast.) Pillans: Esterhuysen 3587 (K).
C. rectum (Mast.) Pillans: Parker, R. N. 3482 (K).
C. tectorum (L.f.) Pillans: (i) Burchell 3996, An. 1813 (K). (ii) Cheadle, V. I. CA907; Cape (S).

SPECIES DESCRIPTIONS

***C. aggregatum* (Mast.) Pillans.** Culm diameter 4 mm.

Culm surface

Epidermis: cells irregular, 4–7-sided, walls moderately thickened; frequent gaps between transverse walls filled with granular substance. **Stomata:** subsidiary cells with perpendicular ends.

Culm T.S.

Epidermis: cells in 2 layers, outer slightly higher than wide, cells of inner layer 3–4 times higher than outer. **Stomata** slightly sunken. **Parenchyma sheath**

mainly 2-layered but 3-layered on either side of scl. ribs. **Sclerenchyma:** sheath *c.* 10-layered. **Vascular bundles:** medullary vb's as for genus but phloem abutting directly on xylem.

***C. andreaeanum* Pillans.** Culm diameter 3·5 mm.

Culm surface (Fig. 28. G)

Epidermis: cells more or less regularly pentagonal or hexagonal, moderately thick-walled, walls straight; no gaps between cells. **Stomata:** see Fig. 28. G. **Cuticular marks** granular, slight longitudinal striations also present.

Culm T.S. (Fig. 28. C)

Epidermis: cells 1-layered, 2½–4 times higher than wide. **Stomata** superficial; subsidiary cells twice height of guard cells. **Chlorenchyma** 2-layered; apertures between protective cells elongated, situated about ½ way down length of cells. **Parenchyma sheath:** outline slightly ridged, mainly 3-, sometimes 4-layered, cells of outer layer narrower than others; walls moderately thickened.

***C. capitatum* (Steud.) Pillans** (Type). Culm diameter 2 mm.

Culm surface (Fig. 28. F)

Epidermis: cells 5–7-sided, longer than wide; walls thick; cells next to stomata smaller and with thicker walls. **Stomata.** See Fig. 28. F. **Cuticular marks** granular; wavy lines present above anticlinal walls of epidermal cells.

Culm T.S.

Epidermis: cells 1-layered; *c.* 3 times higher than wide; anticlinal walls occasionally wavy. **Stomata** superficial; guard cells with slight lip at outer aperture. **Chlorenchyma** 2-layered, cells *c.* 6 times higher than wide; pegs and air-spaces fewer and larger in inner layer; protective cells extending only partly into inner layer. **Parenchyma sheath** 1–3-layered, mainly 1-layered over peripheral vb's. **Sclerenchyma:** sheath 10–12-layered, fibres thick-walled, distinct from ground tissue. **Vascular bundles:** medullary, phloem abutting directly on xylem. **Silica:** fine granules in some cells of parenchyma sheath.

***C. chartaceum* (Pillans) Pillans.** Culm diameter 3·5 mm.

Culm surface

Epidermis: cells irregular, 4–7-sided, stomata often associated with smaller cells; walls moderately thickened; lumina rounded to oval. **Stomata** frequent; similar to those of *C. andreaeanum.*

Culm T.S.

Epidermis: cells 1-layered, cells 4–5 times higher than wide; outer walls slightly corrugate, v. thick; anticlinal walls v. infrequently slightly wavy. **Stomata** superficial; similar to those of *C. capitatum.* **Chlorenchyma** mainly 2-layered; cells of inner layer occasionally divided tangentially in half; protective cells in 2 layers. **Parenchyma sheath** 2–3-layered; cell walls thick. **Sclerenchyma:** sheath 8–10 cells thick, ridged opposite each peripheral vb. **Central ground tissue:** parenchymatous cells of various sizes, walls slightly thickened; no central cavity.

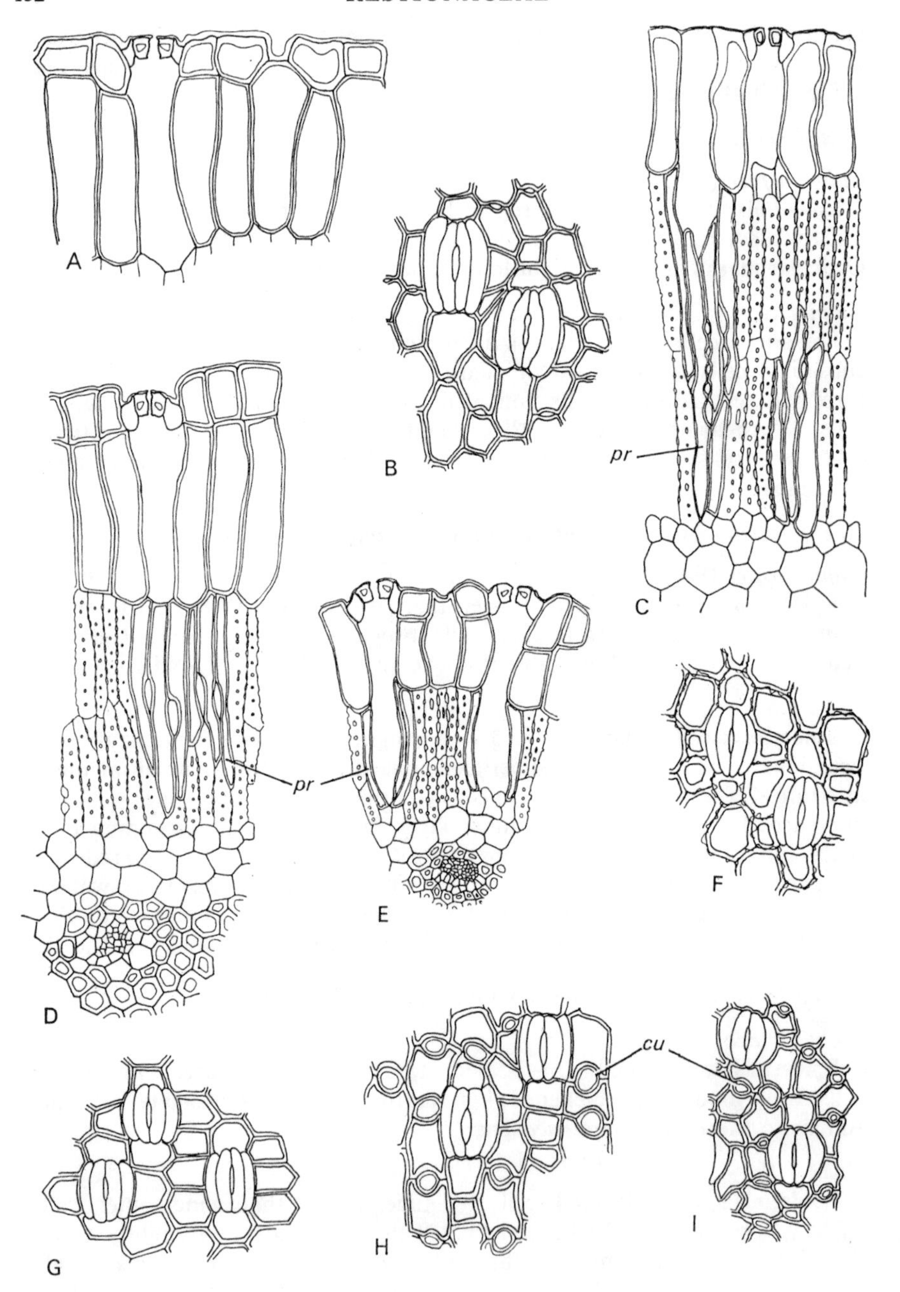

FIG. 28. Restionaceae. *Chondropetalum*. A, H, *C. ebracteatum*: A, detail of culm T.S.; H, culm epidermis. B, D, *C. marlothii*: B, culm epidermis; D, detail of culm T.S. C, G, *C. andreaeanum*: C, detail of culm T.S.; G, culm epidermis. E, I, *C. nudum*: E, detail of culm T.S.; I, culm epidermis. F, *C. capitatum*, culm surface. (All $\times 185$.)
cu = cuticle; pr = protective cell.

***C. deustum* Rottb.** Culm diameter 1·0 mm.

Leaf: see genus description.

Culm surface

Epidermis: cells square, walls more or less straight at low focus, wavy at high focus; gaps present between end walls of cells. **Stomata:** subsidiary cells with acute ends.

Culm T.S.

Epidermis: cells 2-layered; cells of outer layer *c.* 1½ times wider than high; depressions between cells with square corners; cells of inner layer more or less square. **Stomata** superficial; subsidiary cells with ridge on outer wall, guard cells with pronounced outer and smaller inner lips. **Chlorenchyma:** cells in 2 layers of equal height, pegs infrequent, air-spaces wide; protective cells normally 1 layer deep and extending ½ way into inner palisade layer. **Parenchyma sheath** 1-layered, cells almost square, exhibiting a range of sizes. **Vascular bundles:** medullary vb's with phloem abutting directly on xylem. **Bundle sheaths** frequently reduced to 2-layered caps at xylem and phloem poles.

***C. ebracteatum* (Kunth) Pillans.** Culm diameter 2·5 mm.

Culm surface (Fig. 28. H)

Epidermis: cells square to oblong; end walls concave, with circular gaps between them. **Stomata:** subsidiary cells with perpendicular or acute ends.

Culm T.S. (Fig. 28. A)

Epidermis: cells in 2 layers: outer cells about as high as wide; depressions between some cells as deep as straight anticlinal walls of adjacent cells; cells of inner layer up to *c.* 4 times higher than wide; outer walls dovetailing with inner walls of outer layer. **Stomata** as in *C. deustum*. **Chlorenchyma** 2-layered; pegs of outer layer smallest and most numerous; protective cells in 1 or 2 layers, extending part way into inner palisade layer, or almost to parenchyma sheath. **Parenchyma sheath** 2-, occasionally 3-layered, cell walls slightly thickened.

***C. hookerianum* (Mast.) Pillans.** Culm diameter 1·1–1·2 mm.

Culm surface

Epidermis: cells oblong; anticlinal walls wavy at surface, straight at slightly lower focus, end walls concave; frequent gaps present between end walls. **Stomata:** subsidiary cells slightly longer than guard cells, end walls rounded.

Culm T.S.

Epidermis: cells 2-layered; as in *C. ebracteatum*; cells of outer layer slightly higher than wide; those of inner layer *c.* twice as high as wide. **Stomata** as in *C. ebracteatum*. **Chlorenchyma** 2-layered. **Parenchyma sheath** 1-layered.

***C. macrocarpum* (Kunth) Pillans.** Culm diameter 3 mm.

Culm surface

Epidermis: cells mostly hexagonal, sometimes 4- or 5-sided, slightly wider than long or nearly square. Cells at either side of stoma produced into 2 lobes or papillae overarching sunken stomatal apparatus. Anticlinal walls straight. **Stomata:** subsidiary cells slightly longer than guard cells, with rounded ends.

Culm T.S.

Cuticle thick, outer part clear, inner part granular and staining red with safranin. **Epidermis:** cells in 2–3 layers; outer cells more or less square to upright; intermediate cells sporadic, normally short; inner cells *c.* 3 times higher than wide. **Papillae** extending over stomata. **Stomata** sunken; subsidiary cells with ridge on outer wall; guard cells with pronounced lip at outer aperture and ridge on inwardly facing wall. **Chlorenchyma** 2-layered; pegs wide, well separated; air-spaces large. **Parenchyma sheath** 1–2-layered opposite peripheral vb's, 3–4-layered between them.

***C. marlothii* (Pillans) Pillans.** Culm diameter 3·5 mm.

Culm surface (Fig. 28. B)

Epidermis: cells irregular, 4–6-sided, walls thickened, not wavy; gaps between transverse walls small, oval. **Stomata:** subsidiary cells with rounded ends; guard cell pair lemon-shaped, equal in length to subsidiary cells.

Culm T.S. (Fig 28. D)

Epidermis: cells in 2 layers; cells of outer layer slightly higher than wide, those of inner layer up to 3–4 times higher than wide. **Stomata** slightly sunken; guard cells with outer lip upturned, inner lip small. **Chlorenchyma** 2-layered; protective cells extending to $\frac{2}{3}$ depth of inner layer. **Parenchyma sheath** 2–3–(4)-layered, widest on either side of peripheral vb's.

***C. microcarpum* (Kunth) Pillans.** Culm diameter 0·8 mm.

Culm surface not seen.

Culm T.S.

Epidermis: cells in 2 layers; outer cells slightly to $1\frac{1}{2}$ times higher than wide, inner cells of same width, but twice as high (seen to alternate with cells of outer layer, in L.S.). **Stomata** deeply sunken, attached to cells of inner epidermal layer; facing walls of pairs of guard cells close to one another only at outer aperture and sloping apart to inside. (L.S. stoma: one end mounted higher on epidermal cell than other; all stomata inclined in same direction; guard cells with marked constriction at centre.) **Chlorenchyma** 2-layered; cells broad, *c.* 3–4 times higher than wide. **Parenchyma sheath** mainly 1-, occasionally 2-layered.

Rhizome T.S.: see genus description.

***C. mucronatum* (Nees) Pillans.** Culm diameter 7 mm.

Culm surface

Epidermis: cells mainly hexagonal, slightly longer than wide; walls thick, straight; gaps present between transverse walls of some cells. **Stomata:** guard cells thick-walled, outline of pair lemon-shaped, slightly longer than subsidiary cells.

Culm T.S.

Epidermis: cells in 2 layers; outer cells as high as to twice as high as wide; inner cells 4–6 times higher than wide; cells of irregular heights, some intruding into chlorenchyma. Outermost wall v. thick, often with concave face. **Stomata** superficial; subsidiary cells extending inwards much further than in most spp., being twice as high as guard cells; guard cells with outer lip accentuated by cuticle, and ridge on inner wall. **Chlorenchyma:** 2 layers. **Parenchyma sheath:** cells thick-walled, in 2 layers.

***C. nitidum* (Mast.) Pillans.** Culm diameter 2·1 mm.

Culm surface

Epidermis: cells mainly hexagonal, as long as to 1½ times longer than wide; cells next to stomata with thick, slightly wavy walls, walls of remaining cells straight, moderately thickened. **Stomata:** guard cell pair with lemon-shaped outline.

Culm T.S.

Epidermis: cells 1-layered; 2–2½ times higher than wide; anticlinal walls slightly wavy. **Stomata** superficial; guard cells with slight lip at outer aperture. **Chlorenchyma** 2-layered. Protective cells similar in shape to peg-cells, but with thick walls and fewer pegs; apertures numerous, in series along sides of tube. **Parenchyma sheath** 1–2-layered.

***C. nudum* Rottb.** Culm diameter 1·0 mm.

Culm surface (Fig. 28. I)

Epidermis: cells mostly hexagonal, as long as or slightly longer than wide; end wall concave, gaps present; other walls slightly wavy. **Stomata:** subsidiary cells with obtuse end walls.

Culm T.S. (Fig. 28. E)

Epidermis: cells 1–2-layered; cells of outer layer more or less square in median section, cells of inner layer 2–2½ times higher than wide. Deep depressions present above shorter cells. **Stomata** superficial; guard cells with lips at outer and inner apertures. **Chlorenchyma** 2-layered. **Parenchyma sheath.** (1)–2–3 cells wide.

Root T.S.

Root-hairs not seen. **Epidermis:** cells thin-walled, as high as wide. **Cortex:** single outer layer of thin-walled cells, enclosing 3 layers of thick-walled cells, these 4-sided rectangular in outline, 1–1½ times as wide as high; middle and inner cortex as for genus description.

***C. paniculatum* (Mast.) Pillans.** Culm diameter 2·5 mm.

Culm surface

Epidermis: cells mainly hexagonal, mostly as long as wide; walls thick, particularly in cells bordering on stomata. **Stomata:** guard cell pair with lemon-shaped outline.

Culm T.S. (Pl. V. A)

Epidermis: cells 1-layered, 2–3 times higher than wide, anticlinal walls straight, except those lining substomatal cavity. **Stomata** superficial; subsidiary cells with ridge on outer wall; guard cells with lips at inner and outer apertures. **Chlorenchyma** 2-layered. **Parenchyma sheath** 1–2-layered, cells large, walls thick.

***C. rectum* (Mast.) Pillans.** Culm diameter 1·0 and 1·2 mm.

Culm surface

Epidermis: cells rectangular or elongated-hexagonal, *c.* twice as long as wide, gaps present, walls thick, wavy at outer surface. **Stomata:** subsidiary cells variable, guard cell pair with lemon-shaped outline.

Culm T.S.

Epidermis: cells 2–3–(4)-layered; when 2-layered, outer cells *c.* 1½ times wider than high, inner cells *c.* 2½ times higher than outer layer; when 3- or 4-layered, outer layer as above, middle layer(s) as outer, and inner 1½ times higher than outer layer. **Stomata** sunken, attached to cells of outer and middle layers of epidermis; guard cells much constricted in median section, expanded at either end, ridge present on outer and inner walls. **Chlorenchyma** 2–3-layered, protective cells of same length as cells of outer palisade, apertures between protective cells elongated, reaching almost to their apices and bases. **Parenchyma sheath** consisting of 1 layer of wide cells.

Rhizome and *root*; see genus description.

***C. tectorum* (L.f.) Pillans.** Culm diameter 1·2 and 1·7 mm.

Culm surface (3996 only)

Epidermis: cells rectangular, slightly longer than wide, anticlinal walls wavy; frequent, rounded gaps present between transverse walls. **Stomata:** subsidiary cells wide, with acute ends, as long as guard cells.

Culm T.S.

Epidermis: cells 1–2-layered, cells of outer layer slightly wider than high, inner layer either reaching to the outer surface, the cells being more than twice as high as wide, or reaching to the outer layer and then 1–2 times as high as wide. (The two different heights occur in the same cell, according to the plane of section, since cells of inner layer alternate with those of outer.) **Stomata** superficial; guard cells without lips. **Chlorenchyma** 2–3-layered, consisting of peg-cells (3996 basal; cells lobed); protective cells extending inwards from half-way up substomatal tube formed by epidermal cells to depth of outer layer of palisade cells. **Parenchyma sheath** of 1–2 layers.

Rhizome and *root*: see genus description.

Taxonomic Notes

This is a genus with a complex history. It was incorporated into *Restio* a short time after it was described (Jussieu 1789); it was separated again by Beauvois (1828) but shortly afterwards it was included in *Restio* once again by Nees (1830). In 1853, Lindley removed it from *Restio* and included it in *Elegia*; it was retained there by Masters (1878) and Bentham and Hooker (1883). Bentham and Hooker recognized *Dovea* Kunth as a separate genus. Gilg (1890) noticed that *Elegia* L. (including *Chondropetalum* Rottb.) was anatomically similar to *Dovea* Kunth.

Pillans (1928) separated *Chondropetalum* Rottb. once more and included in it all *Dovea* species as synonyms; he recognized *Elegia* L. as a distinct genus. Many of the *Chondropetalum* species are cited by Pillans with synonyms from

Elegia, as would be expected from the previous history. The genus *Askidiosperma* Steud. was included by Pillans as a synonym for *Chondropetalum.*

Chondropetalum was found to be divisible into 2 groups as a result of this present work, with either 1 or 2–(3–4) layers of epidermal cells respectively. None of the species in the first group was found to have synonyms in *Elegia* (see p. 174) in contradistinction to those of the second group in which there are only 2 species to which this does not apply. Individual members of the second group are almost indistinguishable from *Elegia.*

It is possible that individual *Chondropetalum* species with 2 epidermal layers could be identifiable from the relative heights of the cells of the 2 layers as seen in T.S., but more evidence to test this is still needed.

C. capitatum (Steud.) Pillans is the species synonymous with *Askidiosperma capitatum* Steud. Gilg-Benedict (1930) did not agree that this species belonged to *Chondropetalum* and reinstated the genus *Askidiosperma* on morphological and anatomical grounds. During the present investigation the type specimen of *C. capitatum* (Steud.) Pillans was found to belong to the group with a 1-layered epidermis. Since species belonging to this group are similar anatomically, it seems unreasonable to regard *Askidiosperma* as an independent genus. The species with 1 layer of epidermal cells are anatomically very similar to certain South African species of *Restio* (here included in group 3) and *Leptocarpus*, as well as some species of *Staberoha. C. macrocarpum* (Kunth) Pillans is anatomically distinctive; it has diagnostic bilobed subsidiary cells which overarch the sunken stomata.

An unusual feature shared by many *Elegia* and *Chondropetalum* species with more than one epidermal layer is the presence of gaps between the end walls of some epidermal cells of the outer layer, through which the outer, exposed wall of a cell from the inner epidermal layer can be seen.

The species of the 2 groups are as follows.

(i) Epidermis 1-layered. *C. andreaeanum* Pillans, *C. capitatum* (Steud.) Pillans, *C. chartaceum* (Pillans) Pillans, *C. nitidum* (Mast.) Pillans, and *C. paniculatum* (Mast.) Pillans.

(ii) Epidermis 2–(3–4–) layered. (*a*) Stomata sunken: *C. macrocarpum* (Kunth) Pillans, *C. marlothii* (Pillans) Pillans, *C. microcarpum* (Kunth) Pillans, and *C. rectum* (Mast.) Pillans; (*b*) stomata superficial: *C. aggregatum* (Mast.) Pillans, *C. deustum* Rottb., *C. ebracteatum* (Kunth) Pillans, *C. hookerianum* (Mast.) Pillans, *C. mucronatum* (Nees) Pillans, *C. nudum* Rottb., and *C. tectorum* (L.f.) Pillans.

COLEOCARYA S. T. Blake

No. of spp. 1; examined 1.

Genus and Species Description

Leaf not seen.

Culm surface

Hairs absent. **Epidermis:** cells 4-sided, as long, or up to twice as long as wide, those next to stomata of shorter type; walls moderately thick to thick, wavy. **Stomata:** subsidiary cells crescentiform or with obtuse ends; guard cell pair with almost circular outline. **Cuticular marks** granular, with distinct, wide, longitudinal striations. **Silica** and **tannin:** none seen.

Culm T.S. (Fig. 29. A)

Outline circular. **Epidermis:** cells 4-sided; those between stomata as high as or slightly higher than wide, cells next to stomata extending further into chlorenchyma, with walls bordering on the adjacent epidermal cells equal to them in height, and walls bordering on the substomatal cavity produced inwards, being 1½–2 times longer than the others. Outer walls v. thick, anticlinal walls v. thick at outer ends, the inner half of these walls and the inner walls being slightly thickened; anticlinal walls, particularly those lining substomatal cavity, wavy. **Stomata** superficial, guard cells with slight outer and inner lips. **Chlorenchyma:** 1 layer of peg-cells; cells opposite stomata 2–3 times higher than wide, those between stomata *c.* 4–6 times higher than wide; pegs well developed, in files of 8–12. **Parenchyma sheath** 1-layered, cells oval, up to about twice as wide as high, walls slightly thickened; some stegmata present. **Sclerenchyma:** sheath circular in outline, composed of 5–6 layers of fibres, those of outer 1–3 layers thick- or v. thick-walled, narrow, those of inner layers wider, fibres of innermost layer widest, walls moderately thickened to thick.

Vascular bundles. (i) Peripheral vb's: xylem principally composed of single-layered arc of tracheids; flanking tracheids medium-sized, middle tracheids narrow, all thin-walled and angular; phloem situated in concavity of arc. (ii) Medullary vb's: outline rounded; flanking mxv's medium-sized, angular, with slightly thickened walls, separated from one another by 3–5 rows of narrow, parenchymatous cells; lateral walls of mxv's with scalariform pitting; perforation plates oblique, scalariform; px poles present in all medullary vb's. Phloem poles radially oval in outline, separated from xylem by single layer of narrow parenchyma cells with slightly thickened walls. Most medullary vb's free from culm scl. sheath, those in ground tissue grouped in twos or threes, their sheaths being in contact at the flanks.

Bundle sheaths 1–2-layered, fibres at poles narrower than those on flanks, those next to phloem pole v. thick-walled, those next to xylem pole v. narrow, with slightly thickened walls, others moderately thick-walled. **Central ground tissue** parenchymatous, outer cells moderately thick-walled, innermost cells wider, with slightly thickened walls, some breaking down. **Silica:** spheroidal-nodular bodies present in stegmata in parenchyma sheath. **Tannin:** none seen.

Rhizome and *root* not seen.

MATERIAL EXAMINED

Coleocarya gracilis S. T. Blake: Blake, S. T. 14300, An. 1940; Stradbroke Is., Austr. (Dupl. Type) (K). Culm diameter 1·2 mm.

TAXONOMIC NOTES

This monospecific genus has not been described previously from the anatomical point of view. The unspecialized anatomy is similar to that of *Restio confertospicatus* Steud., without pillar cells or protective cells, and with the epidermal cells surrounding the stomata extending into the 1-layered chlorenchyma. There are no hairs in *Coleocarya*, but these are present in *R. confertospicatus*.

Several genera or species from SW. Australia show anatomical resemblance to *Coleocarya*. Of these, *Harperia* Fitz. (see p. 175) is the most similar, but the elongated epidermal cells of *Coleocarya* are not as long as those of *Harperia*, and it lacks the mounds formed by the outwardly projecting epidermal cells in *Harperia*. *Onychosepalum* Steud. is anatomically similar to *Coleocarya*, but like *Harperia* it has low mounds of projecting epidermal cells. *Lepidobolus* Nees and *Loxocarya* species (except *L. pubescens* Benth.) also have inwardly elongated epidermal cells surrounding the stomata, together with multicellular hairs, and they also often have low mounds formed by the epidermal cells.

Restio nitens Nees is readily distinguished from *Coleocarya* on the basis of its anatomy, because although it has the inwardly elongated epidermal cells it also has pillar cells.

A fairly close relationship between *Coleocarya* and some of the genera or species just mentioned seems likely from anatomical data. It should be remembered, however, that this apparent similarity may be misleading because we do not know whether the plants are primarily or secondarily simple.

DIELSIA Gilg

No. of spp. 1; examined 1.

Genus and Species Description

Leaf not seen.

Culm surface.

Hairs and **papillae** absent. **Epidermis.** Cells of 2 types: (i) Large, thick-walled, in uniseriate (rarely biseriate) longitudinal files; outline of individual cells 4–6-sided, $1\frac{1}{2}$–2 times wider than long. (ii) Smaller cells with slightly thickened walls, mostly axially elongated and with 6–7 sides; of varying sizes. Walls slightly wavy in some cells. **Stomata:** subsidiary cells variable (Fig. 29. C); guard cells equal in length to subsidiary cells, outline of pair lemon-shaped. **Silica:** bodies absent. **Tannin:** in large cells. **Cuticular marks** granular.

Culm T.S. (Fig. 29. B, D)

Outline polygonal. **Epidermis.** Cells of 2 types: (i) Large cells (see Culm surface), at angles of culm, opposite and joined to wedge-shaped scl. girders from mechanical cylinder. These cells 2–3 times as high as and up to twice as wide as the others; outer walls thick, slightly corrugated, other walls thin to slightly thickened. Anticlinal walls slightly wavy, with 2 facets, lumen widening from the outer wall to a point equal in height with shorter cells of epidermis and narrowing again to inner wall. (ii) Smaller cells (see Culm surface) more or less square in outline; outer walls thick, slightly corrugated, other walls thin to slightly thickened. **Stomata:** subsidiary cells compressed at centre, but extending above and below guard cells; guard cells with ridge

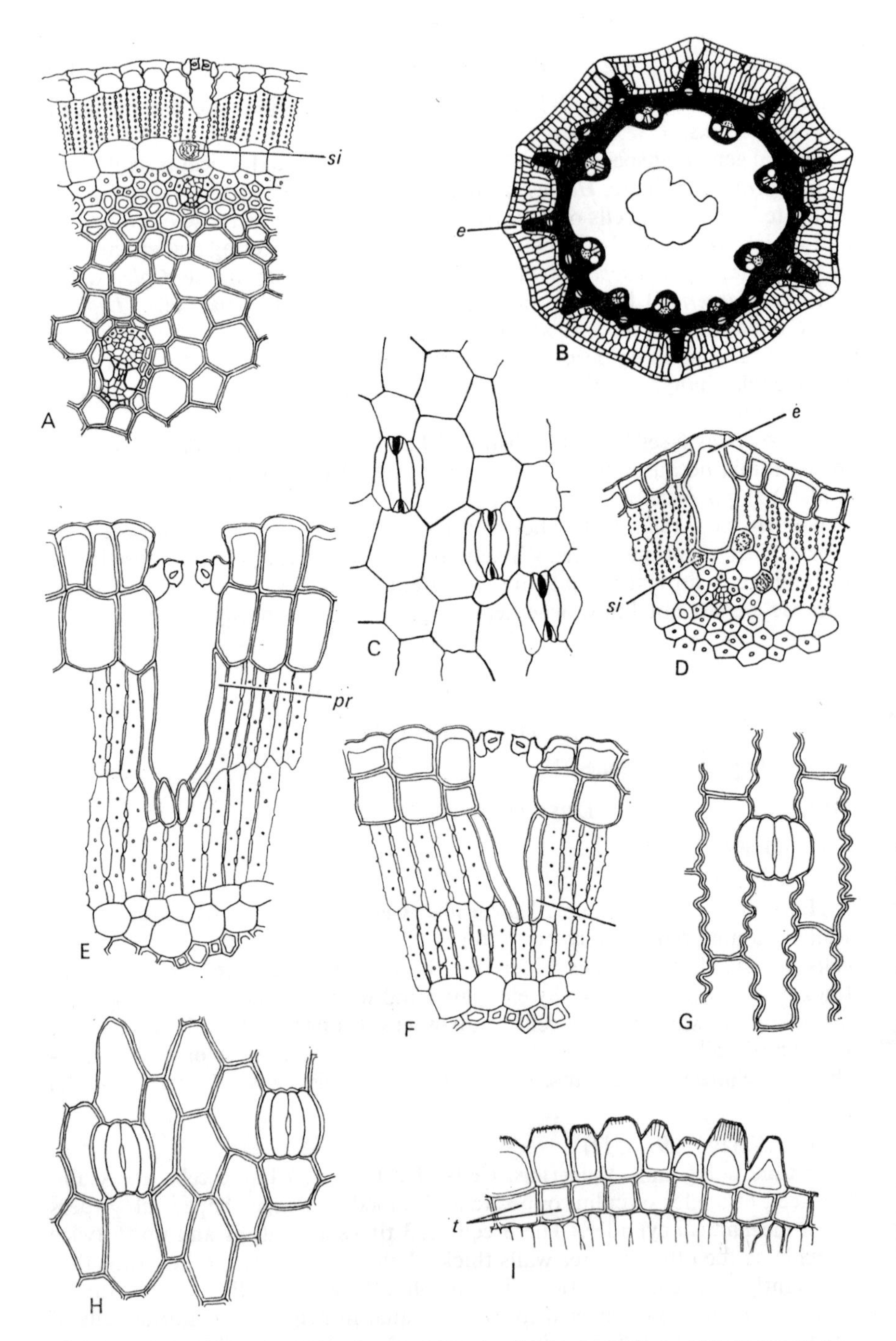

FIG. 29. Restionaceae. A, *Coleocarya gracilis*, detail of culm T.S. (×185). B–D, *Dielsia cygnorum*: B, plan of culm T.S. (×40); C, culm epidermis (×235); D, detail of culm T.S. (×150). E–H, *Elegia obtusiflora* showing two different culm types (E, H, Acocks 17460; F, Drège 2517; G, Vogts 50): E, F, detail of culm T.S. (×200); G, H, culm epidermis (×200). I, *E. neesii*, detail of culm epidermis T.S. (×290).
e = enlarged epidermal cell; pr = protective cell; si = silica-body; t = tannin.

on inner wall. **Hypodermis** absent. **Chlorenchyma** consisting of palisade peg-cells in 2–3 layers; walls thin. **Parenchyma sheath:** cells rounded, in 1 layer; extending on flanks of scl. girders to base of larger epidermal cells. **Sclerenchyma:** sheath composed of 4–6 layers of thick-walled fibres; produced at regular intervals into girders each *c.* 4 cells wide, and terminating in a large epidermal cell.

Vascular bundles. (i) Peripheral vb's: xylem composed of single or double row of narrow tracheids, tracheids of the outer layer being slightly wider than the rest; phloem abutting directly on xylem; bundle embedded in scl. sheath and situated either opposite to or between scl. girders. (ii) Medullary vb's: almost oval in outline, with 1 wide, angular mxv on either flank; lateral walls of vessels with opposite pitting, perforation plates scalariform, oblique. Phloem separated from xylem by single layer of narrow cells having thin walls next to phloem cells and thick walls towards xylem. All medullary vb's enclosed by or in contact with scl. sheath or cylinder.

Bundle sheaths: cells in 1 layer; those at phloem pole narrowest and with thickest walls. **Central ground tissue** parenchymatous, *c.* 5 outer layers of cells with slightly thickened walls, central cells v. thin-walled, frequently breaking down to form central cavity. **Tannin** present in most of larger epidermal cells. **Silica:** spheroidal-nodular bodies numerous, solitary bodies occurring in some cells of parenchyma sheath.

Rhizome and *root* not seen.

Material Examined

Dielsia cygnorum Gilg: (i) Diels, An. 1902; W. Australia (K). (ii) Pritzel, E., No. 304; W. Australia (K). Culm diameter 0·5 and 0·8 mm.

Taxonomic Notes

This monospecific genus is readily distinguishable from all other genera or species in the Restionaceae by its epidermal cells. The culm, which is polygonal in T.S., has an enlarged epidermal cell at each angle and a single layer of smaller epidermal cells on the slightly concave faces between the angles. A girder of fibres extends from the sclerenchyma sheath to each of the large epidermal cells.

The only other member of the Restionaceae known to have an epidermis composed of 2 types of cell is *Restio leptocarpoides* Benth. In this species, however, the large epidermal cells are arranged in numerous longitudinal files, each of which is separated from the next file of the same kind by 1–3 files of small cells. Furthermore the large cells, instead of being almost level with the surface of the culm and extending inwards, as in *Dielsia*, protrude outwards and overarch the smaller cells on either side. *R. leptocarpoides* has sclerenchyma girders extending to some but not all of the large epidermal cells.

In spite of the similarities to which attention has just been drawn, the differences between the epidermal structure of *R. leptocarpoides* and *Dielsia* are probably sufficient to indicate that the two should not be grouped close together in a classification.

Apart from its unusual epidermis, *Dielsia* shares many anatomical characters with other members of the Restionaceae, and is probably correctly assigned to the family.

ELEGIA L.

No. of spp. *c.* 35; examined 21.

Genus Description

Leaf surface (sheathing base)

Seen in *E. asperiflora, E. stipularis,* and *E. verticillaris.*

Hairs absent. **Papillae** present in *E. verticillaris.* **Epidermis:** cells (abaxial only) more or less square, or slightly elongated, or up to 3 times longer than wide; anticlinal walls wavy. **Stomata** (abaxial only): subsidiary cells crescentiform or parallel-sided and with obtuse ends; guard cells with more or less parallel sides and rounded ends, or pair of cells with lemon-shaped outline. **Silica:** bodies absent. **Tannin** in some cells. **Cuticular marks** granular.

Leaf T.S. (sheathing base)

Seen in *E. fistulosa, E. grandis, E. stipularis,* and *E. verticillaris.*

Outline flattened, tapering to margins, encircling culm with 1½ turns. **Epidermis.** (i) Abaxial cells 1-layered, except in *E. grandis*; cells more or less square; outer walls thick or v. thick, sometimes produced into dome-shaped papillae, anticlinal walls slightly to moderately thickened, straight, inner walls thin to slightly thickened. (ii) Adaxial cells 4- or 5-sided, walls thin to slightly thickened. **Stomata** sunken in *E. grandis*, superficial in other spp.; subsidiary cells extending above and below guard cells. **Chlorenchyma** 1–2-layered, composed of peg-cells and of more or less isodiametric and lobed cells, walls thin. **Vascular bundles** of 2 sizes, large and small alternating, phloem well developed, xylem represented by relatively few (4–8) tracheids, or vb's with 1 wide tracheid on either flank (*E. grandis*). **Bundle sheaths** sclerenchymatous, fibres moderately thick- to thick-walled, in 1 layer on flanks and up to 3 layers at bundle poles; parenchymatous sheath, if present, indistinguishable from ground parenchyma. **Sclerenchyma** restricted to bundle sheaths, except in *E. grandis*, where continuous between bundles in 5 layers; fibres v. thick-walled, separated from chlorenchyma by 2 layers of parenchymatous cells and from abaxial epidermis by 2–3 layers of parenchymatous cells. **Ground tissue** parenchymatous in all spp. except *E. grandis* (see sclerenchyma), present between bundles and in 1–3 layers separating chlorenchyma from adaxial epidermis; walls frequently thin, but occasionally slightly thickened. **Air-cavities** absent. **Silica** absent. **Tannin** in cells of abaxial epidermis.

Leaf T.S. (blade)

Seen in *E. neesii.*

Outline oval, major axis 0·75 mm, minor axis 0·3 mm. **Epidermis:** cells rectangular, slightly wider than high; in 1 layer. **Stomata** slightly sunken, as for culm of same sp. **Chlorenchyma:** palisade cells 2–3 times higher than wide, present in 2 layers; pegs present but few. **Parenchyma sheath** mainly 1-, occasionally 2-layered, cells rounded, medium-sized. **Vascular bundles** in an arc of 3; xylem composed of 3 groups of narrow tracheids occupying most of central area of arc. **Sclerenchyma** present as caps 1–2 layers thick, between phloem and parenchyma sheath. **Ground tissue:** none present. **Silica** and **tannin:** none present.

Culm surface

Hairs absent. **Papillae** present in *E. neesii* and *E. obtusiflora*, but never well developed. **Epidermis:** cells frequently elongated-hexagonal or rectangular, sometimes square; anticlinal walls straight or wavy; transverse walls frequently concave, bordering gaps between cells of outer epidermal layer. **Stomata:** subsidiary cells thin-walled, crescentiform or parallel-sided and with obtuse to acute ends; guard cells thick-walled, usually about as long as subsidiary cells. **Silica** absent. **Tannin** frequently present, but normally confined to cells of inner epidermal layer. **Cuticular marks** granular; longitudinal ridges sometimes present.

Culm T.S. (Pl. V. B)

Outline circular. **Cuticle** thick. **Papillae** present in *E. neesii* and *E. obtusiflora*. **Epidermis.** Cells normally in 2 layers but occasionally 1- or 3–4-layered in places in some spp.; cells of both layers of equal height or those of inner layer higher than those of outer layer. Outer epidermal cells of unequal heights in 1 sp. (*E. coleura*), and discontinuous in *E. neesii.* Occasional gaps present between cells of outer layer in some spp., the bases of these being lined by outer walls of cells from inner layer. Outer cell walls frequently thick and curving inwards at cell margins, inner and anticlinal walls slightly to moderately thickened; walls of inner cell-layer frequently thinner than those of outer layer. Anticlinal walls straight or wavy. **Stomata** superficial or sunken; when sunken, often attached to cells of inner epidermal layer. Subsidiary cells exhibiting a wide range of shape; guard cells often with outer lip, inner lip, or ridge on inner wall. **Hypodermis** interpreted as inner epidermis. **Chlorenchyma:** peg-cells in 2 (rarely 3) layers, often of similar height; cells of inner layer frequently having largest pegs; protective cells present, extending from epidermis into culm through 1 or more layers of chlorenchyma according to sp. **Parenchyma sheath** 1- 2–several-layered; walls thin or slightly thickened, cells rounded, wide. **Sclerenchyma:** sheath composed of fibres with thick or moderately thickened walls, in 3–7 layers in most spp.

Vascular bundles. (i) Peripheral vb's with arc of narrow or medium-sized tracheids partially enclosing phloem pole. (ii) Medullary vb's: outline frequently oval, sometimes circular, mostly having 1 wide mxv on either flank; phloem normally separated from xylem by single-layered sheath of narrow cells, these with thickened walls next to xylem and thin walls next to phloem. Lateral wall pitting of vessels scalariform; perforation plates slightly oblique or oblique, scalariform. Px present in all or confined to inner bundles; some or all vb's embedded in culm scl. sheath. **Bundle sheaths:** fibres normally in 1 layer at xylem pole and bundle flanks and in 2–3 layers at phloem pole. **Central ground tissue** parenchymatous, central cells sometimes breaking down to form central cavity. **Silica:** no silica-bodies present. **Tannin** in some epidermal cells.

Rhizome T.S.

Seen in *E. asperiflora*, *E. fistulosa*, *E. obtusiflora*, and *E. stipularis*.

Epidermis: cells 5-sided, outer walls slightly convex, thick; anticlinal walls short, thick; inner walls with 2 faces, moderately thickened. **Hypodermis:** up to 6 layers of sclereids, cells of inner 2 layers having thick inner walls and thinner outer and anticlinal walls. **Cortex** (outer part = hypodermis?); inner part composed of 4–5 layers of thin-walled, more or less isodiametric, lobed, parenchymatous cells. **Endodermoid sheath** normally 1-layered; outer walls thin, other walls thickened. **Vascular bundles.** (i) Peripheral vb's collateral; vessels in an arc. (ii) Medullary vb's amphivasal; vessels medium-sized, thick-walled, angular, lateral wall pitting scalariform, pits wide, bordered; perforation plates oblique, simple. Lumina frequently tannin-filled. **Bundle sheaths:** fibres thick-walled, arranged in 2–3 layers, walls heavily pitted. **Central ground tissue** parenchymatous, forming islands or lacunae between vb's. **Silica:** none present. **Tannin** in many cells of epidermis and cortex, and in some vessels.

Root T.S.

Seen in *E. asperiflora*, *E. fistulosa*, *E. obtusiflora*, and *E. stipularis*.

Root-hairs developed from cells similar to those of remainder of epidermis. **Epidermis:** cells thin-walled, square or rectangular and upright. **Hypodermis:** 1 layer of thin-walled cells, *c.* ½ height of epidermal cells. **Cortex:** (i) outer region consisting of 4–5 layers of fibres in some spp.; (ii) middle part, consisting of thin-walled, parenchymatous cells, either arranged in radial plates 1 cell in thickness, or continuous; (iii) inner part composed of 1–2 layers of smaller, parenchymatous cells. **Endodermis** 1-layered, cells thin-walled, or with thickenings on inner and anticlinal walls. **Pericycle:** cells either in 1–2 layers and parenchymatous, or in 5–6 layers and sclerenchymatous. **Vascular system:** mxv's either in a single ring, or in a ring and also scattered throughout centre of root; in the last instance, sometimes with phloem strands accompanying vessels in root centre. **Central ground tissue** sclerenchymatous and composed of narrow, thick-walled fibres, or parenchymatous (*E. asperiflora*). **Tannin** absent.

MATERIAL EXAMINED

The genus is a South African endemic

Elegia asperiflora (Nees) Kunth: (i) Schlechter 9571; Vogelgat (K). (ii) Burchell 8140 (K). (iii) Schlechter 6017, An. 1894 (K). (iv) Acocks, J. H. P. 21672; 19·8 m W. of Humansdorp (K). (iv) Fourcade, H. G. 2434, An. 1927; Humansdorp (K). (vi) Parker, R. N., 4683, A. 1951; Jonkershoek (K). (vii) Herb. Benth. 106! in Drège's Herb. (K). (viii) Cheadle, V. I. CA709 (S).

E. coleura Nees ex Mast.: Rehmann, A. 2571; Worc. Div. (K).

E. cuspidata Mast.: MacGillivray, Voyage of H.M.S. *Herald*, 1832; Bot. No. 437, Simon's Bay, C.O.G.H. (type) (K).

E. equisetacea Mast.: Fourcade, H. G. 1373, An. 1921 (K).

E. fistulosa Kunth: Cheadle, V. I. CA833; Cape (S).

E. galpinii N.E. Br.: Riversdale Div. (K).

E. glauca Mast.: Esterhuysen 22570 (K).

E. grandis (Spreng. ex Nees) Kunth: Burchell. No number (K).

E. intermedia (Steud.) Pillans: Barker, W. 22454; Kirstenbosch (K).

E. juncea L.: Bolus 4456 ♂, An. 1878; Cape Town (K).

E. muirii Pillans: Esterhuysen 23162, An. 1954 (K).
E. neesii Mast.: (i) Burchell 569; Table Mt. (K). (ii) Esterhuysen 11461, An. 1945; Stellenbosch (K).
E. obtusiflora Mast. (all Kew material as *E. parviflora* Kunth): (i) Cheadle, V. I. CA708; Cape (S). (ii) Godfrey GH1259, An. 1952; Vanrhynsdorp (K). (iii) Esterhuysen 25742, An. 1956; Ceres Div. (K). (iv) Parker, R. N. 4842 ♀; Bettys Bay, Caledon Div. (K). (v) Parker, R. N. 4843 ♂; Bettys Bay, Caledon Div. (K). (vi) Taylor 3029; Cape (K). (vii) Vogts, M. 50; Caledon Div. (K). (viii) Smith, C. A. 5025; Bredasdorp (K). (ix) Kuntze, O. 350; Caledon Div. (K). (x) Kuntze 194 (K). (xi) Drège, 2517! (K). (xii) Burchell 8647 (K). (xiii) Drège 120; Dutoits Kloof (K). (xiv) Esterhuysen 3683; Worc. Div. (K). (xv) Acocks, J. P. H. 17460; Calvinia Dist. (K).
E. racemosa (Poir.) Pers.: Hutchinson, J. 34 (K).
E. spathacea Mast.: Parker, R. N. 4693 ♂, An. 1951; Rooi Els (K).
E. squamosa Mast.: ♀ (K).
E. stipularis Mast.: (i) Pickstone 22 ♂; Paarl and Caledon Div. (K). (ii) Cheadle, V. I. CA825; Cape (S).
E. thyrsoidea (Mast.) Pillans: Fourcade, H. G. 2345, An. 1922; headwaters of Wagenboom R. (K).
E. vaginulata Mast.: Bolus 5480 (K).
E. verreauxii Mast.: Zeyher (K).
E. verticillaris (L.f.) Kunth: Worsdell, W. C., An. 1909 (K).

Species Descriptions

***E. asperiflora* (Nees) Kunth.** Culm diameter 1–2 mm.

Leaf surface (abaxial)

Epidermis: cells elongated, rectangular, walls thick, v. wavy. **Stomata:** long walls of subsidiary cells wavy.

Culm surface

Epidermis: cells square to $1\frac{1}{2}$–2 times as long as wide, walls slightly or moderately thickened, wavy; rounded gaps present between end walls of some cells. **Stomata:** subsidiary cells shorter than guard cells, with ends perpendicular or obtuse, some with slightly wavy long walls.

Culm T.S. (Fig. 30. F)

Epidermis: cells in 2 layers of equal height (only 1 cell thick in places); individual cells slightly higher than wide. **Stomata** superficial; guard cells with outer and inner lips. **Chlorenchyma:** outer cells slightly shorter than inner; protective cells extending part way into inner layer, individual cells slightly swollen at inner and outer ends, apertures elongated. **Parenchyma sheath** mainly 2-layered, 1- or 3-layered in places.

Rhizome T.S.

Cortical region not preserved; other characters as for genus description.

Root T.S.

V. little thickening in any tissue. **Epidermis:** cells upright, $1\frac{1}{2}$ times as high as wide. **Cortex:** outer 2 layers of cells having slightly lignified walls, other cells loosely packed, rounded, parenchymatous, in 8–10 layers, cells of outer layers largest. **Endodermis:** cells thin-walled, more or less square. **Pericycle** 1-layered, cells thin-walled. **Vascular system:** polyarch. **Central ground tissue:** parenchymatous.

E. coleura **Nees ex Mast.** Culm diameter 1·5 mm.

Culm surface

Epidermis: cells mostly rectangular, occasionally wider at middle than ends, *c.* twice as long as wide; anticlinal walls moderately thickened, wavy; rounded gaps, often containing silica-like material, present between cell end walls. **Stomata:** subsidiary cells with perpendicular or rounded end walls; guard cells longer than subsidiary cells and with pincer-shaped ridge at each end of aperture. **Tannin:** none seen.

Culm T.S. (Fig. 30. D)

Epidermis: cells in 2 layers (occasionally 3), cells of each layer of uneven heights, overall height of combined layers more or less constant. Cells square or higher than wide, anticlinal walls straight. Occasional gaps occurring in outer cell layer, bases of which lined by outer walls of cells of inner layer. **Stomata** superficial; guard cells with lip at outer and inner apertures. **Chlorenchyma:** protective cells extending to half-way into inner palisade layer, apertures between protective cells wide and elongated. **Parenchyma sheath:** mainly 2-, occasionally 3-layered. **Silica:** present as granules in pits between cells of outer epidermis (?).

E. cuspidata **Mast.** (type). Culm diameter 4·0 mm.

Culm surface

Epidermis: cell outlines irregular, mainly more or less square or rectangular, anticlinal walls wavy; rounded gaps present between cells. **Stomata** as in *E. coleura*, but more frequent. **Tannin:** none seen.

Culm T.S.

Epidermis: cells mainly in 2 layers, outer cells more or less square, anticlinal walls wavy, inner cells rectangular, 3–4 times higher than outer cells, sometimes reaching to outer surface; depressions present between some cells of outer layer. **Stomata** superficial; guard cells with lip at outer aperture. **Chlorenchyma:** protective cells extending half way into inner chlorenchyma layer, apertures between protective cells elongated, opening to cells of outer chlorenchyma layer. **Parenchyma sheath** 1-, mainly 2-, sometimes 3-layered. **Tannin:** none seen.

E. equisetacea **Mast.** Culm diameter 2·5 mm.

Culm surface

Epidermis: cells mostly rectangular, about twice as long as wide, anticlinal walls wavy. **Stomata** as in *E. coleura.*

Culm T.S.

Epidermis: cells in 2 layers, those of outer layer slightly wider than high and half the height of those of inner layer. **Stomata** superficial; guard cells with lip at outer and inner apertures. **Chlorenchyma:** cells of outer layer the shorter; protective cells as in *E. cuspidata.* **Parenchyma sheath** mostly 2-layered, occasionally 1 or 3 cells thick. **Tannin** in some cells of inner epidermal layer.

***E. fistulosa* Kunth.** Culm diameter 2·5 mm.

Culm surface

Epidermis: cells rectangular or hexagonal and 1½–2 times longer than wide. **Stomata:** subsidiary cells crescentiform. **Tannin:** none seen.

Culm T.S. (Fig. 30. C)

Epidermis: cells in 2–3 layers, outer cells more or less square, inner cells 2–3 times higher than those of outer layer, some reaching culm surface. **Stomata** sunken; subsidiary cells attached to inner epidermal cells; guard cells with lip at outer aperture and ridge on inner wall. **Chlorenchyma:** cells in 2–3 layers; protective cells extending into inner layer of peg-cells, adjacent cells touching one another at few points. **Parenchyma sheath** 1–2-layered.

Rhizome T.S.

As for genus description, but **endodermoid sheath** poorly differentiated and **central ground tissue** restricted to strands of parenchyma between vb's.

Root T.S.

Cortex with 2 outer layers of fibres, middle region occupied by radial plates of thin-walled cells followed by 2 inner layers of narrow, thin-walled cells. **Endodermis:** cells with strongly thickened inner and anticlinal walls and thin outer walls. **Vascular system:** mxv's scattered. **Central ground tissue** sclerenchymatous.

***E. galpinii* N.E. Br.** Culm diameter 2·5 mm.

Culm surface

Epidermis: cell outlines irregular, mainly hexagonal; cells as long as, or up to 1½–2 times longer than wide, some wider than long; walls moderately thickened, straight. **Stomata** solitary or paired, or in groups of up to 4; if in pairs or groups, then common wall between adjacent subsidiary cells oblique (Fig. 30. B). **Tannin** in some cells.

Culm T.S. (Fig. 30. A)

Epidermis: cells in 2 layers; those of outer layer more or less square, those of inner layer elongated, *c.* 3–4 times higher than those of outer layer (Fig. 30. A). **Stomata** slightly sunken; guard cells with pronounced lip at outer aperture and ridge on inner wall. **Chlorenchyma:** protective cells thick-walled, extending almost to parenchyma sheath; apertures between adjacent cells wide and elongated. **Parenchyma sheath** mainly 2-layered, but 1- or 3-layered in places; walls slightly thickened.

***E. glauca* Mast.** Culm diameter 4·0 mm.

Culm surface

Epidermis: cells mainly 6-sided, mostly 1½–2 times longer than wide, but some as long as wide; walls thick, wavy at surface but straight at slightly lower focus. **Stomata** solitary; subsidiary cells with perpendicular end walls.

Culm T.S.

Epidermis: cells in 2 layers, rectangular, up to twice as high as wide, those of inner layer slightly higher than those of outer; anticlinal walls wavy. **Stomata** sunken; attached low down in tube formed by outer epidermal cells; guard cells with outer and inner lips. (Guard cells relatively narrow in median T.S., nearly twice as wide at each end.) **Chlorenchyma:** protective cells thick-walled, extending to *c.* ⅓ of way into inner layer of palisade cells; apertures between protective cells present towards their inner ends. **Parenchyma sheath** mainly 2-layered; cell walls densely covered with simple pits.

***E. grandis* (Spreng. ex Nees) Kunth.** Culm diameter 2 mm.

Leaf T.S.

Epidermis: (i) abaxial cells 2-layered and as in culm; (ii) adaxial cells 4-sided, *c.* 4 times wider than high.

Culm surface

Epidermis: cells 4-sided, mostly slightly wider than long, some up to 1½ times wider than long, walls moderately thickened. **Stomata** smaller than in other *Elegia* spp., visible at lower plane of focus than surface of epidermis and slightly overarched by adjacent epidermal cells; guard cell pair with lemon-shaped outline. **Tannin** in many epidermal cells. **Cuticular marks** granular, with longitudinal striations.

Culm T.S.

Cuticle thick, ridged. **Epidermis:** cells in 2 or, occasionally, 3 layers; cells of outer layer mostly *c.* 1½ times higher than wide, outer walls v. thick, dome-shaped, anticlinal walls with tapering thickening, not wavy; cells of inner layer similar in size to those of outer, rectangular, some divided periclinally into halves. **Stomata** sunken, attached to cells of inner epidermal layer; guard cells with pronounced, upturned lip at outer aperture and lip at inner aperture. **Chlorenchyma:** protective cells extending through outer chlorenchyma layer; adjacent cells touching one another at outer end only, each with concave long walls and slightly expanded ends. **Parenchyma sheath** 2–3-layered, cells rounded-hexagonal, mostly wider than high, those of outer layer the smallest. **Sclerenchyma:** sheath with slight ribs opposite peripheral vb's.

***E. intermedia* (Steud.) Pillans.** Culm diameter 1·2 mm.

Culm surface

Epidermis: cells rectangular to hexagonal, mainly 1½–2 times longer than wide, walls moderately thickened, wavy at surface but straight at lower focus; rounded gaps present between some end walls. **Stomata:** subsidiary cells with perpendicular or obtuse end walls, long anticlinal walls sometimes wavy. **Tannin:** none seen.

Culm T.S.

Epidermis: cells in 2 layers; those of outer layer slightly wider than high, those of inner layer as wide as and 1½–3 times higher than the outer cells, sometimes protruding between them and reaching surface. Anticlinal walls slightly wavy. **Stomata** superficial; guard cells with lips at outer and inner apertures. **Chlorenchyma:** protective cells extending to half way into inner chlorenchyma layer; apertures between protective cells present in inner half of tube. **Parenchyma sheath** mainly 2-layered, cell walls with abundant simple pits.

***E. juncea* L.** Culm diameter 2·0 mm.

Culm surface

Epidermis: cells mainly 6-sided, 1½–3 times longer than wide; anticlinal walls moderately thickened, not wavy. **Stomata:** subsidiary cells compressed centrally.

Culm T.S. (Fig. 30. E)

Epidermis: cells in 2 layers, those of outer layer slightly higher than wide, those of inner layer 1½–2 times higher than the outer, but of same width; anticlinal walls mainly straight (Fig. 30. E). **Stomata** superficial; guard cells with lip at outer aperture and ridge on inner wall. **Chlorenchyma:** protective cells extending to about half-way into inner chlorenchyma layer, having elongated apertures between them. **Parenchyma sheath** 2–3-layered.

***E. muirii* Pillans.** Culm diameter 1·8 mm.

Culm surface

Epidermis: cells mainly 6-sided and 1½–2 times longer than wide, walls thick, v. slightly wavy; rounded gaps present between the end walls of some cells. **Stomata:** subsidiary cells with perpendicular or acute end walls. **Tannin:** none seen.

Culm T.S.

Epidermis: cells in 2 similar layers, individual cells mainly twice as high as wide. **Stomata** sunken, attached to epidermal cells at junction between inner and outer layers; guard cells with beak-like outer lip at either end as seen at lower focus. **Chlorenchyma:** cells mainly in 2, occasionally in 3 layers. Protective cells extending part way into inner palisade layer. **Parenchyma sheath:** cells mainly in 2 layers, individual cells more or less hexagonal.

***E. neesii* Mast.** Culm diameter 2·0 mm.

Leaf: see genus description.

Culm surface

Epidermis: cells in 2 layers; outer layer discontinuous, cells occurring in irregular areas, cells of inner layer being exposed between them; cells of outer layer papillate, papillae appearing rounded at high focus, hexagonal outline of cells visible at lower focus; cells of inner layer hexagonal, mainly 2–3 times longer than wide, walls thick, slightly wavy. **Stomata** located in inner layer of epidermis; subsidiary cells crescentiform. **Cuticular marks:** prominent longitudinal striations.

Culm T.S. (Fig. 29. I)

Epidermis: cells in 2 layers; outer layer discontinuous, cells solitary or in groups of 2–3+, individual cells papillate, with broad basal region, anticlinal walls tapering to outer wall; inner layer continuous, cells 4–6-sided, 1½ times higher than wide or more or less square. **Papillae** frequently diverging on adjacent cells. **Stomata** attached to epidermal cells of inner layer; subsidiary

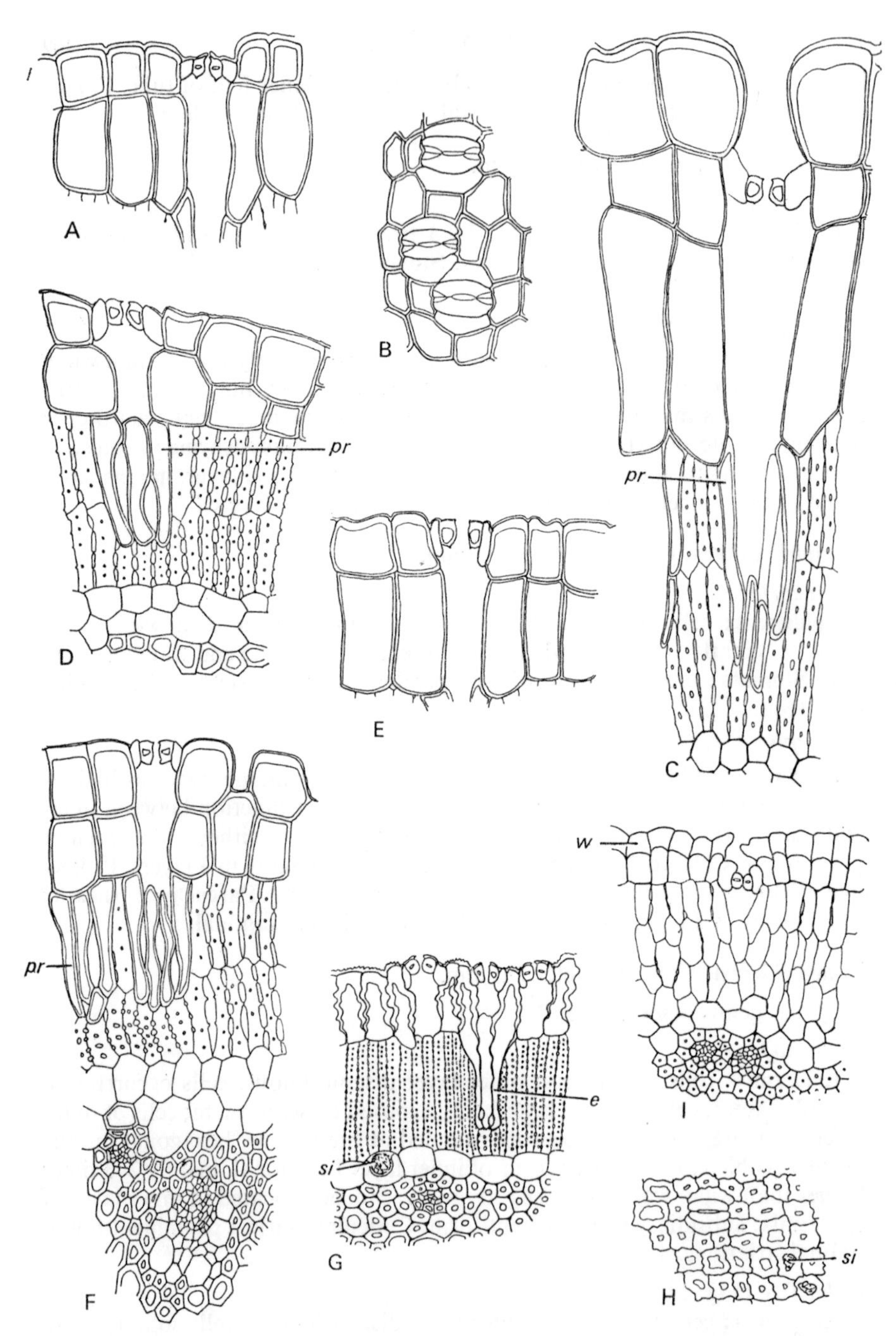

FIG. 30. Restionaceae. A, B, *Elegia galpinii*: A, detail of culm T.S.; B, culm epidermis. C, *E. fistulosa*, detail of culm T.S. D, *E. coleura*, detail of culm T.S. E, *E. juncea*, detail of culm epidermis T.S. F, *E. asperiflora*, detail of culm T.S. G, H, *Harperia lateriflora*: G, detail of culm T.S.; H, culm surface, note silica. I, *Hopkinsia anoectocolea*, detail of culm T.S. (All except I ×200, I ×185.)
e = elongated epidermal cell; pr = protective cell; si = silica-body; w = thick outer wall.

cells with ridge on inner wall; guard cells with outer and inner lips. **Chlorenchyma** mainly 2-, occasionally 3-layered; protective cells as high as cells of outer chlorenchyma layer, walls slightly thickened, apertures elongated. **Parenchyma sheath** mainly 2-, occasionally 1-layered.

***E. obtusiflora* Mast.** Culm diameter 0·8–2·0 mm.

Two anatomical types, A and B, were recognizable in the material examined.

Culm surface (Fig. 29. G, H)

Papillae: outer epidermal cells slightly papillate (in group A only). **Epidermis.** In group A cells normally pentagonal-hexagonal, slightly to 2–3 times longer than wide; walls moderately thickened, not wavy. In group B cells mostly rectangular, 2–5 times longer than wide, some cells more or less square; walls moderately thickened, wavy. **Stomata:** in both groups, subsidiary cells low dome-shaped, end walls mainly obtuse or perpendicular, sometimes acute; guard cell pair frequently with lemon-shaped outline. **Tannin** frequent in epidermal cells.

Culm T.S. (Fig. 29. E, F)

Papillae overarching stomata, and present on other epidermal cells (in group A only). **Epidermis:** cells in 2 layers; those of outer layer more or less square or up to 2 times higher than wide (becoming wider than high in basal material); outer walls convex (group A) or frequently concave (group B); cells of inner layer either of similar dimensions, or slightly taller or shorter than those of outer layer. **Stomata.** In group A sunken, attached to epidermal cells at junction between the 2 layers; most subsidiary cells with ridge on outer wall; guard cells with outer lip and poorly or well-developed ridge on inner wall. In group B superficial; subsidiary cells without outer ridge; guard cells with lip at outer aperture and slight ridge on inner wall. **Chlorenchyma:** both groups, peg-cells in 2 similar layers (1 layer in basal material); some protective cells extending almost to parenchyma sheath, others only as high as cells of outer palisade layer; apertures elongated. **Parenchyma sheath** normally 1-layered with occasional 2-layered areas; mainly 2-layered in one specimen.

Rhizome and *root*: see genus description.

***E. racemosa* (Poir.) Pers.** Culm diameter 2·5 by 2 mm.

Culm surface

Epidermis: cells 4-sided, 1½–2 times longer than wide, except at either end of stomata, where *c.* ½ as long as wide; walls moderately thickened, wavy. **Stomata:** subsidiary cells with perpendicular or acute end walls; guard cell pair occasionally with lemon-shaped outline. **Tannin** in some cells.

Culm T.S.

Epidermis: cells in 2 layers, those of outer layer slightly to 1½ times wider than high, those of inner layer mostly twice as high as wide (*c.* 3 times higher than those of outer layer); some outer walls concave, most curving inwards at cell margins. **Stomata** superficial; guard cells with lips at outer and inner apertures. **Chlorenchyma:** protective cells extending half way into inner chlorenchyma layer, joined to one another at their inner and outer ends only. **Parenchyma sheath** mainly 2-, occasionally 3-layered.

***E. spathacea* Mast.** Culm diameter 1·5 mm.

Culm surface

Epidermis: cells more or less square, or rectangular and up to twice as long as wide, walls wavy, moderately thickened. **Stomata:** subsidiary cells with obtuse or occasionally perpendicular end walls. **Tannin** in some cells.

Culm T.S.

Epidermis: cells in 2 layers, those of outer layer nearly square or 1½ times wider than high, those of inner layer mostly twice as high as wide and 2–3 times higher than those of outer layer; walls moderately thickened, outer walls curving inwards slightly at cell margins. **Stomata** superficial; subsidiary cells with ridge on outer wall; guard cells with outer and inner lips. **Chlorenchyma:** cells of inner layer slightly longer than those of outer layer; protective cells extending to over half-way into inner chlorenchyma layer; apertures elongated. **Parenchyma sheath** 1–2-layered.

***E. squamosa* Mast.** Culm diameter 1 mm.

Culm surface

Epidermis: cells 4-sided, of 2 main types; (i) short, more or less square, at either end of stomata; (ii) mostly twice as long as wide, between files of stomata; walls moderately thickened, wavy. **Stomata:** subsidiary cells variable.

Culm T.S.

Epidermis: cells in 2 layers, cells of individual layers of irregular height (anticlinal walls inclined with respect to culm surface as seen in L.S.). **Stomata** superficial; guard cells thick-walled, with lip at outer and inner apertures. **Chlorenchyma:** protective cells extending almost to parenchyma sheath, apertures between protective cells slit-like, elongated. **Parenchyma sheath** mainly 1-, occasionally 2-layered.

***E. stipularis* Mast.** Culm diameter 1·1, 1·5 mm.

Culm surface

Epidermis: cells rectangular, 2–3 times longer than wide, walls moderately thickened, not wavy. **Stomata:** subsidiary cells with perpendicular end walls. **Tannin:** none seen.

Culm T.S.

Epidermis: cells in 2 layers, more or less square; outer walls frequently convex, curving inwards at cell margins, other walls straight. **Stomata** superficial; guard cells with lip at outer aperture. **Chlorenchyma:** protective cells extending to half way into inner palisade layer, touching one another at mouth and base of tube only. **Parenchyma sheath** 1–2-layered.

Leaf, *rhizome*, and *root*: see genus description.

***E. thyrsoidea* (Mast.) Pillans.** Culm diameter 1·0 mm.

Culm surface

Epidermis: cells more or less rectangular, 1½–3 times longer than wide, cells at ends of stomata shorter; walls moderately thickened, slightly wavy. **Stomata:** subsidiary cells narrow, with acute or obtuse end walls. **Tannin:** occluding lumina of most epidermal cells.

Culm T.S.

Epidermis: cells in 2 layers, more or less square; outer walls curving inwards at margins, some concave. **Stomata** superficial; guard cells with outer and inner lips. **Chlorenchyma** 2-layered, cell walls v. thin, protective cells with slightly to moderately thickened walls, extending almost to parenchyma sheath, joined to one another at inner and outer ends. **Parenchyma sheath** mainly 1-layered. **Tannin** plentiful in cells of outer epidermal layer, present in a few cells of central ground tissue.

E. vaginulata **Mast.** Culm diameter 1 mm.

Culm surface

Epidermis: cells 4-sided, mostly 1½–2 times longer than wide, those at ends of stomata more or less square; walls moderately thickened, slightly wavy. **Stomata:** subsidiary cells with acute end walls; guard cell pair with lemon-shaped outline.

Culm T.S. (Pl. V. B)

Epidermis: cells in 2 layers, those of outer layer slightly to 1½ times higher than wide, outer walls often slightly concave, curving inwards at cell margins; most cells of inner layer similar in size to those of outer layer, but some higher. **Stomata** slightly sunken, but attached to cells of outer epidermal layer; guard cells with lip at outer and inner apertures. **Chlorenchyma:** protective cells extending from ¾ way up sides of substomatal tube formed by inner layer of epidermal cells to half-way into inner palisade layer, inner ends of cells hastate. **Parenchyma sheath** mainly 1-layered, 2-layered at sides of some peripheral vb's, cells wide. **Tannin:** none seen.

E. verreauxii **Mast.** Culm diameter 0·75 mm.

Culm surface

Epidermis: cells 4-sided, more or less square or 1½–2 times longer than wide; those at ends of stomata shorter, more or less circular; walls moderately thickened, v. wavy. **Stomata:** subsidiary cells variable, often with acute end walls.

Culm T.S.

Epidermis: cells in 2, and occasionally 3 or 4 layers; those of outer layer mostly as high as or slightly higher than wide, inner cells as high as or up to 1½ times higher than those of outer layer; outer walls frequently concave, curving inwards at cell margins; walls of inner cells slightly thickened, not wavy. **Stomata** superficial; guard cells with outer lip and ridge on inner wall. **Chlorenchyma:** protective cells extending from inner walls of epidermal cells almost to parenchyma sheath, expanded towards inner end before tapering rapidly to point; apertures between protective cells v. wide, extending almost whole height of cells. **Parenchyma sheath** mainly 1-layered.

E. verticillaris **(L.f.) Kunth.** Culm diameter 5·0 mm.

Leaf: see genus description.

Culm surface

Epidermis: cells 4- and 6-sided, more or less square, or 1½–2½ times as long as wide, walls moderately thickened, not wavy; cells of various lengths intermixed in longitudinal files, but short cells often present at ends of stomata. **Stomata:** subsidiary cells with obtuse or rounded end walls.

Culm T.S.

Epidermis: cells in 1–2 layers, those of outer layer more or less square or slightly papillate, those of inner layer of similar dimensions or nearly twice as high; outer walls curving inwards at cell margins. **Stomata** superficial; subsidiary cells with ridge on outer wall; guard cells with outer and inner lips. **Chlorenchyma:** protective cells extending from inner epidermal cell walls to half-way into inner chlorenchyma layer, apertures elongated. **Parenchyma sheath:** cells in 3 layers opposite peripheral vb's, 4-layered between them. **Tannin:** none seen.

Taxonomic Notes

Elegia has been regarded as a valid genus since it was first described, except for a short period when it was included with *Restio* by Nees (1830).

Lamprocaulos Mast. was included in *Elegia* by Bentham and Hooker (1883) and by Pillans (1928), but Gilg (1890) and Gilg-Benedict (1930) both recognized it as a separate genus. Gilg-Benedict included in *Lamprocaulos* the species *L. neesii* Mast. and *L. grandis* Mast. and named a new species *L. schlechteri* Gilg-Benedict.

The relationship between *Chondropetalum* Rottb. and *Elegia* has been discussed on p. 157.

The present author has examined material of *Elegia* (*Lamprocaulos*) *neesii* Mast. and *Elegia* (*Lamprocaulos*) *grandis* (Spreng. ex Nees) Kunth in an attempt to see if the separation of these species as *Lamprocaulos* has any support from anatomical evidence. Both species have sunken stomata, but so have *E. muirii* Pillans and *E. glauca* Mast. *E. neesii* has pronounced papillae and a discontinuous outer epidermal layer. The Kew material of *E. grandis* does not have papillae, but that examined by Gilg-Benedict may have, according to her genus description. Papillae of a slightly different nature from those in *E. neesii* are to be found in some specimens of *E. obtusiflora*. There is no clear anatomical character which, in linking *E. neesii* and *E. grandis*, separates them from the rest of the *Elegia* species. At present there is little anatomical justification for recognizing *Lamprocaulos* as a separate genus. When the relationship of *Elegia* to *Chondropetalum* is next considered it may be necessary to make new combinations; at such a time the species of *Elegia* with sunken stomata and/or papillae should be examined carefully to see if they do constitute a natural group. If they do, it may be reasonable to give this group generic status and call it *Lamprocaulos*.

Species of *Restio* figure prominently in the lists of synonyms for some *Elegia* species given by Pillans. No species currently recognized as belonging to the genus *Restio* that has been examined during the present investigation has a double epidermis. This supports the view that *Restio* and *Elegia* are probably distinct genera, and that *Elegia* species with synonyms in *Restio* can be regarded as good *Elegia* species. Further taxonomic adjustments may be needed if the relationships between *Elegia* and *Chondropetalum* ever come to be resolved.

Individual *Elegia* species frequently have distinctive epidermal characters. Examples include the proportionate heights of the cells of the epidermal layers, the number of epidermal layers, the position of the stomata relative to the culm surface, and the presence or absence of papillae.

All species of *Elegia* have protective cells; these show a range of variation, and the type of protective cell present may prove to be a character of diagnostic value when more samples of individual species are examined.

E. obtusiflora Mast. as at present represented in the Kew Herbarium appears to consist of specimens with 2 distinct types of anatomical structure and requires further investigation.

HARPERIA W. V. Fitzgerald

No. of spp. 1; examined 1.

Genus and Species Description

Leaf not seen.

Culm surface (Fig. 30. H)

Hairs absent. **Epidermis:** cells as long as wide to 1½ times longer than wide, walls thick, wavy. **Stomata:** subsidiary cells variable, as long as guard cells. **Silica:** granular material and irregularly shaped bodies present in some epidermal cells. **Tannin:** none seen. **Cuticular marks** granular, with fine, longitudinal striations.

Culm T.S. (Fig. 30. G)

Outline circular, with irregular, low mounds. **Cuticle** thick. **Epidermis.** Cells arranged in 1 layer; mainly 2½–3 or 4 times higher than wide, but those next to stomata v. elongated, with a tapering projection extending towards and often reaching parenchyma sheath. Projections from adjacent cells surrounding a stoma united along their anticlinal walls and forming a substomatal tube, with rounded or oval apertures between the side walls, near the inner end of the tube, opening to adjacent chlorenchyma cells. Outer walls of epidermal cells v. thick; anticlinal walls wavy, thick except for inner $\frac{1}{6}$, where slightly thickened; inner walls thin. Inwardly directed projections from cells surrounding stomata with slightly thickened walls lining substomatal cavity but walls next to chlorenchyma cells thin, with numerous small pegs. All epidermal cells inclined at an angle of *c.* 50° to culm surface as seen in L.S.

Stomata superficial; guard cells with pronounced cuticular lip at outer aperture. **Chlorenchyma:** 1 layer of peg-cells; individual cells about 10–12 times longer than wide, with numerous, small pegs. **Parenchyma sheath:** cells 1½–2 times wider than high, with slightly thickened walls. **Sclerenchyma:** sheath 6–8-layered, fibres medium-sized to wide, hexagonal, with slightly rounded corners, walls thick; intercellular spaces filled with extracellular substances. **Vascular bundles.** (i) Peripheral vb's: tracheids narrow, arranged in single-layered arc partially enclosing small, oval phloem pole. (ii) Medullary vb's with 1 narrow, rounded-angular mxv. on either flank; lateral wall pitting of vessels scalariform; perforation plates oblique, fenestrate-scalariform. Vessels separated from one another by 3–4 rows of v. narrow cells.

Phloem poles ensheathed by 1 layer of v. narrow cells with slightly to moderately thickened walls. Px present in all bundles. All medullary bundles free from culm scl. sheath, some rotated from normal radial orientation.

Bundle sheaths: fibres narrow, thick-walled, with rounded corners; arranged in 2 layers at phloem pole, 1-layered on flanks and at xylem pole. **Central ground tissue** parenchymatous, most cells with slightly to moderately thickened walls; small scattered areas of cells with thin walls also present; central cavity absent. **Silica** present as: (i) spheroidal-nodular bodies in stegmata of scl. sheath; (ii) granular material or irregular bodies in some epidermal cells and some cells of central ground tissue. **Tannin:** none seen.

Rhizome and *root* not seen.

MATERIAL EXAMINED

Harperia lateriflora W. V. Fitzgerald: Fitzgerald, W., N.S.W. 48392, An. 1903; Cunderdin, W. Australia (K). Culm diameter 1·0 mm.

TAXONOMIC NOTES

The genus *Harperia* was described in 1904, after the publication of Gilg's work, but in time for an anatomical investigation to be carried out by Gilg-Benedict. Fitzgerald stated that it is morphologically similar to *Loxocarya* and *Lepidobolus* (see pp. 197, 227) and Gilg-Benedict commented that it is anatomically similar to *Loxocarya.* During the present investigation the anatomical similarity to *Loxocarya* (excl. *L. pubescens* Benth.) and *Lepidobolus* has been confirmed. No hairs were found on *Harperia*; they are present in the other genera.

The most striking anatomical feature of *Harperia* is the greatly elongated epidermal cells, which extend inwards almost to the parenchyma sheath, forming a substomatal tube which probably has the same physiological function as similar tubes formed by protective cells in other genera.

Similarity has also been noted between *Harperia, Coleocarya* S. T. Blake, and *Restio confertospicatus* Steud. (see p. 158).

HOPKINSIA W. V. Fitzgerald

No. of spp. 1; examined 1.

GENUS AND SPECIES DESCRIPTION

Leaf surface not seen

Leaf T.S. (sheathing base)

Epidermis. (i) Abaxial cells with thick outer walls, anticlinal and inner walls thin; cells mostly twice as high as wide. (ii) Adaxial cells with all walls thin; cells as high as wide to $1\frac{1}{2}$ times higher than wide. **Stomata** as in culm; present on abaxial surface only. **Hypodermis** absent. **Chlorenchyma** composed of 2–3 layers of more or less isodiametric cells situated below abaxial epidermis. **Vascular bundles** with few cells in the phloem and xylem poles; larger bundles with 1 medium-sized angular tracheid on either flank, smaller bundles with narrow tracheids only. **Bundle sheaths** sclerenchymatous, 1–2-layered. **Sclerenchyma** continuous with bundle sheaths, in 4–5 layers, occupying up to

½ thickness of leaf, situated between adaxial epidermis and chlorenchyma. **Ground tissue:** 1 layer of parenchyma present between scl. and adaxial epidermis. **Air-cavities:** none present. **Silica:** none seen. **Tannin** in some epidermal cells.

Culm surface

Epidermis: cells rectangular, mostly between 2–4 times longer than wide; walls moderately thickened, wavy. **Stomata:** subsidiary cells with acute or perpendicular end walls; guard cell pair with oval outline. **Silica:** none seen. **Tannin** in some epidermal cells. **Cuticular marks** granular.

Culm T.S. (Fig. 30. I)

Epidermis: cells oblong, 2–4 times higher than wide; outer walls convex, heavily thickened, occluding up to ½ lumen. Anticlinal and inner walls straight and thin. Cells next to stomata with papillae overarching suprastomatal cavity. **Stomata** sunken to level of inner wall of epidermal cells and overarched by epidermal cells; guard cells without lips or ridges. **Chlorenchyma:** palisade cells loosely packed, in 3 or occasionally 4 layers; pegs infrequent, short. Cells 1½–4 times higher than wide.

Parenchyma sheath: 1 (occasionally 2) layer(s) of cells, walls slightly thickened. **Sclerenchyma:** sheath strongly developed, *c.* 10-layered; outline ribbed; ribs corresponding to positions of peripheral vb's. Two distinct zones present: (i) outer, fibres narrow, walls v. thick; (ii) inner, fibres wider, walls thick, lumina wider. **Vascular bundles.** (i) Peripheral vb's each with small pole of phloem and xylem. (ii) Medullary vb's with 1 narrow or medium-sized, rounded-angular mxv on either flank; flanking vessels widely spaced. Lateral wall pitting scalariform; perforation plates simple, slightly oblique. Phloem surrounded by scl. sheath. Bundles graded in size: small to medium-sized outer bundles embedded in scl. sheath, and larger, free, central bundles scattered. **Bundle sheaths:** fibres 1-layered on flanks, 2-layered at xylem and 3-layered at phloem poles. **Central ground tissue** parenchymatous; walls moderately thickened except for lacunae of thin-walled cells; no central cavity. **Silica:** none seen. **Tannin** in some epidermal and ground tissue cells.

Rhizome T.S.

Epidermis not seen. **Cortex:** (i) outer part consisting of 6–8 layers of parenchymatous cells with slightly thickened walls; (ii) middle part 8–9-layered consisting of hexagonal parenchymatous and sclerenchymatous cells intermixed; (iii) inner part consisting of 3–4 layers composed of thin-walled rounded cells. **Endodermoid sheath** not developed. **Vascular bundles** scattered; mxv's medium-sized, angular, arranged in horseshoe around or completely encircling phloem. **Bundle sheaths** mostly 4-layered, fibres joining those of ground tissue and indistinguishable from them in places. **Central ground tissue** composed of scl. just described and lacunae filled with parenchymatous, thin-walled cells. **Tannin** and **silica:** none seen.

Root T.S.

Outer and cortical layers not seen. **Endodermis** 1-layered, cells mostly 1½ times higher than wide; outer and anticlinal walls straight, inner walls with 2 faces; all walls heavily thickened. **Pericycle** 7–8-layered, cells mostly 1½–2 times higher than wide, those of outer layers with moderately thick to thick walls, those of inner layers with slightly to moderately thickened walls. **Vascular system:** mxv's wide, mostly solitary, some pairs, radially oval, arranged in 1 ring. **Central ground tissue:** cells of outer 7–8 layers narrow, hexagonal with slightly to moderately thickened walls; central cells narrow, rounded, thin-walled. **Tannin** and **silica:** none seen.

MATERIAL EXAMINED

Hopkinsia anoectocolea (F. v. Muell.) D. F. Cutler, under the names of: *H. scabrida* W. V. Fitzgerald: (i) ♂ and ♀ Fitzgerald, W. V.; Cunderdin, Nov. 1903, '*H. calovaginata* Gilg' det. E. Gilg; dupl. type, Berlin Bot. Mus. (ii) Fitzgerald, W. V.; Cunderdin, Nov. 1903, State Herb. W. Australia.

(*b*) *H. calovaginata* (Gilg) Pilger: (i) Gardner, C. A.; 86–87 mile peg, E. of Meckering, W. Australia, 14 Dec. 1945 (2 sheets), State Herb. W. Australia. (ii) Gardner, C. A., Tammin, Sept. 1939, State Herb. W. Australia. (iii) Gardner, C. A. 6449, 8 Sept. 1942. (iv) Alpin, T. E. H. 617; ½ mile N.W. Wyola Siding, W. Australia, 3 Feb. 1960. State Herb. W. Australia. (v) Cheadle, V. I. CA457; 3 miles E. of Meckering, W. Australia, 1960. Herb. V. I. Cheadle; Univ. of California.

(*c*) *Lepyrodia anaectocolea* F. v. Muell. Drummond, type; Australia (K) (*anaectocolea* is the original, incorrect spelling).

Culm diameter 2 mm.

TAXONOMIC NOTES

The first name given to the plant now known as *Hopkinsia anoectocolea* (F. v. Muell.) D. F. Cutler is *Lepyrodia anaectocolea* F. v. Muell. This has been shown (Cutler 1967) to be synonymous with *Hopkinsia calovaginata* (Gilg) Gilg-Benedict (=*Anarthria calovaginata* Gilg and *Hopkinsia scabrida* W. V. Fitzgerald).

Gilg-Benedict (1930) considered that there is anatomical similarity between *Hopkinsia* and *Lyginia*, but this was not found to be so in the material examined for the present work. Indeed, the anatomical similarity between *Hopkinsia* and *Lepyrodia anaectocolea* led to the discovery that they are synonymous.

HYPODISCUS Nees

No. of spp. 12; examined 9.

GENUS DESCRIPTION

Leaf surface

Abaxial, seen in *H. albo-aristatus*, *H. aristatus*, and *H. binatus*.

Hairs absent. **Epidermis:** cells 4-sided, as long as or up to 3 or more times longer than wide, or hexagonal and up to twice as long as wide; walls moderately thickened, wavy in quadrangular cells, straight in hexagonal cells. **Stomata:** subsidiary cells with acute end walls in all spp.; guard cell types as in culm. **Tannin:** none seen. **Cuticular marks** granular.

Leaf T.S.

Sheathing base, seen in *H. albo-aristatus*, *H. aristatus*, and *H. neesii*.

Hairs absent. **Epidermis:** (i) adaxial cells rectangular, mostly twice as wide as high, outer walls slightly thickened, other walls thin, or all walls thick (*H. neesii*); (ii) abaxial cells 4-sided, as high as wide, with thick, convex outer walls (*H. neesii*) or *c.* 1½–2 times higher than wide, with thick, concave outer walls (*H. albo-aristatus*), or about as high as wide, outer walls concave, moderately thickened (*H. aristatus*). Cells of both surfaces becoming progressively smaller towards the margins. **Stomata** present in abaxial surface only; superficial in *H. aristatus*, sunken in *H. albo-aristatus*, not seen in *H. neesii*; guard cells with lips at outer and inner apertures. **Chlorenchyma** composed of lobed or peg-cells, as high as wide to 2–3 times higher than wide, present in areas between vb sheaths, ground tissue and abaxial epidermis; protective cells similar to bone-shaped type in culm.

Vascular bundles all small; smaller vb's with few narrow tracheids and phloem cells, alternating with larger vb's, characterized by 1 medium-sized tracheid on either flank of arc of narrow or narrow to medium-sized tracheids, and by a larger phloem pole. **Bundle sheaths:** O.S. restricted to single-layered cap of parenchyma cells to abaxial side of bundle; I.S. sclerenchymatous; fibres in 1–2 layers at phloem pole, 3–4 layers at xylem pole and flanks, or joining laterally with those on flanks of adjacent bundles. **Sclerenchyma** represented by bundle sheaths and their lateral bridges; bridging scl. either abutting directly on adaxial epidermis, or separated from it by 1–2 layers of parenchymatous ground tissue. **Ground tissue** parenchymatous, cells with thin or slightly thickened walls. **Air-cavities** absent. **Silica:** individual, spheroidal-nodular bodies present in some cells of outer abaxial layer of scl. bundle sheath. **Tannin** in some abaxial epidermal cells.

Culm surface

Hairs and **papillae** absent. **Epidermis:** cells 4-, 5-, or 6-sided, shorter than or up to 7 times longer than wide, walls moderately thickened or v. thick, straight or wavy. **Stomata:** subsidiary cells variable, guard cells with rounded ends and parallel end walls, or outline of pair lemon-shaped. **Silica** absent. **Tannin** in epidermal cells of some spp. **Cuticular marks** granular, occasionally with superimposed longitudinal or transverse striations.

Culm T.S.

Outline circular or oval. **Cuticle** moderately thick to thick. **Epidermis:** cells in 1 layer; more or less square or up to 2(–3) times higher than wide, outer walls thick or v. thick, straight or concave; anticlinal walls slightly or moderately thickened, straight or wavy; inner walls slightly to moderately thickened. **Stomata** superficial or sunken; subsidiary cells variable in shape; guard cells sometimes with lips at outer and/or inner apertures, sometimes with ridge on inner wall. **Chlorenchyma** composed of peg-cells either (i) in 2–(3) uninterrupted layers or (ii) in (1)–3 layers, interrupted by ribs from scl. sheath and their associated pillar cells. Peg-cells normally with few pegs, cells arranged in chequerboard-like manner as seen in T.L.S., each joining on to its neighbours at its 4 corners only, individual cells alternating with air-spaces. Protective cells present in all spp., normally surrounding substomatal cavity to depth of outer chlorenchyma layer only. Protective cells represented by either (i) little modified chlorenchyma cells, with slightly to moderately thickened walls lining the substomatal tube, and with rounded or oval apertures near to or at outer end of tube, particularly as seen in L.S., or (ii) palisade cells with expanded ends and parallel-sided middle portion, i.e. bone-shaped, or with ends extending laterally in an I-shape or T-shape. Cells of type (ii) in contact with neighbours at outer and inner ends only, apertures between them wide and v. elongated (Fig. 31. C).

Parenchyma sheath 1-layered in all spp., either completely encircling culm, with all cells similar, or interrupted by scl. ribs, some with pillar cells as well as normal sheath cells; the number of pillar cells opposite to a rib normally equalling number of cells making up width of girder. **Sclerenchyma.** Sheath of 2 types: (i) outline circular, 3–5-layered; (ii) outline circular, with short or long rectangular ribs radiating towards epidermis, these either partially or completely dividing chlorenchyma into sectors. Fibres narrow to medium-sized, thick- or v. thick-walled; those surrounding peripheral vb's in spp. with ribs medium-sized, moderately thick-walled.

Vascular bundles. (i) Peripheral vb's: tracheids (*a*) all narrow, medium-sized or wide, arranged in 1–2-layered arc, or (*b*) with 1 wider tracheid on either flank of arc; arc partially enclosing rounded or oval phloem pole; peripheral vb's alternating with ribs. (ii) Medullary vb's: outer bundles smaller than inner; 1 narrow to medium-sized, angular, mxv on either flank, these frequently touching, or separated by 1–4 rows of narrow cells. Lateral wall pitting of mxv's scalariform, perforation plates oblique and scalariform or transverse to nearly transverse, simple. Phloem close to xylem, frequently abutting directly on it and overarching flanking mxv's. Narrow sclereids or sclerified parenchyma cells present in phloem of some spp. (*H. albo-aristatus, binatus, neesii, willdenowia*). Px poles present in all inner, but absent from most outer vb's. Bundles all embedded in, some embedded in, or all free from culm scl. sheath according to sp. **Bundle sheaths:** fibres at poles narrow with moderately thick or thick walls, in 1–2–(4) layers, those on flanks wider, with slightly to moderately thickened walls, often difficult to distinguish from ground tissue. **Central ground tissue** parenchymatous; outer cells, surrounding vb's, with slightly to moderately thickened walls; central cells, in bundle-free area, thin-walled, frequently breaking down to form cavity. **Silica:** spheroidal-nodular bodies present in certain cells of outer layer of scl. sheath in some spp. **Tannin** in some epidermal cells of certain spp.

Rhizome T.S.

Seen in *H. albo-aristatus* and *H. aristatus.*

Epidermis: cells narrow, square or slightly higher than wide, thick-walled. **Cortex:** (i) outer cells mostly 2–3 times wider than, and twice as high as, those of epidermis, walls of all cells v. thick, or those of occasional cells slightly thickened (*H. aristatus*); cells present in 10–12 layers; (ii) inner part composed of 4–6 layers of thin-walled cells; cells slightly smaller than those of outer cortex. **Endodermoid sheath** 1-layered; cells with U-shaped wall thickenings. **Vascular bundles:** most amphivasal, mxv's wide, angular, enclosing rounded phloem pole; bundles scattered. **Bundle sheaths** difficult to distinguish from ground tissue; cells in 1–2 layers, flattened, with moderately thickened walls. **Central ground tissue:** matrix composed of cells with moderately thickened walls, interspersed with strands of thin-walled cells. **Silica** and **Tannin** not seen.

Root T.S.

Seen in *H. albo-aristatus*, *H. aristatus*, and *H. willdenowia*.

Root-hairs arising from cells similar to remainder in epidermis, hair shaft *c.* $\frac{1}{3}$–$\frac{1}{2}$ width of base. **Epidermis:** cells 2–3 times wider than high, thin-walled. **Cortex:** (i) outer part composed of 2 layers of thin-walled cells and 1 layer of thick-walled cells, all similar in size to those of epidermis; (ii) inner part, composed of radiating plates of cells separated from one another by air-spaces. Plates 3–many-celled in radial direction. **Endodermis:** cells in 1 layer, as high as wide or twice as high as wide, outer walls thin or moderately thickened, other walls v. thick; lumina frequently entirely occluded in *H. albo-aristatus*. **Pericycle:** cells parenchymatous, 5–6-sided, 2–3 times higher than wide, or about as high as wide, walls slightly to moderately thickened; cells in *c.* 4 layers in *H. albo-aristatus*, 6–9 in *H. aristatus*, and 5–6 in *H. willdenowia*. **Vascular system:** mxv's wide, oval, arranged in 1 or 2 rings immediately to inside of ring formed by phloem strands. **Central ground tissue** composed of narrow, 5–6-sided cells with moderately thickened walls. **Tannin:** none seen.

Material Examined

Hypodiscus albo-aristatus (Nees) Mast.: Cheadle, V. I. CA770; Cape (S).
H. alternans Pillans: Parker, R. N. 3520 (K).
H. argenteus (Thunb.) Mast.: Parker, R. N. 4770 (K).
H. aristatus (Thunb.) Nees: (i) Cheadle, V. I. CA819; Cape (S); (ii) CA824; Cape (S).
H. binatus (Steud.) Mast.: Burchell (M).
H. neesii Mast.: Esterhuysen 3750 ♀, An. 1940; 4700 ft, S. slopes Bonteberg Mts. (K).
H. striatus (Kunth) Mast.: Esterhuysen, An. 1941 (K).
H. synchroolepis (Steud.) Mast.: '816 *Restio*, = 9608 Drège!' (K).
H. willdenowia (Nees) Mast.; Parker, R. N., 4089 ♀; sandy soil (K).

Species Descriptions

***H. albo-aristatus* (Nees) Mast.** Culm diameter 1 mm.

Culm surface (Fig. 31. E)

Epidermis: cells 4-sided, mainly 2–3 times longer than wide, but some, particularly those next to stomata, as long as or shorter than wide; walls wavy. **Stomata:** subsidiary cells with wavy walls; guard cell pair with lemon-shaped outline. **Tannin** in some epidermal cells. **Cuticular marks:** single longitudinal striation near to each longitudinal wall.

Culm T.S. (Fig. 31. D)

Outline circular. **Epidermis:** cells mostly about as high as wide; outer walls slightly concave; anticlinal walls sometimes wavy. **Stomata** superficial; guard cells with slight ridge on inner wall; walls to pore convex. **Chlorenchyma:** cells 4–5 times higher than wide; protective cells of generic type (ii) see p. 180. **Parenchyma sheath** interrupted between peripheral vb's by broad scl. ribs

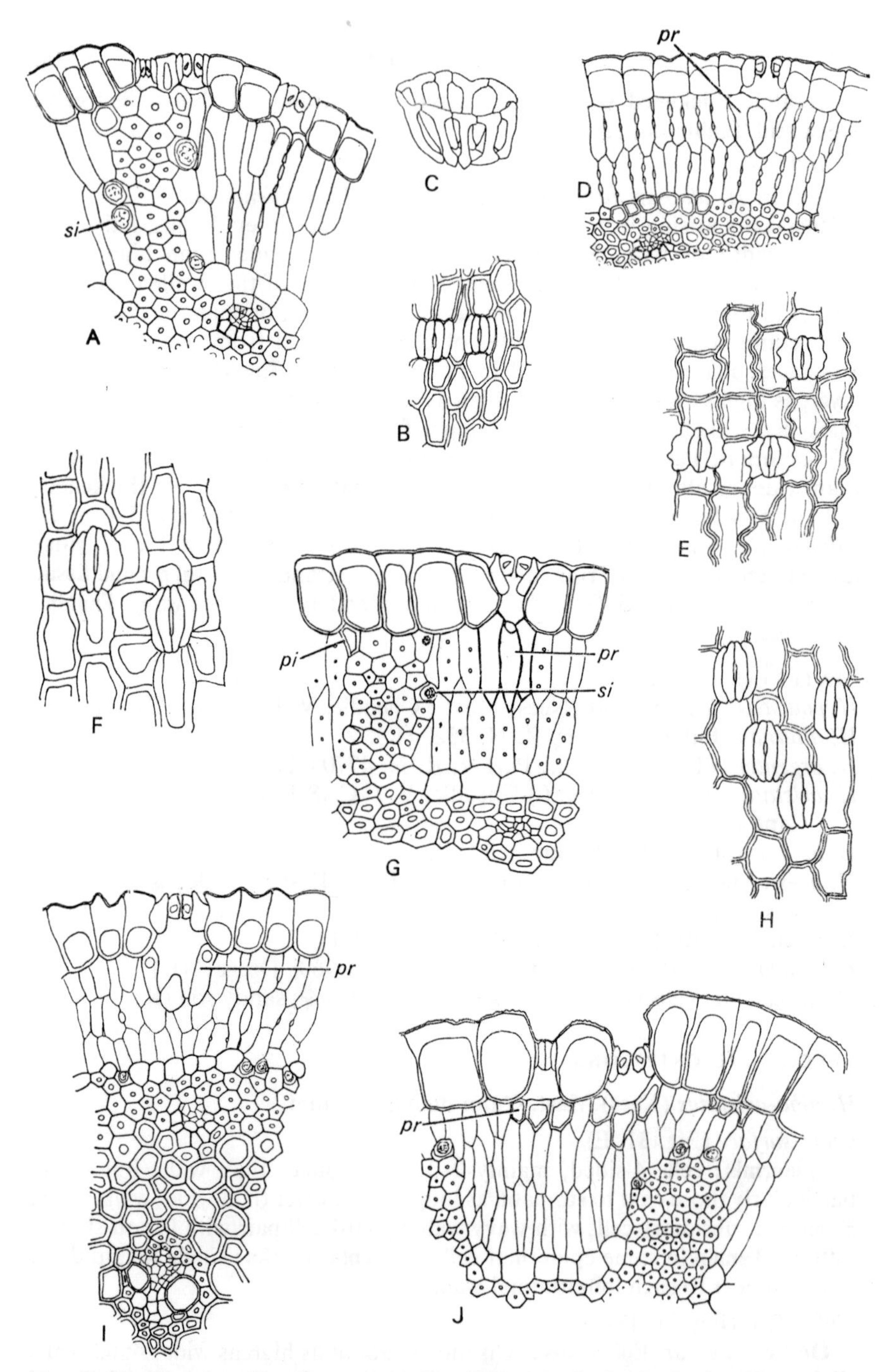

FIG. 31. Restionaceae. *Hypodiscus*. A, B, *H. willdenowia*: A, detail of culm T.S.; B. culm surface. C, *Hypodiscus*, diagram of arrangement of protective cells as seen in 3 dimensions. D, E, *H. albo-aristatus*: D, detail of culm T.S.; E, culm surface. F, *H. binatus*, culm surface. G, H, *H. striatus*: G, detail of culm T.S.; H, culm surface. I, *H. alternans*, detail of culm T.S. J, *H. neesii*, detail of culm T.S. (A–H × 200, I, J × 185.)
pi = pillar cell; pr = protective cell; si = silica-body.

from scl. sheath. **Sclerenchyma:** sheath outline rounded, with low ribs 1 fibre high and up to 7–9 fibres wide. **Silica:** none seen. **Tannin** in some epidermal cells.

Leaf, *rhizome*, and *root*: see genus description.

***H. alternans* Pillans.** Culm diameter 1 by 0·75 mm.

Culm surface

Epidermis: cells mostly 4-sided, frequently 1½–2½ times longer than wide, but some shorter than or as long as wide; walls moderately thickened, axial walls wavy. **Stomata:** subsidiary cells with wavy walls. **Tannin:** none seen. **Cuticular marks:** with short lines running transversely across walls.

Culm T.S. (Fig. 31. I)

Outline oval. **Epidermis:** cells 4-sided, mostly twice as high as wide; outer walls v. thick, normally concave at outer surface, lumina reduced to about half height of cells. **Stomata** superficial; guard cells with slight lip at outer and inner apertures. **Chlorenchyma:** 3 layers of similar cells; individual cells 2–4 times higher than wide. Protective cells of generic type (ii), see p. 180. **Parenchyma sheath** interrupted in places by 1–4 fibres from scl. sheath. **Sclerenchyma:** sheath 5–7-layered, outline oval, with low ribs between peripheral vb's. **Silica:** bodies present in stegmata of outer scl. layer, particularly in cells of ribs. **Tannin:** none seen.

***H. argenteus* (Thunb.) Mast.** Culm diameter 1·5 mm.

Culm surface

Epidermis: cells of unequal sizes; 4–6-sided, ranging from wider than long to 1½ times longer than wide; walls moderately thickened to thick, slightly wavy. **Stomata:** subsidiary cells variable, normally with obtuse end walls, slightly shorter than guard cells; guard cell end walls slightly pointed. **Tannin** in some epidermal cells.

Culm T.S. (Fig. 32. A)

Outline circular. **Epidermis:** cells 2–2½ times higher than wide; outer walls v. thick, outer surface straight or slightly concave; anticlinal walls moderately thickened to thick, wavy. **Stomata** superficial; subsidiary cells with ridge on outer wall, elongated, about twice height of guard cells; guard cells narrow, with lip at inner aperture. **Chlorenchyma:** cells in 2 or 3 layers, divided into 21 sectors by ribs from scl. sheath and their associated pillar cells; chlorenchyma cells from 3–8 times but mostly 5–6 times higher than wide; cells arranged as in a chequerboard as seen in T.L.S.; protective cells of generic type (i) (see p. 179). **Parenchyma sheath** composed of moderately thick-walled pillar cells extending between scl. ribs and epidermis, and 1 discontinuous layer of wide to medium-sized, slightly thickened to thin-walled, rounded-hexagonal cells, between and on inner flanks of ribs. **Sclerenchyma:** sheath 4–6-layered, with 21 rectangular ribs radiating from it at more or less regular intervals; each rib 2–5 cells wide and 8–10 cells high. **Silica:** bodies present in stegmata in outer fibre layer of flanks of scl. ribs. **Tannin** in some epidermal cells.

***H. aristatus* (Thunb.) Nees.** Culm diameter 1·5, 2 mm.
(Culm material not typical; from within leaf sheath.)

Culm surface not seen.

Culm T.S.

Outline circular. **Epidermis:** cells more or less square, outer walls v. thick, anticlinal walls not wavy. **Stomata** superficial; guard cells with lip at inner aperture. **Chlorenchyma:** cells isodiametric (atypical?). **Parenchyma sheath** interrupted by scl. ribs in CA824. **Silica** and **tannin:** none seen.

Leaf, rhizome, and *root*: see genus description.

***H. binatus* (Steud.) Mast.** Culm diameter 1·3 mm.

Leaf surface: see genus description.

Culm surface (Fig. 31. F)

Epidermis: cells mostly 4-sided, those next to stomata slightly wider than or as wide as long, others mainly 2–3 times longer than wide; walls thick, slightly wavy. **Stomata:** subsidiary cells with wavy or slightly wavy walls; guard cell pair with lemon-shaped outline. **Tannin** in some epidermal cells.

Culm T.S.

Outline circular. **Epidermis:** cells 4-sided, as high as to 1½ times higher than wide; outer walls v. thick, outer surface irregular; anticlinal walls not wavy. **Stomata** superficial; subsidiary cells with slight ridge on outer wall, but nearly twice as high as guard cells; guard cells with v. small lips at outer and inner apertures. **Chlorenchyma:** 2 similar layers; cells mainly 2½–3 times higher than wide; protective cells of generic type (ii) (see p. 180). **Parenchyma sheath** interrupted in places by cells from scl. sheath. **Sclerenchyma:** sheath cells present in *c.* 8 layers; outline circular with low intrusions, 1 cell high and 2–4 cells wide, into parenchyma sheath between peripheral vb's. **Silica:** bodies present in stegmata of outer layer of scl. sheath, particularly in the ribs from the sheath. **Tannin** in some epidermal cells.

***H. neesii* Mast.** Culm diameter 0·8 mm.

Leaf T.S.: see genus description.

Culm surface

Epidermis: cells of unequal sizes and irregular shapes, 4–6-sided, walls straight, thick. **Stomata:** subsidiary cells variable; guard cell pair with lemon-shaped outline. **Tannin** in many epidermal cells.

Culm T.S. (Fig. 31. J)

Outline circular. **Cuticle** thick, with slight ridges. **Epidermis:** cells mainly twice as high as wide, some 1½–3 times higher than wide, outer walls v. thick, frequently slightly convex; anticlinal walls not wavy, thick at outer ends, thickening tapering, inner half of walls moderately thickened. **Stomata** sunken; mounted about half-way down anticlinal walls of epidermal cells; subsidiary cells extending further to inside of guard cells than normal; guard cells with ridge on inner wall. **Chlorenchyma** composed of 1, 2, or 3 layers of palisade cells with few short pegs; cells mainly 4–6 times higher than wide; cells of outermost layer completely encircling culm, those of inner layer(s) interrupted by strong rectangular ribs from scl. sheath. Protective cells of generic type (i) (see p. 179). **Parenchyma sheath** interrupted by ribs from scl. sheath, but extending part way up flanks of ribs. **Sclerenchyma:** sheath circular in outline, with 12 evenly spaced ribs radiating into chlorenchyma, but stopping at inner face of outer chlorenchyma layer. Ribs rectangular in outline, or with slightly expanded, rounded, outer ends, each *c.* 12 cells high, 4–8 cells wide at inner end and 6–9 cells wide at broadest point, near outer end. Sheath between ribs *c.* 7 layers wide. **Vascular bundles:** peripheral vb's alternating with ribs; occasional fusions observed between peripheral vb's and small, outer medullary vb's. **Silica:** bodies present in stegmata in outer layer of scl. sheath, particularly in rounded ends of ribs. **Tannin** in many epidermal cells.

***H. striatus* (Kunth) Mast.** Culm diameter 1 mm.

Culm surface (Fig. 31. H)

Epidermis: cells 4–6-sided, as long as wide to 4 times longer than wide; walls moderately thickened to thick, wavy at surface, straighter at slightly lower focus. **Stomata:** subsidiary cells with acute, obtuse, or slightly rounded ends; guard cell pair with lemon-shaped outline. **Tannin:** none seen.

Culm T.S. (Fig. 31. G)

Outline circular. **Epidermis:** cells all of similar height, 2–2½ times higher than wide; outer walls thick to v. thick, slightly convex; some anticlinal walls wavy. **Stomata** superficial; subsidiary cells about twice as high as guard cells; guard cells with slight lip on inner wall. **Chlorenchyma** divided into 17 sectors by radiating ribs from scl. sheath and their associated pillar cells; individual peg-cells mostly 4–5 times higher than wide. Protective cells mostly of generic type (ii) (see p. 180), but with pointed inner ends. **Parenchyma sheath** interrupted by scl. ribs; short pillar cells sometimes present opposite to ends of ribs. **Sclerenchyma:** sheath 4–5-layered, with 17 fairly regularly spaced ribs radiating towards epidermis; ribs parallel-sided or wedge-shaped, and widest at outer ends, ranging from 1–3 cells wide, or 1–2 cells wide at inner end and 3–5 cells wide at outer end; normally 12–14 cells high, either reaching epidermis, or separated from it by short pillar cells. **Vascular bundles:** peripheral vb's alternating with scl. ribs. **Silica:** bodies present in stegmata on flanks of scl. ribs. **Tannin:** none seen.

***H. synchroolepis* (Steud.) Mast.** Culm diameter 1 mm.

Culm surface

Epidermis: cells 4–6-sided, mostly 1½–3 times longer than wide, walls thick, wavy. **Stomata:** subsidiary cells with acute end walls; guard cells narrow. **Tannin** in some epidermal cells.

Culm T.S.

Outline circular. **Epidermis:** cells all of similar height, 2½–3 times higher than wide; outer walls v. thick, outer surface concave in some cells; anticlinal walls wavy, v. thick at outer ends, thickening tapering rapidly to half-way down walls, and inner halves of walls evenly, slightly to moderately thickened. **Stomata** superficial; subsidiary cells nearly twice height of guard cells; guard cells with lips at outer and inner apertures. **Chlorenchyma** composed of 2 similar, uninterrupted layers of cells; individual cells mostly 3–4 times higher than wide; protective cells of generic type (ii) (see p. 180). **Parenchyma sheath** interrupted in places by 1–2 scl. fibres or stegmata. **Sclerenchyma:** sheath mainly 2-layered, fibres v. thick-walled, with narrow lumina; 1–2 fibres interrupting parenchyma sheath at intervals. **Silica:** bodies present in isolated stegmata interrupting parenchyma sheath. **Tannin:** none seen.

***H. willdenowia* (Nees) Mast.** Culm diameter 2·5 by 1 mm.

Culm surface (Fig. 31. B)

Epidermis: cell outlines and sizes variable; 4–6-sided, some cells wider than long, and some up to 7 times longer than wide, but most between 1½ and 2 times longer than wide; walls moderately thickened to thick, not or only slightly wavy. **Stomata:** subsidiary cells with perpendicular or acute end walls. **Tannin** in many epidermal cells.

Culm T.S. (Fig. 31. A)

Outline flattened-oval. **Epidermis:** cells 4-sided, mostly *c.* 1½ times higher than wide, some, particularly those opposite to scl. ribs, about twice as high as wide; outer walls v. thick, outer surface sometimes slightly concave; outermost ends of anticlinal walls thick, but thickening tapering rapidly and remainder of anticlinal walls slightly to moderately thickened; anticlinal walls not wavy. **Stomata** superficial; subsidiary cells twice as high as guard cells; guard cells with ridge on inner wall. **Chlorenchyma:** 2 layers of cells, layers interrupted at either angle of flattened culm by rib from scl. sheath; protective cells of generic type (i) (see p. 179), but with pegs on anticlinal walls, and slit-like apertures; some occasional protective cells elongated, and extending to parenchyma sheath. **Parenchyma sheath** interrupted at angles of flattened stem by long scl. ribs, and at other places by low ribs. Pillar cells of normal type not present. **Sclerenchyma:** sheath 1–3-layered, sometimes 4–5-layered opposite larger medullary vb's, with 1 rib radiating to epidermis at either angle of flattened culm, and several low ribs interrupting parenchyma sheath in other places; flanking ribs mainly 3 cells wide, and *c.* 10 cells high. **Vascular bundles:** peripheral vb's not opposite ribs. **Silica:** bodies present in stegmata on flanks of ribs. **Tannin** in many epidermal cells.

Root; see genus description.

TAXONOMIC NOTES

Shortly after it was first described, this genus was included in *Willdenowia* Thunb. by Endlicher (1836). Kunth (1841) separated it once again but regarded it as a genus of doubtful status. It has continued to be recognized since that time.

Boeckhia Kunth and *Lepidanthus* Nees were included in it by Masters (1878), and these with the addition of *Leucoploeus* Nees were recognized as synonyms by Bentham and Hooker (1883), Pillans (1928), and Gilg-Benedict (1930). Pillans recognized the synonymy of some *Willdenowia* species with certain *Hypodiscus* species.

Both Gilg and Gilg-Benedict gave a general account of the anatomy of *Hypodiscus*, but they did not examine enough species to realize that they fall into 2 distinct groups on the basis of their anatomical characters. These groups are: (1) Most spp. with pillar cells and all with protective cells and well-developed scl. ribs radiating at regular intervals from the scl. sheath to the pillar cells (if the latter are present). Group (1) can further be divided into 2 subgroups: (*a*) with superficial stomata in *H. argenteus* (Thunb.) Mast., *H. striatus* (Kunth) Mast., and *H. willdenowia* (Nees) Mast.; (*b*) with sunken stomata in *H. neesii* Mast. (2) With protective cells but no pillar cells, and with only weak ridges interrupting the parenchyma sheath and not penetrating the chlorenchyma, in *H. albo-aristatus* (Nees) Mast., *H. alternans* Pillans, *H. binatus* (Steud.) Mast., and *H. synchroolepis* (Steud.) Mast.

Most of the species in group 1 above are anatomically very similar to the spp. of *Willdenowia* here described in group 1 of that genus. *H. argenteus* (Parker 4770) (synonymous with *W. simplex* N. E. Br. and *Leucoploeus argenteus* Nees, according to Pillans) is anatomically indistinguishable from *W. striata* Thunb. forma *monstrosa* Mast. labelled 'W. *ecklonii* Endl.', '*Nematanthus ecklonii* Nees ab Esinb.' in the Kew Herbarium.

H. willdenowia (Nees) Mast. (Parker 4089) (synonymous with *W. striata* Spreng. and *Lepidanthus willdenowia* Nees, according to Pillans) is anatomically very similar to *H. argenteus* (Thunb.) Mast.

H. neesii Mast. (synonymous with *Leucoploeus striatus* Nees ex Drège, according to Pillans), whilst basically similar to *H. willdenowia* (Nees) Mast., has shorter girders and lacks pillar cells. In these respects it is similar to *W. lucaeana* Kunth (from which it is readily distinguished by its sunken stomata).

H. striatus (Kunth) Mast. (synonymous with *Boeckhia striata* Thunb., according to Pillans) has girders or partial girders of scl. which may terminate at short pillar cells.

Since the *Hypodiscus* species of group 1 are anatomically so similar to *Willdenowia* species of group 1 of this work, it is probable that they collectively constitute a natural taxon.

The *Hypodiscus* species of group 2 include 1 species, *H. albo-aristatus* (Nees) Mast., which is said by Pillans to be synonymous with *Boeckhia laevigata* Kunth. It has been noted that the other *Boeckhia* species is a synonym of a species of *Hypodiscus* belonging to group 1. This would be unexpected if *Boeckhia* itself constituted a natural genus and the significance of this discontinuity is not clear at present. The possibility that the spirit material of *H. albo-aristatus* which was examined anatomically may be incorrectly named should not be overlooked. This species needs further investigation.

The individual species of *Hypodiscus* belonging to group 2 are anatomically very similar to one another, but it already seems probable that characters of the epidermal cells and stomata that can be seen in surface preparations could be used for identification at the species level.

In the species of group 2 the short scl. ridges present between the peripheral vb's which interrupt the parenchyma sheath are worthy of note. They may be the equivalent of the larger girders in species belonging to group 1. If they are, this could indicate an affinity between the two groups. The parenchyma sheath is entire in species of *Willdenowia* belonging to group 2, suggesting a discontinuity between this and *Hypodiscus* group 2 (see p. 323).

The pillar cells that are to be found in some species of *Hypodiscus* and *Willdenowia* are unique among South African Restionaceae. They occur, however, in species from other countries (Australia, New Zealand, S. Vietnam (Cochin China), Malaysia, Chile). Whereas the pillar cells in species from countries other than South Africa appear to be modified cells of the parenchyma sheath, the corresponding cells in *Willdenowia* and *Hypodiscus* species seem, like the protective cells, to be derived from palisade chlorenchyma cells. In some instances where stomata are very close to scl. girders, the thickened palisade cell acts as both pillar and protective cell (see Fig. 32. A of *H. argenteus*). If this difference in the origin of the pillar cells is valid, it follows that pillar cells in the true sense are unknown in African species of the Restionaceae. This would indicate a major discontinuity in the family.

HYPOLAENA R. Br.

No. of spp. *c.* 12[1]; examined 7.

Genus Description

The spp. fall into 2 distinct geographical and anatomical groups. (1) Spp. from South Africa. (2) Spp. from Australia and Tasmania.

GROUP 1

Leaf surface not seen.

Leaf T.S.

Blade seen in *H. graminifolia* (Fig. 32. C).

Outline parallel-sided, with rounded margins; *c.* 10 times wider than thick. **Hairs** and **papillae** absent. **Epidermis:** (i) abaxial cells mostly as high as or slightly higher than wide, *c.* ¾ height of adaxial cells; (ii) adaxial cells slightly to 1½ times higher than wide. Outer walls moderately thickened to thick, other walls slightly to moderately thickened, in cells of both surfaces. **Stomata** as in culm of same sp. **Chlorenchyma** as in culm of same sp.; filling all space between vb sheaths and epidermis; protective cells similar to peg-cells, but with slightly thickened walls. **Vascular bundles** 13, arranged in 1 row; medium-sized or large alternating with small vb. Large vb's with 1 wide, angular tracheid with slightly thickened walls on either flank of arc composed of up to 5 rows of medium-sized and narrow tracheids, phloem pole with curved outer face and flattened face next to xylem.

Bundle sheaths: O.S. parenchymatous, cells in 1 layer, similar to those surrounding free culm bundles. I.S. sclerenchymatous: (i) in small and medium bundles; fibres narrow, with moderately thickened walls, in 1–2 layers at phloem pole and 1 layer on flanks and at xylem pole; (ii) in large bundles; fibres narrow, with moderately thickened walls, in 3–4 layers at phloem pole, fibres wider, 1–2-layered on flanks and around xylem pole. **Sclerenchyma** restricted to bundle sheaths. **Air-cavities** absent. **Silica:** spheroidal-nodular bodies present in stegmata in outer layer of scl. bundle sheaths.

Culm surface

Hairs and **papillae** absent. **Epidermis:** cells 4-sided (sometimes 5–6-sided),

[1] Including spp. described as *Mastersiella*; see p. 237.

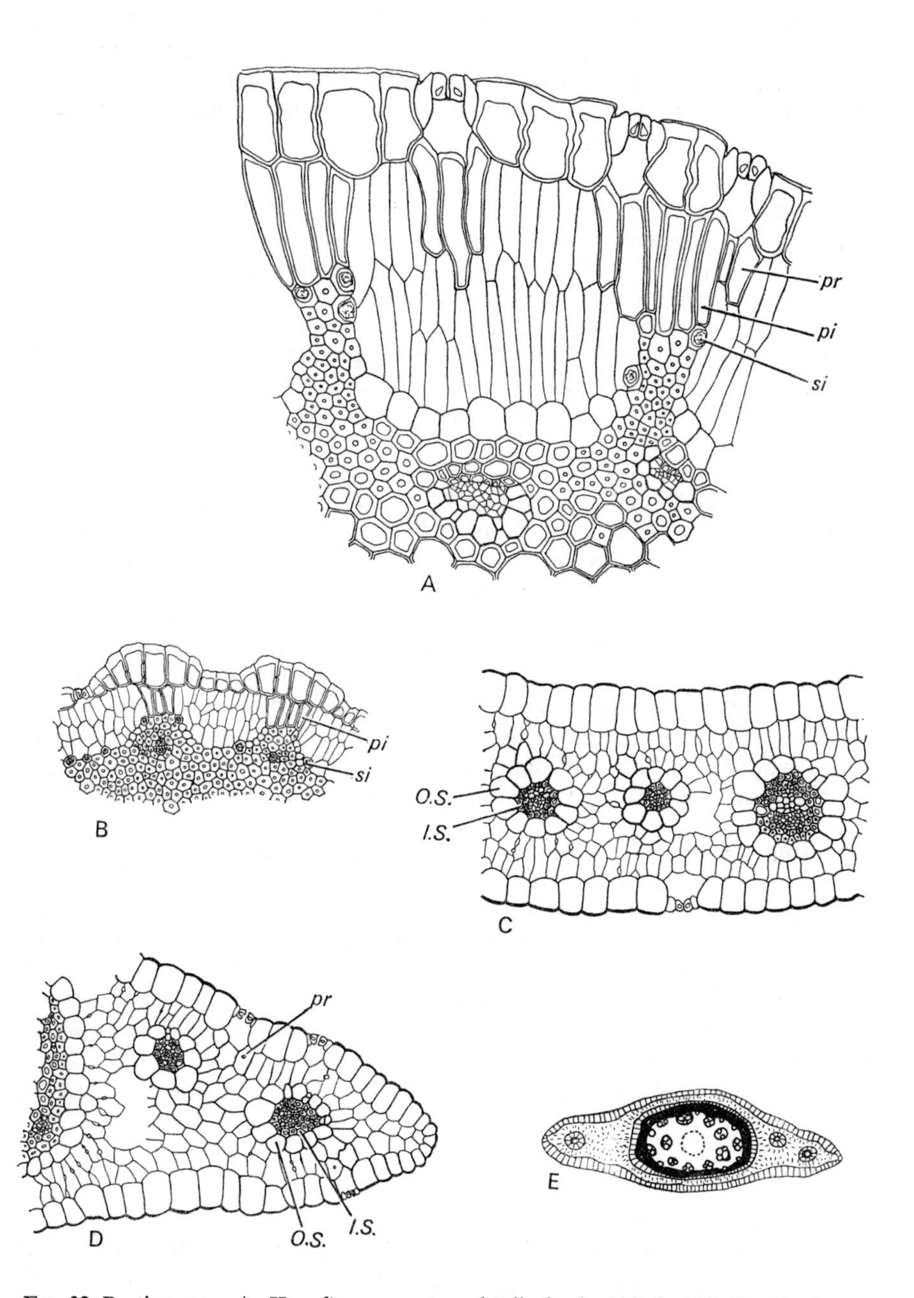

FIG. 32. Restionaceae. A, *Hypodiscus argenteus*, detail of culm T.S. (×185). B, *Hypolaena fastigiata*, detail of culm T.S. (×100). C–E *Hypolaena graminifolia*: C, part of leaf T.S. (×100); D, detail of part of culm T.S. (×100); E, diagram of culm T.S. (×25).
I.S. = inner bundle sheath; O.S. = outer bundle sheath; pi = pillar cell; pr = protective cell; si = silica-body.

as long as or up to 2–4 times longer than wide; walls moderately thickened, normally wavy. **Stomata:** subsidiary cells variable; guard cell pair with circular outline, or ends of individual guard cells rounded. **Silica** absent. **Tannin** in some or most epidermal cells. **Cuticular marks:** longitudinal striations in all spp. except *H. spathulata.*

Culm T.S.

Outline circular or oval. **Cuticle** slightly to moderately thickened, with slight ridges. **Epidermis:** cells present in 1 layer, individual cells 4-sided, slightly to 2 times higher than wide. Outer walls thick; anticlinal walls not wavy, moderately thickened; inner walls moderately thickened. **Stomata** superficial; guard cells with lips at outer and inner apertures, or with ridge on inner wall and without inner lip. **Chlorenchyma** composed of peg-cells in 1 or 2(–3) layers; protective cells present; lobed cells present in *H. graminifolia.* **Parenchyma sheath** 1–2-layered; cells mostly slightly to $1\frac{1}{2}$ times wider than high. **Sclerenchyma:** sheath 3–8-layered, fibres thick- to v. thick-walled, those of outer layers narrowest.

Vascular bundles. (i) Peripheral vb's: tracheids angular, medium-sized or wide, thin-walled, arranged in arc partially enclosing rounded or oval phloem pole. (ii) Medullary vb's: outline oval or circular; some outer bundles with 1 central, medium-sized to wide, rounded mxv, capped by a small phloem pole and without px; inner bundles with 1 wide, circular-angular or many-sided mxv on either flank, with scalariform lateral wall pitting and simple, more or less transverse, or oblique scalariform-reticulate perforation plates. Mxv's separated from one another by 1, 2, or 3 rows of narrow cells. Phloem overarching and extending partly between mxv's, but separated from xylem by 1 layer of narrow cells with slightly or moderately thickened walls in all spp. but *H. spathulata.* Px present in inner bundles; some bundles free from culm scl. sheath.

Bundle sheaths: fibres in 1–2 layers at poles, narrow, with moderately thickened to thick walls; those on flanks in 1 layer, somewhat wider, with slightly or moderately thickened walls; weak in *H. spathulata,* walls slightly thickened. **Central ground tissue** parenchymatous; cells surrounding vb's with moderately thickened walls, those of culm centre thin-walled. **Silica** as spheroidal-nodular bodies, in stegmata situated in outer layer of culm scl. sheath. **Tannin** in some epidermal cells. Note: *H. graminifolia* has unusual features; see species description.

Rhizome and *root* not seen.

GROUP 2

Leaf surface

Abaxial, *H. fastigiata* only.

Hairs: none seen. **Epidermis:** cells 4-sided, mostly 5–8(–10) times longer than wide, some slightly longer than wide; end walls sometimes oblique; walls moderately thickened, wavy. **Stomata:** subsidiary cells with perpendicular or acute end walls, long anticlinal walls frequently slightly wavy; guard cell pair with more or less oval outline. **Silica:** none seen. **Tannin** in many epidermal cells. **Cuticular marks** granular.

Leaf T.S.

Scale leaf, *H. fastigiata* only.

Epidermis: (i) abaxial cells 4–5-sided, slightly to $1\frac{1}{2}$ times wider than high, outer walls thick, other walls moderately thickened; (ii) adaxial cells 4-sided, flattened, mostly twice as wide as high and *c.* $\frac{1}{3}$ height of abaxial cells; all walls slightly to moderately thickened. **Stomata** present in abaxial surface only, superficial; guard cells without lips or ridges. **Chlorenchyma** restricted to small areas of more or less isodiametric cells below stomata. **Vascular bundles** v. small, each composed of several narrow tracheids to adaxial side of small phloem pole. **Bundle sheaths** sclerenchymatous; fibres narrow, thick-walled, extending in several layers to abaxial epidermis at phloem pole, indistinguishable from ground scl. on bundle flanks and at xylem pole. **Sclerenchyma** represented by that of bundle sheaths, and by 5 layers of fibres between bundles and 2–3 layers to adaxial side of xylem poles; separated from adaxial epidermis by 1–2 layers of parenchymatous cells and from abaxial epidermis by 1 layer of narrow parenchyma cells or by chlorenchyma. **Air-cavities** absent. **Silica:** small spheroidal-nodular bodies in stegmata in outer layer of bundle sheaths at phloem poles. **Tannin** in many epidermal cells.

Culm surface

Hairs of 2 types: (i) in *H. exsulca*; similar to those in Australian *Leptocarpus* spp.; (ii) multicellular, peltate, or sub-sessile (Fig. 33. G, p. 200). **Epidermis:** cells 4–6-sided, as long as wide or up to twice or 4 times as long as wide; walls slightly to moderately thickened or thick, wavy. **Stomata:** subsidiary cells variable; guard cells with rounded end walls. **Silica:** none seen. **Tannin** in *H. fastigiata* only, in some epidermal cells. **Cuticular marks** granular, with or without longitudinal striations.

Culm T.S.

Outline circular. **Cuticle** granular, slightly to moderately thickened, with or without small ridges. **Hairs:** see above. **Epidermis:** cells (i) all more or less of uniform size, as high as wide, outer walls moderately thickened, other walls slightly thickened, or (ii) of varying sizes, those next to stomata smallest, *c.* $1\frac{1}{2}$ times higher than wide, those opposite to pillar cells $1\frac{1}{2}$–3 times higher than smallest cells, twice as high as wide. Cells of intermediate sizes between the two extremes. Outer walls thick, other walls moderately thickened. **Stomata** superficial or sunken; guard cells either without lips, or with lips at outer aperture, with or without ridge on inner wall. **Chlorenchyma** composed of 2–3 layers of palisade cells without or with few pegs; layers divided into sectors by ribs from scl. sheath and their associated pillar cells; individual cells mostly $1\frac{1}{2}$–2 or 2–3 times higher than wide. Protective cells absent. **Parenchyma sheath** composed of cells of 2 types: (i) elongated pillar cells, each 2–3 times higher than wide, extending between ribs from scl. sheath to epidermis; (ii) rounded-hexagonal cells situated in 1 layer on flanks of ribs from culm scl. sheath; individual cells frequently slightly wider than high; these cells occasionally present in place of pillar cells over scl. ribs. **Sclerenchyma:** sheath outline circular, with low, dome-shaped ribs each opposite to and enclosing a single peripheral vb. Main part of sheath *c.* 7–8-layered;

fibres of outer layers narrow, those of innermost layers wider, all v. thick-walled.

Vascular bundles: (i) peripheral vb's with 1 wide tracheid on either flank of arc of narrow tracheids, or tracheids narrow and medium-sized, arranged in 1-layered arc partially enclosing rounded or oval phloem pole; (ii) medullary vb's: outer bundles small, inner large; each with 1 mxv on either flank, those of outer bundles narrow, those of inner bundles wide, angular, thin-walled. Flanking mxv's separated from one another by 2–6 layers of narrow cells with slightly thickened walls. Phloem pole oval, separated from xylem by 1 layer of narrow cells with slightly thickened walls. Px present in all but smallest bundles. Always some of the bundles free from culm scl. sheath, except in *H. fasciculata.* **Bundle sheaths:** fibres narrow, thick-walled, in 1 or 2–(3) layers at phloem pole, in 1 layer on flanks and round xylem pole. **Central ground tissue** parenchymatous; cells surrounding vb's with moderately thickened walls, those at culm centre thin-walled, breaking down. **Silica:** spheroidal-nodular bodies present in stegmata in outer layer of scl. sheath. **Tannin** in some epidermal cells.

Rhizome T.S.

H. exsulca only

Outer tissues not seen. **Cortex:** inner part composed of several layers of loosely packed, lobed, isodiametric and palisade, thin-walled cells. **Endodermoid sheath:** cells in 1 layer, hexagonal, mostly as high as wide, thick-walled. **Vascular bundles** scattered, mostly amphivasal but some outer bundles with arc of wide, angular, moderately thick-walled mxv's, with scalariform lateral wall-pitting and simple, more or less transverse perforation plates. Wide tracheids present in 1 layer round most mxv's, those next to phloem narrow; phloem poles circular. **Bundle sheaths** several-layered, boundaries merging imperceptibly with ground tissue. **Ground tissue** composed of matrix of sclereids with thick walls, and scattered regions of lobed, more or less isodiametric, thin-walled, parenchymatous cells. **Silica:** small, spheroidal-nodular bodies present, frequently several per cell, in parenchyma cells bordering lacunae in ground tissue, and also in some inner cortical cells. **Tannin:** none seen.

Root T.S.

H. exsulca only.

Root-hairs and **epidermis** not seen. **Cortex:** (i) outer part composed of *c.* 3 layers of narrow cells with slightly to moderately thickened walls; (ii) middle part consisting of radiating plates of parenchymatous cells separated from one another by lysigenous air-spaces; each plate 2–3-layered and 15–20 cells high; (iii) inner part 1–2-layered, cells narrow, parenchymatous, with thin walls. **Endodermis** 1-layered, cells $1\frac{1}{2}$ times higher than wide, all walls v. thick. **Pericycle** 1–2-layered, cells slightly wider than high, *c.* $\frac{1}{4}$ height of endodermal cells, walls slightly thickened. **Vascular tissue:** mxv's angular, oval, wide, situated in 1 ring to inside of phloem and px poles; each mxv surrounded by 1 layer of flattened tracheids. **Central ground tissue** composed of narrow to medium-sized cells with moderately thickened to thick walls. **Tannin:** none seen.

Material Examined

GROUP 1. South Africa.

Hypolaena crinalis (Mast.) Pillans: Esterhuysen 16530; Slanghoek Mts. (K).

H. graminifolia (Kunth) Pillans: (i) Parker, R. N. 4592 ♂ (K). (ii) Schlechter 9255, An. 1896 (K).

H. purpurea Pillans: (i) Esterhuysen 4592 (K). (ii) Muir 3214 ♂ Dupl. (K).

H. spathulata Pillans: Esterhuysen 16358 (K).

GROUP 2. Australia, Tasmania.

H. exsulca R. Br.: Cheadle CA82; Perth, W. Aust. (S).

H. fasciculata W. V. Fitzgerald: (i) Andrews, C., 1st Coln. 1106; Albany, W. Aust. (ii) Baudin, An. 1801; W. Aust. (K).

H. fastigiata R. Br.: Curtis, W. M.; Blackman's Bay, Hobart, Tasmania (S).

Species Descriptions

GROUP 1

***H. crinalis* (Mast.) Pillans.** Culm diameter 0·5 mm.
(basal material)

Culm surface

Epidermis: cells mostly 2–4 times longer than wide. **Stomata:** subsidiary cells with acute or rounded end walls.

Culm T.S.

Outline circular. **Epidermis:** cells slightly higher than wide. **Chlorenchyma:** cells in 1 or 2 layers; individual cells as high as or up to twice as high as wide; protective cells extending in 1 layer to cells of parenchyma sheath, or inner layer of chlorenchyma; walls moderately thickened, cells otherwise similar to peg-cells.

***H. graminifolia* (Kunth) Pillans.** Culm diameter 1·5 by 0·5, 0·75 by 0·25 mm

Leaf T.S.: see group 1 description (Fig. 32. C).

Culm surface

Epidermis: cells as long as wide or up to 4 times longer than wide. **Stomata:** subsidiary cells with perpendicular or acute end walls; guard cell pair with more or less oval outline.

Culm T.S. (Fig. 32. D, E)

Outline oval. **Epidermis:** cells mostly about twice as high as wide, graded in size, those on flattened faces largest; outer walls thick, with slightly ridged outer surface. **Stomata:** subsidiary cells with slight ridge on outer wall. **Chlorenchyma** composed of peg-cells and lobed cells; peg-cells mostly 1½–2 times higher than wide, in 1, 2, or 3 layers immediately inside epidermis; lobed cells 1½ times higher than wide or more or less isodiametric (Fig. 32. D), filling all space between bundle parenchyma sheaths and culm parenchyma sheath. Protective cells short, walls next to substomatal cavity slightly thickened, other walls thin. **Parenchyma sheath:** cells as high as wide, or slightly higher than wide; smallest ⅓–½ height of largest, next to palisade chlorenchyma cells; largest next to lobed chlorenchyma cells. **Sclerenchyma:** sheath 3–5-layered; fibres of outer 1–3 layers narrow, lumina filled with silica. **Vascular**

bundles: distribution unlike that of any other member of family; most bundles embedded in, or to inner side of, culm scl. sheath, as usual in spp. belonging to group 1 of the genus, but 1 or 2 free bundles present in either wing of lobed chlorenchyma cells, outside culm scl. sheath. Bundles outside scl. sheath each with arc of narrow to medium-sized, angular, thin-walled tracheids arranged in 2–3 layers, partially enclosing rounded phloem pole; orientated with phloem facing culm surface. Bundles within culm scl. sheath as for group 1. **Bundle sheaths:** (*a*) present around free, outer bundles: O.S. parenchymatous, cells in 1 layer, similar to those of culm parenchyma sheath; I.S. sclerenchymatous, composed of narrow fibres with moderately thick to thick walls, 2–4-layered at phloem pole, 1-layered on flanks and at xylem pole; (*b*) sheaths around medullary vb's, as for group 1 (see p. 190). **Silica** present as (i) granular material in protective cells and outer cells of scl. sheath; (ii) bodies in stegmata in outer layer of culm scl. sheath.

***H. purpurea* Pillans.** Culm diameter 1·2, 1·5 mm.

Culm surface

Epidermis: cells 4–6-sided, cells at ends of stomata mostly as long as wide, others mostly 1½–2 times longer than wide. **Stomata:** subsidiary cells variable; guard cell pair with more or less oval outline.

Culm T.S.

Outline circular. **Epidermis:** cells mostly *c.* 1½ times higher than wide, outer walls slightly convex. **Stomata:** subsidiary cells with ridge on outer wall; some cells about twice as high as wide and extending below guard cells; guard cells with slight lip at inner aperture. **Chlorenchyma:** 2 layers of peg-cells; cells of outer layer 2–4 times higher than wide, those of inner layer 3–5 times higher than wide and *c.* ⅓ wider than outer cells. Protective cells extending in 1 layer ¼–½ way into inner chlorenchyma layer; walls slightly to moderately thickened; apertures between cells oval, elongated, present in inner half of substomatal tube. **Parenchyma sheath** 1-layered, following slightly ribbed outline of scl. sheath and interrupted at intervals by groups of 1 or 2 or up to 8 fibres at the ribs; cells opposite ribs smaller than those between them. **Sclerenchyma:** sheath 3–4- or 4–8-layered, wider at ribs; ribs 3–8 cells wide, extending in 2–3 layers beyond circular outline of main part of sheath, not opposite to peripheral vb's. **Silica:** bodies present in stegmata in outer layer of culm scl. sheath.

***H. spathulata* Pillans.** Culm diameter 1·5 mm.

Culm surface

Epidermis: cells mostly 6-sided, twice as long as wide, some cells 4-sided, half as long as others; walls slightly to moderately thickened, straight or slightly wavy. **Stomata:** subsidiary cells variable; guard cells elongated. **Cuticular marks** coarsely granular (not striate).

Culm T.S.

Outline circular. **Epidermis:** cells 4-sided, mostly *c.* 1½ times higher than wide, outer walls slightly convex. **Stomata:** subsidiary cells with ridge on outer wall. **Chlorenchyma:** 2 similar layers of peg-cells; individual cells *c.* 4 times higher than wide. Protective cells as in *H. purpurea,* extending through outer chlorenchyma layer only. **Parenchyma sheath** 1-layered, cells slightly wider than high, those opposite to short ribs from scl. sheath smallest. **Sclerenchyma:** sheath outline circular, with low ribs between most peripheral vb's; 5–6-layered. **Vascular bundles:** peripheral and medullary vb's, as in species of the genus included in group 1, but phloem abutting directly on xylem. **Silica:** bodies present in stegmata in outer layer of culm scl. sheath.

GROUP 2

***H. exsulca* R. Br.** Culm diameter 1·5 mm.

Culm surface

Hairs of type (i) described for species in the genus included in group 2. **Stomata:** subsidiary cells more or less crescentiform; guard cells shorter and more rounded than normal. **Tannin:** none seen. **Cuticular marks** granular.

Culm T.S.

Epidermis: cells mostly as high as wide. **Stomata** not seen in T.S. **Silica:** bodies present in stegmata in outer layer of culm scl. sheath, particularly on flanks of scl. ribs.

Rhizome and *root*: see description for spp. belonging to group 2.

***H. fasciculata* W. V. Fitzgerald.** Culm diameter 1·1 mm.

Culm surface

Hairs multicellular, sparsely distributed, peltate to sub-sessile, each consisting of 1 basal cell and an oval rosette of thick-(cellulose) walled cells, or wedge-shaped and composed of 5–6 rows of short, 4-sided cells, the largest being at the wide (free) end of the wedge; only 1 cell thick throughout. **Epidermis:** cells 4-sided, slightly to 4 times longer than wide. **Stomata:** subsidiary cells variable. **Tannin:** none seen. **Cuticular marks** finely granular, with superimposed longitudinal striations.

Culm T.S.

Cuticle with slight ridges. **Epidermis:** cells mostly as high as wide, outer walls moderately thickened, outer surface with slight ridges corresponding to those in cuticle; other walls slightly to moderately thickened. **Stomata** small, v. slightly sunken; guard cells with slight lip at outer aperture.

***H. fastigiata* R. Br.** Culm diameter 1·5 mm.

Leaf: see description of spp. in group 2.

Culm surface

Hairs multicellular, flattened, plate-like, each with 1 thick-walled, short basal cell; plate-like structure composed of cells with irregular outlines, but mostly *c.* 3 times longer than wide, with pointed ends; margins of hair indistinct. **Epidermis:** cells as long as wide to twice as long as wide, those at ends of stomata shortest. **Stomata:** subsidiary cells with obtuse end walls. **Cuticular marks** finely granular.

Culm T.S. (Fig. 32. B)

Outline circular, with slight ribs. **Epidermis:** cells 4-sided, those opposite pillar cells largest, about twice as high as wide, those next to stomata smallest, slightly higher than wide, *c.* $\frac{1}{2}$–$\frac{1}{3}$ height of tallest; intermediate sizes present. Conspicuous, circular pits present in anticlinal walls. **Stomata** superficial, small; guard cells with outer lip and ridge on inner wall.

Taxonomic Notes

The species here recognized as members of this genus do not include all of those placed in the genus by Pillans. This is a point of difference in the mode of treatment of *Hypolaena*, for in dealing with South African species of other genera Pillans's classification has been followed. Gilg-Benedict (1930) recognized all of the *Hypolaena* species mentioned by Pillans as anatomically distinct from the Australian representatives of the genus, so she gave them generic status under the new name *Mastersiella*. The present author agrees with Gilg-Benedict's treatment of the genus.

Hypolaena has been regarded as a valid genus since it was first described. Endlicher (1836) and Lindley (1853) included *Cucullifera* species in it as synonyms, but these were later moved into *Cannomois* by Masters (1878). Masters also included *Calorophus* Labill., *Calostrophus* F. v. Muell., *Loxocarya* Benth., and *Desmocladus* Nees in *Hypolaena*. Gilg (1890) and Gilg-Benedict (1930) found the Australian species to be anatomically similar to *Leptocarpus* species from Australia, and Gilg-Benedict described the new genus, *Mastersiella* (see above). Gilg-Benedict regarded *Loxocarya* R. Br. as a good genus and included in it *Desmocladus*. She also re-separated *Calorophus* Labill. (incl. *Calostrophus* F. v. Muell.) from *Hypolaena*.

Pillans did not admit the synonymy with other genera as proposed by Masters. Pillans continued to publish new South African species of *Hypolaena* after Gilg-Benedict's *Mastersiella* had been proposed, thus indicating that he did not recognize the distinction between Australian and South African members of the genus *Hypolaena*.

The present work indicates that the genus, including Australian species, and South African species named subsequently to Gilg-Benedict's work or not seen by her, falls into 2 groups on the basis of anatomy.

(1) Without pillar cells or hairs, with protective cells. This group comprises the South African species *H. crinalis* (Mast.) Pillans, *H. graminifolia* (Kunth) Pillans, *H. purpurea* Pillans, and *H. spathulata* Pillans.

(2) With pillar cells and hairs, without protective cells. This group includes Australian species *H. exsulca* R. Br. and *H. fasciculata* W. V. Fitzgerald and the Tasmanian species *H. fastigiata* R. Br.

The species of group 2 share the possession of pillar cells with Australian *Leptocarpus* species, *Meeboldina* Suess., *Loxocarya pubescens* Benth., *Chaetanthus* R. Br., and Australian *Restio* species here included in group 7, see p. 283; many of these also have hairs. The anatomical similarity may indicate affinities between these genera or species.

The species belonging to group 2 are anatomically distinct from *Calorophus* Labill., which lacks pillar cells and hairs, and one species of which has protective cells. It is evident that Masters was wrong to group *Hypolaena* and *Calorophus* together.

The species of *Hypolaena* belonging to group 1 are all similar in most anatomical characters to *Mastersiella* Gilg-Benedict and should be included in that genus if there is supporting evidence from other sources. They also resemble anatomically South African species of *Restio* belonging to groups 3 and 4 and South African *Leptocarpus* species. It would be necessary to make sure that they could not be included in either of these genera before classifying them with *Mastersiella.*

H. graminifolia (Kunth) Pillans is unique among members of the Restionaceae; the culm has 2 wings of tissue, giving it an oval outline. The mechanical ring is cylindrical and encloses vb's in the way that is normal for the family, but there are also free bundles in the wing tissue. The wing tissue is chlorenchymatous; the free bundles are enclosed by an inner scl. and an outer parenchyma sheath similar to those found round the leaf bundles of the same species.

Apart from the abnormal vb arrangement and wing development there are no anatomical facts which provide reasons for separating this species from the other South African *Mastersiella* (*Hypolaena*) species.

LEPIDOBOLUS Nees

No. of spp. 4; examined 3.

GENUS DESCRIPTION

Leaf surface

Not seen in *L. chaetocephalus.*

Hairs absent. **Papillae** present as outpushings from occasional cells. **Epidermis:** (i) abaxial cells 4-sided, slightly longer than wide, walls thick, not wavy; (ii) adaxial cells not seen. **Stomata** showing same range as in culm. **Silica** and **tannin:** none seen. **Cuticular marks** none; smooth.

Leaf T.S.

Not seen in *L. chaetocephalus.*

Papillae: see leaf surface. **Epidermis:** (i) abaxial cells 4–5-sided, about as high as wide; outer and anticlinal walls v. thick, inner walls slightly or moderately thickened; lumina square to slightly flask-shaped; (ii) adaxial cells more or less hexagonal, outer walls 2-faced, all walls slightly thickened or thin. **Stomata** superficial or slightly sunken; as in culm. **Chlorenchyma:** cells lobed, more or less isodiametric, in 1 layer to inner side of abaxial epidermis. **Vascular bundles** small, with few narrow tracheids and phloem cells. **Bundle sheaths:** O.S. consisting of parenchyma cells in 1 layer; I.S. composed of fibres arranged in 1–2 layers. **Sclerenchyma** present as I.S. only. **Ground tissue** parenchymatous. **Air-cavities** and **tannin:** none present. **Silica:** spheroidal-nodular bodies present in some cells of parenchyma bundle sheaths.

Culm surface (Fig. 33. B, C, F)

Hairs present in 2 spp., multicellular, each composed of a stalk of 2 or 3 square–slightly oblong cells, the lower ones with moderately thickened walls, the end one with thin walls, and several (usually 2–4) radiating branches, each branch consisting of 1 thin-walled, contorted cell, 6–12 times longer than wide, and with a rounded or pointed free end. **Epidermis:** cells as long as to 2–3 times longer than wide; walls thick or v. thick, wavy. **Stomata:** subsidiary cells variable; guard cells with rounded end walls or guard cell pair with compressed oval outline. **Silica** and **tannin:** none seen. **Cuticular marks** granular.

Culm T.S. (Fig. 33. A, D, E)

Outline circular in *L. drapetocoleus*, with frequent, low mounds in other spp. **Cuticle** thick. **Hairs:** see culm surface. **Epidermis:** cells in 1 layer; of equal or unequal heights, ranging from as high as wide to 6 times higher than wide; cells bordering stomata with the anticlinal walls lining substomatal cavity up to twice as long as the others, extending into chlorenchyma (Fig. 33. E). Outer walls v. thick, anticlinal walls wavy, v. thick at their outer ends, thickening tapering near inner ends of walls, inner walls slightly thickened; lumina flask-shaped, base of flask directed inwards, neck narrow to v. narrow, wavy. **Stomata** superficial; subsidiary cells sometimes with ridge on outer wall; guard cells often with ridge on outer wall, or with lip at outer aperture and ridge on inner wall, outer lip becoming a ridge on outer wall at each end. **Hypodermis** absent. **Chlorenchyma** composed of 1 layer of peg-cells, cells mainly between (6), 8, and 15 times higher than wide, with the shorter cells opposite to stomata; pegs small, numerous, mostly arranged in 6 longitudinal files; protective cells absent. **Parenchyma sheath:** cells in 1 layer, rounded, as wide as to 1½–(2) times wider than high, sometimes 1½–2 times higher than wide (*L. preissianus*); walls slightly thickened; some stegmata present, with moderately thickened inner and anticlinal walls and slightly thickened outer wall. **Sclerenchyma:** sheath well developed, outline circular; fibres v. thick-walled, in 6–8 layers, those of outer layers narrow, those of inner layers wider, clearly distinct from ground tissue except in *L. preissianus*.

Vascular bundles. (i) Peripheral vb's with single-layered arc of narrow or medium-sized tracheids partially enclosing rounded or oval phloem pole; flanking tracheids wider than rest in *L. drapetocoleus*. (ii) Medullary vb's mostly with 1 medium-sized to wide, rounded or oval-angular, thick-walled mxv on either flank, separated by 2–4 rows of narrow cells. Lateral wall pitting of mxv scalariform; perforation plates simple, slightly oblique. Phloem poles oval, ensheathed by 1 layer of narrow cells with moderately thickened walls. All except smallest, outer bundles with px pole; some bundles embedded in culm scl. sheath, but most scattered in outer region of ground tissue. **Bundle sheaths:** fibres narrow to medium-sized, with moderately thickened walls, in 1–2 layers at poles, and 1 layer on flanks. **Central ground tissue** parenchymatous, cells of outer layers surrounding vb's with slightly or moderately thickened walls; central cells thin-walled, breaking down to form a wide central cavity. **Silica** present as (i) spheroidal-nodular bodies in stegmata in parenchyma sheath, and occasional cells of central ground tissue; (ii) granular material in some cells of central ground tissue. **Tannin:** none seen.

Rhizome not seen.

Root T.S.

Not seen in *L. chaetocephalus*.

Root-hairs: same width as cells from which they arise. **Epidermis:** cells thin-walled, about as high as wide. **Cortex:** (i) outer part 4-layered, cells slightly larger than those of epidermis, with slightly thickened walls; (ii) middle part composed of 6–8 layers of wide, thin-walled cells, with intercellular spaces; (iii) inner part 2-layered, cells similar in width to those of outer cortex, more or less isodiametric, walls slightly thickened. **Endodermis** 1-layered, cells *c.* 1½ times higher than wide, all walls v. thick, lumina medium-sized or narrow. **Pericycle** composed of 2–3 layers of hexagonal, thick-walled cells. **Vascular tissue:** mxv's wide, circular, present in outer ring and scattered throughout central ground tissue composed of narrow, thick-walled cells. **Tannin:** none seen.

MATERIAL EXAMINED

Lepidobolus chaetocephalus F. v. Muell.: Pritzel 829; Avon, Australia (K).
L. drapetocoleus F. v. Muell.: Cheadle CA362; Adelaide, Australia (S).
L. preissianus Nees: (i) Blake, S. T. 18106; S.W. Div. 10–20 mls. N. of Northampton, Australia (K). (ii) Helms, An. 1891; Elder Exploring Expd., Victoria Desert, Australia (K). (iii) Cheadle CA495; Perth, Australia (S).

SPECIES DESCRIPTIONS

***L. chaetocephalus* F. v. Muell.** Culm diameter 1·5 mm.

Culm surface (Fig. 33. B)

Hairs as for genus. **Epidermis:** cells mostly as long as wide. **Stomata:** subsidiary cells with obtuse end walls; guard cell pair with compressed oval outline but slightly longer than usual.

Culm T.S. (Fig. 33. A)

Epidermis: cells ranging from twice to 6 times higher than wide. **Silica:** spheroidal-nodular bodies present in stegmata of parenchyma sheath.

***L. drapetocoleus* F. v. Muell.** Culm diameter 1·5 mm.

Culm surface not seen.

Culm T.S.

Hairs: none seen (material from within leaf sheath). **Epidermis:** cells as high as to 1½ times higher than wide. **Stomata:** subsidiary cells with ridge on outer wall. **Silica** present as large bodies in stegmata of parenchyma sheath.

Leaf and *root*: see genus description.

***L. preissianus* Nees.** Culm diameter 0·8–1·2 mm.

Culm surface (Fig. 33. C, F)

Hairs: see genus description. **Epidermis:** cells as long as to twice as long as wide. **Stomata:** subsidiary cells variable; guard cells with rounded end walls.

Culm T.S. (Fig. 33. D, E)

Epidermis: cells 1½–2 times higher than wide, the taller cells at ends of stomata. **Silica:** spheroidal-nodular bodies in stegmata in parenchyma sheath.

Leaf and *root*: see genus description.

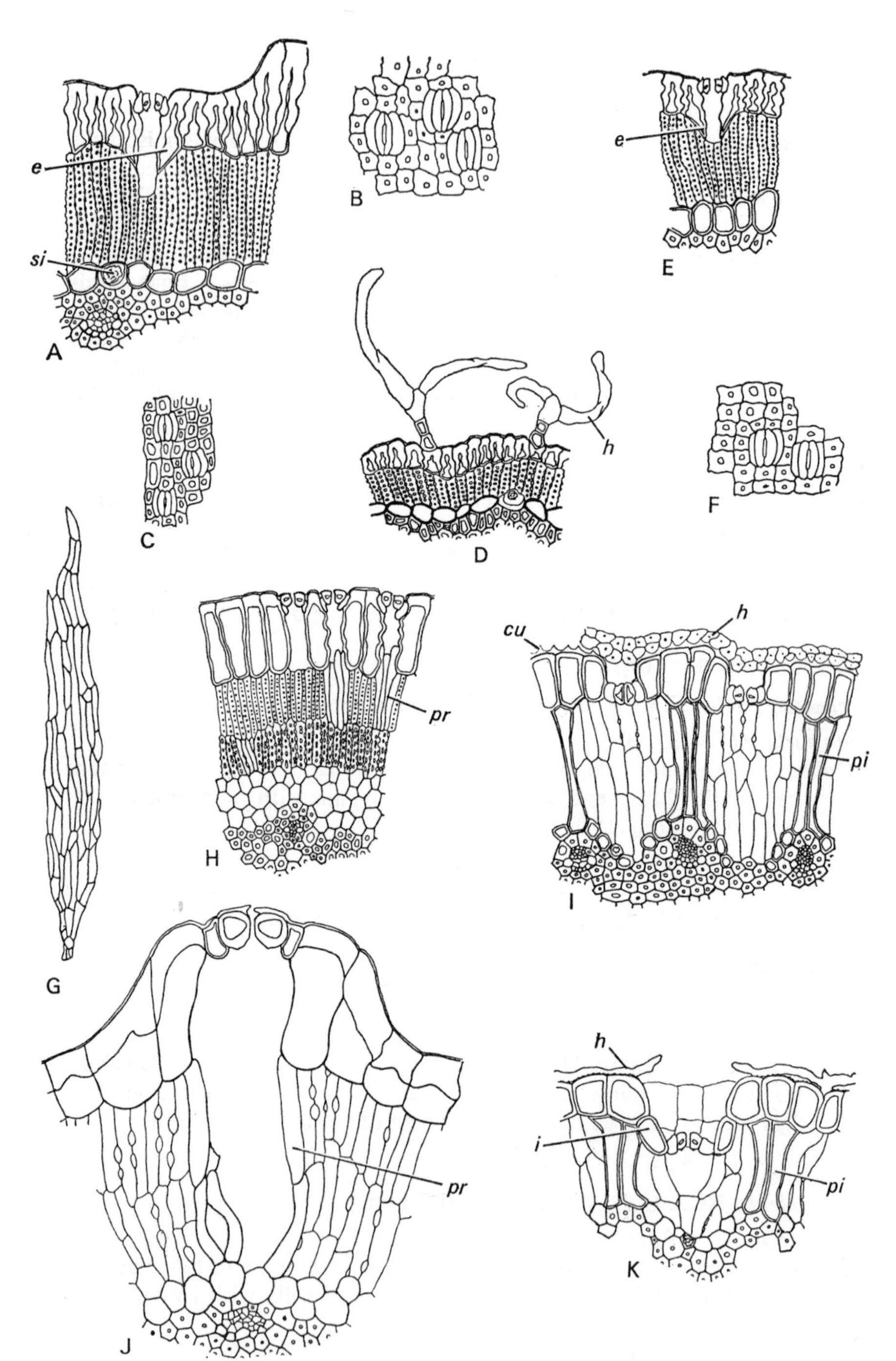

FIG. 33. Restionaceae. A–F, *Lepidobolus*. A, B, *L. chaetocephalus*: A, detail of culm T.S.; B, culm surface. C–F, *L. preissianus*: C, culm surface (Helms); D, detail of culm T.S., note hairs (Helms); E, detail of culm T.S. (Blake 18106); F, culm surface (Blake 18106)(A–F ×265). G–K, *Leptocarpus*. G, K, *L. tenax*: G, hair, surface view (×95); K, detail of culm T.S., note inwardly directed epidermal cells (i) (×185). H, *L. parkeri*, detail of culm T.S. (×100). I, *L. chilensis*, detail of culm T.S., note hair (×185). J, *L. impolitus*, detail of culm T.S. (×185).

cu = cuticle; e = elongated epidermal cell; h = hair; pi = pillar cell; pr = protective cell; si = silica-body.

Taxonomic Notes

The genus has been accepted by all authorities since it was first described. Gilg (1890) saw one species (*L. preissianus* Nees) in which he noted the elongated cells that surround the stomata and extend into the 1-layered chlorenchyma. The 'mucilage-cells' which he recorded in the parenchyma sheath are stegmata containing spheroidal-nodular silica-bodies. His description matches that given in this present account, except that he did not see the hairs. Gilg-Benedict (1930) described the anatomy of the genus as being similar to that of *Onychosepalum* Steud. Her brief account of *Onychosepalum* contains no mention of the elongated epidermal cells that surround the stomata (see p. 243).

The members of the genus *Lepidobolus* exhibit a simple form of culm organization, similar to that of *Coleocarya*. There is a full discussion of the possible interrelationships between these and other genera on p. 159. The hair type found in *Lepidobolus* may serve to distinguish it anatomically from the other genera.

LEPTOCARPUS R. Br.

No. of species 40; examined 19.

Genus Description

The spp. form 2 distinct geographical and anatomical groups:

1. spp. from South Africa.
2. spp. from Australia, New Zealand, Malaysia, South Vietnam, and South America.

GROUP 1

Leaf surface

L. rigoratus only (3911), abaxial.

Hairs: none seen. **Epidermis:** cells 4-sided, mostly as long as wide, some *c.* 1½ times longer than wide; walls moderately thickened, v. wavy. **Stomata** *c.* ⅔ of size of those in culm of same sp.; subsidiary cells with obtuse end walls, long anticlinal walls slightly wavy; guard cell pair with lemon-shaped outline but shorter than normal for the family. **Silica** and **tannin:** none seen. **Cuticular marks** granular.

Leaf T.S.

L. rigoratus (3911).

Epidermis. (i) Abaxial cells 4-sided, mostly as high as wide, outer walls v. thick; outer halves of anticlinal walls thick, thickening tapering rapidly, inner halves of walls moderately thickened; inner walls moderately thickened. (ii) Adaxial cells 4-sided, v. flattened, *c.* 4 times wider than high and *c.* ⅛ of height of abaxial cells; all walls moderately thickened. **Stomata** present in abaxial surface only, superficial; guard cells with lips at outer and inner apertures. **Chlorenchyma** composed of peg-cells in 1 layer below abaxial epidermis; individual cells about twice as high as wide; protective cells absent.

Vascular bundles: large and small alternating; small with 5–10 narrow tracheids and 8–12 phloem cells; large with 1 medium-sized tracheid on either flank of 4–5 narrow to medium-sized tracheids and 8–20 phloem cells. **Bundle**

sheaths. O.S. parenchymatous, as 1-layered cap at phloem, extending laterally and forming continuous sheet below chlorenchyma. I.S. sclerenchymatous, completely ensheathing the vb; fibres narrow, thick-walled, arranged in 1 layer; joined on flanks by ground scl. **Sclerenchyma** present in bundle sheaths and in 1–2 layers between bundles, bounded abaxially by parenchyma layer and adaxially by 1 layer of wide parenchyma cells, these separating it from adaxial epidermis. Scl. continuous to adaxial side of bundle sheaths; fibres either as wide as high, or up to 1½ times higher than wide, with v. thick walls and narrow lumina. **Ground tissue:** single layer of wide parenchymatous cells next to adaxial epidermis, as described above. **Air-cavities** absent. **Silica** and **tannin:** none seen.

Culm surface

Hairs absent. **Papillae** present on basal material of *L. asper*, short, rounded. **Epidermis:** cells 4- or 6-sided, as long as or up to 6 times longer than wide; smallest cells at ends of stomata; walls slightly to moderately thickened, usually wavy. **Stomata:** subsidiary cells variable; guard cells with rounded end walls or guard cell pair with compressed oval outline. **Silica** absent. **Tannin** occasionally present in epidermal cells. **Cuticular marks** granular, occasionally with superimposed longitudinal striations.

Culm T.S.

Outline circular, sometimes with irregularly distributed, low, flat-topped mounds, or with shallow depressions, e.g. *L. esterhuysianae*, *L. membranaceus*. **Cuticle** slightly to moderately thickened, granular, occasionally with slight ridges. **Papillae:** see under culm surface above. **Epidermis.** Cells in 1 layer, either (i) all of similar heights, or (ii) with irregular areas of tall cells amongst shorter cells, giving rise to flat-topped mounds. Cells as high as wide or up to 5 times higher than wide; outer walls thick; anticlinal and inner walls moderately thickened, anticlinal walls straight or wavy. **Stomata** superficial, frequently raised on mounds formed by elongated epidermal cells; subsidiary cells variable; guard cells with outer lip only, or with outer and inner lips, or with outer lip and ridge on inner wall. **Chlorenchyma** composed of peg-cells in 1, 2, or 3–(4) uninterrupted layers; protective cells present around substomatal cavities in 1 or 2 layers or extending from *c.* ½ way up substomatal tube lined by epidermal cells, right to or near to parenchyma sheath. **Parenchyma sheath** 1- or 2–4-layered. **Sclerenchyma:** sheath well developed in most spp., outline circular, or with low, dome-shaped ribs opposite peripheral vb's. Wide, square-ended ribs alternating with peripheral vb's and extending well into chlorenchyma in 1 species. Sheath with all peripheral vb's and all, some, or no medullary vb's embedded in it.

Vascular bundles. (i) Peripheral vb's small, with narrow tracheids arranged in single-layered arc partially enclosing rounded or oval phloem pole (flanking tracheids occasionally wider than others). (ii) Medullary vb's: outer bundles small with 1 medium-sized vessel on either flank or, occasionally, with 1 central mxv; inner bundles larger, with 1 wide mxv on either flank. Lateral wall pitting of vessels scalariform; perforation plates simple, transverse. Vessels separated from one another by 2–4 narrow cells. Phloem pole ensheathed by 1 layer of narrow cells with moderately thickened walls. **Bundle**

sheaths: fibres at poles narrow, thick-walled, in 1–2–(3) layers, those on flanks wider, with slightly or moderately thickened walls, in 1 layer; or sheath weak, all fibres with slightly thickened walls. **Central ground tissue** parenchymatous, outer cells surrounding vb's with moderately thickened walls, central cells thin-walled, frequently breaking down to form central cavity; occasional areas of thin-walled cells scattered in outer layers in some spp. **Silica:** spheroidal-nodular bodies present in some spp. in inner chlorenchyma layer or parenchyma sheath; irregularly shaped bodies present in protective cells in *L. impolitus*. **Tannin** occasionally present in some epidermal cells.

Rhizome and *root* not seen.

GROUP 2

Leaf surface not seen.

Leaf T.S.

Seen in *L. simplex*, *L. spathaceus*, and *L. tenax*.

Hairs: none seen. **Epidermis:** (i) adaxial cells 5-sided, slightly or 1½ times wider than high, walls v. thick, lumina narrow; (ii) abaxial cells mostly 4-sided, as high as wide to twice as high as wide, walls slightly or moderately thickened; about twice height of adaxial cells. **Stomata** present in abaxial surface only, superficial; subsidiary cells sometimes with ridge on outer wall; guard cells without lips or ridges. **Chlorenchyma** composed of lobed, more or less isodiametric cells present in 1 or 2 layers next to abaxial epidermis; divided into longitudinal strands by bundle sheaths.

Vascular bundles of 2 main types: (i) with few or several, narrow to medium-sized tracheids and phloem cells; (ii) with 1 wider tracheid on either flank of group of several narrow tracheids and phloem cells. Bundles of the 2 types alternating. **Bundle sheaths:** O.S. parenchymatous, present as 1 layer of cells at phloem pole in *L. spathaceus* only; I.S. sclerenchymatous, fibres narrow, thick-walled, in 2–3 layers. **Ground tissue:** cells wide in *L. simplex* and *L. tenax*, narrow or medium-sized in *L. spathaceus* (the narrowest being next to chlorenchyma), all thick- or moderately thick-walled; cells filling all space not occupied by bundles and sheaths or chlorenchyma. **Air-cavities** absent. **Silica:** small spheroidal-nodular bodies present in stegmata in outer layer of scl. sheath in *L. spathaceus*. **Tannin:** none seen.

Culm surface

Hairs (Fig. 33. G, I, K) often present; either (i) multicellular, simple, unbranched, uniseriate, basal cell(s) with slightly thickened walls, other cells thin-walled; or (ii) multicellular, plate-like, with 1 short basal cell and flattened, 2-layered plate. Cells all with thick or v. thick (cellulose) walls, individual cells up to 10 times longer than wide. Plate 4–5 times longer than wide, one end pointed, other end truncated; basal cell attached near truncated (proximal) end, on median axis. Margins of adjacent hairs frequently slightly overlapping one another and completely covering culm in some spp. **Epidermis:** cells 4–6-sided, as long as to 5 times longer than wide, walls straight or v. wavy, slightly or moderately thickened. **Stomata** mainly confined to regions between files of pillar cells; subsidiary cells variable; guard cell pair with lemon-shaped or compressed oval outline, or guard cell ends rounded.

Silica and **tannin** present in epidermal cells of some spp. **Cuticular marks** granular, frequently with longitudinal striations.

Culm T.S.

Hairs: see culm surface. **Outline** circular. **Cuticle** moderately thickened or thick, frequently ridged, ridges matching grooves between cells of hairs. **Epidermis:** cells in 1 layer; as high as to 3 times higher than wide, either all of similar height, or some, next to sunken stomata, shorter; cells next to stomata sometimes v. much modified (in *L. tenax*). Outer walls thick or moderately thickened, other walls moderately thickened, anticlinal walls not wavy. **Stomata** superficial or sunken; subsidiary cells v. variable, see individual spp. Guard cells usually without lips or ridges, but occasionally with slight lip at inner aperture or slight ridge on inner wall. **Chlorenchyma** normally 2–3-, or occasionally 4- or 5-layered, layers divided into longitudinal sectors by ribs from scl. sheath and their associated pillar cells. Chlorenchyma cells palisade-like, occasionally with few pegs; protective cells absent. Palisade cells arranged in transverse plates or chequerboard pattern as seen in T.L.S.

Parenchyma sheath composed of 2 types of cell: (i) thin-walled, narrow, rounded cells in 1 layer between chlorenchyma and scl. sheath, often containing stegmata, and frequently interrupted, occurring only on flanks of scl. ribs; (ii) pillar cells, extending between rounded ribs from scl. sheath to epidermis; pillar cells mostly 2–3–(4 or 5) times higher than wide, with slightly or moderately thickened walls. Pillar cells arranged in longitudinal files, normally 2–3 cells wide as seen in T.L.S. **Sclerenchyma:** sheath well developed, 3–4- or 8–10-layered, outline circular, with dome-shaped ribs (partial girders) each opposite to and enclosing 1 peripheral vb; ribs better developed in some spp. than others. Fibres of ribs and outer layers narrow, with thick or v. thick walls, those of inner layers wider, with moderately thickened walls.

Vascular bundles. (i) Peripheral vb's: tracheids all narrow, forming 1–2-layered arc, or arc with 1 medium-sized tracheid on either flank; phloem pole rounded or oval, situated in concavity of arc. (ii) Medullary vb's: outer bundles small, with 1 narrow mxv on either flank; usually lacking px pole; inner, larger bundles with 1 medium-sized to wide, circular or angular mxv on either flank. Vessels with scalariform lateral wall-pitting and slightly transverse, simple perforation plates. Mxv's normally separated from one another by 2–3 files of narrow cells. Phloem pole rounded or oval, abutting directly on xylem or separated from it by 2 layers of narrow cells with moderately thickened walls next to xylem and thin walls next to phloem. Px present in all inner, larger bundles. Some, all, or no bundles free from culm scl. sheath. **Bundle sheaths:** fibres at poles narrow, thick-walled, in (1)–2–3 layers; those on flanks wider, with slightly or moderately thickened walls, in 1 layer. **Central ground tissue** parenchymatous; outer layers composed of cells with slightly to moderately thickened walls, cells of inner layers with thin walls, often breaking down to form central cavity; inner layers separated from outer layers by 1 or 2 layers of narrow, thick-walled cells in some spp. **Silica:** spheroidal-nodular bodies present in stegmata in parenchyma or scl. sheath; sometimes also as irregular bodies or granules in pillar cells, chlorenchyma or central ground tissue. **Tannin** in some epidermal cells of some spp.

Rhizome T.S.

Seen in *L. simplex* and *L. tenax*.

Hairs: (i) unicellular, thin-walled, as high as to 1½ times higher than wide, rounded; and (ii) multicellular, short, biseriate, composed of more or less isodiametric cells with thin walls, hair about 3–4 cells long. **Epidermis:** cells 4-sided, as high as wide, walls thin. **Cortex:** (i) outer part composed of 4–6 layers of moderately thick- or thick-walled sclereids, individual sclereids mostly 1½ times wider than high; (ii) inner part, 5–7–(10)-layered, cells parenchymatous, lobed, loosely packed, more or less isodiametric. **Endodermoid sheath** 1–2-layered, cells with moderately thickened walls in *L. simplex*, outer walls moderately thickened and other walls thick in *L. tenax*. **Vascular bundles:** outer with arc of mxv's, inner amphivasal; vessels narrow to medium-sized, angular, with thick walls; phloem pole rounded. **Bundle sheaths** indistinguishable from ground tissue. **Central ground tissue:** cells around bundles narrow, with moderately thickened walls; areas of thin-walled cells dispersed amongst thicker cells. **Silica:** small spheroidal-nodular bodies present in some thin-walled cells of central ground tissue. **Tannin:** none seen.

Root T.S.

Seen in *L. aristatus*, *L. canus*, *L. simplex*, and *L. tenax*.

Root-hairs infrequent, arising from cells similar to those of remainder of epidermis. **Epidermis:** cells 5-sided; outer wall convex; all walls thin. **Cortex:** (i) outer part composed of 4 peripheral layers of 4–5-sided more or less square cells with thin walls (slightly lignified), and 3–5 inner layers of cells with moderately thickened walls; (ii) middle part consisting of radiating plates of parenchymatous cells, plates separated by air-spaces; (iii) inner part 1-layered, cells narrow, as high as wide, thin-walled. **Endodermis** 1-layered, cells slightly or 1½–2 times higher than wide, walls v. thick, lumina narrow. **Pericycle** composed of 1 layer of narrow, parenchymatous cells. **Vascular system:** mxv's wide, angular, radially oval, mostly in 1 ring. **Central ground tissue:** cells narrow to medium-sized, hexagonal, with moderately thickened walls. **Tannin:** none seen.

MATERIAL EXAMINED

GROUP 1. Spp. from South Africa.

Leptocarpus asper (Mast.) Pillans: (i) Burchell, An. 1815; Niewe Kloof (K). (ii) Burchell, no number (M).

L. burchellii Mast.: Burchell 7185, An. 1874 (K).

L. esterhuysianae Pillans: Esterhuysen 8770, paratype, An. 1942 (K).

L. impolitus (Kunth) Pillans: Bolus, H. 12895, ♀; Hopefield (K).

L. membranaceus Pillans: Schlechter 7765 ♀ (K).

L. parkeri Pillans: Parker, R. N. 4695 ♂, An. 1951; Caledon Div. (K).

L. rigoratus Mast.: (i) Parker, R. N. 3911 ♀, An. 1944; Faure, Stellenbosch Div. (K). (ii) Parker, R. N. 3914 ♂, An. 1944; Faure, Stellenbosch Div. (K).

L. vimineus (Rottb.) Pillans: (i) Zeyher 4989 ♂ (K). (ii) Matching 63a! in Drège's Herb. (K). (iii) Received from Reichenbach fil. in 1865; Worcester Div. (K). (iv) Parker, R. N. 3456 ♂, An. 1940; Somerset West. (K).

GROUP 2. Spp. not from South Africa.

L. aristatus R. Br.: (i) Andrews, C., 1st coln. 1100, An. 1902; Torbay Junction, Nr. Albany, Aust. (K). (ii) Cheadle, V. I. CA71; Perth, Aust. (S). (iii) Gardner, C. A., An. 1936, ♂; Kenwick, Aust. (K). (iv) Mueller, F., An. 1876 (in Kew Herb. as *L. anomalus*, mss. name, C. A. Gardner) (K).

L. brownii Hk. f.: (i) Willich ♂; Boston Pt., S.W. Aust. (K). (ii) Hicks, A. J. 47 ♀; Little Desert, Victoria, Aust. (K).

L. canus Nees: (i) Andrews, C. 1097 ♂; Midland Junction, Perth, Aust. (K). (ii) Andrews, C. 1098 ♀; Midland Junction, Perth, Aust. (K).

L. chilensis (Steud.) Mast.: Hook. Herb. 853, An. 1867; Valdivia, S. America (K).

L. coangustatus Nees: (i) 2662 ♂; Aust. (K). (ii) 2662 ♀; Aust. (K).

L. disjunctus Mast.: (i) Vestal, An. 1917; Trengganu, Malaya (K). (ii) Poilane, M. E. 23177; Cochin China (S. Vietnam) (K).

L. erianthus Benth.: (i) Gardner, C. A., An. 1924, ♀; Hill River, Aust. (K). (ii) Meebold, A., An. 1928; Kenwick, Aust. (K).

L. ramosus R. Br.: Specht, R. L. 932 ♂ An. 1948; Arnhem Land, Aust. (K).

L. simplex A. Rich.: (i) Moore, G. 3704; Spencerville, Canterbury, Salt Marsh, New Zealand (S). (ii) Anderson, A. W. 199 ♂; New Zealand (K). (iii) Wallace, E. H. 4551 ♀; Timani, S. Island, New Zealand (K).

L. spathaceus R. Br.: (i) Specht, R. L. 1953 ♂, An. 1948; Aust. (K). (ii) No. 428; Aust. (K).

L. tenax (Labill.) R. Br.: (i) Rodway 334, An. 1931; Aust. (K). (ii) Cheadle, V. I. CA 94; Grampians, Aust. (S). (iii) Curtis, W. M.; Blackman's Bay, Hobart, Tasmania (S). (iv) Drummond, An. 1843 (in Kew Herb. as *Restio amblycoleus*); Swan R., Aust. (K). (v) Hubbard, C. E. 2631 (in Kew Herb. as *Restio gracilis* R. Br.); Bribie Is., Aust. (K). (vi) Main and Constable, N. H., N.S.W. 6416 (in Kew Herb. on sheet of *Calorophus minor* Hk. f.); Aust. (K).

SPECIES DESCRIPTIONS

GROUP 1

***L. asper* (Mast.) Pillans.** Culm diameter 2 mm.

Culm surface

Papillae: 1 present on each of several isolated cells, short, rounded. **Epidermis:** cells 1½–2 times longer than wide; those around stomata longer than others. **Stomata:** subsidiary cells with acute end walls, long anticlinal wall sometimes slightly wavy; guard cell pair with compressed oval outline. **Tannin** in few epidermal cells. **Cuticular marks:** longitudinal striations present over papillae.

Culm T.S.

Epidermis. Cells of 2 types: (i) those in files between stomata, slightly wider than high; (ii) those in files with stomata, 2–3 times higher than wide, and 2–3 times higher than shorter cells. **Stomata** raised on tall cells; subsidiary cells narrow; guard cells with ridge on inner wall. **Chlorenchyma** mostly 2-layered; cells 2–3 times higher than wide. **Parenchyma sheath** 1-layered, cells mostly slightly wider than high. **Silica:** none seen. **Tannin** in few epidermal cells.

***L. burchellii* Mast.** Culm diameter 0·7 mm.

Culm surface

Epidermis: cells 4(–6)-sided, as long as or up to twice as long as wide. **Stomata:** subsidiary cells with acute end walls. **Tannin** in few epidermal cells. **Cuticular marks** granular.

Culm T.S.

Epidermis: cells 1½–2 times higher than wide, those next to stomata tallest; outer walls thick; anticlinal walls thick at outer ends, thickening tapering rapidly, and inner ⅔ of walls slightly thickened, wavy. **Stomata:** subsidiary cells with slight ridge on outer wall; guard cells with outer lip and ridge on inner wall, walls to pore concave. **Chlorenchyma** 2-layered; cells of outer layer *c.* 10 times higher than wide, cells of inner layer *c.* 8 times higher than wide, as high as, but slightly wider than outer cells and with fewer, larger pegs. Protective cells extending in 2 layers, each layer corresponding in height to adjacent peg-cells; walls slightly thickened; apertures irregular in shape, occurring mainly between cells of inner layer. **Parenchyma sheath** 1–2-layered. **Silica:** none seen. **Tannin** in few epidermal cells.

***L. esterhuysianae* Pillans.** Culm diameter 2·0 mm.

Culm surface

Epidermis: cells 4-sided, mostly as long as wide, those flanking stomata frequently up to twice as long as wide, those at ends of stomata frequently twice as wide as long; walls not wavy. **Stomata** closer together than usual for the genus; subsidiary cells with acute end walls; guard cells wider and shorter than usual, with rounded ends. **Tannin** in some epidermal cells.

Culm T.S.

Outline circular, with frequent, shallow depressions. **Epidermis:** cells 5–6 times higher than wide, shorter cells occurring irregularly in small groups between stomata; outer surface irregular, roughly granular, particularly in shorter cells; anticlinal walls wavy, with thick outer ends, thickening tapering gradually, inner ends moderately thickened. **Stomata:** subsidiary cells short and narrow, with slightly thickened walls; guard cells with pronounced lip at outer aperture. **Chlorenchyma:** 2 layers of cells; those of outer layer *c.* 8–9 times higher than wide; those of inner layer *c.* ⅔ height of, and ⅓ wider than, cells of outer layer, each being 5–6 times higher than wide, with fewer, wider pegs. Protective cells extending to parenchyma sheath; walls moderately thickened, cells of inner layer only slightly modified peg-cells; apertures between cells rounded, occurring mainly between cells of inner layer, and at junction between outer and inner layers. **Parenchyma sheath** 3-layered on either side of peripheral vb's, but only 2-layered between them. **Silica:** none seen. **Tannin** in some epidermal cells.

***L. impolitus* (Kunth) Pillans.** Culm diameter 1·0 mm.

Culm surface

Epidermis: cells as long as wide; those next to stomata with irregular outlines. **Stomata**: subsidiary cells more or less crescentiform. **Tannin** in some epidermal cells.

Culm T.S. (Fig. 33. J)

Epidermis: cells of 2 main sizes: (i) *c.* 1½ times higher than wide, between stomata; (ii) taller, surrounding stomata, up to 3½ times higher than wide; cells of intermediate heights present, with sloping outer walls, linking the 2 types. Anticlinal walls not or only slightly wavy. **Stomata** raised on mounds; subsidiary cells with moderately thickened walls; guard cells with lip at outer aperture and slight ridge on inner wall; substomatal cavities unusually wide at outer end. **Chlorenchyma**: 2–(3) similar layers of cells; individual cells mostly 8–10 times higher than wide; pegs large, sparse; protective cells in 2 layers. **Silica**: irregularly shaped silica-bodies present in many protective cells.

L. membranaceus **Pillans.** Culm diameter 1·5 mm.

Culm surface

Epidermis: cells mostly hexagonal, 1½ times wider than long, those flanking stomata slightly longer than wide. **Stomata**: subsidiary cells with acute end walls. **Tannin** in some epidermal cells. **Cuticular marks** with slight longitudinal striations.

Culm T.S.

Outline circular with shallow depressions. **Epidermis**: cells mostly 3–4 times higher than wide, occasional cells only twice as high as wide; outer walls v. thick, more or less straight or concave; anticlinal walls wavy, v. thick at outer ends, thickening tapering and inner ⅓ of walls only slightly thickened. **Stomata**: guard cells with pronounced outer lip. **Chlorenchyma**: cells in 3–4 layers; those of outer 2 layers *c.* 5 times higher than wide, those of inner layer(s) slightly wider, ranging from slightly higher than wide to 4 or 5 times higher than wide. Protective cells well developed, extending in 2 or 3 layers to inner chlorenchyma layer, or to cells of parenchyma sheath; apertures between cells rounded, most frequent at inner end of substomatal tube, but also occurring along its whole length. **Parenchyma sheath** 1–2-layered; some cells, particularly in outer layer, with lignified thickenings on inner and anticlinal walls. **Central ground tissue** as for group 1, but with scattered areas of thin-walled cells in outer region. **Silica**: bodies present in v. few cells of inner chlorenchyma layer.

L. parkeri **Pillans.** Culm diameter 1 mm.

Culm surface

Epidermis: cells 3–6 times longer than wide. **Stomata**: subsidiary cells with perpendicular or acute end walls. **Tannin** in some epidermal cells.

Culm T.S. (Fig. 33. H)

Outline circular. **Epidermis**: cells mostly *c.* 5 times higher than wide, outer walls v. thick, dome-shaped; anticlinal and inner walls moderately thickened. **Stomata**: subsidiary cells with ridges on outer wall; guard cells with weak outer and inner lips. **Chlorenchyma** composed of peg-cells in 2 layers, cells 6–8 times higher than wide; protective cells extending to parenchyma sheath. **Parenchyma**

sheath 2–3-layered, narrowest opposite ribs of fibres from scl. sheath; cells as high as to 1½ times higher than wide, with rounded corners. **Silica:** bodies present in some cells of parenchyma sheath.

L. rigoratus **Mast.** Culm diameter 1 and 1·2 mm.

Leaf surface and *T.S.*: see genus description for spp. belonging to group 1.

Culm surface

Epidermis: cells 4-, 5-, and 6-sided, as long as or up to 1½ times longer than wide, walls moderately thickened, slightly wavy. **Stomata:** subsidiary cells with rounded or acute end walls. **Tannin** in some epidermal cells. **Cuticular marks:** irregularly striate.

Culm T.S.

Epidermis: cells mostly 1½ times higher than wide; outer part of anticlinal walls thick, thickening tapering and inner halves of walls slightly to moderately thickened. **Stomata:** guard cells with lip at outer and inner apertures. **Chlorenchyma:** 2 similar layers of cells, each cell *c.* 4 times higher than wide; protective cells penetrating 1 or both layers. **Parenchyma sheath** 2- or, occasionally, 3-layered, cells of unequal sizes, mostly slightly wider than high. **Silica:** none seen.

L. vimineus **(Rottb.) Pillans.** Culm diameter 1, 0·8 mm.

Culm surface

Epidermis: cells mostly 6-sided, those next to stomata about as long as wide, others about 1½ times longer than wide. **Stomata:** subsidiary cells with acute end walls. **Tannin** in epidermal cells of some samples.

Culm T.S.

Outline circular. **Epidermis:** cells slightly higher or 1½ times higher than wide in 3456, twice as high as wide in other specimens. **Stomata** superficial, except in 63a! where slightly sunken; guard cells with lip at outer aperture and ridge on inner wall. **Chlorenchyma** 2-layered, cells of outer layer 6–10 times higher than wide, cells of inner layer mostly 5 times higher than wide, with fewer, larger pegs, except in 3456, where cells of both layers *c.* 5 times higher than wide. Protective cells extending in 2 layers to parenchyma sheath. **Parenchyma sheath** 1–2-layered, cells mostly slightly wider than high. **Silica:** none seen.

GROUP 2

L. aristatus **R. Br.** Culm diameter 0·5–1·0 mm.

Culm surface

Hairs of group 2 type (ii) (see p. 203). **Epidermis:** cells as long as wide to 3–5 times longer than wide; walls slightly thickened, v. wavy. **Stomata:** subsidiary cells shorter than guard cells, with acute or obtuse end walls; guard cell pair with lemon-shaped outline. **Tannin** in some epidermal cells in 1 specimen. **Cuticular marks:** longitudinal striations.

Culm T.S.

Cuticle moderately thickened, with ridges. **Epidermis:** cells largest over pillar cells. **Stomata** small, slightly sunken; guard cells with no pronounced lips or ridges. **Chlorenchyma** 2–3-layered, cells 1½–2 times higher than wide. **Central ground tissue** without inner scl. sheath. **Silica** present as (i) bodies, in stegmata of outer layer of scl. sheath, and (ii) granular particles, in some pillar cells.

Root: see genus description for spp. belonging to group 2.

***L. brownii* Hk. f.** Culm diameter 1·0, 0·8 mm.

Culm surface

Hairs of group 2 type (ii) (see p. 203). **Epidermis:** cells mostly 6-sided, 2–4 times longer than wide; walls thick, end walls sometimes oblique. **Stomata:** subsidiary cells crescentiform; guard cell pair outline compressed oval. **Silica:** granular material present in some hair-cells. **Tannin:** none seen. **Cuticular marks:** wavy, longitudinal striations.

Culm T.S.

Cuticle moderately thickened, with pronounced ridges. **Epidermis:** cells all of about same height, 1½–2 times higher than wide (according to plane of section). **Stomata** sunken; attached to innermost end of anticlinal wall of epidermal cell; guard cells without conspicuous lips or ridges. **Chlorenchyma** 2-layered. **Vascular bundles** as for spp. belonging to group 2 (see p. 204), but some outer medullary vb's with 1 narrow, central mxv only. **Central ground tissue** without inner scl. sheath. **Silica** present as (i) bodies in stegmata in outer layer of scl. sheath, and (ii) irregularly shaped particles in some pillar cells.

***L. canus* Nees.** Culm diameter 1 mm.

Culm surface

Hairs of group 2 type (ii) (see p. 203). **Epidermis:** cells 4- or 6-sided, mostly 1½–2 times longer than wide, those above pillar cells slightly larger than others; walls wavy. **Stomata:** subsidiary cells with acute end walls; guard cell pair with lemon-shaped outline. **Tannin:** none seen. **Cuticular marks** granular.

Culm T.S.

Epidermis: cells mostly 1½ times higher than wide, but taller over pillar cells and shorter next to guard cells. **Stomata** slightly sunken; subsidiary cells with slight ridge on outer wall; guard cells with ridge on inner wall. **Chlorenchyma:** cells 2–3 times higher than wide, in 3–4 layers; as for group 2. **Vascular bundles** as for spp. belonging to group 2, but peripheral vb's with 1 wide tracheid on either flank. **Central ground tissue** with 1-layered, inner scl. sheath separating outer from inner cells. **Silica:** bodies present in stegmata of culm scl. sheath and also in some pillar cells.

Root T.S. as for group 2.

***L. chilensis* (Steud.) Mast.** Culm diameter 1·5 mm.

Culm surface not seen.

Culm T.S. (Fig. 33. I, p. 200)

Cuticle thick, with pronounced ridges. **Hairs** of group 2 type (ii) (see p. 203). **Epidermis:** cells next to stomata *c.* 1–2 times higher than wide, those over pillar cells slightly more than twice as high as wide. **Stomata** sunken, attached to inner ends of anticlinal walls of adjacent epidermal cells; guard cells with lip at inner aperture. **Chlorenchyma** 2–3-layered, some cells with few, short pegs; individual cells 3–7 times higher than wide; occasional areas of cells with moderately thickened, lignified walls, otherwise as for group 2. **Vascular bundles** as for spp. belonging to group 2, but peripheral vb's with medium-sized tracheids flanking arc of narrow tracheids. **Central ground tissue** without inner scl. sheath. **Silica** present as: (i) bodies in stegmata in outer layer of culm scl. sheath, and (ii) granular material in some cells of chlorenchyma and central ground tissue. **Tannin:** none seen.

***L. coangustatus* Nees.** Culm diameter 1–1·5 by 0·75 mm.

Culm surface

Hairs of group 2 type (ii) (see p. 203). **Epidermis:** cells 4-, 5-, and 6-sided, mostly as long as wide, those over pillar cells frequently up to 1½ times longer than wide. **Stomata:** subsidiary cells variable, but mainly with obtuse or acute end walls; guard cell pair with lemon-shaped outline. **Tannin:** not seen. **Cuticular marks** finely granular.

Culm T.S.

Outline semicircular to kidney-shaped. **Epidermis:** cells next to stomata *c.* 1½ times higher than wide, others reaching 2–3 times higher than wide; outer walls of cells adjacent to stomata curving down sharply towards stomata. **Stomata** sunken, inner walls level with inner walls of epidermal cells; guard cells without pronounced lips or ridges. **Chlorenchyma:** cells in 4–5 layers, without pegs, cells mostly twice as high as wide. **Central ground tissue** without inner scl. sheath. **Silica:** bodies present in stegmata in outer layer of scl. sheath and also in some pillar cells. **Tannin:** none seen.

***L. disjunctus* Mast.** Culm diameter 1·5 mm.

Culm surface

Hairs of group 2 type (ii) (see p. 203). **Epidermis:** cells mostly 4-sided, 1–2 times longer than wide; walls slightly to moderately thickened, wavy. **Stomata:** subsidiary cells slightly longer than guard cells, with acute end walls, long anticlinal walls sometimes slightly wavy; guard cell pair with lemon-shaped outline. **Tannin** in some epidermal cells.

Culm T.S.

Epidermis: cells as high as to 1½ times higher than wide, those over pillar cells up to twice as wide as, and 1½ times higher than, those next to stomata. **Stomata** superficial; subsidiary cells with slight ridge on outer wall; guard cells without pronounced lips or ridges; walls to pore concave. **Chlorenchyma** mainly 4-layered, as for spp. belonging to group 2; cells 2–4 times higher than wide; infrequent, short pegs on anticlinal walls of some cells. **Central ground tissue** as for spp. belonging to group 2, with inner scl. sheath, 1 layer wide, separating outer from inner region. **Silica:** bodies present in stegmata in outer layer of scl. sheath.

***L. erianthus* Benth.** Culm diameter 1·1, 1·0 mm.

Culm surface

Hairs of group 2 type (ii) (see p. 203). **Epidermis:** cells 4–6-sided, as long as wide to 3 times longer than wide, those over pillar cells sometimes larger than others; walls wavy. **Stomata:** subsidiary cells with obtuse end walls; guard cells sometimes slightly longer than subsidiary cells, with rounded end walls or guard cell pair with lemon-shaped outline. **Tannin:** none seen.

Culm T.S.

Epidermis: cells mostly 1½ times higher than wide, largest opposite pillar cells. **Stomata** superficial; subsidiary cells wider than usual, with ridge on outer wall; guard cells without lips or ridges. **Chlorenchyma** 2–3-layered. **Central ground tissue** as for spp. belonging to group 2, without inner scl. sheath. **Silica:** bodies in stegmata in outer layer of scl. sheath. **Tannin:** none seen.

***L. ramosus* R. Br.** Culm diameter 0·8 mm.

Culm surface

Hairs: none seen. **Epidermis:** cells mostly hexagonal and mainly 2–3 times longer than wide, those at ends of stomata often shorter than wide; walls wavy at surface only. **Stomata:** subsidiary cells with obtuse or rounded end walls. **Tannin:** none seen. **Cuticular marks** granular, with close, fine, longitudinal striations.

Culm T.S.

Epidermis: cells as high as to slightly higher than wide. **Stomata** superficial; guard cells without conspicuous lips or ridges; walls to pore concave. **Chlorenchyma** 3–4-layered, as for spp. belonging to group 2; cells of outer layers 2–3 times higher than wide, those of inner layers mostly 1½ times higher than wide. **Central ground tissue** with inner scl. sheath. **Silica:** (i) as bodies in stegmata in outer layer of culm scl. sheath; (ii) as granular material filling lumina of some pillar cells. **Tannin:** not seen.

***L. simplex* A. Rich.** Culm diameter 1·5–2 mm.

Culm surface

Hairs of group 2 type (ii) (see p. 203 and Pl. VII. A); some cells containing silica. **Epidermis:** cells mostly 4-sided, 1½–2–(3) times longer than wide, those at ends of stomata frequently as long as wide. **Stomata**: subsidiary cells with obtuse end walls; guard cell pair with compressed oval outline. **Tannin:** none seen. **Cuticular marks:** longitudinal striations.

Culm T.S. (Pl. VII. A)

Cuticle thick, with pronounced ridges. **Epidermis:** cells over pillar cells slightly higher than wide, those over chlorenchyma about as high as wide; outer walls slightly concave. **Stomata** sunken; guard cells without pronounced lips or ridges. **Chlorenchyma** 2–3-layered, as for spp. belonging to group 2; cells 2–3 times higher than wide. **Central ground tissue:** inner scl. sheath present in 199 only. **Silica** present as (i) bodies in stegmata in outer layer of culm scl. sheath; (ii) irregular, granular inclusions in some hair-cells. **Tannin:** none seen.

Leaf, *rhizome*, and *root*: see description for spp. belonging to group 2.

***L. spathaceus* R. Br.** Culm diameter 1, 1·2 mm.

Leaf T.S. See description for spp. belonging to group 2.

Culm surface

Hairs of group 2 type (i) (see p. 203). **Epidermis:** cells mostly 4- or 6-sided, as long as or up to twice as long as wide, those over pillar cells *c.* 3 times longer and 1½ times wider than other cells. Cells at ends of stoma rounded, as long as wide, with thick walls; walls of other cells slightly to moderately thickened, wavy. **Stomata:** subsidiary cells with acute end walls and irregularly wavy long walls; guard cell pair with lemon-shaped outline. **Tannin:** in some epidermal cells. **Cuticular marks:** longitudinal striations.

Culm T.S.

Epidermis: cells above pillar cells *c.* 1½ times wider and higher than others in 428; all cells more or less square or slightly wider than high; outer walls straight in smaller cells, convex in larger cells. **Stomata** superficial; subsidiary cells with slight ridge on outer wall; guard cells without pronounced lips or ridges. **Chlorenchyma** 3-layered, cells 2–3 times higher than wide. **Silica:** bodies present in stegmata in outer layer of culm scl. sheath.

***L. tenax* (Labill.) R. Br.** Culm diameter 0·8–1·4 mm.

Culm surface

Hairs (Fig. 33. G) as for group 2 type (ii) (see p. 203). **Epidermis:** cells mostly hexagonal, as long as to twice as long as wide, those flanking stomata often *c.* 3 times longer than wide, those at ends of stomata frequently about twice as wide as long; walls not wavy. **Stomata:** subsidiary cells crescentiform. **Tannin** in many epidermal cells. **Cuticular marks:** longitudinal striations, and pronounced wavy lines above anticlinal walls of epidermal cells.

Culm T.S. (Fig. 33. K, p. 200)

Epidermis. Cells of 2 main types: (i) Those not next to stomata mostly slightly to 1½–2½ times higher than wide. (ii) Those next to stomata 2–3 times wider than high, with rounded corners, *c.* ¼–½ height of cells of type (i); attached at one side to adjacent epidermal cells at angle between anticlinal walls, inclined inwards towards culm centre at angle of 160–180° to culm surface and attached at other side to subsidiary cell of stoma. **Stomata** deeply sunken; guard cells with slight ridge on inner wall. **Chlorenchyma** as for spp. belonging to group 2, but with air-cavity opposite stomata; cells surrounding substomatal cavity with few, long pegs, remaining cells with few, short pegs. **Central ground tissue** as for spp. belonging to group 2; without inner scl. sheath. **Silica:** bodies present in stegmata in parenchyma sheath.

Leaf, *rhizome*, and *root*: see description for spp. belonging to group 2.

Taxonomic Notes

Leptocarpus is the most widely distributed genus in the family. It has not always been considered to be a 'good' genus. Nees (1830) put it with *Restio*, but it was reconstituted by Endlicher in 1836 and remained unchanged except for the addition or subtraction of individual species until the time of Gilg (1890). He recorded that there is anatomical discontinuity between South African and other species, and suggested that the earlier name of *Calopsis* Beauv. ex Desv. should be given to the South African species. Pillans (1928) did not appear to be aware of Gilg's work because he continued to recognize the name *Leptocarpus* for South African species, including *Calopsis* in the synonymy without comment. Gilg-Benedict (1930) affirmed once more that on anatomical grounds the name *Calopsis* Beauv. ex Desv. should be used for the South African members of this genus.

The present work confirms that there are clear anatomical distinctions between South African and other species of *Leptocarpus*, and the proposal that South African species should be called *Calopsis* Beauv. ex Desv. is endorsed. South African species (as recognized by Pillans) all have protective cells and the chlorenchyma is composed of peg-cells. They all lack hairs. Some species have stomata raised on mounds, in others all epidermal cells are of uniform height. On the basis of this last character the South African species described in this work can be divided into 2 groups as an aid to identification. (i) Stomata raised on conspicuous mounds in *L. asper* (Mast.) Pillans, *L. esterhuysianae* Pillans, and *L. impolitus* (Kunth) Pillans. (ii) Stomata not raised on conspicuous mounds in *L. burchellii* Mast., *L. membranaceus* Pillans, *L. parkeri* Pillans, *L. rigoratus* Mast., and *L. vimineus* (Rottb.) Pillans.

The South African species of *Leptocarpus* and species of *Restio* here included in groups 3 and 4 are similar anatomically, although certain species have distinctive characters. This similarity is of interest because some *Leptocarpus* species have synonyms in *Restio* according to Pillans. Since the 2 genera are included in different tribes by Pillans it might be argued that they are not closely related. However, the anatomical evidence for close affinity is quite strong, and it is possible that further investigations may lead to the formation of new combinations.

The second, widely distributed group of *Leptocarpus* species is remarkably uniform anatomically. This provides additional evidence for the taxonomic reliability and stability of certain anatomical characters in classifying the Restionaceae. Each species has pillar cells radiating from dome-shaped ridges from the scl. sheath. The chlorenchyma is divided into sectors by the pillar cells, and the chlorenchyma cells themselves are arranged in transverse plates, or in a chequerboard pattern as seen in tangential longitudinal section. They have occasional pegs in certain species, but are not true 'peg-cells'. Protective cells are absent. Flattened 'fan hairs' are characteristic of this group (see p. 203); they are present in all species except *L. spathaceus* R. Br., which has erect, multicellular, uniseriate hairs, and *L. ramosus* R. Br., in which no hairs were seen.

Some species of group 2 have a 1–2-layered inner scl. ring separating the thin-walled central cells from the thicker-walled cells of the outer part of the ground tissue. This ring is not a constant feature of the species concerned, but when present, it is diagnostic for members of the genus.

Primary divisions of *Leptocarpus* for the purpose of identification can be made on a geographical basis. Some species can be further characterized. *L. tenax* (Labill.) R. Br. in Australia and Tasmania has modified epidermal cells surrounding the sunken stomata (see p. 213). Other species have sunken stomata, but do not exhibit the modified type of epidermal cell found in *L. tenax*; these are *L. chilensis* (Steud.) Mast., *L. coangustatus* Nees, *L. canus* Nees, *L. brownii* Hk. f., and *L. simplex* A. Rich.

Some species have sunken and superficial stomata (*L. spathaceus* R.Br.). More specimens need to be examined in order to determine the taxonomic value of the character of stomatal position relative to the culm surface in this genus.

Except for the variation in characters outlined above, the species, excluding those in South Africa, appear to constitute a natural group on the basis of their anatomy.

There are other Australian genera which have pillar cells and hairs. Their possible affinity to *Leptocarpus* should be considered by future workers. These genera are discussed under *Chaetanthus* (p. 147.)

The subgeneric divisions of Australian species of *Leptocarpus* recognized by Bentham and Hooker in 1883 (Diplantherae, Homoeantherae) are not reflected in anatomical structure. The pillar cell appears to be an 'Australian' character (see p. 188). On this basis, the species or genera sharing this character may be considered to have a common ancestry (unless the character evolved several times), but they may have diverged sufficiently in other characters to warrant their recognition as distinct modern groups.

LEPYRODIA R. Br.

No. of species *c.* 20; examined 17.

Genus Description

The genus is divided into 3 groups for clarity of description.

Group 1

Culm surface

Hairs and **papillae** absent. **Epidermis:** cells square, rectangular, or hexagonal, as long as wide or up to twice as long as wide; walls moderately thick or thick, wavy or only wavy at surface but straight at slightly lower focus; lumina rectangular, circular, or oval. **Stomata:** subsidiary cells with perpendicular, obtuse, or acute end walls, as long as guard cells in all spp. except *L. tasmanica*, where slightly longer; guard cells with rounded end walls. **Silica** granular, filling entire lumen of some epidermal cells in *L. stricta*, and present to a lesser extent in *L. tasmanica* and *L. caudata*. **Tannin** seen only in some cells of *L. caudata*. **Cuticular marks** granular; straight or slightly wavy longitudinal striations present in all spp. except *L. tasmanica*.

Culm T.S.

Outline circular or oval, or with slight mounds (*L. tasmanica*). **Cuticle** thick, frequently with slight ridges. **Epidermis:** cells in 1 layer, 4-sided, mostly as high as to 1½–2 times higher than wide, those next to stomata often higher (up to 4 times higher than wide in *L. caudata*), projecting into chlorenchyma; outer walls thick or v. thick, outer surface concave in *L. stricta* and *L. caudata*. Anticlinal walls wavy, moderately thickened or thick at outer ends, thickening tapering rapidly and inner halves of walls and inner walls slightly or moderately thickened. **Stomata** superficial; guard cells with outer lips in all spp., inner lips in *L. caudata* and *L. tasmanica*, ridge on inner wall in *L. interrupta*. **Chlorenchyma** composed of palisade cells with occasional pegs; cells arranged in 2 layers or, v. occasionally, in 3–4 layers. Adjacent cells separated from one another by air-spaces equivalent to slightly less than cell width; cells of inner and outer layers shortly interdigitating (cells arranged in longitudinal, 1-layered plates as seen in T.L.S.). Protective cells moderately thick-walled, extending in 2–3 layers to parenchyma sheath, those of outer layer frequently twice as long as those of other layers. As seen in T.L.S., walls of substomatal tube lined by 1–3 layers of protective cells at either end of stoma and 1–5 layers on flanks, substomatal tubes of adjacent stomata frequently touching. Apertures between protective cells of 2 types: (i) small, rounded, at junctions between inner and outer cell layers; (ii) narrow, elongated, between cells of inner layers. **Parenchyma sheath** 1–3-layered, continuous, cells rounded hexagonal, with slightly thickened walls. **Sclerenchyma**: sheath 2–8-layered; outline circular, with low, dome-shaped ribs opposite to peripheral vb's. Fibres of outer 2–5 layers thick- to v. thick-walled, except those to outer side of peripheral vb's, where fibres moderately thick-walled, all narrow. Fibres of inner layers wider, with wider lumina.

Vascular bundles. (i) Peripheral vb's: long diameter tangential, each vb with 1 medium-sized, angular, thin-walled tracheid on either flank, and 2–3 narrow (or, infrequently, medium-sized) tracheids in 1 or 2–3 layers between them; outline of phloem pole oval. (ii) Medullary vb's: outer smaller than inner, frequently with 1 central mxv and without px poles; inner bundles with 1 wide (medium-sized in *L. interrupta*), many-sided, thin-walled vessel on either flank, these separated by 1–3 files of narrow parenchymatous cells. Mxv's with scalariform wall-pitting and oblique, scalariform or transverse, simple perforation plates. Phloem pole overarching and abutting directly on xylem. Px present in most larger bundles. Most medullary vb's free from culm scl. sheath. **Bundle sheaths** poorly developed, 1–(2)-layered; fibres at poles narrow, with moderately thick to thick walls, those on flanks wider, with slightly to moderately thickened walls. **Central ground tissue** parenchymatous; cells surrounding vb's with slightly to moderately thickened (lignified) walls, but cells of inner bundle-free layers wider, with cellulose wall-thickening; innermost cells frequently breaking down to form central cavity. **Silica** present in epidermal cells of some spp. as amorphous material. **Tannin** in some mxv's of *L. stricta* (possibly pathological) and some epidermal cells of *L. caudata*.

Leaf, *rhizome*, and *root* not seen.

GROUP 2*a*

Leaf surface

Seen only in *L. scariosa* (abaxial).

Hairs absent. **Epidermis**: cells hexagonal, 2–3 times wider than long, end walls wide, longitudinal walls composed of 2 short sides; all walls v. thick, slightly wavy. **Stomata** sparse; subsidiary cells with acute end walls, mostly longer than guard cells; guard cell pair with lemon-shaped outline. **Silica**: solitary, spheroidal-nodular bodies present in many cells. **Tannin**: none seen.

Leaf T.S.

Seen only in *L. scariosa*.

Epidermis: (i) abaxial cells 1½ times higher than wide, all walls thick, anticlinal walls v. wavy; (ii) adaxial cells not seen. **Stomata** superficial; subsidiary cells with slight ridge on outer wall; guard cells with lip at outer aperture and ridge on inner wall. **Chlorenchyma** composed of 1 layer of peg-cells to inside of abaxial epidermis; cells 4–5 times higher than wide. Other, more or less isodiametric cells with pegs (possibly chlorenchyma) present in 1–2 layers between palisade chlorenchyma and parenchymatous ground tissue. **Vascular bundles** each with 1–2 wide, angular, thin-walled tracheids on either flank, large bundles also with several narrow, central tracheids; phloem poles circular, ensheathed by 1 layer of narrow cells with thin walls next to phloem and thick walls next to xylem. **Bundle sheaths.** O.S. parenchymatous, 1–2-layered, extending over phloem pole and round flanks, but not round xylem pole; also extended laterally and forming layer between adjacent bundles. I.S. sclerenchymatous, completely encircling bundles; fibres narrow, thick-walled, in 1–2 layers; sheath continuous laterally with 1–2 layers of fibres lying to adaxial side of parenchyma sheath extensions. **Sclerenchyma** present as described above. **Ground tissue** parenchymatous, comprising: (*a*) that of bundle sheath extensions, (*b*) 1–2 layers of thin-walled, flattened cells between scl. and adaxial epidermis. **Air-cavities** absent. **Silica:** see abaxial epidermis, surface view. **Tannin:** none seen.

Culm surface

Hairs and **papillae** absent. **Epidermis:** cells 4-sided, most frequently as long as wide, occasionally 1½–5 times longer than wide and sometimes 1½–2 times wider than long; walls slightly to moderately thickened, wavy. **Stomata:** subsidiary cells variable; guard cells mostly with rounded end walls, guard cell pairs with lemon-shaped outline in some spp. **Silica** present as solitary, spheroidal-nodular bodies in specialized cells; cells as long as wide with thick anticlinal walls, either solitary or arranged in longitudinal files of up to 5. **Tannin** in some epidermal cells. **Cuticular marks** granular; longitudinal striations often present.

Culm T.S.

Outline circular or oval. **Cuticle** slightly to moderately thickened, often with small ridges. **Epidermis:** cells in 1 layer, as high as wide to 1½ times higher than wide; outer walls moderately thickened to thick, curving inwards slightly at cell margins; anticlinal walls moderately thickened, wavy, characteristically with few, conspicuous undulations (Pl. VIII. B); inner walls moderately thickened. Some cells modified, with v. thick inner and anticlinal walls and thin or slightly thickened outer walls, each cell containing 1 spheroidal-nodular silica-body (Pl. VIII. B). **Stomata** superficial (slightly sunken in *L. muirii*); subsidiary cells variable, guard cells with pronounced, beak-like lip at outer aperture. **Chlorenchyma** composed of 2 dissimilar layers of peg-cells (Pl. VIII. A, B); those of outer layer *c.* 8 times higher than wide, with numerous, small pegs arranged in 4–6 longitudinal files; cells of inner layer frequently 2–3 times wider than and only ¾ height of outer cells, individual cells normally 2–3 times higher than wide and having 4–6 longitudinal files of 3–4 wide pegs. Air-spaces between cells of inner layer much wider than those between cells of outer layer. Cells arranged in reticulate manner as seen in T.L.S. Protective cells present in some spp. as slightly modified cells of outer chlorenchyma layer, with slightly thickened walls; substomatal tube opening to chlorenchyma near its inner end by circular apertures; linings of substomatal tube consisting of 1 layer of protective cells as seen in T.L.S. **Parenchyma sheath** 1–3-layered, uninterrupted, cells up to twice as high as wide, walls normally slightly thickened. **Sclerenchyma:** sheath 3–9-layered, outline circular with slight, dome-shaped ribs opposite peripheral vb's; fibres of outer 2–7 layers thick- or v. thick-walled, narrow, those of inner layers wider, with moderately thickened walls; innermost layer frequently difficult to distinguish from central ground tissue.

Vascular bundles. (i) Peripheral vb's: outline more or less circular, all tracheids narrow, or some vb's with 1 or 2 medium-sized, angular, thin-walled tracheids on either flank of arc of narrow tracheids; phloem pole rounded, situated in concavity of arc. (ii) Medullary vb's: outer bundles small, with no px, inner bundles larger, circular or oval in outline, each with 1 narrow or medium-sized, angular mxv on either flank. Mxv's separated from one another by 2–3 rows of narrow cells. Lateral wall-pitting of vessels scalariform; perforation plates oblique, scalariform (scalariform-fenestrate in *L. flexuosa* and *hermaphrodita*). Phloem poles circular, separated from xylem by single layer of narrow cells with slightly thickened walls, or with thin walls next to phloem and slightly thickened walls next to xylem. Px poles well developed; no bundles free from culm scl. sheath in most spp. **Bundle sheaths:** fibres at bundle poles narrow, thick-walled, normally in 1–2 layers; those on flanks wider, with moderately thickened walls, in 1 layer. **Central ground tissue** parenchymatous; cells surrounding vb's with moderately thickened walls; central cells thin-walled, frequently breaking down to form central cavity. **Silica** present as spheroidal-nodular bodies, as described above, in some epidermal cells, and as granular material in some cells of central ground tissue. **Tannin** in some epidermal cells.

Rhizome T.S.

Seen in *L. muirii* and *L. scariosa.*

Epidermis: cells 4–5-sided, slightly wider than high, narrow; walls slightly thickened. **Cortex.** (i) Outer part composed of 3 layers of moderately thick-walled, hexagonal cells in *L. muirii* and 6–7 layers of hexagonal cells with slightly thickened walls in *L. scariosa.* (ii) Inner part parenchymatous; cells hexagonal, thin-walled in outer 4–5 layers and rounded in inner 7–8 layers in *L. muirii*, hexagonal, v. thin-walled, in 6–8 layers in *L. scariosa*; cells of innermost 1–2 layers in both spp. hexagonal, walls slightly thickened. **Endodermoid sheath** 1-layered; cells with thin outer walls and slightly thickened inner and anticlinal walls in *L. scariosa*, and with thin outer walls but other walls thick in *L. muirii*. **Vascular bundles:** outer bundles with arc of mxv's and circular phloem pole; inner bundles amphivasal, those of *L. muirii* with several layers of narrow mxv's, *L. scariosa* with 1–(2) layer(s) of wide, angular mxv's. **Central ground tissue** composed of thick-walled sclerenchymatous cells surrounding vb's in *L. muirii*, parenchymatous, cells with moderately thickened walls in *L. scariosa*; some thin-walled cells present in lacunae in both spp. **Silica** and **tannin**: none seen.

Root T.S.

Seen in *L. muirii* only.

Epidermis and **root-hairs** not seen. **Cortex.** (i) Outer part consisting of 2 peripheral layers of rectangular, thin-walled cells each *c.* 1½ times wider than high, followed to inside by 1–2 layers of similar cells with slightly thickened walls. (ii) Middle part composed of single-layered, radiating plates of rounded cells, 6–8 cells deep, with wider cells between them. (iii) Inner part 2–3-lavered, cells mainly 1½ times wider than high, with rounded corners and intercellular spaces. **Endodermis** mostly 1-, occasionally 2-layered, cells nearly square, outer walls thin, inner and anticlinal walls heavily thickened; lumina v. reduced. **Pericycle** 1-layered, cells narrow, hexagonal, walls moderately thickened. **Vascular system :** mxv's mostly in 1 ring to inside of that formed by phloem and px, vessels solitary or in radial pairs, oval-angular. **Central ground tissue** sclerenchymatous, cells hexagonal, thick-walled. **Tannin :** none seen.

GROUP 2*b*

Similar in most points to group 2*a*, but differing in the following respects.

Culm surface

Stomata : subsidiary cells; wall next to anticlinal wall of guard cell produced into 2 short papillae extending over guard cell.

Culm T.S.

Chlorenchyma composed of 1 layer of palisade peg-cells; individual cells *c.* 6 times higher than wide, with numerous pegs; protective cells v. weak, seen only in *L. glauca.* **Parenchyma sheath** 1-layered.

MATERIAL EXAMINED

GROUP 1

Lepyrodia caudata L. Johnson and O. Evans: no number; Aust. (K).

L. gracilis R. Br.: Forsyth W., Kneucker distrib. No. 117; Aust. (K).

L. interrupta F. v. Muell.: Hubbard, C. E. 3904, An. 1930; Aust. (K).

L. tasmanica Hk. f.: Rodway 2480; Tasmania (K).

GROUP 2*a*

L. anarthria F. v. Muell. ex Benth.: Johnson and Constable 17760, An. 1951; Aust. (K).

L. drummondiana Steud.: Drummond 347; Aust. (K).

L. flexuosa (Benth.) L. Johnson and O. Evans: Sutton 847, An. 1904; Grampians, Aust. (K).

L. hermaphrodita R. Br.: Andrews, 1st colln., 1080, An. 1902; Aust. (K).

L. leptocaulis L. Johnson and O. Evans: Constable, E. F., N.S.W. 56903, An. 1962; Aust. (K).

L. macra Nees: Drummond. No number; Swan R., W. Aust. (K).

L. monoica F. v. Muell.: Drummond 447; Aust. (K).

L. monoica var. *foliosa* F. v. Muell.: Mueller, F., An. 1876; Porongerup. (K).

L. muelleri Benth.: Constable 22088, An. 1953; Hat Head, N.S.W., Aust. (K).

L. muirii F. v. Muell.: Blake, S.T. 18039, An. 1947; Kirup, S.W. Div., Aust. (K).

L. scariosa R. Br.: (i) Cheadle CA178, An. 1961; Aust. (S). (ii) Stephenson 338; Aust. (K).

L. valliculae J. M. Black: Inman, R. S. A., An. 1928; Aust. (K).

GROUP 2*b*

L. glauca (Nees) F. v. Muell.: Drummond. No number; Swan R., W. Aust. (K).

L. heleocharoides Gilg: Pritzel, E. 172, An. 1900; Aust. (K).

SPECIES DESCRIPTIONS

GROUP 1

***L. caudata* L. Johnson and O. Evans.** Culm diameter 2 by 1·5 mm.

Culm surface

Stomata: subsidiary cells with obtuse end walls, long anticlinal walls wavy at culm surface. **Tannin** in some epidermal cells.

Culm T.S.

Outline oval. **Epidermis:** cells mostly 1½ times higher than wide, those next to stomata projecting into chlorenchyma, being 2½–4 times higher than wide; outer walls occasionally with shallow depressions, outer surface irregular; anticlinal walls moderately thickened. **Stomata:** guard cells with outer and inner lips. **Silica** granular, in occasional epidermal cells.

***L. gracilis* R. Br.** Culm diameter 1·0 mm.

Culm surface

Stomata: subsidiary cells with obtuse end walls. **Tannin** absent.

Culm T.S.

Outline circular. **Epidermis:** cells mostly as high as to 1½ times higher than wide, those next to stomata twice as high as wide. **Stomata:** guard cells with outer lip.

***L. interrupta* F. v. Muell.** Culm diameter 1 by 0·75 mm.

Culm surface

Epidermis: cells hexagonal, mostly about as long as wide, walls slightly to moderately thickened. **Stomata:** subsidiary cells with obtuse end walls.

Culm T.S.

Outline oval. **Epidermis:** cells mostly slightly or 1½ times higher than wide; those next to stomata often twice as high as wide; outer walls moderately thickened, occasionally with shallow, irregular depressions in outer surface; anticlinal walls slightly to moderately thickened. **Stomata:** guard cells with lip at outer aperture and v. slight ridge on inner wall.

***L. tasmanica* Hk. f.** Culm diameter 0·8 mm.

Culm surface (Fig. 34. C)

Epidermis: cells next to stomata mostly as long as wide, others up to twice as long as wide. **Stomata:** subsidiary cells frequently longer than guard cells, with acute end walls and slightly expanded ends. **Silica:** granular particles, or spheroidal-nodular bodies in some epidermal cells.

Culm T.S. (Fig. 34. D, E)

Outline circular, with slight mounds. **Epidermis:** cells 1½–2½ times higher than wide; those between stomata frequently extended outwards and forming low mounds; anticlinal walls v. wavy, with tapering thickening. **Stomata:** guard cells with lips at outer and inner apertures. **Silica** in some epidermal cells.

GROUP *2a*

***L. anarthria* F. v. Muell. ex Benth.** Culm diameter 0·8 mm.

Culm surface

Epidermis: cells as long as to 2 or 5 times longer than wide, shorter cells at ends of stomata. **Stomata:** subsidiary cells with perpendicular or acute end walls and wavy anticlinal walls.

Culm T.S.

Outline circular. **Stomata:** subsidiary cells with slight ridge on outer wall.

***L. drummondiana* Steud.** Culm diameter 0·1 mm.

Culm surface

Epidermis: cells mostly slightly wider than long, some nearly as wide as long. **Stomata:** subsidiary cells wide, with obtuse end walls.

Culm T.S.

Outline circular. **Epidermis:** cells *c.* 1½ times higher than wide, all walls moderately thickened. **Chlorenchyma** as for group *2a*, but with 1 or 2 inner layers; protective cells absent.

***L. flexuosa* (Benth.) L. Johnson and O. Evans.** Culm diameter 0·8 mm.

Culm surface

Epidermis: cells mostly as long as wide, some, particularly those next to stomata, slightly wider than long. **Stomata:** subsidiary cells with acute end walls, but long anticlinal walls slightly wavy at high focus; guard cell pair with lemon-shaped outline. **Tannin:** none seen.

Culm T.S.

Outline circular. **Epidermis:** cells *c.* 1½ times higher than wide. **Stomata:** subsidiary cells with ridge on outer wall. **Chlorenchyma:** protective cells with slightly thickened walls, often containing irregularly shaped silica-bodies.

***L. hermaphrodita* R. Br.** Culm diameter 0·6 mm.

Culm surface

Epidermis: cells mostly as wide as long. **Stomata:** subsidiary cells with acute end walls, long anticlinal wall of many cells slightly concave; long anticlinal walls of guard cells with 3 faces.

Culm T.S.

Outline circular. **Epidermis:** cells *c.* 1½ times higher than wide. **Stomata:** guard cells with lip at outer aperture and slight ridge on inner wall.

***L. leptocaulis* L. Johnson and O. Evans.** Culm diameter 0·4 mm.

Culm surface

Epidermis: cells mostly as wide as long, those next to stomata smaller than others. **Stomata:** subsidiary cells with acute end walls, long anticlinal walls slightly wavy at high focus. **Tannin:** none seen.

Culm T.S.

Outline circular. **Cuticle** with minute ridges. **Epidermis:** cells mostly as high as wide; anticlinal walls with tapered thickening. **Stomata:** guard cells with lip at outer aperture and slight ridge on inner wall.

***L. macra* Nees.** Culm diameter 1 mm.

Culm surface

Epidermis: cells mostly as long as wide. **Stomata:** subsidiary cells wide, with obtuse ends. **Tannin:** none seen.

Culm T.S.

Outline circular. **Cuticle** with slight ridges. **Epidermis:** cells slightly higher than wide. Outer surface of outer walls with minute ridges. **Stomata:** guard cells with outer lip. **Silica** as for spp. belonging to group 2*a*, also some granular material in certain cells of central ground tissue.

***L. monoica* F. v. Muell.** Culm diameter 2 by 1·2 mm.

Culm surface

Epidermis: cells as long as wide or slightly wider than long. **Stomata:** subsidiary cells with acute end walls. **Tannin:** none seen.

Culm T.S.

Outline oval. **Cuticle** with marked ridges. **Epidermis:** cells mostly as wide as high. **Stomata:** guard cells with outer lip. **Chlorenchyma** without protective cells.

***L. monoica* var. *foliosa* F. v. Muell.** Culm diameter 0·7 by 0·5 mm.

Similar to *L. monoica*, but **tannin** in some epidermal cells.

***L. muelleri* Benth.** Culm diameter 1·5 by 1 mm.

Culm surface

Epidermis: cells mostly as long as wide, some wider than long, particularly next to stomata. **Stomata:** subsidiary cells with acute end walls; guard cell pair with more or less lemon-shaped outline. **Tannin:** none seen.

Culm T.S.

Outline oval. **Cuticle** with small ridges. **Epidermis:** cells nearly 1½ times higher than wide. **Stomata:** guard cells with outer lip and ridge on inner wall. **Chlorenchyma:** protective cells not developed.

***L. muirii* F. v. Muell.**

Culm surface

Epidermis: cells mostly as long as wide. **Stomata:** subsidiary cells wide, with obtuse end walls. **Tannin:** none seen.

Culm T.S.

Outline circular. **Cuticle** without pronounced ridges. **Epidermis:** cells mostly slightly higher than wide. **Stomata:** slightly sunken; subsidiary cells with v. pronounced ridge on outer wall; guard cells with slight ridge on inner wall. **Chlorenchyma** 3-layered in places; cells of all layers of similar heights.

Rhizome and *Root*: see group 2*a* genus description on p. 218.

***L. scariosa* R. Br.** Culm diameter 1·2 and 2·1 mm.

Culm surface (Fig. 34. B)

Epidermis: cells 4-sided, mostly twice as wide as long; those next to stomata smaller than remainder. **Stomata:** subsidiary cells with acute or obtuse end walls, long anticlinal walls occasionally slightly wavy. **Tannin** in occasional epidermal cells.

Culm T.S. (Fig. 34. A; Pl. VIII. A, B)

Outline circular. **Cuticle** with slight ridges. **Epidermis:** cells mostly *c.* 1½ times higher than wide. **Stomata:** guard cells with outer lip and slight ridge on inner wall. **Silica** present as bodies, similar to those in other spp. belonging to group 2*a*, and also as granular material in some cells of central ground tissue. **Tannin** in occasional epidermal cells.

Leaf and *rhizome*: see group 2*a* genus description on pp. 216, 218.

***L. valliculae* J. M. Black.** Culm diameter 0·7 mm.

Culm surface

Epidermis: cells mostly as long as wide, those next to stomata sometimes shorter than wide. **Stomata:** subsidiary cells with obtuse end walls, long anticlinal walls slightly wavy. **Tannin:** none seen.

Culm T.S.

Outline circular. **Cuticle** not ridged. **Epidermis:** cells mostly as high as wide. **Stomata:** guard cells with outer lip. **Chlorenchyma** without protective cells.

GROUP 2*b*

***L. glauca* (Nees) F. v. Muell.** Culm diameter 5 mm.

Culm surface

Epidermis: cells irregular, but mostly as long as wide or slightly shorter than wide. **Tannin:** none seen.

Culm T.S

Outline circular. **Cuticle** without ridges. **Epidermis:** cells slightly higher than wide. **Stomata:** subsidiary cells reaching surface, but with guard cells

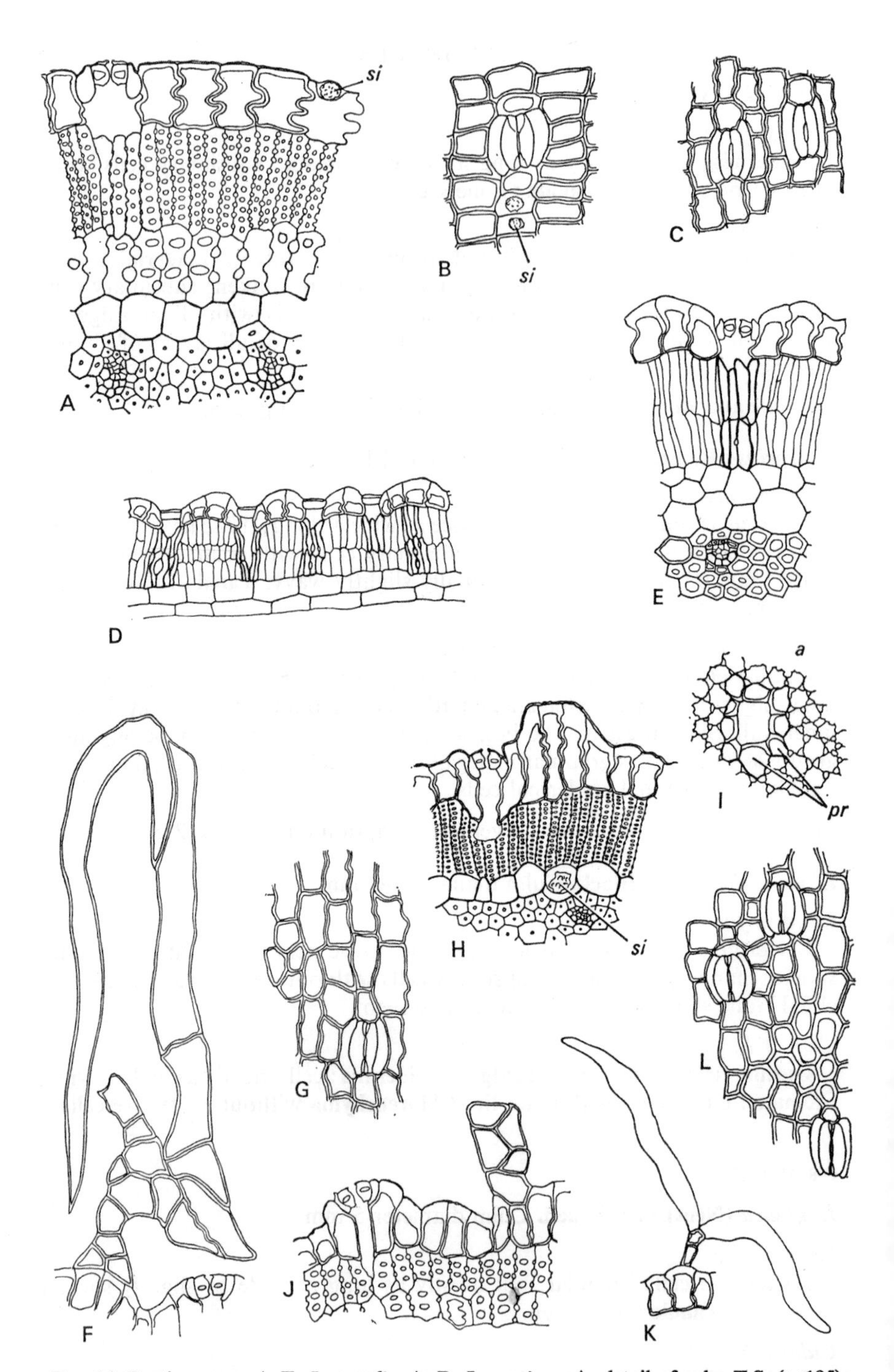

FIG. 34. Restionaceae. A–E, *Lepyrodia*. A, B, *L. scariosa*: A, detail of culm T.S. (×185); B, culm surface, note silica (×185). C–E, *L. tasmanica*: C, culm surface (× 185); D, detail of culm L.S. (×100); E, detail of culm T.S. (×185). F–L, *Loxocarya*. F, G, J, *L. fasciculata*: F, hair on culm; G, culm surface; J, detail of culm T.S. (all ×185). H, I, K, L, *L. cinerea*: H, detail of culm T.S. (×185); I, culm T.L.S. just below epidermis showing peg-cells (a) and epidermal cells (pr) (×290); K, hair on culm (×185); L, culm surface; note area of smaller cells corresponding to tall epidermal cells of T.S. (×185).
si = silica-body.

mounted near their inner ends; subsidiary cells with ridge on outer wall projecting over guard cells; guard cells without lips or ridges. **Silica** present as bodies similar to those in spp. belonging to group 2*a*, and granular material in some cells of parenchyma sheath and central ground tissue.

L. heleocharoides **Gilg.** Culm diameter 0·8 mm.

Culm surface as *L. glauca.*

Culm T.S.

Outline circular. **Cuticle** without ridges. **Epidermis:** cells mostly 1½ times higher than wide. **Stomata:** subsidiary cells with ridge on outer wall overarching adjacent guard cells; guard cells without lips or ridges. **Silica** as bodies similar to those in spp. belonging to group 2*a*, or as granular material present in some cells of central ground tissue.

Taxonomic Notes

Lepyrodia R. Br. has been accepted by all authors since its original description. The genus *Sporadanthus* F. v. Muell. was included in it as a synonym by Hieronymus (1888), where it has remained until recently. The genus *Sporadanthus* is currently regarded as a distinct genus in New Zealand (Campbell 1964) and is treated as such in the present work (p. 300). Johnson and Evans (1966) state that they have the question of the relationship of *Sporadanthus* with *Lepyrodia* under review. Gilg-Benedict (1930) found *Lepyrodia* to be composed of 2 anatomical types. Johnson and Evans (1963*c*) worked on the anatomy and morphology of some *Lepyrodia* species and incorporated Briggs's (1963) cytological data into their paper. They reported that the combined anatomical and cytological data show that the genus comprises 2 distinct groups, but they were not ready to divide it on this evidence alone, since they could find no characters of gross morphology by which they could distinguish between the groups.

The results of the present work are in agreement on most points with the conclusions of Johnson and Evans. Three anatomical groups are recognized here, 2 of which are clearly distinct and 1 may be only a subdivision of one of the major groups. The chromosome numbers which are quoted in the tables that follow are those recorded by Briggs (1963) and the groups to which species are assigned by Johnson and Evans (J. and E.) are also indicated in the tables.

Table 1

GROUP 1. Chlorenchyma not composed of peg-cells, but of palisade cells arranged in longitudinal plates; protective cells present in all species; silica present as granular material in some epidermal cells of certain species; stomata superficial.

	2n	J+E
L. caudata L. Johnson and O. Evans	18	B
L. gracilis R. Br.	18	B
L. interrupta F. v. Muell.	18	B
L. tasmanica Hk. f.	..	B

(*Sporadanthus traversii* F. v. Muell. belongs to this anatomical group; see p. 300).

TABLE 2

GROUP 2*a*. Chlorenchyma composed of 2 layers of peg-cells, cells of outer layer longer and with smaller pegs than those of inner layer; protective cells present in some species; silica present as solitary, spheroidal-nodular bodies in some epidermal cells; stomata superficial.

	2n	J+E
L. anarthria F. v. Muell. ex Benth.	14	A
L. drummondiana Steud.	..	..
L. flexuosa (Benth.) L. Johnson and O. Evans	..	A
L. hermaphrodita R. Br.	..	A
L. leptocaulis L. Johnson and O. Evans	14	A
L. macra Nees	..	A
L. monoica F. v. Muell.	..	..
L. monoica var. *foliosa* F. v. Muell.	..	..
L. muelleri Benth.	14	A
[1]*L. muirii* F. v. Muell.	..	B
L. scariosa R. Br.	14	A
L. valliculae J. M. Black	..	A

TABLE 3

GROUP 2*b*. As group 2*a*, but with 1 layer of chlorenchyma, and subsidiary cells produced into papillae.

	2n	J+E
L. glauca (Nees) F. v. Muell.	..	B
L. heleocharoides Gilg	..	..

The 2*b* species are so similar anatomically that they may be synonymous. *L. glauca* needs further investigation; the chlorenchyma of the specimen examined is unlike that of the other members of Johnson and Evans's group B.

Although Johnson and Evans feel unable to suggest a division of this genus on the evidence that they present, the present author suggests that the evidence now available is overwhelmingly in favour of a division into two major groups. The fact that *Sporadanthus* F. v. Muell. corresponds in so many anatomical details with species of group 1 suggests an affinity between them. The differences in gross morphology between the vegetative parts of *Sporadanthus* and *Lepyrodia* species of group 1 could easily be manifestations of adaptation to widely differing habitats. It is suggested that the name *Sporadanthus* may be suitable for all these species.

Epidermal silica-bodies (recognized first in this genus by Solereder and Meyer, 1929) are unknown in other members of the Restionaceae except *Thamnochortus bachmannii* Mast. and *Harperia*.

[1] Johnson and Evans (1966) record that this species was inadvertently included in the wrong list in the 1963*c* publication.

LOXOCARYA R. Br.

No. of spp. 8; examined 6.

Genus Description

Leaf surface not seen.

Leaf T.S.

Seen in *L. cinerea* and *L. flexuosa.*

Hairs: none seen. **Cuticle:** thick with small ridges on abaxial surface, thin on adaxial surface. **Epidermis:** (i) abaxial cells 4-sided, as wide as to twice as wide as high; outer walls thick, other walls slightly thickened or thin; (ii) adaxial cells 4-sided, 8–10 times wider than high, *c.* $\frac{1}{4}$–$\frac{1}{5}$ of the height of abaxial cells; walls slightly thickened. **Stomata** present in abaxial epidermis only, superficial; guard cells with lip at outer aperture and slight lip at inner aperture; walls to pore concave. **Chlorenchyma:** cells in 1–3 layers, next to abaxial epidermis; cells mostly as wide as high, some slightly higher than wide, all with wide, oval pegs.

Vascular bundles each with oval phloem pole and group of narrow and medium-sized tracheids; some bundles small, with few phloem and xylem cells, these alternating with larger bundles. **Bundle sheaths.** O.S. parenchymatous, cells wide with slightly thickened walls; arranged in 1 layer at phloem pole, and with lateral extensions in *L. cinerea*; arranged in 1 layer on bundle flanks, interrupted at poles by fibres from I.S. in *L. flexuosa.* I.S. sclerenchymatous; fibres narrow, in 1–3 layers at phloem pole; wider and arranged in 1–2 layers on flanks; 2–3-layered at xylem pole, fibres of outermost layer wide, those of inner layers narrow. All fibres thick-walled. **Sclerenchyma** restricted to that of bundle sheaths. **Ground tissue** parenchymatous, cells wide, thin-walled, in 1–2 layers filling space between bundle sheaths and adaxial epidermis, and in 3–4 layers between individual bundles. **Air-cavities** absent. **Silica:** spheroidal-nodular bodies present in stegmata in outer layer of scl. of bundle sheath at phloem poles. **Tannin:** none seen.

Culm surface

Hairs multicellular, composed of short stalk and 1 or several elongated branch cells; see individual spp. **Epidermis.** Cells 4–6-sided, of 2 main sizes: (i) those next to stomata, and in groups of 4–30, about as long as wide; (ii) those not next to stomata and not in groups of short cells, 1$\frac{1}{2}$–2 or 2–6 times longer than wide; walls slightly to moderately thickened (occasionally thick), straight or wavy, or wavy at surface only. **Stomata:** subsidiary cells and guard cells variable, see species. **Silica** absent. **Tannin** in epidermal cells of *L. pubescens* only. **Cuticular marks** granular, with or without longitudinal striations.

Culm T.S.

Outline circular or kidney-shaped, with low mounds formed by taller epidermal cells. **Cuticle** slightly or moderately thickened, occasionally thick. **Hairs:** see individual spp. **Epidermis.** Cells 4-sided, short or elongated. (i) Short cells normally as high as to 1$\frac{1}{2}$ times higher than wide, present near to stomata; cells next to stomata in all spp. except *L. pubescens* extending inwards into chlorenchyma, forming sides of substomatal cavity. (ii) Taller cells, situated between and occasionally next to the stomata, often up to 5 times higher than wide, forming mounds (appearing as groups of smaller cells as seen in surface view). Cells in centre of mounds tallest, those on flanks of intermediate height, with inclined outer walls (occasional multicellular, round-topped emergences present in some spp.). Outer walls v. thick; anticlinal walls often wavy, moderately or slightly thickened, or thick at outer ends and slightly thickened at inner ends, thickening tapering gradually. Inner walls slightly or moderately thickened. **Stomata** superficial, raised on tall cells in 1 species; subsidiary cells variable, see species; guard cells with outer lips and inner lips. **Chlorenchyma:** (*a*) 1- or 2-layered in most spp., peg-cells varying from slightly to 12 times higher than wide, those between stomata normally tallest, layers uninterrupted; (*b*) 3–4-layered in *L. pubescens*, cells with few pegs, mostly slightly to 2–(3) times higher than wide, layers interrupted by scl. ribs and their associated pillar cells. Protective cells absent. **Parenchyma sheath** 1-layered, cells all as wide as to slightly or 1½–2 times wider than high, except *L. pubescens*, where of 2 types: (i) circular, more or less as high as wide, on flanks of ribs from scl. sheath; (ii) pillar cells, extending from outer ends of ribs from scl. sheath to epidermis; individual cells 2–4–(5) times higher than wide, with slightly to moderately thickened walls. **Sclerenchyma:** sheath well developed in all spp., outline circular with low ribs opposite peripheral vb's in all spp. except *L. pubescens*, this species having pronounced scl. ribs opposite peripheral vb's. Outer fibres narrow, v. thick-walled; inner fibres frequently wider, with thick or v. thick walls.

Vascular bundles. (i) Peripheral vb's each with arc of narrow or medium-sized tracheids, sometimes with 1 wider tracheid on either flank; phloem pole situated in concavity of arc. (ii) Medullary vb's: outer bundles smallest, inner largest; outline oval; each vb with 1 mxv on either flank. Flanking mxv's circular or angular, narrow, medium-sized, or wide, with slightly or moderately thickened walls. Phloem poles rounded or oval, separated from xylem by 1-layered sheath of narrow cells with thin walls next to phloem, and slightly or moderately thickened walls next to xylem. **Bundle sheaths** 1–2-layered at poles, usually 1-layered on flanks; fibres narrow, with thick or v. thick walls at poles; wider, with slightly thickened, moderately thickened, or thick walls on flanks. **Central ground tissue** parenchymatous, outer cells around vb's with slightly or moderately thickened walls, those in culm centre sometimes thin-walled, and breaking down. **Silica:** spheroidal-nodular bodies present in stegmata in parenchyma sheath or outer layer of scl. sheath. **Tannin** in epidermal cells of *L. pubescens*.

Material Examined

All spp. Australian.

Loxocarya cinerea R. Br.: (i) Andrews, C., 1st colln., 1112 ♂, An. 1900; 100 miles E. of Perth (K). (ii) Mueller, F., ♂; Upper Hay River (K). (iii) Blake, S. T. 18153, An. 1947; open Eucalypt, on granite hillside (K).

L. fasciculata (R. Br.) Benth.: (i) Koch, M. 1887 ♂, An. 1911; Lowden, S. W. Aust. (K). (ii) Carter, C. B., ♂, An. 1915; Westbourne (K).

L. flexuosa (R. Br.) Benth.: (i) Brown, R., Bennet Distrib., No. 5849 (type) (K). (ii) Andrews, C., 1st colln., 1110 ♂; limestone, rocky ground, Cottesloe, nr. Perth (K). (iii) Drummond, 96; Swan River (K).

L. pubescens (R. Br.) Benth.: (i) Brown, R., Bennet Distrib. No. 5852; King George's Sound (type) (K). (ii) Pries ♂; Busselton (K). (iii) Koch, M. 1880 ♀, An. 1911; Lowden (K).

L. vestita Benth.: Oldfield; Murchison Estuary (K).

L. virgata Benth.: Steward, F. 63, An. 1914; Narrogin Exptal. Farm (K).

Species Descriptions

***L. cinerea* R. Br.** Culm diameter 0·9, 0·9, 0·5 mm.

Leaf T.S.: see genus description.

Culm surface (Fig. 34. K, L.)

Hairs multicellular (Fig. 34. K). **Epidermis:** cells as long as wide to 1½–(2) times longer than wide, smaller cells in clusters of 8–30 (corresponding to mounds formed by elongated cells as seen in T.S.); walls sometimes wavy. **Stomata:** subsidiary cells variable. **Cuticular marks:** longitudinal striations.

Culm T.S. (Fig. 34. H, I)

Outline circular, with low mounds. **Epidermis:** cells smallest between mounds, where mainly as high as or up to 1½ times higher than wide; tallest in centre of mounds, where up to 5 or 6 times higher than wide. Cells surrounding stomata extending *c.* ⅕–½ of way into chlorenchyma. **Stomata** normally amongst shorter cells; guard cells with pronounced lip at outer and slight lip at inner aperture. **Chlorenchyma:** see Fig. 34. I.

***L. fasciculata* (R. Br.) Benth.** Culm diameter 0·7 by 0·5, 1·2 by 1·0–1·1 mm.

Culm surface (Fig. 34. F, G)

Hairs: see Fig. 34. F. **Epidermis.** Cells mostly 4-sided, of 2 types: (i) as long as or slightly longer than wide, frequently grouped around stoma (elongated as seen in T.S.), and (ii) 2–6 times longer than wide, normally between stomata (shorter cells as seen in T.S.); walls wavy. **Stomata** often raised on taller cells; subsidiary cells mainly with acute end walls. **Cuticular marks:** fine, longitudinal striations.

Culm T.S. (Fig. 34. J)

Outline circular, with pronounced mounds and short, club-shaped emergences (possibly hair bases?). **Epidermis.** Cells of 2 main types: (i) elongated, 3–4 times higher than wide; sometimes associated with stomata and then also extending slightly into chlorenchyma; sometimes producing multicellular hair bases (Fig. 34. J); (ii) shorter cells, mostly as high as to 1½ times higher than wide, normally present between stomata, but occasionally associated with them (cells mostly of this type in Koch specimen). **Stomata:** subsidiary cells variable; guard cells with slight lips at outer and inner apertures. **Chlorenchyma:** see Fig. 34. J.

***L. flexuosa* (R. Br.) Benth.** Culm diameter 1 by 0·5, 1·0, 0·5 mm.

Leaf T.S.: see genus description.

Culm surface

Hairs: basal cells only seen in (1110) and (96). **Epidermis**: cells mostly 4-sided, as long as wide and up to 3 or 4 times longer than wide; groups of smaller, short cells present in (1110) and (96), forming mounds as seen in T.S.; walls wavy. **Stomata**: subsidiary cells with acute end walls, long anticlinal walls sometimes slightly wavy; guard cell pair occasionally with lemon-shaped outline. **Cuticular marks**: longitudinal striations.

Culm T.S.

Outline circular, with mounds. **Epidermis**: cells mostly slightly to 1½ times higher than wide in (5849) and (96), those next to stomata 2–4 times higher than wide, extending into chlorenchyma, forming substomatal tube; some elongated, 3–4 times higher than wide, in (1110), forming mounds. **Stomata**: guard cells with lips at outer and inner apertures. **Chlorenchyma**: cells in 1 or 2 layers, individual cells mostly 5–6 or 3–10(–12) times higher than wide; cells opposite substomatal cavities shortest.

L. pubescens **(R. Br.) Benth.** Culm diameter 1·1, 1·1, 0·5 mm.

Culm surface

Hairs: stalk uni- or bi-seriate, composed of 3–6 cells with moderately thickened walls; flattened, free branching distal part consisting of many elongated, moderately thick-walled, loosely associated cells. **Epidermis**: cells mostly 4-sided, sometimes 5–6-sided, as long as to 4–5 times longer than wide, but mostly 1½–2 times longer than wide; walls wavy, except in (1880), where irregular in outline and mostly short; cells over pillar cells usually longest. **Stomata**: subsidiary cells variable. **Tannin** present in many epidermal cells. **Cuticular marks**: longitudinal striations.

Culm T.S.

Outline circular, or sometimes with mounds. **Epidermis**: cells all about as high as wide except in (1880), where as high as, or up to 1½ times higher than wide next to stomata, and reaching 2½–3½ times higher than wide between stomata. **Stomata**: guard cells with lips at outer and inner apertures. **Chlorenchyma** mostly 3–(4)-layered, see genus description.

L. vestita **Benth.** Culm diameter 0·75 by 0·5 mm.

Culm surface

Hairs similar to those in *L. fasciculata*. **Epidermis**: cells 4-sided, 1½–4 times longer than wide, walls wavy. **Stomata**: subsidiary cells with wavy long walls; guard cell pair with lemon-shaped outline.

Culm T.S.

Outline kidney-shaped, with low mounds. **Epidermis**: cells mostly as high as or slightly higher than wide, some up to twice as high as wide, forming low, rounded mounds; those next to stomata extending a short distance into chlorenchyma and lining substomatal cavity. **Stomata**: subsidiary cells elongated, with pronounced ridge on outer wall; guard cells with lips at outer and inner apertures. **Chlorenchyma** 2-layered; cells of outer layer 3–5 times higher than wide, those of inner layer 2–3 times higher than wide, slightly wider than outer cells.

L. virgata **Benth.** Culm diameter 1 mm.

Culm surface

Hairs as in *L. fasciculata*; see Fig. 34. F. **Epidermis:** cells mostly 4-sided, as long as wide near stomata and in groups (corresponding to mounds as seen in T.S.), and slightly to $1\frac{1}{2}$–(2) times longer than wide between stomata; walls slightly wavy at surface, straight at lower focus. **Stomata:** subsidiary cells with irregular outlines; guard cell pair with lemon-shaped outline.

Culm T.S.

Outline circular, with irregular mounds. **Epidermis:** cells mostly slightly to $1\frac{1}{2}$ times higher than wide, some elongated, forming mounds, largest of these *c.* 5 times higher than wide, those on flanks of mounds of intermediate size, with sloping outer walls; cells surrounding stomata elongated inwards, being *c.* 4 times higher than wide, anticlinal walls lining substomatal cavity about twice as long as others, inner walls sharply inclined. **Stomata:** guard cells with pronounced outer lip and lip at inner aperture. **Chlorenchyma** 1-layered, peg-cells mostly *c.* 10 times higher than wide, those opposite stomata 5–7 times higher than wide.

Taxonomic Notes

This genus has been recognized by all authorities since its description, up to the time of Masters (1878), who split it into 2 groups, one of which he included with *Restio* as *R. loxocarya* Mast. = *L. cinerea* R. Br. The species remaining from Bentham's treatment of *Loxocarya* were put by Masters under *Hypolaena* R. Br. in a division on their own 'Stylus unicus indivisus'. Bentham and Hooker (1883) did not recognize this classification and retained *Loxocarya* as published by R. Brown (1810), including in it Bentham's species. They thought that *Haplostigma* F. v. Muell. and *Desmocladus* Nees should be placed with *Loxocarya*.

Loxocarya was not examined by Gilg, but Gilg-Benedict (1930) gave a brief, partly accurate anatomical account of it, following Bentham and Hooker's interpretation of the content of the genus.

The present study shows that the genus falls into 2 distinct anatomical groups. The first includes only *L. pubescens* (R. Br.) Benth., which differs from the other species in having pillar cells. In order to ensure correct identity, type material of this species was examined. It is similar to Australian species of *Leptocarpus*. If the complex of Australian species showing pillar cells is further investigated, the taxonomic position of this species will need to be examined critically.

The second group contains species that lack pillar cells. The epidermal cells surrounding the stomata penetrate into the chlorenchyma forming a sub-stomatal tube, as in *Lepidobolus* species and *Coleocarya*, amongst others. For a discussion of the possible interrelationships of these groups, see pp. 159, 201.

There is a variety of hair form, and scale-like emergences are present in some *Loxocarya* species but the genus, apart from *L. pubescens* (R. Br.) Benth., constitutes a relatively uniform anatomical group.

It may eventually be possible to distinguish individual species by using combinations of anatomical characters; but more samples should first be examined to make sure that the data are reliable for this purpose.

LYGINIA R. Br.

No. of spp. 1 (2?); examined 2.

Genus Description

Leaf surface only, seen in *L. tenax*.

Epidermis: cells hexagonal, 4–5 times wider than long, each with long transverse walls and 2 short walls on either side; all walls thick. Cells next to stomata produced into overarching papillae (see Culm surface). **Stomata:** subsidiary cells wide, with obtuse end walls; guard cell pair with lemon-shaped outline.

Culm surface

L. tenax only (Fig. 35. B).

Hairs: none seen. **Epidermis.** Cells consisting of 2 main types: (i) cells not associated with stomata, 4-sided, mostly 2–4 times longer than wide, walls thick, wavy; (ii) cells surrounding stomata, mostly 2–4 times longer than wide, but with walls bordering stoma produced into overarching papillae. Papillae from adjacent cells frequently fused at their inner ends, but free at their outer ends. **Stomata** sunken and overarched by papillae; subsidiary cells irregular in outline, but mostly dome-shaped, with rounded or perpendicular ends (see Fig. 35. F); guard cells narrower than normal for the family. **Silica:** none seen. **Tannin** in many epidermal and some subsidiary cells. **Crystals:** none seen. **Cuticular marks:** none seen.

Culm T.S. (Fig. 35. A)

Outline circular, sometimes with irregularly distributed, low dome-shaped mounds. **Cuticle** moderately thickened. **Epidermis.** Cells in 1 layer, all more or less level at culm surface, except for raised papillate cells next to stomata, but extending inwards to differing degrees, the tallest being up to 7 times higher than wide, the shortest about twice as high as wide, with a graded series between the extremes. As seen in L.S., all cells inclined at angle of *c.* 15° from perpendicular. Outer walls v. thick, anticlinal walls v. wavy, thick at outer end, but thickening tapering gradually to thin inner end, inner walls thin. **Stomata** sunken, attached to inner end of anticlinal walls of epidermal cells (as seen in L.S., Fig. 35. D, stomata attached to short cells at one end and long cells at other, and thus inclined at *c.* 45° to culm surface, all stomata inclined in same direction). Guard cells without lips or ridges. **Chlorenchyma** composed of palisade cells arranged in 2 layers; cells of 2 types, one without pegs: (i) cells opposite stomata, thin-walled; (ii) cells not opposite stomata, with moderately thickened walls. Cells of type (ii) as seen in T.L.S. forming irregular reticulum with cells of type (i) in the lacunae, the thinner-walled cells joined to one another at adjacent angles, thus alternating with air-spaces and looking like a chequerboard (Fig. 35. C). **Parenchyma sheath:** cells in 1 layer, individual cells 1½–2 times higher than wide, with slightly thickened walls; outline of parenchyma sheath following that of scl. sheath. **Sclerenchyma**: sheath 4–7-layered, outline circular with low, dome-shaped ribs each accommodating 1 peripheral vb; fibres of outer layers thick-walled, with medium-sized lumina; those occurring in the 1 layer to outer side of each peripheral vb and in the inner sheath layers with moderately thickened walls.

Vascular bundles. (i) Peripheral vb's with 4–6 narrow, angular, thin-walled tracheids arranged in an arc partially enclosing a small phloem pole. (ii) Medullary vb's: (*a*) Outer, smaller vb's, embedded in culm scl. sheath, each having on either flank 1 narrow or medium-sized, many-sided mxv with slightly thickened walls. Mxv's separated from one another by narrow parenchymatous cells arranged in 2–3 rows; px poles present in some bundles. (*b*) Inner, larger vb's, free from culm scl. sheath, each with 1 (occasionally 2) medium-sized, many sided mxv(s) on either flank separated from one another by narrow parenchymatous cells arranged in 2–3 rows. Lateral wall-pitting of vessels scalariform; perforation plates scalariform or reticulate-scalariform. Phloem pole circular, separated from xylem by 1 layer of narrow cells with thin walls facing the phloem and with slightly thickened walls facing the xylem. Px poles well developed, separated from bundle sheath on either side by 1–2 layers of thin-walled parenchymatous cells. **Bundle sheaths**: fibres with moderately thickened walls present in 3–4 layers at xylem and phloem poles and in 1, frequently discontinuous, layer on bundle flanks. **Central ground tissue** parenchymatous, most cells rounded, with slightly thickened walls, some with thin walls; intercellular spaces frequent; no central cavity formed. **Silica**: none seen. **Tannin** in many epidermal cells, some subsidiary cells, and some cells of parenchyma sheath. **Crystals** square, slab-shaped, occurring in some specimens in cells of central ground tissue (Fig. 35. E).

Rhizome T.S.

Seen in *L. tenax* only.

Epidermis: cells as wide as high, or up to twice as wide as high; outer walls slightly thickened, anticlinal and inner walls thin. **Stomata** superficial; guard cells without lips or ridges, walls slightly thickened. **Hypodermis** composed of 2–5-layered groups of 4–6-sided cells with slightly thickened walls, dispersed among parenchymatous, thin-walled cells. **Cortex**: parenchymatous, thin-walled, rounded cells present in 8–10 layers, interspersed with occasional strands of 6–8 narrow, thick-walled fibres. **Endodermoid sheath** not apparent. **Vascular bundles** collateral, scattered, outermost with 1–2, innermost with 8–10 angular, thick-walled vessels at xylem pole. Lateral wall-pitting of vessels scalariform; perforation plates simple, oblique; phloem poles oval. **Central ground tissue** with several layers of sclerenchymatous, thick-walled cells surrounding individual vb's, the scl. from adjacent sheaths frequently confluent, and lacunae filled with parenchymatous, thick-walled, lobed, more or less isodiametric cells. **Silica** and **tannin**: none seen.

Root T.S.

Seen in *L. tenax* only.

Cortex not seen. **Endodermis** 1-layered, cells small, as wide as high or slightly wider than high; outer walls thin, other walls heavily thickened, lumina narrow. **Pericycle** 2–3-layered, cells more or less hexagonal, thick-walled, as high as wide. **Vascular tissue:** phloem strands restricted to 1 ring immediately to inside of pericycle, approximately 3 times as numerous as mxv's; mxv's wide, oval-angular, thick-walled, many arranged in single ring just inside that formed by phloem strands. **Central ground tissue:** outer 6–7 layers thick- or v. thick-walled, inner cells thin-walled, breaking down. **Tannin:** none seen.

MATERIAL EXAMINED

Lyginia barbata R. Br.: (i) ♀ An. 1913; Narrogin, W. Aust. (K). (ii) ♀, Grimwade Expd., An. 1947; Pallings Creek, W. Aust. (K). (iii) Pritzel 101; Avon, W. Aust. (K).

L. tenax (Labill.) C. A. Gardner: Cheadle, V. I. CA73; Perth, W. Aust. (S).

SPECIES DESCRIPTIONS

***L. barbata* R. Br.** Culm diameter 0·8, 1, and 1 mm.

Culm T.S. (Fig. 35. A, C, E)

Similar to genus description in all but following details. **Epidermis:** cells next to stomata produced into larger papillae in Grimwade than in other material. **Vascular bundles:** some of larger medullary vb's attached to culm scl. sheath by fibres at phloem pole.

***L. tenax* (Labill.) C. A. Gardner.** Culm diameter 1·5 mm.

See genus description.

TAXONOMIC NOTES

The genus has remained unchanged since its original description. It is unusual in the family because its species have 2 anthers to each stamen. Hieronymus (1888) included it with *Anarthria* and *Ecdeiocolea* in his *Diplantherae*. The anatomy of *Lyginia* is, however, quite unlike that of *Anarthria* or *Ecdeiocolea* (see Cutler and Shaw 1965), but there are no good grounds for excluding it from the Restionaceae.

Gilg (1890) noticed that the stomata are inclined relative to the culm surface as seen in L.S. and this feature was commented on by Solereder and Meyer (1929). This unusual arrangement is described on p. 232. Gilg mentioned and illustrated a small 'corona' present on the walls of some epidermal cells surrounding the stomata. He observed that it was present only in plants which had been collected at a certain time of the year. This unusual structure was not seen in any of the material examined for the present work, and its nature remains uncertain.

Besides being readily recognizable by the orientation of its stomata, *Lyginia* also has an unusual chlorenchyma structure. Chambers of palisade cells opposite the stomata are bounded by cells with thickened walls (see p. 232). These cells are not in the normal position for protective cells and they appear to be mainly mechanical. They are derived from palisade chlorenchyma cells, and therefore cannot be regarded as true pillar cells.

Crystals, present in this genus, are unknown in any other members of the Restionaceae.

While certain details of the anatomy indicate that *Lyginia* is rather isolated from other genera in the Restionaceae, the basic arrangement of tissues is very similar to that in the other restionaceous genera.

It is probable that *L. barbata* R. Br. is a synonym of *L. tenax* (Labill.) C. A. Gardner.

MASTERSIELLA Gilg-Benedict

No. of spp. *c.* 6; examined 5.

Genus Description

Leaf surface not seen.

Leaf T.S.

M. digitata only.

Outline: encircling culm with $1\frac{1}{2}$ turns, with a longitudinal rib on abaxial surface opposite to each vb. **Hairs** and **papillae:** none seen. **Epidermis:** (i) adaxial cells 5–6-sided, $1\frac{1}{2}$–2 times wider than high, the largest being *c.* $\frac{1}{3}$ height of abaxial cells; walls all moderately thickened; (ii) abaxial cells 4-sided, as high as or slightly higher than wide, those next to stomata narrower and shorter than others; outer walls thick, other walls slightly thickened. **Stomata** present in abaxial surface only, superficial; guard cells with ridge on inner wall; walls to pore concave. **Chlorenchyma** represented by flattened, lobed cells in 1 layer beneath abaxial epidermis.

Vascular bundles of 2 sizes, alternating: (i) larger vb's, with arc of medium-sized, angular tracheids partially enclosing rounded phloem pole; (ii) smaller vb's, with xylem pole composed of 2–4 narrow tracheids and phloem pole of 4–6 cells. **Bundle sheaths:** O.S. parenchymatous, cells in 1 layer, at phloem pole only; I.S. sclerenchymatous, present around larger bundles only, fibres in 1–2 layers, narrow, with thick or moderately thickened walls. **Sclerenchyma** represented by that of I.S. only. **Ground tissue** parenchymatous, cells wider than those of O.S.; arranged in 2–3 layers between bundles, separating chlorenchyma from adaxial epidermis, and in 1–2 layers to adaxial side of vb's. **Air-cavities** absent. **Silica:** spheroidal-nodular bodies present, 1 to a cell, in isolated cells of O.S. **Tannin** in some epidermal cells.

Culm surface

Hairs absent. **Papillae** present in 1 sp. (*M. diffusa*). **Epidermis:** cells 4- or 6-sided, as long as or up to 2 or 2–5 times longer than wide; walls moderately thickened, thick or v. thick, straight or wavy. **Stomata:** subsidiary cells variable; guard cells with rounded ends. **Silica** absent. **Tannin** in some epidermal cells. **Cuticular marks:** granular; longitudinal striations present in 1 sp. (*M. diffusa*).

Culm T.S.

Outline circular, with or without low mounds. **Cuticle** slightly thickened to thick, smooth, or with minute ridges. **Papillae** present, 1 to a cell, on some cells in *M. diffusa*. **Epidermis.** Cells in 1 layer, and arranged in 2 alternative ways as follows. (i) All of similar dimensions, 4-sided, 1–2 times higher than wide. Outer walls thick or v. thick, flattened dome-shaped, or with raised rim. Outer surface rough, with minute granular appearance, or smooth. Anticlinal walls either (*a*) thick or v. thick at outer ends, thickening tapering, and inner part of walls moderately or slightly thickened, or (*b*) slightly to moderately thickened, straight or wavy; inner walls slightly or moderately thickened. (ii) Some cells taller than others (usually next to or near to stomata), forming low mounds; taller cells 2–3 times higher than wide, shorter cells as high as to *c.* 1½ times higher than wide. Wall thickenings as for type (i).

Stomata superficial, sometimes raised on tall cells; subsidiary cells variable; guard cells with lip at outer aperture and, in some spp., with inner lip. **Chlorenchyma** composed of 2 layers of peg-cells; cells of outer layer usually taller and narrower than those of inner layer; cells arranged in lattice as seen in T.L.S., each cell joined to its neighbours by pegs at 4, 5, or 6 angles. Protective cells present in all spp., extending from epidermis either in 1 layer stretching to junction between chlorenchyma layers, or in 2 layers reaching the parenchyma sheath or only part way into inner chlorenchyma layer. Apertures between cells wide, short or elongated, occurring mainly in inner half of substomatal tube. **Parenchyma sheath** 1- (2-) layered, interrupted in places in *M. diffusa* and *M. laxiflora* by 1 layer of fibres from scl. sheath. Fibres single or in groups of up to 6. Cells of sheath as wide as high or 1½–2 times wider than high. **Sclerenchyma:** sheath 2–3- or 8–10-layered, outline circular, with occasional ribs as described in 'parenchyma sheath' above. Fibres v. thick-walled, with narrow lumina.

Vascular bundles. (i) Peripheral vb's: phloem pole circular or oval, partially enclosed by single-layered arc of medium-sized and narrow, angular tracheids, but separated from tracheids by 1 layer of narrow parenchyma cells with thin to slightly thickened walls. (ii) Medullary vb's: some outer bundles with 1 central mxv. All other bundles with 1 or 2 mxv's on either flank; flanking vessels medium-sized to wide, angular, circular or oval, with moderately thick to thick walls. Flanking vessels of some outer bundles touching one another, those of inner bundles separated by 1–3 rows of narrow, parenchymatous cells with slightly thickened walls. Lateral wall-pitting of flanking mxv scalariform; perforation plates frequently simple, nearly transverse, occasionally oblique, scalariform. Phloem poles circular or oval, abutting directly on xylem in *M. digitata*, ensheathed by 1 layer of narrow cells with all walls, or only those next to xylem, thickened in all other spp. Px present in all but smallest, outer bundles. Most bundles free from culm scl. sheath.

Bundle sheaths: fibres at phloem pole narrow, v. thick-walled, in 1 or 2 layers; those at xylem pole 1-layered, narrow, thick- or v. thick-walled; flanking fibres wider, walls thick or moderately thickened; in 1 layer. **Central ground tissue** parenchymatous; outer cells, surrounding medullary vb's, medium-sized or wide, with slightly or moderately thickened walls; inner cells, in area free from vb's, wide, thin-walled, often breaking down to form a central cavity. **Silica:** (i) as granular particles, in some cells of central ground tissue; (ii) as spheroidal-nodular bodies, in certain cells of outer layer of scl. sheath in some spp. **Tannin** in some epidermal cells.

Rhizome T.S.

M. digitata only.

Epidermis: cells present in 1 layer; 4-sided, 1½–2 times wider than high; outer walls slightly thickened, other walls thin. **Hypodermis:** cells in 1 layer, up to twice the size of epidermal cells, all walls thin. **Cortex.** (i) Outer part 6–7-layered; cells 2–3 times wider than high, walls moderately thickened or thick, individual cells *c.* 2–3 times higher than epidermal cells. (ii) Inner part 4–5-layered, cells thin-walled. **Endodermoid layer:** cells similar to those of inner cortex, but inner and anticlinal walls moderately thickened. **Vascular bundles:** outline circular, all more or less amphivasal; mxv's angular, wide, walls slightly thickened; phloem pole circular. **Bundle sheaths** 3–4-layered; fibres with moderately thick or thick walls; sheaths of adjacent bundles joined together, particularly in outer region of stele. **Central ground tissue** composed of lacunae of lobed, thin-walled, more or less isodiametric cells forming an aerenchyma between bundle sheaths. **Silica:** spheroidal-nodular bodies present in some cells of ground tissue.

Root T.S. (Fig. 35. N)

M. digitata only.

Root-hairs arising from square, thin-walled cells; shaft of hair *c.* ⅓ of width of base. **Epidermis:** cells similar to those bearing root-hairs. **Cortex.** (i) Outer part composed of 3–4 layers of thin-walled cells similar to those of piliferous layer. (ii) Middle part made up of radiating plates of cells, each plate 1 cell wide and up to 11 cells long, separated from its neighbours by lysigenous air-spaces. (iii) Inner part composed of 2–3 layers of flattened, thin-walled cells, each cell 1½–2 times wider than high. **Endodermis** 1-layered, cells more or less square, outer walls thin, inner walls v. thick, anticlinal walls thin at outer ends, v. thick at inner ends, thickening tapering rapidly. **Pericycle** composed of 3–4 layers of hexagonal, thick- or v. thick-walled cells. **Vascular system:** mxv's v. wide, oval in outline, each enclosed in a layer of flattened tracheids, and arranged in 1 ring to inside of that formed by phloem poles and px strands. **Central ground tissue** sclerenchymatous; fibres hexagonal, thick- or v. thick-walled. **Tannin:** none seen.

MATERIAL EXAMINED

Note: The arrangement of material in the Kew Herbarium follows Pillans's classification of 1928; *Mastersiella* was described in 1930 by Gilg-Benedict.

Mastersiella browniana (Mast.) Gilg-Benedict: Stokoe 21499 (in Kew Herb. as *Restio debilis* Nees var. *subulatus* Pillans); S. Afr. (K).

M. diffusa (Mast.) Gilg-Benedict: Parker, R. N. 4647 ♀, An. 1951 (in Kew Herb. as *Hypolaena diffusa* Mast.); Caledon, Elgin Forest, S. Afr. (K).

M. digitata (Thunb.) Gilg-Benedict: (i) Taylor, H. C. 3610, An. 1962 (in Kew Herb. as *Hypolaena digitata* Pillans); Caledon, S. Afr. (K). (ii) Cheadle, V. I. CA 838 (S).

M. hyalina (Mast.) Gilg-Benedict: Pillans, N. S. 17699, Sheet II, An. 1924 (in Kew Herb. as *Leptocarpus hyalinus*); Caledon, Zwart Berg, S. Slopes, S. Afr. (K).

M. laxiflora (Nees) Gilg-Benedict: Acocks, J. P. H. 21097 (in Kew Herb. as *Hypolaena laxiflora* Nees); S. Afr. (K).

Species Descriptions

***M. browniana* (Mast.) Gilg-Benedict.** Culm diameter 0·8 mm.

Culm surface

Epidermis: cells 4-sided, as long as or up to twice as long as wide, with shortest cells next to stomata; walls v. thick, some wavy. **Stomata:** subsidiary cells sometimes with many-faceted long walls, end walls often acute.

Culm T.S.

Outline circular. **Epidermis:** cells all of similar sizes, slightly to 1½ times higher than wide, outer walls slightly convex; anticlinal walls wavy, thickening tapering. **Stomata:** guard cells with pronounced lip at outer aperture. **Bundle sheaths:** scl. caps only. **Silica** as granular particles, in some cells of central ground tissue.

***M. diffusa* (Mast.) Gilg-Benedict.** Culm diameter 0·8 mm.

Culm surface

Papillae present, 1 to a cell, on occasional cells. **Epidermis:** cells 4- or 6-sided, 1½–3 times longer than wide, frequently with shorter cells next to stomata; walls moderately thickened, slightly wavy. **Stomata:** subsidiary cell end walls acute or rounded; guard cells with wavy long walls. **Cuticular marks** granular, and with longitudinal striations superimposed.

Culm T.S. (Fig. 35. M)

Cuticle with conspicuous ridges. **Papillae** present, 1 to a cell, on occasional cells. **Epidermis:** cells mostly about as high as wide, those next to stomata frequently twice as high as wide, lifting stomata above other epidermal cells; anticlinal walls not wavy. **Stomata:** guard cells with pronounced lip at outer aperture. **Chlorenchyma:** protective cells extending from epidermis to about half-way into inner chlorenchyma layer. **Parenchyma sheath** 1-layered, infrequently interrupted by occasional scl. fibres. **Silica:** none seen.

***M. digitata* (Thunb.) Gilg-Benedict.** Culm diameter 1·0 mm.

Culm surface (Fig. 35. J)

Epidermis: cells 4-, 5-, and 6-sided, as long as wide to nearly twice as long as wide, walls slightly to moderately thickened, not wavy.

Culm T.S. (Fig. 35. I)

Cuticle slightly thickened, with slight ridges. **Epidermis:** cells mostly twice as high as wide; outer walls curving inwards at margins, outer face more or less flat or slightly concave; anticlinal walls not wavy. **Stomata:** guard cells with lips at outer and inner aperture. **Chlorenchyma:** protective cells extending in 1 (occasionally 2) layer(s) to about half-way into inner chlorenchyma layer. **Silica:** bodies present in stegmata in outer layer of scl. sheath.

Leaf, rhizome, and *root*: see genus description on p. 235.

***M. hyalina* (Mast.) Gilg-Benedict.** Culm diameter 1 mm.

Culm surface (Fig. 35. H)

Epidermis. Cells up to 6-sided, of 2 sizes: (i) those next to stomata, slightly longer than wide, *c.* ⅓–½ length of larger cells; (ii) those not next to stomata

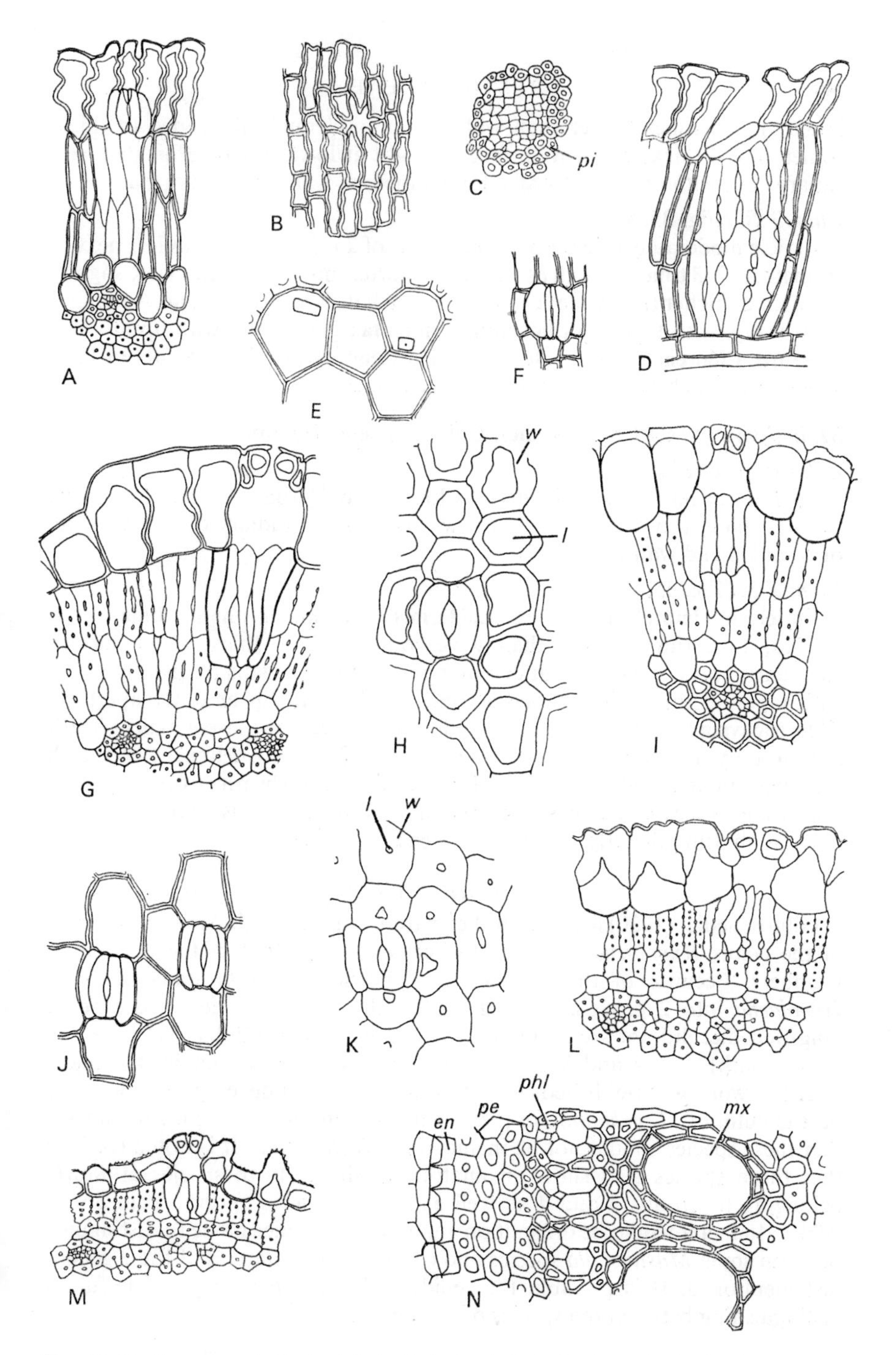

FIG. 35. Restionaceae. A–F, *Lyginia* (×185). A, C, E, *L. barbata*: A, detail of culm T.S., note inclined stoma; C, T.L.S. culm chlorenchyma just below epidermis (pi = pillar cells); E, crystals in cells of ground tissue. B, D, F, *L. tenax*: B, culm surface, note papillate epidermal cells overhanging stomatal pit; D, detail of culm L.S.; F, stoma, surface view. G–N, *Mastersiella* (×200). G, H, *M. hyalina*: G, detail of culm T.S.; H, culm surface. I, J, N, *M. digitata*: I, detail of culm T.S.; J, culm surface; N, part of root T.S.: en = endodermis, pe = pericycle, phl = phloem. K, L, *M. laxiflora*: K, culm surface; L, detail of culm T.S. M, *M. diffusa*, detail of culm T.S.
l = lumen; w = cell wall.

larger, 1½–2½ times longer than wide; walls moderately thickened to thick, sometimes v. thick, wavy at surface, straight at lower focus. **Stomata:** subsidiary cells with acute end walls, long anticlinal walls sometimes wavy.

Culm T.S. (Fig. 35. G)

Cuticle moderately thick. **Epidermis.** Cells of 2 main sizes: (i) taller, next to stomata, *c*. 2–3 times as high as wide; (ii) shorter, not next to stomata, slightly to 1½ times higher than wide; cells of intermediate size also present. Anticlinal walls wavy, with tapering thickening. **Stomata:** guard cells with lip at outer aperture. **Chlorenchyma:** protective cells extending in 1 layer to about halfway into inner chlorenchyma layer. **Silica:** none seen.

M. laxiflora **(Nees) Gilg-Benedict.** Culm diameter 1·0 mm.

Culm surface (Fig. 35. K)

Epidermis: cells mainly 4-sided, as long as to 1½ times longer than wide, walls v. thick, wavy, lumina v. narrow. **Stomata:** subsidiary cells with acute or rounded end walls.

Culm T.S. (Fig. 35. L)

Cuticle moderately thickened. **Epidermis:** cells all of similar dimensions, 4-sided, 1½–2 times higher than wide; outer wall v. thick, outer surface irregularly wavy, curving in at cell margins; anticlinal walls frequently wavy, with tapering thickening. **Stomata:** guard cells with lip at outer aperture. **Chlorenchyma:** protective cells extending in 2 layers to parenchyma sheath (1 layer only shown in Fig. 35. L, due to plane of section). **Parenchyma sheath** 1-layered, interrupted in places by low ribs from scl. sheath, 1 cell high, 1–5 cells wide. **Vascular bundles:** peripheral vb's as for genus, but some with phloem abutting directly on xylem. **Silica:** none seen.

Taxonomic Notes

Gilg-Benedict (1930) found that the anatomy of the South African members of the genus *Hypolaena* available to her was unlike that of the Australian members. Because of this and also on morphological grounds she split *Hypolaena* into 2 groups, naming the South African species *Mastersiella*. Gilg-Benedict did not see material of *H. crinalis* (Mast.) Pillans, *H. graminifolia* (Kunth) Pillans, and *H. purpurea* Pillans since they were named by Pillans after her work was published. Apart from the unusual development of wings on the culm of *H. graminifolia* there is little anatomical evidence to suggest that these species should not be included in *Mastersiella*. It is noted (p. 197) that these species also show an anatomical similarity to species of *Restio* belonging to groups 3 and 4 as well as to *Leptocarpus* species from South Africa. In fact it would be difficult on anatomical evidence alone to distinguish between some *Mastersiella* species and certain representatives of the 2 genera just mentioned. It is possible that epidermal characters may be useful in distinguishing between the species of this genus.

MEEBOLDINA Suess.

No. of spp. 1; examined 1.

GENUS AND SPECIES DESCRIPTION

Leaf not seen.

Culm surface (Fig. 36. C–F, p. 244)

Hairs v. complex in structure; outline of individual hairs difficult to determine; point of attachment to basal cell near one end. Each hair probably composed of the following cells: (i) 2 thin-walled cells lying side by side, each *c.* 8 times longer than wide, with rounded or somewhat acute ends; (ii) v. thick-walled, narrow, elongated cells of various lengths, but mostly $\frac{1}{3}$–$\frac{1}{2}$ as wide as thin-walled cells; arranged around the margins of the plate formed by the thin-walled cells; the outer (free) edge of the thick-walled cells irregularly papillate, the papillae being small and often with recurving branches. The hairs so formed overlapping one another at their edges, and orientated with their long axes parallel to the long axis of the culm (Fig. 36. D–F).

Epidermis: cells mostly 4-sided, some 5- or 6-sided, frequently 2–3 times and occasionally 4 times longer than wide; walls slightly to moderately thickened, wavy. **Stomata:** subsidiary cells narrow, as long as, or slightly longer than, guard cells, with rounded ends; guard cell pair with oval outline. **Silica:** none seen. **Tannin** in epidermal cells opposite pillar cells (q.v.). **Cuticular marks:** faint, longitudinal striations.

Culm T.S. (Fig. 36. A, B)

Outline irregularly 7-sided. **Hairs:** see surface view. Consisting of scale-like units attached to epidermis near hair edge by a single, short, thick-walled, basal cell (Fig. 36. G). **Epidermis:** cells 4-sided, as wide as or slightly wider than high, the squarer cells being taller than the others. Outer walls thick to v. thick, slightly convex, other walls moderately thickened, anticlinal walls not wavy. **Stomata** superficial; guard cells with outer lip and ridge on inner wall. **Chlorenchyma** divided into 7[1] longitudinal chambers by pillar cells; chlorenchyma cells shortly palisade-like, *c.* 1½–2 times higher than wide, present in 1–3 layers. No pegs seen. Intercellular spaces between anticlinal walls of cells wide (particularly as seen in L.S.). Protective cells absent.

Parenchyma sheath. Cells of 2 types: (i) pillar cells, mostly *c.* 3 times longer than wide, ends wider than shaft, walls slightly thickened; extending from ribs of scl. sheath to inner wall of epidermal cells; occurring in pairs in longitudinal files as seen in L.S. (some pillars composed of 2 or 3 shorter cells); (ii) rounded cells, each *c.* 1½ times wider than high, with slightly thickened walls, occurring mainly on flanks of low ribs from scl. sheath. **Sclerenchyma:** sheath 4–5-layered with evenly spaced, rounded ribs, each containing 1 peripheral vb and opposite to pillar cells; fibres of outer 3 layers narrow, with v. thick walls and narrow lumina, those of next nner layers wider, with moderately thickened to thick walls.

Vascular bundles. (i) Peripheral vb's present in ribs from scl. sheath, each with an arc of 3–6 narrow to medium-sized, thin-walled, angular tracheids, the tracheids at either end of the arc sometimes wider than the others.

[1] In this specimen; number may vary from specimen to specimen.

Phloem poles partially enclosed by arc of tracheids and composed of a total of 8–12 sieve-tubes and companion cells. (ii) Medullary vb's: flanking mxv's narrow or medium-sized, angular, with slightly thickened walls, vessels separated from one another by 3–4 rows of narrow, parenchymatous cells with slightly thickened walls; lateral wall-pitting of mxv's scalariform; perforation plates oblique, scalariform. Phloem pole oval in outline, separated from xylem by 1 layer of narrow cells with thin walls next to phloem cells and slightly thickened walls next to xylem. Px present in all bundles. All bundles attached by fibres at phloem pole to culm scl. sheath. **Bundle sheaths:** fibres narrow, 2-layered at phloem pole; those at xylem pole and on flanks wider, all with moderately thickened walls. **Central ground tissue** parenchymatous, cells wide, with moderately thickened walls; central cells not breaking down. **Silica:** spheroidal-nodular bodies present in stegmata in outer layer of culm scl. sheath. **Tannin** in some epidermal cells.

Rhizome and *root* not seen.

Material Examined

Meeboldina denmarkica Suess.: Meebold A. 1389, An. 1928, ♂ from type (Munich Herb.); Australia. Culm diameter 0·5 mm.

Taxonomic Notes

This is the most recently described genus in the Restionaceae. Its anatomy is very similar to that of Australian species of *Leptocarpus* and to species of *Restio* belonging to group 7 as well as to *Chaetanthus leptocarpoides* R. Br., *Loxocarya pubescens* Benth., and *Hypolaena* R. Br. (Australian species). It shares with them multicellular hairs and chlorenchyma divided into sectors by pillar cells. It has no protective cells; these are not usually to be found in species with pillar cells. The hair type is diagnostic for the genus and is so different from that encountered in the other genera mentioned above that this alone may indicate a real discontinuity with them. These genera are discussed again on p. 147 under *Chaetanthus*.

ONYCHOSEPALUM Steud.

No. of spp. 1; examined 1.

Genus and Species Description

Leaf not seen.

Culm surface (Fig. 36. J)

Hairs and **papillae** absent. **Epidermis:** cells 4-sided, as long as to 1½ times longer than wide; walls thick, wavy. **Stomata:** subsidiary cells variable; guard cell pair with almost lemon-shaped outline. **Silica** and **tannin:** none seen. **Cuticular marks** granular with longitudinal striations.

Culm T.S. (Fig. 36. H–I)

Outline circular with low, irregularly distributed mounds. **Cuticle** moderately thickened, with slight ridges. **Epidermis.** Cells present in 1 layer, of irregular sizes, those next to stomata being the highest, 2–3 times higher than wide, extending further to both the inside and the outside of the culm than the neighbouring cells, the latter being *c.* 1½–2 times higher than wide. Outer walls v. thick; anticlinal walls wavy, thick at outer ends, thickening tapering, and inner ⅙ of walls slightly thickened. Anticlinal walls lining substomatal cavities longer than others in the same cells, part extending into chlorenchyma slightly thickened; inner walls slightly thickened (Fig. 36 I). Cell lumina flask-shaped, with wavy necks. **Stomata** superficial; guard cells with lip at outer aperture. **Hypodermis** absent. **Chlorenchyma** composed of peg-cells in 1 layer; individual cells 5–6 times higher than wide, with numerous, medium-sized pegs; cells opposite stomata with fewer, slightly larger pegs. **Parenchyma sheath:** cells present in 1 layer, 1½–2 times wider than high, most with slightly thickened walls, some with moderately thickened inner and anticlinal walls and each containing a silica-body. **Sclerenchyma:** sheath 5–7-layered, fibres thick-walled, inner layers of cell walls gelatinous.

Vascular bundles. (i) Peripheral vb's with 1 medium-sized, angular tracheid on either flank of 1–2-layered arc of narrower tracheids; phloem pole situated in concavity of arc. (ii) Medullary vb's: outline oval, each with 1 medium-sized, moderately thick-walled, angular mxv on either flank, these separated by 2–3 rows of narrow tracheids with moderately thickened walls. Phloem poles oval in outline, separated from xylem by 1 layer of narrow cells with thin walls next to phloem and slightly thickened walls next to xylem. Px present in all but outermost, smallest bundles. Some bundles free from culm scl. sheath. **Bundle sheaths:** fibres narrow, thick-walled, in 1 layer at xylem and phloem poles, the whole bundle enclosed by 1 layer of wider, thick-walled fibres. **Central ground tissue** parenchymatous; intercellular spaces present between inner cells; all cells with moderately thickened walls; no central cavity present. **Silica** in stegmata. **Tannin** absent.

Rhizome and *root* not seen.

MATERIAL EXAMINED

Onychosepalum laxiflorum Steud.; Drummond 325 (Type); Swan River, Australia (K). Culm diameter 0·8 mm.

TAXONOMIC NOTES

The genus has been accepted by all authors since its description. It was not examined by Gilg, but Gilg-Benedict described the morphology and anatomy in 1930. She noted the simple anatomy which she said was similar to that of *Lepidobolus*. She did not comment on the lack of hairs in *Onychosepalum* or make observations on the epidermal cells surrounding the stomata, which are elongated as in *Lepidobolus*, *Coleocarya*, and *Harperia*. The relationships of these genera are discussed in the section on *Coleocarya* (p. 159), where it is noted that *Onychosepalum* has low mounds formed by areas of epidermal cells extending above the general level of the culm surface.

PHYLLOCOMOS Mast.

No. of spp. 1; examined 1.

GENUS AND SPECIES DESCRIPTION

Leaf sheath

Leaf surface (abaxial only seen). All characters as for culm surface.

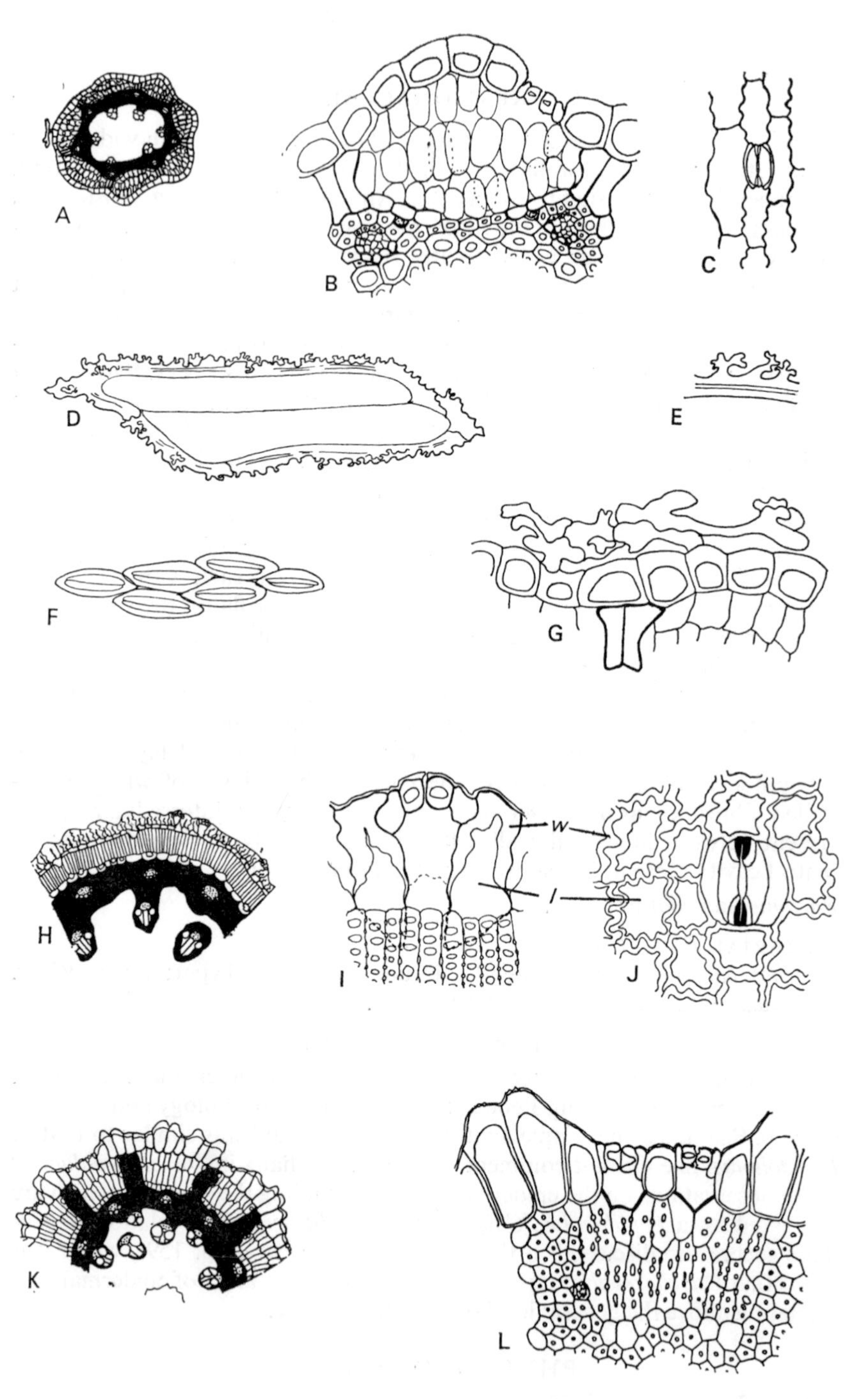

FIG. 36. Restionaceae. A–G, *Meeboldina denmarkica*: A, diagram of culm T.S. (×40); B, detail of culm T.S. (×200); C, culm surface (×200); D, hair, surface view (×150); E, detail of papillate hair-margin cell (×200); F, plan showing arrangement of hairs (×30); G, 2 hairs in T.S. (×200). H–J, *Onychosepalum laxiflorum*: H, diagram of sector of culm T.S. (×40); I, detail of stoma and elongated epidermal cells (×235); J, culm surface (×235). K, L, *Phyllocomos insignis*: K, diagram of sector of culm T.S. (×40); L, detail of culm T.S. (×150).
l = lumen; w = cell wall.

Leaf T.S.

Epidermis: (i) abaxial cells as in culm; (ii) adaxial cells pentagonal, outer walls convex, anticlinal and 2 inner walls straight; all walls slightly thickened. **Stomata** present in abaxial epidermis only; of same type as in culm. **Hypodermis** absent. **Chlorenchyma** present between vb's in thicker part of leaf; composed of peg-cells; all walls thin, except in cells lining the shallow substomatal cavities, when slightly thickened. **Vascular bundles** of 2 main sizes: (i) larger, having 1–2 medium-sized tracheids on either flank; (ii) smaller, alternating with larger, with a group of narrow tracheids at the xylem pole. Phloem poles developed in both types, but restricted to v. few sieve-tubes and companion cells in the smaller bundles.

Bundle sheaths double; O.S. parenchymatous, 1-layered, present around all larger and some smaller bundles, better developed to abaxial side of the bundle; I.S. sclerenchymatous, present around all bundles, smaller vb's with 1-layered and larger vb's with up to 3-layered I.S. Fibres heavily thickened, particularly at phloem poles. **Sclerenchyma** present as bundle sheaths only. **Ground tissue** parenchymatous, up to 4 layers thick in thickest part of leaf, not present at margins, where adaxial epidermis faces abaxial epidermis. (Extreme margins only 1 cell thick.) **Air-cavities** absent. **Silica:** spheroidal-nodular bodies present in cells of O.S. **Tannin** in larger cells of abaxial epidermis.

Culm surface

Hairs: none present. **Papillae** present, 1 to a cell, on certain cells occurring in an unevenly distributed reticulum. **Epidermis.** Cells of 2 main types; (i) papillate cells more or less hexagonal in outline; (ii) cells without papillae, rectangular to elongated-hexagonal in outline. Walls of both types of cell slightly thickened, straight or somewhat wavy. **Stomata** present among non-papillate cells; subsidiary cells equal in length to guard cells, with obtuse or rounded end walls; guard cells with rounded ends. **Silica:** bodies absent. **Tannin** in papillate cells. **Cuticular marks** granular, with superimposed longitudinal striations.

Culm T.S. (Fig. 36. K, L)

Outline more or less circular, with irregular ridges. **Cuticle** thick, ridged. **Epidermis:** tallest cells papillate, *c.* 80μm high; shortest non-papillate; outer walls thick, other walls moderately thickened; anticlinal walls straight. **Stomata** superficial; subsidiary cells extending above and below guard cells; guard cells with lips at outer and inner apertures. **Hypodermis** absent. **Chlorenchyma:** peg-cells in 2 layers, divided into 10 sectors by scl. ribs; walls of peg-cells slightly thickened; cells lining shallow substomatal cavities moderately thick-walled on side lining cavity, other walls thin. **Parenchyma sheath** consisting of 1 layer of cells interrupted by ribs of scl.; and also as occasional cells lining sides of the scl. ribs. **Sclerenchyma:** sheath produced into strong, rectangular ribs, each *c.* 4 cells wide, extending to the epidermis and dividing the chlorenchyma into more or less equal sectors; cylindrical part of sheath *c.* 4 cells wide.

Vascular bundles. (i) Peripheral vb's small, with 2–4 narrow tracheids arranged in an arc, and a well-defined phloem pole; alternating with scl. ribs. (ii) Medullary vb's with 1 medium-sized mxv on either flank; these vessels angular, with oblique, scalariform perforation plates and scalariform lateral wall-pitting. Phloem abutting directly on xylem. **Bundle sheaths** weak, 1 cell thick at bundle poles and discontinuous round the flanks of some bundles; walls of fibres at phloem pole more strongly thickened than those of other fibres. **Central ground tissue** parenchymatous, cells breaking down at centre of culm to form cavity. **Silica:** bodies spheroidal-nodular, present in some cells of parenchyma sheath. **Tannin** in larger epidermal cells.

Rhizome and *root* not seen.

MATERIAL EXAMINED

Phyllocomos insignis Mast.: (i) Schlechter 9956, An. 1896; Mitchell's Pass, Ceres Div., S. Afr. (isotype) Edinburgh. (ii) Esterhuysen 13851; Oliphants River Mts., Ceres Div., S. Afr. (K). Culm diameter 1·1 mm.

TAXONOMIC NOTES

This genus was described after Gilg's anatomical work had appeared, but it was examined anatomically by Gilg-Benedict (1930). It has been found that her description is accurate from the study of isotype material made for this present investigation. Gilg-Benedict noticed that some of the scl. ribs, which typify this genus, do not extend right to the epidermis. They were, however, all found to do so in the material examined during the present investigation, thus demonstrating that slight variation is to be expected in this character.

The genus is anatomically distinctive; it is similar to *Dielsia* Gilg (see p. 160) and *Restio leptocarpoides* Benth. (see p. 288) among Australian genera in having scl. ribs extending from the scl. sheath to the epidermis, but it is easily distinguishable from these species because it lacks their specialized, enlarged epidermal cells. There are areas of larger epidermal cells among the smaller ones in *Phyllocomos*, but the larger cells are not arranged in longitudinal files. The scl. ribs alternate with the peripheral vb's in *Phyllocomos*; they are opposite to them in *Dielsia* and in the Australian species of *Restio* belonging to group 10 (see p. 290).

The other genera in which species with scl. ribs occur are South African (*Willdenowia*, *Hypodiscus*), but, unlike *Phyllocomos*, they have well-developed protective cells.

Phyllocomos shares the main histological culm characters with other members of the family Restionaceae, but the anatomy indicates that it occupies an isolated taxonomic position. This is further borne out by the morphology, since the flowers are hermaphrodite and not dioecious as in the other South African genera.

RESTIO Rottb.

No. of spp. *c.* 125; examined 84.

The genus is divided into 10 groups for clarity of description. A v. brief outline of the diagnostic anatomical characters of the culm of each of the groups is given below, together with a list of the material examined for the groups. This is followed by detailed anatomical accounts of the groups.

MATERIAL EXAMINED

GROUP 1. Australia, Tasmania.

Culm T.S.

Chlorenchyma composed of peg-cells normally arranged in 1, but sometimes in 2 or 3 layers. Protective cells, pillar cells, hairs, and papillae absent. Sclerenchyma sheath without pronounced ribs. Epidermal cells of more or less uniform size. See under group 10 (p. 250) for characters distinguishing it from group 1.

Restio australis R. Br.: (i) Cheadle, V. I. CA 379; Tasmania (S). (ii) Constable, E. F. 7281, An. 1948; Mt. Cameron, Newnes Distr., N.S.W., swamp, sandstone (K). (iii) ? 422; mountain side (K). (iv) Burbridge 3388; swamp (K).

R. dimorphus R. Br.: Salisbury, E. J., An. 1949.; West Point Nature reserve, N.S.W. (K).

R. fimbriatus L. Johnson and O. Evans: (i) Rodway, F. A. 249, An. 1930; Fitzroy Falls, swampy area, N.S.W. (K). (ii) Constable, E. F. 36568, An. 1956; Wingello State Forest, swampy soil, sandstone, N.S.W. (K).

R. gracilis R. Br.: (i) Mair, H. K., and Constable, E. F. ♂ 16462; damp, swampy ground, sandstone, N.S.W. (K). (ii) Mair, H. K., and Constable, E. F. ♀ 16464; waterfall 24 miles S.W. Sydney, N.S.W. (K). (iii) Rodway, F. A. 1750A; Bulli Dams, N.S.W. (K).

R. longipes L. Johnson and O. Evans: Constable, E. F., N.S.W., 35056; Clyde Mts. (K).

R. monocephalus R. Br.: Curtis, W. M.; Blackman's Bay, Hobart, Tasmania (S).

R. oligocephalus F. v. Muell.: Rodway, F. A. 97 (K).

R. pallens R. Br.: (i) Hubbard, C. E. 2249, An. 1930; Stradbroke Is. (K). (ii) Hubbard, C. E. 2586, An. 1930; Bribie Is. (K). (iii) Constable, E. F. 22082, An. 1953 (K). (iv) Constable, E. F. 19362, An. 1952; Bombah Pt. (K).

R. stenocoleus L. Johnson and O. Evans: Williams, J. B., N.S.W., 53480; Bullock Creek, Armidale to Ebor Rd. (K).

GROUP 2. Australia.

As for group 1, but with multicellular hairs, and epidermal cells next to stomata extending inwards into chlorenchyma.

R. confertospicatus Steud.: Drummond ♂, Herb. F. Mueller, An. 1876 (K).

GROUP 3. South Africa.

As for group 1, but with protective cells. See under groups 4, 5, and 6 for characters distinguishing them from group 3.

R. aridus Pillans: Esterhuysen 8476, An. 1942; Worcester Div. ('a dwarf and slender growth form') (K).

R. bifarius Mast.: Parker, R. N. 4885 (K).

R. bifidus Thunb.: Parker, R. N. 4759, An. 1952 (K).

R. bifurcus Nees ex Mast.: (i) Parker, R. N. 4876 ♀, An. 1953 (K). (ii) Burchell, no number (M).

R. bolusii Pillans: Stokoe 1357, An. 1926 (K).

R. brunneus Pillans: Esterhuysen 8070 ♂, An. 1942; Boontjieskraal, Cederberg Mts., Clanwilliam Div. (K).
R. callistachyus Kunth: Fourcade 2526 (K).
R. compressus Rottb.: (i) Esterhuysen 24471, An. 1955 (K). (ii) ? ♂ 7535 (K).
R. curviramis Kunth: Schlechter 7466, An. 1896 (K).
R. cuspidatus Thunb.: Parker, R. N. 34860 ♀ (K).
R. dispar Mast.: Parker, R. N. 4326; in stony stream bed (K).
R. dodii Pillans: Pillans, N. S. 7302 (K).
R. egregius Hochst.: Schlechter 7240 (K).
R. festuciformis Nees ex Mast.: Schlechter 2600 (K).
R. foliosus N.E. Br.: Acocks, J. P. H. 21091; N. side arid Fynbos (K).
R. fourcadei Pillans: Thode, J. A2426; Cape (K).
R. fraternus Kunth: Esterhuysen 9650; nr. Mt. Summit (K).
R. galpinii Pillans: Esterhuysen 8820, An. 1943; Natal (K).
R. giganteus (Kunth) N.E. Br.: Burchell (M).
R. gossypinus Mast.: Pillans, N. S. 7320 (K).
R. laniger Kunth: Acocks, J. P. H. 17076 (K).
R. leptocladus Mast.: Schlechter 9720 (K).
R. major (Mast.) Pillans: Acocks, J. P. H. 21671 (K).
R. micans Nees: Bolus 4448 ♀ (K).
R. miser Kunth: Esterhuysen 7374 ♀ (K).
R. multiflorus Spreng.: Garside 5446 (K).
R. nodus Pillans: Esterhuysen 8717; Worcester Div. (K).
R. ocreatus Kunth: Pillans, N. S. 6128 ♀ (K).
R. pachystachyus Kunth: = Pillans 17700 ♂ in Bolus Herb. (K).
R. pedicellatus Mast.: Parker, R. N. 4853 ♀ (K).
R. praeacutus Mast.: ? previously under *R. bifurcus* Nees (K).
R. quinquefarius Nees: Burchell 188; Cape (K).
R. rhodocoma Mast.: (i) Drège 2016 ♂; Alexandria Div. (K). (ii) Burchell 3373 (K). (iii) Schlechter 6021; nr. river (K).
R. rottboellioides Kunth: Esterhuysen 5812, An. 1941; Clanwilliam Div. (K).
R. sarocladus Mast.: Stokoe 2482 (K).
R. setiger Kunth: Esterhuysen 5804 (K).
R. sieberi Kunth var. *sieberi*: Phillips 11107 (K).
R. sieberi var. *venustulus* (Kunth) Pillans: Esterhuysen 24395, An. 1955 (K).
R. stereocaulis Mast.: Masters, M. T., 9303 (K).
R. strobilifer Kunth: Bolus 5484 (K).
R. tenuissimus Kunth: (i) Parker, R. N. ♂ 4614; Stellenbosch Div., wet, sandy flats (K). (ii) Schlechter 7790 (*R. ludwigii* Steud. in Kew Herb.) (K). (iii) Schlechter 2053, An. 1907 (*R. ludwigii* Steud. in Kew Herb.) (K).
R. triflorus Rottb.: (i) Parker, R. N. 3545 ♀, An. 1940; Somerset West (K). (ii) Esterhuysen 24202, An. 1955 (K). (iii) Parker, R. N. 3657, An. 1942; Helderberg, Stellenbosch Div. (K).
R. vilis Kunth: Schlechter 8470 (K).
R. virgeus Mast.: Phillips 11109 (K).

GROUP 4. South Africa.

As group 3, but with some epidermal cells taller than others, forming low mounds on culm surface; some spp. with papillae.

R. cincinnatus Mast.: Schlechter 348, Sheet 2, An. 1892; Table Mt. (K).
R. duthieae Pillans: Esterhuysen 22082; Cape, moist sand, E. slopes of rocky plateau (K).
R. filiformis Poir.: Parker, R. N. 4660 ♂; Caledon Div. (K).
R. fruticosus Thunb.: (i) Schlechter 10462 (K). (ii) Esterhuysen 24585; nr. stream (K). (iii) Schonland 3347SS, An. 1919, Nat. Herb. Pretoria (in Kew Herb. as *R. macowani* Pill.) (K).
R. macer Kunth: Pillans, N.S. 7604; Piquetberg Mts. (K).
R. marlothii Pillans: Compton 3155 (K).
R. monanthus Mast.: Acocks, J. P. H. 19717 ♀, An. 1958; Calvinia, Cape (K).
R. patens Mast.: Stokoe, T. P., An. 1929; Worcester Div. (K).
R. purpurascens Nees ex Mast.: Stokoe, T. P. 1330 ♂ (K).
R. pygmaeus Pillans: Pillans, N.S. 15892 (K).
R. scaber Mast.: Stokoe, T. P. 2844 (K).
R. scaberulus N.E. Br.: Burchell 5732 and Bolus 11394 ♂ (K).
R. triticeus Rottb.: (i) Garside, S. 1039, Sheet II (K). (ii) Acocks, J. P. H. 11061; sandy bush (K).
R. zwartbergensis Pillans: Parker, R. N. 4860; nr. stream (K).

GROUP 5. Madagascar.

As group 3, but with 1 papilla on each culm epidermal cell.

R. madagascariensis H. Chermez.: Humbert, H. 3892, An. 1924 (K).

GROUP 6. South Africa.

As group 3, but with square culm T.S. and with 1 scl. rib extending between mechanical cylinder and epidermis at each angle.

R. quadratus Mast.: Burchell 408 (K).
R. tetragonus Thunb.: Schlechter 10289 (K).

GROUP 7. Australia, Tasmania.

As for Australian *Leptocarpus* spp., with pillar cells, some with hairs.

R. amblycoleus F. v. Muell.: Drummond ♂ (type) (K).
R. chaunocoleus F. v. Muell.: Drummond 967; Swan R. (type) (K).
R. complanatus R.Br.: (i) Cheadle, V. I. CA 99; Grampians (S). (ii) Cheadle, V.I. CA 300; Coff's Harbour (S). (iii) Cheadle, V. I. CA 413; Tasmania (S).
R. gracillior F. v. Muell. ex Benth.: Drummond 68; Swan R. (K).
R. grispatus R. Br.: Brown, R.; Lucky Bay (K).
R. laxus R. Br.: Andrews, C. 1st colln. 1087; Albany (K).
R. megalotheca F. v. Muell.: Drummond ♀; Swan R. (K).
R. nitens Nees: Drummond; Swan R. (K).
R. ornatus Steud.: Drummond 339 (type) (K).
R. sphacelatus R. Br.: Willis, J. H., An. 1950; Duke of Orleans Bay (K).

GROUP 8. Australia

Culm T.S.

Epidermal cells of 2 sizes, large and small, alternating in longitudinal rows;

ribs from scl. sheath extending to some of rows of larger cells. Pillar cells and protective cells absent.

R. leptocarpoides Benth.: (i) Diels 6135 (K). (ii) ? ♀; King's Sound (K).

R. leptocarpoides var. *monostachyus* F. v. Muell. ex Benth.: F. Mueller ♀ An. 1876; Stirling Range (type) (K).

GROUP 9. Australia.

Culms flattened, with scl. ribs extending from mechanical cylinder to epidermis; epidermal cells not grossly dimorphic.

R. applanatus Spreng.: Drummond ♀; Swan R. (K).

R. tremulus R. Br.: (i) Harvey, W. H., An. 1854 (K). (ii) Drummond 106 (K).

GROUP 10. Australia.

As group 1, but with papillae, and stomata sunken in longitudinal grooves.

R. fastigiatus R. Br.: (i) Rodway, F. A. 2591; Upper Clyde R., N.S.W. (K). (ii) Salisbury, E. J., An. 1949; W. Pt. Nat. Res. N.S.W. (K). (iii) ? N.S.W. H. 347, An. 1950; Blackheath, Blue Mts., damp boggy ground, sandstone (K). (iv) Constable, E. F. 7583; Linden, Blue Mts., swampy ground, sandstone, N.S.W. (K). (v) Forsyth, W. 118, An. 1902; Port Jackson (K). (vi) Rodway, F. A. 198, An. 1928; Jarvis Bay, N.S.W. (K). (vii) Morrison, D. A. 5197; Sydney, Nat. Park, N.S.W. (K). (viii) Stephenson; Sydney, N.S.W. (K). (ix) Rodway, F. A. 1754; Sassafras Falls, 25 miles W. of Nowra, sandstone boulders, N.S.W. (K).

RESTIO GROUP 1. Australia

Leaf surface

Scale leaf, seen in *R. monocephalus*

Epidermis: (i) abaxial cells 4-sided, 3–6 times longer than wide, walls slightly to moderately thickened, wavy; (ii) adaxial cells 4-sided, end walls transverse or oblique, cells 4–6 times longer than wide, *c.* $\frac{1}{3}$ as wide again as abaxial cells; walls slightly to moderately thickened, not wavy. **Stomata** present in abaxial surface only; subsidiary cells with perpendicular or acute end walls; guard cell pair with lemon-shaped outline. **Silica:** none seen. **Tannin** in some epidermal cells. **Cuticular marks:** longitudinal striations in cuticle of abaxial surface.

Leaf T.S.

Scale leaf, seen in *R. australis* and *R. monocephalus.*

Outline: encircling culm by $1\frac{1}{2}$ turns, central portion with slight abaxial ribs opposite vb's; blade tapering to margins. **Cuticle** thick, granular on abaxial surface, thin on adaxial surface. **Hairs** absent. **Papillae** present as short dome-shaped emergences from some abaxial epidermal cells in *R. australis*. **Epidermis.** (i) Abaxial cells 4-sided, as high as to $1\frac{1}{2}$ times higher than wide in *R. australis*, about $1\frac{1}{2}$ times wider than high in *R. monocephalus*; outer walls moderately thickened (*R. monocephalus*) or v. thick (*R. australis*); other walls slightly or slightly to moderately thickened. (ii) Adaxial cells 4–5-sided, about twice as wide as high, slightly wider than, but *c.* $1\frac{1}{2}$ times height of abaxial cells; all walls moderately thickened. **Stomata** present in abaxial surface only, similar to those in culm of *R. monocephalus*; subsidiary cells with ridge on outer wall; guard cells with beak-like lip at outer aperture. **Chlorenchyma:** cells lobed, mostly as high as wide, restricted to areas opposite to stomata, between vb's in *R. australis*, but present in 1 or 2 more or less continuous layers next to abaxial epidermis in *R. monocephalus*.

Vascular bundles each with 1–2-layered arc of narrow tracheids (occasionally with wider ones on flanks) partially enclosing circular phloem pole; bundles of 2 main sizes, large and small, alternating. **Bundle sheaths:** O.S. parenchymatous, present in *R. monocephalus* only, as 1–2 layers of cells at phloem pole; I.S. sclerenchymatous, completely encircling bundles; fibres narrow, thick-walled, in 2 layers at phloem pole, wider, in 2–6 layers on flanks and 1 layer at xylem pole (fibres gelatinous in *R. australis*). **Sclerenchyma** present in bundle sheaths and in 1–2 layers between bundles, frequently joining sheaths of adjacent bundles. **Ground tissue** parenchymatous, cells in 1 layer to inside of adaxial epidermis, and also occasionally present between vb's. **Air-cavities** absent. **Silica:** spheroidal-nodular bodies in stegmata in outer scl. layer at phloem poles. **Tannin** in many epidermal cells.

Culm surface

Hairs and **papillae** absent. **Epidermis:** cells 4–6-sided, as long as or up to 4 times longer than wide; walls moderately thickened, wavy or straight. **Stomata** variable; see spp. descriptions. **Silica** absent. **Tannin** in epidermal cells of some spp. **Cuticular marks** granular, occasionally with superimposed longitudinal striations.

Culm T.S. (Pl. IV. B).

Outline circular or oval. **Cuticle** slightly thickened–thick. **Epidermis.** Cells present in 1 layer, 4-sided, frequently as high as wide to 1½ times higher than wide or, in some spp. with some cells 3 or 5–6 times higher than wide. Outer walls thick or v. thick; anticlinal walls straight or wavy, slightly or moderately thickened, or with tapering thickening; inner walls slightly or moderately thickened. **Stomata** superficial or sunken; variable, see individual spp. **Chlorenchyma** composed of peg-cells in 1, 2, or 3 layers; cells mostly 2–4 times higher than wide, but in some spp. reaching 10 times higher than wide; pegs small, numerous. Protective cells absent. **Parenchyma sheath** 1-, 2-, or 1–2-layered; cells rounded, slightly to 1½ times higher than wide or up to 1½–(2) times wider than high, with slightly thickened walls. **Sclerenchyma:** sheath well developed in all spp., outline circular, frequently with low, dome-shaped ribs opposite to peripheral vb's; fibres of outer layers narrow, with thick or v. thick walls, those of inner layers wider, with thick walls.

Vascular bundles. (i) Peripheral vb's frequently large, with wider tracheids flanking several-layered arc of narrower tracheids; phloem pole circular or oval, situated in concavity of xylem arc. (ii) Medullary vb's: outline oval; each with 1 narrow, medium-sized, or wide, angular (occasionally rounded or oval) mxv on either flank. Lateral wall-pitting of vessels scalariform; perforation plates usually oblique, scalariform or scalariform-fenestrate. Phloem poles oval, ensheathed by 1–2 layers of narrow cells with slightly or moderately thickened walls. Px present in all, or only in larger, inner bundles. Some or no bundles free from culm scl. sheath. **Bundle sheaths:** fibres narrow, thick- or v. thick-walled, in 2 layers at phloem pole; wider with moderately or slightly thickened walls, in 1 layer on flanks; narrow or medium-sized, in 1 or 2 layers at xylem pole. **Central ground tissue** parenchymatous, outer cells around vb's with slightly or moderately thickened walls; central cells thin-walled, breaking down to form central cavity, or with moderately thickened walls and not breaking down. **Silica:** spheroidal-nodular bodies present in stegmata in parenchyma sheath or outer layer of scl. sheath in most spp. **Tannin** in epidermal cells of some spp.

Rhizome T.S.

Seen in *R. australis* and *R. monocephalus*.

Outline circular. **Hairs** multicellular, each composed of 1 small, basal cell and a single file of several thin-walled cells, each cell 2–4 times longer than wide. **Epidermis:** cells 4–5-sided, as high as or slightly to 1½ times higher than wide, walls thin or slightly thickened. **Cortex.** (i) Outer part; cells with slightly thickened (lignified) walls with numerous, simple pits, cells of outer 7–8 layers 6-sided, 2–4 times wider than high (slightly to 1½ times higher than wide in *R. monocephalus*), those of inner 1–2 layers 2–3 times higher than wide with slightly or moderately thickened walls. (ii) Inner part composed of 3–4 layers of loosely packed, parenchymatous, thin-walled cells, each 3–4 times higher than wide. **Endodermoid sheath** not developed in *R. australis*, 2-layered in *R. monocephalus*, cells 6-sided, as high as wide, walls slightly to moderately thickened.

Vascular bundles: outer vb's small, inner larger, all more or less amphivasal; mxv's angular, medium-sized to wide; perforation plates simple, lateral wall-pitting scalariform and alternate. **Bundle sheaths:** cells medium-sized, with moderately thickened to thick walls; merging into cells of ground tissue. **Central ground tissue** composed of a matrix of moderately thick- to thick-walled sclereids and of small lacunae filled with thin-walled, parenchymatous cells. **Silica:** spheroidal-nodular bodies present in stegmata in outer layer of ground tissue in *R. australis*, none seen in *R. monocephalus*. **Tannin** in isolated vessels (and in many epidermal cells in *R. monocephalus*).

Root T.S.

Seen in *R. monocephalus*.

Epidermis not seen. **Cortex:** (i) outer part 5–6-layered, cells 6-sided, slightly wider than high, with slightly thickened walls; (ii) middle part composed of wide, thin-walled cells (often breaking down) and smaller cells forming 1–2-layered, radiating plates; (iii) inner part consisting of 1–2-layers of rounded, narrow, thin-walled cells. **Endodermis:** 1 layer of large, 4-sided cells, cells 3 times higher than wide, v. thick-walled, lumina occluded. **Pericycle:** 1–2 layers of cells each *c.* ⅛ height of endodermal cells; walls moderately thickened. **Vascular system:** mxv's wide, oval-angular, 10 arranged in 1 ring inside that formed by phloem and px. Each mxv ensheathed by 1 layer of flattened tracheids. **Central ground tissue** sclerenchymatous, cells v. thick-walled. **Tannin:** none seen.

SPECIES DESCRIPTIONS

***R. australis* R. Br.** Culm diameter 1·2–3·0 mm.

Culm surface

Epidermis: cells 4–6-sided, as long as, or up to twice as long as wide, some of shorter cells only *c.* ½ width of longer cells; walls wavy at culm surface. **Stomata:** subsidiary cells with acute or obtuse end walls; guard cell pair with lemon-shaped outline. **Tannin** in some epidermal cells. **Cuticular marks** granular, granules arranged in longitudinal files.

Culm T.S.

Epidermis: cells next to stomata *c.* 1½ times higher than wide, others normally 1½–3 times higher than wide, those farthest away from stomata up to 5–6 times higher than wide in (7281); anticlinal walls with tapering thickening, slightly wavy. **Stomata** sunken; with slight lip at outer aperture and ridge on inner wall. **Chlorenchyma** (1)–2–3-layered. **Central ground tissue:** inner cells separated from outer cells by 1–2 layers of narrow, thick-walled fibres in specimen 3388.

Leaf and *rhizome*: see group 1 genus description.

***R. dimorphus* R. Br.** Culm diameter 1·1 mm.

Culm surface

Epidermis: cells 4–6-sided; 1½–4 times longer than wide, walls slightly to moderately thickened. **Stomata** sunken, slightly overarched by surrounding epidermal cells, leaving the stomata visible through longitudinally orientated, oval holes, subsidiary cells with rounded end walls, longer than guard cells; guard cell pair narrow with lemon-shaped outline. **Tannin:** none seen.

Culm T.S.

Epidermis: cells mostly as high as wide; (stomata overarched by cells next to them); outer walls v. thick, irregular, often slightly concave; anticlinal and inner walls slightly thickened. **Stomata** small, sunken, attached to inner end of anticlinal walls of adjacent epidermal cells; guard cells with slight lip at outer aperture. **Chlorenchyma:** cells in 1 layer. **Tannin:** none seen.

***R. fimbriatus* L. Johnson and O. Evans.** Culm diameter 0·8 mm.

Culm surface

Epidermis: cells 4-sided; as long as to 4 times longer than wide, walls moderately thickened to thick, wavy. **Stomata:** subsidiary cells with acute end walls, long anticlinal walls often slightly wavy; guard cells with acute ends. **Tannin** in some epidermal cells.

Culm T.S.

Epidermis: cells mainly 1½ times higher than wide; anticlinal walls not wavy. **Stomata** superficial; subsidiary cells with slight ridge on outer wall; guard cells with slight lip at outer aperture and ridge on inner wall. **Chlorenchyma:** cells in 1 and 2 layers. **Tannin** in some epidermal cells.

***R. gracilis* R. Br.** Culm diameter 0·9–1·1 mm.

Culm surface

Epidermis: cells 4–6-sided; 2–3 times longer than wide, walls moderately thickened, wavy. **Stomata:** subsidiary cells with rounded end walls and slightly wavy long walls; guard cell pair with lemon-shaped outline. **Tannin** in many epidermal cells.

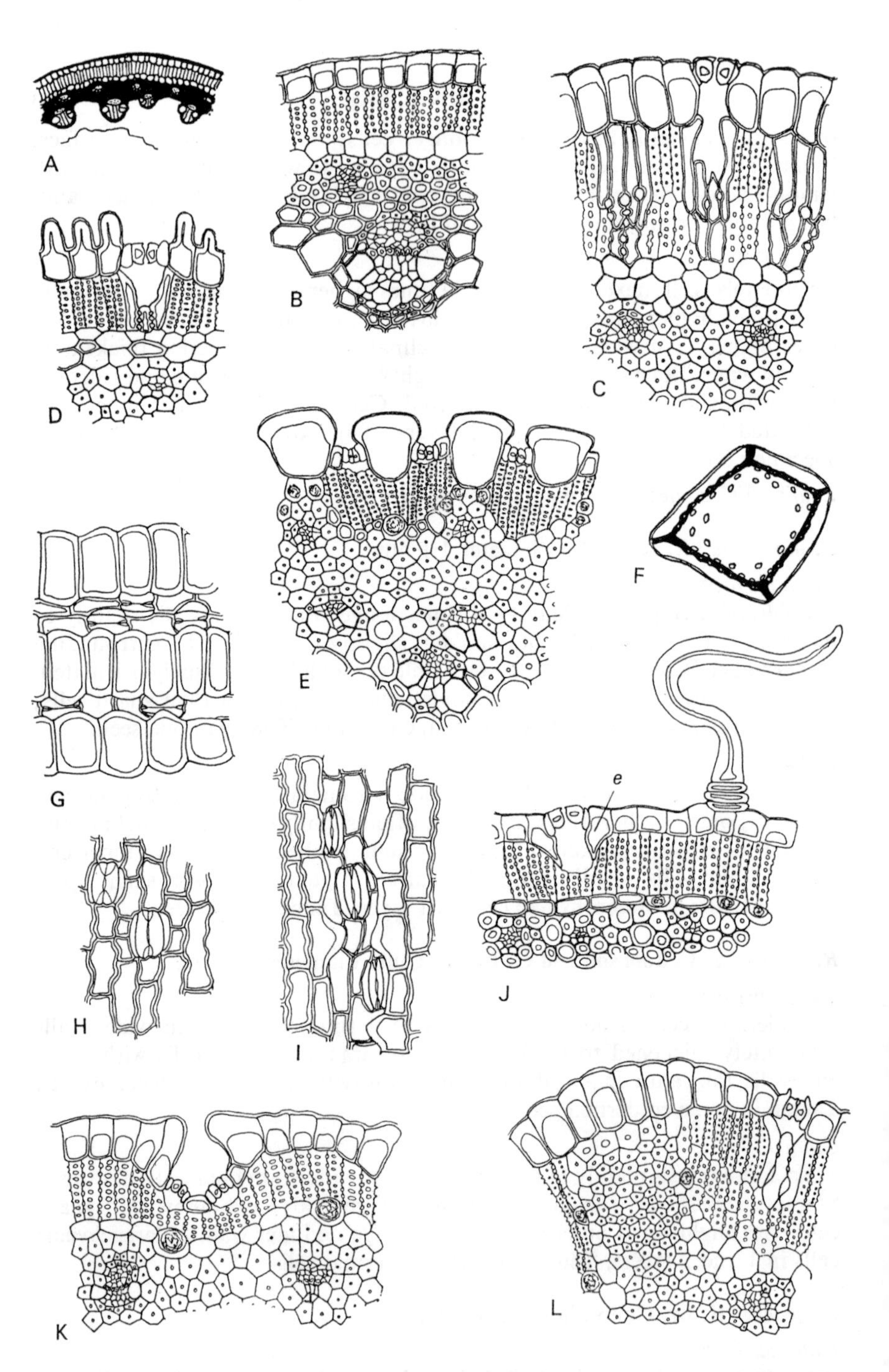

FIG. 37. Restionaceae. *Restio*. A, B, *R. gracilis*: A, diagram of sector of culm T.S. (×50); B, detail of culm T.S. (×185). C, *R. saroclados*, detail of culm T.S. (×185). D, *R. madagascariensis*, detail of culm T.S. (×185). E, G, *R. leptocarpoides*: E, detail of culm T.S. (×185); G, culm surface (×185). F, L, *R. tetragonus*: F, diagram of culm T.S. (×12); L, detail of angle of culm T.S. (×185). H, J, *R. confertospicatus*: H, culm surface (×200); J, detail of culm T.S.; note hair and elongated epidermal cells next to stoma (e); (×200). I, K, *R. fastigiatus*: I, culm surface, stomata in groove (×200); K, detail of culm T.S. (×200).

Culm T.S. (Fig. 37. A, B; Pl. IV. B)

Epidermis: cells as high as to slightly higher than wide; outer walls v. thick, surface irregular. **Stomata** superficial; guard cells with ridge on inner wall. **Chlorenchyma:** cells in 1 layer. **Tannin** in many epidermal cells.

***R. longipes* L. Johnson and O. Evans.** Culm diameter 2·0 mm.

Culm surface

Epidermis: cells 4-sided, 1½–2 times longer than wide, walls moderately thickened, wavy. **Stomata:** subsidiary cells with acute or rounded end walls, long anticlinal walls slightly wavy; guard cell pair narrow. **Tannin:** none seen.

Culm T.S.

Epidermis: cells 1½ times higher than wide; outer walls v. thick; anticlinal walls with tapering thickening. **Stomata** sl. sunken; subsidiary cells with ridge on outer wall; guard cells with v. slight lips at outer and inner apertures. **Chlorenchyma:** cells in 2–(3) layers. **Silica:** small bodies in some outer cells of scl. sheath. **Tannin** absent.

***R. monocephalus* R. Br.** Culm diameter 1·1 by 0·9 mm.

Culm surface

Epidermis: cells 4-sided; as long as to twice as long as wide, walls slightly to moderately thickened, wavy. **Stomata:** subsidiary cells with obtuse end walls; guard cell pair with lemon-shaped outline. **Tannin** in many epidermal cells.

Culm T.S.

Outline oval. **Epidermis:** cells mostly twice as high as wide; outer walls v. thick, frequently concave, and curving in slightly at margins. **Stomata** superficial; guard cells with v. slight lip at outer aperture. **Chlorenchyma** 2–(3)-layered. **Central ground tissue:** outer cells with slightly to moderately thickened walls, central cells with slightly thickened walls. **Tannin** in many epidermal cells.

Leaf, *rhizome*, and *root*; see group 1 genus description (pp. 250, 252).

***R. oligocephalus* F. v. Muell.** Culm diameter 1·0 by 0·7 mm.

Culm surface

Epidermis: cells 4-sided; as long as to twice as long as wide, walls moderately thickened, wavy. **Stomata:** subsidiary cells variable; guard cells with narrow rounded ends. **Tannin** in some epidermal cells. **Cuticular marks** granular, with discontinuous longitudinal striations.

Culm T.S.

Outline oval. **Epidermis:** cells 1½–2 times higher than wide; outer walls v. thick, outer surface granular. **Stomata** superficial; subsidiary cells with ridge on outer wall; guard cells without lip at outer or inner apertures. **Chlorenchyma** 2- or 3-layered. **Vascular bundles:** peripheral vb's larger than in most spp., tracheids narrow to wide, arranged in 3–4-layered arc with wider ones flanking; arc partially enclosing circular phloem pole. **Central ground tissue:** outer 4–5 layers of cells narrow, central cells wider, all with slightly thickened walls. **Tannin** in some epidermal cells.

***R. pallens* R. Br.** Culm diameter 2–3 mm.

Culm surface

Epidermis: cells 4-sided; as long as to twice as long as wide, walls moderately thickened, wavy. **Stomata:** subsidiary cells variable; guard cells with rounded ends. **Tannin** in some epidermal cells.

Culm T.S.

Epidermis: cells mostly 1½ times higher than wide; outer walls v. thick, frequently convex; anticlinal walls sometimes slightly wavy. **Stomata:** subsidiary cells sometimes with ridge on outer wall; guard cells with slight lip at outer aperture and ridge on inner wall. **Chlorenchyma** 1- or 2-layered. Protective cells absent, but peg-cells opposite to stomata with larger, fewer pegs than their neighbours, and v. occasionally with slightly lignified but unthickened walls. **Silica:** large bodies present in many cells of parenchyma sheath, and smaller ones frequently present in some cells of outer layer of culm scl. sheath. **Tannin** in some epidermal cells.

***R. stenocoleus* L. Johnson and O. Evans.** Culm diameter 1·1 mm.

Culm surface

Epidermis: cells 4–6-sided; as long as to 1½–2 times longer than wide, walls moderately thickened to thick, straight at surface, wavy at lower focus. **Stomata:** subsidiary cells with acute end walls; guard cells with more or less rounded walls. **Tannin** in many epidermal cells.

Culm T.S.

Epidermis: cells between stomata 1½–2 times higher than wide; those next to stomata 2½–3 times higher than wide, extending inwards further than other cells; outer walls v. thick, outer surface granular; anticlinal walls wavy. **Stomata** superficial; guard cells with slight outer lip and ridge on inner wall. **Chlorenchyma** 2–(3)-layered. **Tannin** in many epidermal cells.

RESTIO GROUP 2. Australia

Genus descriptions as for group 1, but with hairs, and epidermal cells surrounding stomata elongated and extending into chlorenchyma.

SPECIES DESCRIPTION

***R. confertospicatus* Steud.** Culm diameter 1·5 mm.

Culm surface (Fig. 37. H)

Hairs uniseriate, see Fig. 37. J. **Epidermis:** cells mostly 4-sided; as long as wide to 4 times longer than wide, walls moderately thickened, wavy. **Stomata:** subsidiary cells narrow, with acute or obtuse end walls; guard cells with rounded ends. **Tannin:** none seen.

Culm T.S. (Fig. 37. J)

Epidermis. Cells as high as to 1½ times higher than wide; those next to stomata with walls lining substomatal cavity about 1½ times longer, and extending further into chlorenchyma than other walls. Outer walls v. thick; anticlinal walls thick, with tapering thickening. **Stomata** superficial; subsidiary cells with outer ridge, and extending inwards further than in most spp. of group 1; guard cells with lip at outer aperture. **Chlorenchyma:** cells in 1 layer. **Parenchyma sheath** 1-layered. **Tannin:** none seen.

RESTIO GROUP 3. South Africa

Leaf surface

Seen in *R. sieberi* var. *venustulus* (Kunth) Pillans (abaxial, sheathing base).

Hairs and **papillae** absent. **Epidermis:** cells 4-sided, some up to twice as long as wide, others slightly wider than long; walls moderately thickened, wavy. **Stomata** similar to those of culm, but *c.* ⅔ as large. **Tannin** in many cells. **Cuticular marks** coarsely granular.

Leaf T.S.

Seen in *R. sieberi* var. *venustulus* (Kunth) Pillans (sheathing base).

Epidermis: (i) abaxial cells 4-sided, mostly 1½–2 times wider than high; outer walls thick, frequently convex; anticlinal and inner walls slightly thickened; (ii) adaxial cells 6-sided, slightly narrower than and about half the height of abaxial cells; all walls moderately thickened; anticlinal walls v. short, each cell with 3 facets to inner wall. **Stomata** superficial, infrequent, in abaxial surface only; guard cells as in culm. **Chlorenchyma:** cells as wide as high or slightly wider than high, with pegs, occurring in 1 layer between scl. or parenchyma O.S. and abaxial epidermis; no protective cells seen. **Vascular bundles** all small, each with several narrow tracheids and phloem cells. **Bundle sheaths:** O.S. parenchymatous, restricted to 1 layer of cells at phloem pole; I.S. sclerenchymatous, fibres narrow, thick-walled, in 1 layer; joining at xylem pole and on flanks with sclerenchymatous ground tissue. **Sclerenchyma:** in bundle sheaths just described and as wide, v. thick-walled fibres occupying all of area between chlorenchyma and adaxial epidermis; individual fibres reaching twice as high as wide. **Air-cavities** absent. **Ground tissue** sclerenchymatous. **Silica:** spheroidal-nodular bodies present in some cells of parenchymatous O.S. **Tannin:** none seen.

Leaf T.S.

Seen in *R. rottboellioides* Kunth (blade).

Outline semicircular, the flattened face being adaxial. **Hairs** and **papillae** absent. **Epidermis:** (i) abaxial cells 4-sided, slightly to 1½ times wider than high, outer walls thick, other walls moderately thickened; (ii) adaxial cells 4-sided, 2–3 times wider than high, outer walls thick, other walls moderately thickened. **Stomata** superficial, present on abaxial surface only; similar to those in culm. **Chlorenchyma** present in 2 layers beneath abaxial epidermis; peg-cells of outer layer 3–4 times higher than wide, with numerous medium-sized pegs; inner layer composed of shorter, slightly wider cells ranging from as high as to twice as high as wide, similar to those of inner layer in culm. Protective cells present as slightly thickened, palisade peg-cells. **Parenchyma sheath:** completely encircling, in 1 or 2 layers next to chlorenchyma or adaxial epidermis.

Vascular bundles: 9 present, arranged in 1 group of 4, 2 groups of 2, and 1 singly, in arc; each with numerous medium-sized to wide, angular tracheids. **Bundle sheaths** sclerenchymatous; fibres narrow, moderately thick- or thick-walled, in 1 layer at phloem pole and on flanks and 2–3 layers at xylem poles.

Air-cavities: 1 wide air-cavity present to inside of that part of parenchyma sheath next to adaxial epidermis, formed by breakdown of thin-walled cells. **Ground tissue** parenchymatous, cells with moderately thickened or thick walls.

Culm surface

Hairs and **papillae** absent apart from papillae on epidermis enclosed by sheathing leaf base in *R. giganteus* and *R. stereocaulis.* **Epidermis.** Cells 4–6-sided, most frequently as long as or up to 1½ times longer than wide, occasionally 2–(9) times longer than wide, rarely slightly wider than long; shorter cells often next to stomata. Walls usually moderately thickened, sometimes slightly thickened, thick, or v. thick, wavy or straight, or wavy at surface and straight at lower focus. **Stomata:** subsidiary cells variable; guard cells usually with rounded ends. **Silica** absent. **Tannin** in epidermal cells of all spp. except *R. bolusii, curviramis, multiflorus, rottboellioides, saroclados, stereocaulis, strobilifer, vilis,* and *virgeus.* **Cuticular marks** granular, rarely with longitudinal striations superimposed.

Culm T.S.

Outline usually circular, rarely oval or kidney-shaped. **Cuticle** thick, occasionally slightly or moderately thickened, rarely v. thick. **Epidermis.** Cells present in 1 layer (2-layered in places in *R. callistachyus*); 4-sided; cells usually all of similar height, occasionally some shorter than rest, as high as to 5 times higher than wide. Outer walls thick or v. thick, outer surface convex to concave, sometimes irregular; anticlinal walls straight or wavy, usually moderately thickened, sometimes thick at outer end, thickening tapering to inner end or tapering rapidly, and inner part of walls slightly or moderately thickened; inner walls usually moderately, sometimes slightly thickened. **Stomata** usually superficial, rarely sunken; subsidiary cells variable; guard cells usually with small lip at outer aperture, occasionally with inner lip or ridge on inner wall. **Chlorenchyma** normally composed of 2 layers of peg-cells, some spp. with 1, 1–2, or 2–3 layers; protective cells present in all spp. **Parenchyma sheath** 1-, 1–2- or, occasionally 2–5-layered; cells as wide as high, or up to twice as wide as high or twice as high as wide; walls usually slightly thickened. **Sclerenchyma:** sheath well-developed in all spp., ranging from 2–6-layered to 9–10-layered; outline circular, often with low, dome-shaped ribs opposite to peripheral vb's (ribs between peripheral vb's in *R. callistachyus* and *festuciformis*); fibres of outer layers usually narrow, with thick or v. thick walls; fibres of inner layers wider, usually thick-walled, sometimes difficult to distinguish from groundtissue.

Vascular bundles. (i) Peripheral vb's each with arc of tracheids partially enclosing circular or oval phloem pole; tracheids all of same size, or flanking or central tracheids wider than others. (ii) Medullary vb's: outer bundles smallest, often with 1 central, narrow or medium-sized mxv, or with 1 narrow or medium-sized mxv on either flank; inner bundles larger, each with 1 narrow, medium-sized, or wide, rounded or angular mxv on either flank, these separated by 1–2 or 2–4 rows of narrow cells. Phloem poles oval or circular, usually ensheathed by 1 layer of narrow cells with slightly or moderately thickened walls (those next to phloem sometimes thin). Px poles absent from smaller, outer bundles. No or several to many vb's free from culm scl. sheath.

Bundle sheaths: fibres normally in 2 layers at phloem pole and 1 layer on flanks and at xylem pole, sometimes in 1 or 3 layers at phloem pole and 2 layers at xylem pole; those at poles normally narrow, with thick or v. thick walls; those on flanks normally wider, with moderately or slightly thickened walls. **Central ground tissue** parenchymatous; outer cells, surrounding vb's, usually with moderately thickened walls; inner cells thin-walled and breaking down, or with moderately thickened walls; some spp. with scattered areas of thin-walled cells among outer cells. **Silica** present as (i) spheroidal-nodular bodies in some cells of parenchyma sheath in certain spp. (and sometimes in chlorenchyma cells of inner layer); (ii) granular material in occasional cells of ground tissue, or parenchyma sheath or both, or in protective cells or chlorenchyma cells in some spp. **Tannin** in some epidermal cells in most spp. (see culm surface description above).

Rhizome and *root* not seen.

Species Descriptions

***R. aridus* Pillans.** Culm diameter 0·4 mm.

Culm surface

Epidermis: cells 4-sided; 1–1½ times longer than wide, walls moderately thickened, slightly wavy. **Stomata:** subsidiary cells with obtuse end walls.

Culm T.S.

Epidermis: cells next to stomata twice as high as wide, others *c.* 1½ times higher than wide; outer walls v. thick, outer surface slightly irregular. **Stomata:** guard cells with lip at outer aperture and ridge on inner wall. **Chlorenchyma** 2-layered. Protective cells extending from *c.* ⅓ of way up tube formed by epidermal cells surrounding stomata, to inner chlorenchyma layer. **Parenchyma sheath** 1-layered, cells slightly wider than high. **Silica:** bodies present in occasional cells of parenchyma sheath.

***R. bifarius* Mast.** Culm diameter 1·0 mm.

Culm surface

Epidermis: cells 4-sided; more or less square, walls thick, wavy at surface only. **Stomata:** subsidiary cells with rounded ends.

Culm T.S.

Epidermis: cells all of same height, 1½ times higher than wide; outer walls thick to v. thick, anticlinal walls with tapering thickening, sometimes wavy. **Stomata:** subsidiary cells short; guard cells with lip at outer aperture. **Chlorenchyma:** cells in 2 similar layers; protective cells extending in 2 layers to parenchyma sheath. **Parenchyma sheath** 1–2-layered, cells as high as or up to twice as high as wide. **Vascular bundles:** medullary, phloem pole circular, abutting directly on xylem, or partially ensheathed. **Silica** granular, present in most protective cells.

***R. bifidus* Thunb.** Culm diameter 0·8 mm.

Culm surface

Epidermis: cells mostly 4-sided; those next to stomata frequently as long as wide, others 1½–2 times longer than wide, walls slightly wavy, thick.

Stomata : subsidiary cells with acute end walls, long anticlinal walls sometimes wavy.

Culm T.S.

Epidermis : cells all of same height, 2–2½ times higher than wide; outer walls v. thick, outer surface slightly uneven; anticlinal walls sometimes wavy, with tapering thickening. **Stomata :** guard cells with outer lip. **Chlorenchyma :** cells in 2 layers; protective cells extending in 1 layer from half way up substomatal tube formed by epidermal cells, to inner chlorenchyma layer; walls slightly thickened. **Parenchyma sheath** 1-layered, cells mostly as wide as high. **Silica** granular, present in some protective cells and some cells of central ground tissue.

R. bifurcus **Nees ex Mast.** Culm diameter 1·5, 3·0 mm.

Culm surface

Epidermis : cells mostly 4- and 6-sided; those next to stomata as long as wide, others up to 1½ times longer than wide, walls wavy at surface only, thick. **Stomata :** subsidiary cells with rounded or acute end walls, long anticlinal walls slightly wavy; guard cell pair sometimes slightly lemon-shaped.

Culm T.S.

Epidermis : cells all of same height or occasional cells slightly shorter than others, 2–3 times higher than wide; outer walls v. thick, curving in at cell margins, some anticlinal walls wavy. **Stomata :** subsidiary cells sometimes with ridge on outer wall; guard cells with lip at outer aperture and slight ridge on inner wall. **Chlorenchyma :** cells in 2 layers; protective cells in 2 layers, extending from part way into tube formed by epidermal cells lining sub-stomatal cavity, to parenchyma sheath. **Parenchyma sheath** 1–2–(3)-layered, cells as wide as high or up to 1½ times higher than wide. **Silica** granular, present in some cells of central ground tissue.

R. bolusii **Pillans.** Culm diameter 1·5 by 2·3 mm.

Culm surface

Epidermis : cells 4–6-sided; most as long as wide, some at ends of stomata twice as wide as long, walls thick or v. thick. **Stomata :** subsidiary cells with acute end walls.

Culm T.S.

Outline oval. **Epidermis :** cells mostly 3 times higher than wide; outer walls v. thick, flat or slightly concave; anticlinal walls wavy, with tapering thickening. **Stomata :** guard cells with lip at outer aperture. **Chlorenchyma :** cells in 2 layers; protective cells extending in 2 layers from inner wall of epidermis to parenchyma sheath. **Parenchyma sheath** 1–2-layered, cells mostly up to 1½ times higher than wide. **Silica :** none seen.

R. brunneus **Pillans.** Culm diameter 2·0 mm.

Culm surface

Epidermis : cells 4–6-sided; mostly about as long as wide, walls moderately thick to thick, not wavy. **Stomata :** subsidiary cells with acute end walls. **Cuticular marks** coarsely granular, with slight longitudinal striations superimposed.

Culm T.S.

Epidermis: cells 3½–4 times higher than wide; outer walls v. thick; anticlinal walls wavy, outer end v. thick, thickening tapering. **Stomata:** guard cells with lip at outer aperture. **Chlorenchyma:** cells in 2 layers; protective cells with slightly thickened walls, extending in 2 layers from epidermis to parenchyma sheath; cells of outer layer elongated, projecting about half-way into inner chlorenchyma layers. **Parenchyma sheath** 1–3-layered, cells mostly as high as wide, some larger than others. **Silica:** none seen.

***R. callistachyus* Kunth.** Culm diameter 1·5 by 1·1 mm.

Culm surface

Epidermis: cells 4-sided; mostly twice as long as wide, walls slightly to moderately thickened, wavy. **Stomata:** subsidiary cells with obtuse end walls or crescentiform.

Culm T.S.

Outline oval. **Epidermis:** cells present in 1 and 2 layers; 4-sided, larger cells 1½–2 times higher than wide, some of these being divided periclinally to produce 2 more or less equal cells or with the inner cell the larger. **Stomata:** subsidiary cells sometimes with ridge on outer wall; guard cells with slight lip at outer and inner apertures. **Chlorenchyma:** cells in 2 similar layers; protective cells extending in 1 layer from epidermis to half way into inner chlorenchyma layer, or reaching parenchyma sheath. **Parenchyma sheath** 1–2-layered, cells rounded hexagonal, mostly as high as wide. **Silica:** none seen.

***R. compressus* Rottb.** Culm diameter 2·0 by 1·2; 2·0 by 1·1 mm.

Culm surface

Epidermis: cells 4–6-sided; as long as to 1½–2 times longer than wide, walls moderately thickened to thick, slightly wavy. **Stomata** small; subsidiary cells narrow, end walls perpendicular or obtuse. **Silica:** none seen.

Culm T.S.

Outline oval. **Epidermis:** cells as high as to slightly higher than wide, or (in 24471) twice as high as wide. Outer walls v. thick, some slightly concave. **Stomata** sunken, attached to inner end of anticlinal walls of adjacent epidermal cells; guard cells unusual, with lips on wall to pore. **Chlorenchyma:** cells mainly in 1 layer; protective cells with moderately thickened walls, extending in 1–(2) layer(s) to parenchyma sheath. **Parenchyma sheath** 1–2-layered, cells rounded, as high as to 1½ times higher than wide. **Silica:** none seen.

***R. curviramis* Kunth.** Culm diameter 0·7 mm.

Culm surface

Epidermis: cells 4–6-sided; slightly wider than long or as long as wide, walls v. thick, wavy. **Stomata:** subsidiary cells with obtuse end walls: guard cells wide.

Culm T.S.

Epidermis: cells mostly twice as high as wide; outer walls v. thick; anticlinal walls v. wavy, thickening tapering rapidly. **Stomata:** guard cells with pronounced outer lip. **Chlorenchyma:** cells in 2 layers; protective cells composed of somewhat modified peg-cells, with slightly thickened walls, extending in 2 layers from *c.* ⅓ of way up substomatal cavity lined by epidermal cells, to parenchyma sheath; cells of outer layer elongated, extending about half-way into inner chlorenchyma layer, cells of inner layer short. **Parenchyma sheath** 1-layered, cells as high as to slightly higher than wide. **Silica:** spheroidal-nodular bodies in some cells of parenchyma sheath.

R. cuspidatus **Thunb.** Culm diameter 1·0 mm.

Culm surface

Epidermis: cells 4-sided; those next to stomata as long as wide, others as long as to twice as long as wide, walls moderately thickened, wavy. **Stomata:** subsidiary cells with wavy long walls.

Culm T.S.

Epidermis: cells mostly slightly wider than high, those next to stomata as high as wide. **Stomata:** guard cells with lip at outer aperture and slight ridge on inner wall. **Chlorenchyma:** cells in 2 layers. Protective cells 2-layered, with slightly thickened walls, extending from epidermis to parenchyma sheath; cells of outer layer extending about half-way into inner chlorenchyma layer. **Parenchyma sheath** 1-layered, cells mostly 1½ times wider than high. **Silica:** bodies present in some cells of parenchyma sheath.

R. dispar **Mast.** Culm diameter 2·0 mm.

Culm surface

Epidermis: cells 4–6-sided; mostly about as long as wide, walls moderately thickened. **Stomata:** subsidiary cells variable.

Culm T.S.

Epidermis: cells mainly twice as high as wide; outer walls v. thick; anticlinal walls wavy, with tapering thickening. **Stomata:** guard cells with lip at outer and inner apertures. **Chlorenchyma:** cells in 2 layers; protective cells consisting of slightly modified peg-cells, extending from ¼ of way up substomatal cavity lined by epidermal cells, to parenchyma sheath, in 2–(3) layers. **Parenchyma sheath** 1–2-layered, cells mostly 1½ times higher than wide, with some smaller, rounded cells also present. **Silica:** granular material present in some chlorenchyma cells.

R. dodii **Pillans.** Culm diameter 0·8 mm.

Culm surface

Epidermis: cells 4-sided; as long as wide to 1½ times longer than wide, walls thick to v. thick, wavy. **Stomata:** subsidiary cells variable.

Culm T.S.

Cuticle v. thick. **Epidermis:** cells mostly as high as wide; outer walls v. thick, frequently concave; anticlinal walls wavy, with abruptly tapering thickening. **Stomata:** guard cells with lip at outer aperture. **Chlorenchyma:** peg-cells and protective cells in 2 layers. **Parenchyma sheath** 1-layered, cells circular or slightly higher than wide. **Silica** present as (i) granular material filling lumina of most protective cells; (ii) spheroidal-nodular bodies in some cells of inner chlorenchyma layer and parenchyma sheath. **Tannin** in many vessels (material possibly pathological).

***R. egregius* Hochst.** Culm diameter 1·1 mm.

Culm surface

Epidermis: cells 4–6-sided; some as long as wide, others up to 1½ times longer than wide, walls thick, slightly wavy. **Stomata:** subsidiary cells with acute end walls.

Culm T.S.

Epidermis: cells 2–2½ times higher than wide; outer walls v. thick; anticlinal walls slightly wavy. **Stomata:** subsidiary cells with ridge on outer wall; guard cells with lip at outer aperture and ridge on inner wall. **Chlorenchyma:** cells in 2 layers; protective cells extending in 2 layers from *c.* ⅓ of way up substomatal cavity lined by epidermal cells, to parenchyma sheath. (All chlorenchyma cells inclined at slight angle to surface as seen in L.S.) **Parenchyma sheath** 3–5-layered, cells circular or slightly higher than wide. **Silica** granular, present in some cells of ground tissue.

***R. festuciformis* Nees ex Mast.** Culm diameter 1·0 mm.

Culm surface

Epidermis: cells 4-sided; mostly 6–9 times longer than wide, walls moderately thickened, not wavy. **Stomata:** subsidiary cells with acute end walls.

Culm T.S.

Epidermis: cells 1½–2 times higher than wide; outer walls v. thick, flattened-convex. **Stomata** raised on cells slightly taller than others; guard cells with slight lip at outer aperture and slight ridge on inner wall. **Chlorenchyma:** cells in 1 or 2 layers; protective cells in 1 layer. **Parenchyma sheath** 1-layered, interrupted at irregular intervals by fibres from scl. sheath; cells mainly 1½ times wider than high. **Sclerenchyma:** sheath 4–6-layered; outline circular with irregular 1-layered ribs between peripheral vb's, interrupting parenchyma sheath. **Silica:** spheroidal-nodular bodies present in some cells of parenchyma sheath and occasionally in cells of inner chlorenchyma layer.

***R. foliosus* N. E. Br.** Culm diameter 2·0 mm.

Culm surface

Epidermis: cells 4–6-sided; as long as wide, walls v. thick, wavy. **Stomata** sunken; subsidiary cells with acute end walls; guard cells wide.

Culm T.S.

Cuticle v. thick. **Epidermis:** cells 3½ times higher than wide; outer walls v. thick, flattened convex, cells free from one another (separated by cuticular in-pushings) to depth equal in thickness to outer wall; anticlinal walls wavy, thickening tapering abruptly. **Stomata** sunken, mounted ¾ of way down anticlinal walls of surrounding epidermal cells; guard cells with lip at outer aperture. **Chlorenchyma:** cells in 2 layers; protective cells formed from slightly thickened but otherwise little modified peg-cells of both layers. **Parenchyma sheath** 1–2-layered, cells circular to oval and twice as high as wide. **Central ground tissue** parenchymatous, outer cells moderately thick-walled, interspersed by wide strands of thin-walled cells; central cells thin-walled. **Silica :** granular material in some cells of central ground tissue.

***R. fourcadei* Pillans.** Culm diameter 1·0 mm.

Culm surface not seen.

Culm T.S.

Outline circular, with 1 flattened face. **Cuticle** thick, granular. **Epidermis :** cells as high as to 1½ times higher than wide; outer walls v. thick, frequently convex and with slightly ridged outer surface. **Stomata :** subsidiary cells with ridge on outer wall; guard cells with lip at outer aperture. **Chlorenchyma :** cells in 2 layers; protective cells v. weak, in 1 layer extending from *c.* ¼ of way up substomatal cavity lined by epidermal cells, to junction between inner and outer palisade layers. **Parenchyma sheath** 1-layered, cells slightly to 1½ times wider than high, walls thin. **Silica :** bodies present in some cells of parenchyma sheath.

***R. fraternus* Kunth.** Culm diameter 1·2 mm.

Culm surface

Epidermis : cells 4–6-sided; mostly as long as wide, some slightly wider than long, walls thick, wavy at surface only. **Stomata :** subsidiary cells with acute end walls.

Culm T.S.

Epidermis : cells about twice as high as wide; outer walls v. thick; anticlinal walls wavy, with tapering thickening. **Stomata :** guard cells with lip at outer aperture and slight ridge on inner wall. **Chlorenchyma :** cells in 2 layers. Protective cells extending in 2 layers from about half-way up substomatal cavity lined by epidermal cells, to parenchyma sheath. **Parenchyma sheath** 1–2-layered, cells rounded hexagonal, mostly as wide as high. **Silica :** bodies present in some cells of parenchyma sheath.

***R. galpinii* Pillans.** Culm diameter 0·9 mm.

Culm surface

Epidermis : cells 4–6-sided; of varying widths, as long as wide to *c.* 1½ times longer than wide, walls moderately thickened, not wavy. **Stomata** sunken; subsidiary cells with acute end walls.

Culm T.S.

Epidermis : cells mostly 1½ times higher than wide; outer walls v. thick, outer surface granular. **Stomata** sunken, mounted at inner end of anticlinal walls of adjacent epidermal cells; subsidiary cells with guard cells mounted towards inner end; guard cells with lip at outer aperture and ridge on inner wall. **Chlorenchyma :** cells in 2 or, occasionally, 3 layers; protective cells extending in 1 layer from epidermis to junction between chlorenchyma layers. **Parenchyma sheath** 1–3-layered, cells circular to 1½ times higher than wide. **Silica :** none seen.

***R. giganteus* (Kunth) N. E. Br.** Culm diameter 4·5 mm.

Culm surface (Material examined taken from part of culm enclosed by leaf sheath.)

Papillae short, 1 present on few epidermal cells. **Epidermis:** cells 4–6-sided; 2–2½ times longer than wide, walls slightly to moderately thickened, wavy. **Stomata:** subsidiary cells with acute end walls, slightly longer than guard cells.

Culm T.S.

Papillae: see above under Culm surface. **Epidermis:** cells as high as wide; outer walls v. thick; anticlinal and inner walls moderately thickened. **Stomata:** guard cells with lip at outer aperture and ridge on inner wall. **Chlorenchyma:** cells in 2 layers; protective cells poorly developed, represented by slightly thickened palisade cells of both layers. **Parenchyma sheath** 1–2-layered, cells 1½–2 times wider than high. **Central ground tissue** parenchymatous, outer cells with slightly to moderately thickened walls, accompanied by scattered areas of thin-walled cells. **Silica:** none seen.

***R. gossypinus* Mast.** Culm diameter 0·9 mm.

Culm surface

Epidermis: cells 4–6-sided; as long as to 2½ times longer than wide, walls moderately thickened to thick, wavy. **Stomata:** subsidiary cells irregular, but mainly with acute end walls, long walls frequently wavy; guard cells shorter and wider than usual in this group of species.

Culm T.S.

Epidermis: cells slightly to 1½ times wider than high; outer walls v. thick. **Stomata:** guard cells with lip at outer aperture. **Chlorenchyma:** cells in 2 layers; protective cells thick-walled, extending in 1 layer from epidermis to junction between chlorenchyma layers, or part way into inner chlorenchyma layer. **Parenchyma sheath** 1-layered, cells mostly 1½ times wider than high. **Silica:** occasional bodies in some cells of parenchyma sheath.

***R. laniger* Kunth.** Culm diameter 1·0 mm.

Culm surface

Epidermis: cells 4–6-sided; as long as to 1½ times longer than wide, walls thick, wavy. **Stomata:** subsidiary cells with obtuse or acute end walls; guard cells wide.

Culm T.S.

Epidermis: cells slightly to 1½ times higher than wide; outer walls thick, outer surface granular; anticlinal walls wavy, thick; inner walls moderately thickened. **Stomata:** guard cells with lip at outer aperture. **Chlorenchyma:** cells in 2 layers; protective cells extending in 2 layers from *c.* ⅓ of way up substomatal cavity lined by epidermal cells, to parenchyma sheath, cells of outer layer extending *c.* ½ way into inner chlorenchyma layer. **Parenchyma sheath** mainly 1-, occasionally 2-layered, cells mostly twice as high as wide. **Silica:** silica-bodies in some cells of parenchyma sheath.

***R. leptocladus* Mast.** Culm diameter 1·0 mm.

Culm surface

Epidermis: cells 4-sided; as long as to 1½–2 times longer than wide, walls moderately thickened, wavy. **Stomata:** subsidiary cells with acute or rounded end walls.

Culm T.S.

Epidermis: cells as high as to slightly higher than wide; outer walls thick, frequently slightly concave. **Stomata:** guard cells with lips at outer and inner apertures. **Chlorenchyma:** cells in 2 layers; protective cells differing from other chlorenchyma cells only in shape, inner cells with longer pegs and separated by wider air-spaces than their neighbours. **Parenchyma sheath** 1-layered, cells circular or slightly wider than high. **Silica:** large bodies in many cells of parenchyma sheath.

***R. major* (Mast.) Pillans.** Culm diameter 1·5 mm.

Culm surface

Epidermis: cells 6-sided; mostly as long as wide, walls moderately thickened, straight. **Stomata** overarched by epidermal cells, visible through cross-shaped hole; subsidiary cells with obtuse end walls.

Culm T.S.

Epidermis: cells mostly 1½–2 times higher than wide; outer walls v. thick, frequently slightly concave. **Stomata** sunken; subsidiary cells twice as high as guard cells, outer walls sloping to guard cell; guard cells with ridge on inner wall. **Chlorenchyma:** cells mainly in 1 layer, occasionally 2-layered by periclinal division; protective cells present in 1 layer, extending from epidermis to parenchyma sheath. **Central ground tissue** parenchymatous, all cells with slightly thickened walls. **Silica:** none seen.

***R. micans* Nees.** Culm diameter 1·3 mm.

Culm surface

Epidermis: cells 4-sided; as long as, to twice as long as wide; walls moderately thickened to thick, wavy. **Stomata:** subsidiary cells with perpendicular or acute end walls.

Culm T.S.

Epidermis: cells slightly to 1½ times higher than wide; outer walls v. thick, irregular, frequently concave. **Stomata:** subsidiary cells with ridge on outer wall; guard cells with well-developed lip at outer and slight lip at inner apertures. **Chlorenchyma:** cells in 2–3 layers; protective cells 1-layered, resembling slightly modified peg-cells, extending from epidermis to junction between outer and second layers of chlorenchyma. **Parenchyma sheath** 1–2-layered, cells narrow and as wide as high, or medium-sized and slightly wider than high. **Central ground tissue** parenchymatous, outer cells with slightly to moderately thickened walls, islands of thin-walled cells also present; central cells thin-walled, breaking down to form central cavity. **Silica:** spheroidal-nodular bodies in some cells of parenchyma sheath.

***R. miser* Kunth.** Culm diameter 0·5 mm.

Culm surface

Epidermis: cells 4-sided; as long as to 1½ times longer than wide; walls moderately thickened, wavy. **Stomata:** subsidiary cells with obtuse end walls.

Culm T.S.

Epidermis: cells 1½ times higher than wide; outer walls thick to v. thick. **Stomata:** guard cells with lip at outer aperture. **Chlorenchyma:** cells in 1 layer;

protective cells extending in 1 layer from half-way up substomatal tube lined by epidermal cells to parenchyma sheath. **Parenchyma sheath** 1–2-layered, cells of 2 sizes: (i) smaller, rounded; (ii) larger, *c.* 1½ times higher than wide. **Silica** granular, present in most protective cells and some cells of central ground tissue.

***R. multiflorus* Spreng.** Culm diameter 2·0 mm.

Culm surface

Epidermis: cells mostly 6-sided, those next to stomata 4-sided; as long as to 1½ times longer than wide; walls moderately thickened, wavy at surface only, straight at lower focus. **Stomata:** subsidiary cells with acute or obtuse end walls.

Culm T.S.

Epidermis: cells mostly 2½ times higher than wide; outer walls v. thick; some anticlinal walls wavy. **Stomata:** guard cells with lip at outer aperture. **Chlorenchyma:** cells in 2 or 3 layers; protective cells extending in 2–3 layers from epidermis to parenchyma sheath. **Parenchyma sheath** mostly 2–3-layered, cells narrow to medium-sized, rounded to slightly wider than high. **Silica:** none seen.

***R. nodus* Pillans.** Culm diameter 1·1 by 0·8 mm.

Culm surface

Epidermis: cells 4–6-sided; as long as wide, walls wavy, v. thick. **Stomata:** subsidiary cells with perpendicular or acute end walls.

Culm T.S.

Outline kidney-shaped. **Epidermis:** cells 3½–4 times higher than wide; outer walls v. thick; anticlinal walls with tapering thickening. **Stomata:** guard cells with lip at outer aperture. **Chlorenchyma:** cells in 2 layers. **Parenchyma sheath** 1–2–(3)-layered, cells higher than wide. **Central ground tissue** parenchymatous, outer cells with thick or moderately thickened walls, with occasional strands of thin-walled cells, central cells thin-walled. **Silica:** none seen.

***R. ocreatus* Kunth.** Culm diameter 2·0 mm.

Culm surface

Epidermis: cells 4–6-sided; mostly as long as wide, but irregular in outline; walls thick. **Stomata:** subsidiary cells with acute end walls; guard cells wider and shorter than usual.

Culm T.S.

Epidermis: cells twice as high as wide; outer walls v. thick, straight but curving in at cell margins; anticlinal walls wavy, with tapering thickening. **Stomata:** subsidiary cells with outer part compressed; guard cells with lip at outer aperture. **Chlorenchyma:** cells in 2 layers; cells not perpendicular to epidermis as seen in L.S.; protective cells in 2 layers, composed of peg-cells with slightly thickened walls. **Parenchyma sheath** 1–2-layered, cells circular or slightly higher than wide. **Silica** in most protective cells and some cells of central ground tissue, in granular form.

***R. pachystachyus* Kunth.** Culm diameter 2·0 mm.

Culm surface

Epidermis: cells mostly 6-sided; slightly wider than long and sometimes about as long as wide, walls v. thick; cells irregularly orientated, the largest cells surrounding stomata. **Stomata:** subsidiary cells with acute end walls.

Culm T.S.

Epidermis: some cells between stomata shorter than others, 2½ times higher than wide, remainder *c.* 2½–3 times higher than wide; outer walls v. thick, outer surface irregular; anticlinal walls frequently wavy, with tapering thickening. **Stomata:** subsidiary cells with outer part compressed; guard cells with lips at outer and inner apertures. **Chlorenchyma:** cells in 2 (rarely 3) layers; protective cells present in 2 layers, extending from *c.* ⅓ of way up substomatal cavity lined by epidermal cells, to parenchyma sheath. **Parenchyma sheath** 1–2-layered, cells rounded or slightly to 1½ times wider than high. **Silica** absent.

***R. pedicellatus* Mast.** Culm diameter 0·9 mm.

Culm surface

Epidermis: cells 4–6-sided; mostly slightly wider than long, some 1½ times longer than wide, walls thick. **Stomata:** subsidiary cells with obtuse end walls; guard cells wide.

Culm T.S.

Epidermis: cells mostly 1½ times higher than wide; outer walls v. thick, more or less flat, anticlinal walls with tapering thickening. **Stomata:** guard cells with outer lip and slight ridge on inner wall. **Chlorenchyma:** cells in 2 layers; protective cells well developed, extending in 1 or 2 layers from *c.* ½ to ¼ of way up substomatal cavity lined by epidermal cells, to junction between outer and inner chlorenchyma layers, or right to parenchyma sheath. **Parenchyma sheath** mainly 1-, occasionally 2-layered, most cells nearly twice as high as wide, some smaller cells rounded. **Silica** granular, present in many protective cells and some cells of central ground tissue.

***R. praeacutus* Mast.** Culm diameter 1·5 by 2·0 mm.

Culm surface

Epidermis: cells 4–6-sided; as long as wide to *c.* 1½ times longer than wide, the smaller cells being next to stomata; walls moderately thickened to thick, not wavy. **Stomata:** subsidiary cells mostly with acute end walls.

Culm T.S.

Outline oval. **Epidermis:** cells mostly 2½ times higher than wide; outer walls v. thick, frequently slightly convex; anticlinal walls v. occasionally wavy. **Stomata:** subsidiary cells with outer part compressed; guard cells with lip at outer aperture and v. slight ridge on inner wall. **Chlorenchyma:** cells in 2 layers; protective cells well developed, with pointed inner ends, extending in 1 or 2 layers from *c.* ¼ of way up substomatal cavity lined by epidermal cells, to junction between outer and inner chlorenchyma layers, or to parenchyma sheath. **Parenchyma sheath** mainly 2-, occasionally 1- or 3-layered, cells rounded, mostly as wide as high. **Silica:** none seen.

R. quinquefarius **Nees.** Culm diameter 1·1 mm.

Culm surface

Epidermis: cells 4-sided; some as long as wide, others mostly slightly to 1½ times longer than wide, walls moderately thickened, wavy. **Stomata:** subsidiary cells with acute end walls, longer than guard cells.

Culm T.S.

Epidermis: cells large, mostly just over twice as high as wide; outer walls thick, outer surface slightly irregular; anticlinal walls wavy. **Stomata:** subsidiary cells elongated and compressed; guard cells with lip at outer and inner aperture. **Chlorenchyma:** cells in 2 layers; protective cells extending in 2–(3) layers from *c.* ¼ of way up walls of epidermal cells lining substomatal cavity to parenchyma sheath. **Parenchyma sheath** mainly 2-layered, cells wide, mostly slightly higher than wide. **Silica** absent.

R. rhodocoma **Mast.** Culm diameter 1·5 and 2 mm.

Culm surface

Epidermis: cells 4-sided; those next to stomata as long as wide, others up to twice as long as wide but often only as long as wide; walls moderately thickened, wavy at surface. **Stomata:** subsidiary cells with acute end walls, slightly longer than guard cells, often with wavy long anticlinal walls; guard cells shorter and wider than in most spp.

Culm T.S.

Epidermis: cells as high as wide in (3373), 2–2½ times higher than wide in other specimens; outer walls thick, slightly convex; anticlinal walls sometimes wavy, with tapering thickening. **Stomata:** guard cells with lip at outer and inner apertures. **Chlorenchyma:** peg-cells and protective cells in 2 layers. **Parenchyma sheath** 1-, mainly 2-layered, cells circular to 1½ times higher than wide, some larger than others. **Central ground tissue** parenchymatous, most outer cells with slightly to moderately thickened walls, with islands of thin-walled cells scattered amongst them, central cells thin-walled. **Silica:** none seen.

R. rottboellioides **Kunth.** Culm diameter 1·5 mm.

Culm surface

Epidermis: cells 4-sided; most as long as, some slightly shorter than wide; walls v. thick, lumina rounded. **Stomata:** subsidiary cells with obtuse end walls and with slightly wavy long anticlinal walls.

Culm T.S.

Epidermis: cells mostly as high as wide; outer walls v. thick, with irregular outer surface; anticlinal walls v. wavy, with tapering thickening. **Stomata:** guard cells with lip at outer aperture and ridge on inner wall. **Chlorenchyma:** cells in 2 layers; protective cells composed of slightly thickened cells of outer chlorenchyma layer, with wavy walls; cells of inner chlorenchyma layer opposite to these less regular than the remainder. **Parenchyma sheath** mostly 1-layered, cells large, being slightly higher than cells of inner chlorenchyma layer and 1½ or 2 times higher than wide. **Silica:** spheroidal-nodular bodies in some cells of inner chlorenchyma layer and parenchyma sheath.

***R. saroclados* Mast.** Culm diameter 1·1 mm.

Culm surface

Epidermis: cells 4- or 6-sided; those next to stomata wider than long or as long as wide, others between stomata slightly longer than to twice as long as wide; walls moderately thickened, not wavy. **Stomata:** subsidiary cells with acute end walls.

Culm T.S. (Fig. 37. C, p. 254)

Epidermis: cells all of same height, mostly twice as high as wide; outer walls v. thick, slightly convex; anticlinal walls not wavy. **Stomata:** subsidiary cells narrow; guard cells with lip at outer aperture and slight ridge on inner wall. **Chlorenchyma:** cells in 2 similar layers; protective cells with moderately thickened walls, extending in 2 layers from *c.* ¼ of way into substomatal tube lined by epidermal cells, to parenchyma sheath. **Parenchyma sheath** 1–3-layered, cells polygonal, of various sizes, largest where in 1 layer. **Silica:** none seen.

***R. setiger* Kunth.** Culm diameter 2·0 mm.

Culm surface

Epidermis: cells 4- or 6-sided; mostly as long as wide, some slightly longer than wide, others slightly wider than long; walls v. wavy, v. thick. **Stomata:** subsidiary cells with acute end walls and wavy long anticlinal walls, guard cell pair wide, sometimes with lemon-shaped outline.

Culm T.S.

Cuticle v. thick. **Epidermis:** cells mostly 2–3 times higher than wide; outer walls v. thick, outer surface granular; anticlinal walls v. wavy, with tapering thickening. **Stomata:** guard cells with outer lip. **Chlorenchyma:** cells in 2 layers; protective cells consisting of 2 layers of cells, with slightly thickened walls, extending from epidermal cells to parenchyma sheath. **Parenchyma sheath** 1-, 2-, or 3-layered, cells wide, rounded or up to 1½ or 2 times higher than wide. **Silica:** large bodies present in some cells of parenchyma sheath and granular material present in some cells of central ground tissue.

***R. sieberi* Kunth var. *sieberi*.** Culm diameter 1·1 mm.

Culm surface

Epidermis: cells 4–6-sided, as long as wide to 1½–2 times longer than wide, walls thick, not wavy. **Stomata:** subsidiary cells with acute end walls; guard cell pair wide.

Culm T.S.

Epidermis: cells mostly twice as high as wide; outer walls v. thick, outer surface granular; anticlinal walls frequently wavy, with tapering thickening. **Stomata:** guard cells with lip at outer aperture and slight ridge on inner wall. **Chlorenchyma:** cells in 2 layers; protective cells present in 2 layers, cells reaching from *c.* ⅓ of way up substomatal cavity lined by epidermal cells, to parenchyma sheath. **Parenchyma sheath** 1-layered, cells rounded, as high as wide or slightly wider than high. **Silica:** irregular, more or less spheroidal-nodular bodies present in some cells of parenchyma sheath and central ground tissue.

***R. sieberi* var. *venustulus* (Kunth) Pillans.** Culm diameter 1·5 mm.

Culm surface

Epidermis: cells 4–6-sided; as long as to twice as long as wide; walls thick, straight at surface, wavy at lower focus. **Stomata:** subsidiary cells mainly with obtuse end walls.

Culm T.S.

Epidermis: cells mostly *c.* $1\frac{1}{2}$ times higher than wide; outer walls v. thick, granular at surface; anticlinal walls wavy, with rapidly tapering thickening. **Stomata:** guard cells with lip at outer aperture and slight ridge on inner wall; walls to pore concave. **Chlorenchyma:** protective cells present as slightly modified cells of outer chlorenchyma layer, extending in 1 layer from *c.* $\frac{1}{3}$ of way up substomatal cavity lined by epidermal cells to $\frac{1}{5}$–$\frac{1}{4}$ of way into inner chlorenchyma layer. **Parenchyma sheath** mainly 1-layered, cells as wide as high or up to twice as wide as high. **Silica** present as (i) granular material filling most protective cells, and (ii) bodies in some cells of parenchyma sheath.

***R. stereocaulis* Mast.** Culm diameter 0·8 by 0·4 mm.

Culm surface

Material taken from part of culm enclosed by leaf sheath.

Papillae low, 1 to a cell on occasional cells. **Epidermis:** cells 4-sided; 4–9 times longer than wide, walls slightly to moderately thickened, wavy. **Stomata:** subsidiary cells narrow, with acute end walls. **Cuticular marks:** with regularly spaced transverse bands of coarse granules, giving the appearance of longitudinal striations.

Culm T.S.

Outline oval. **Cuticle** slightly thickened, with low ridges. **Papillae** short, on cells towards regions of sharpest curvature. **Epidermis:** papillate cells largest, those on flatter areas smallest, mostly as high as wide, some slightly higher than wide; outer walls thick, convex, with low ridges. **Stomata:** subsidiary cells with low ridges as on epidermal cells; guard cells with lip at outer aperture. **Chlorenchyma:** cells in 1 layer; most cells appearing only slightly higher than wide (inclined to culm surface as seen in L.S.. and really twice as high as wide); protective cells I-shaped as seen in L.S. **Parenchyma sheath** 1-layered, interrupted at intervals by areas of scl. 1–2 cells thick, cells slightly wider than high. **Sclerenchyma:** sheath 5–6-layered, outline oval, with low, wide intrusions into parenchyma sheath. **Silica** absent.

***R. strobilifer* Kunth.** Culm diameter 1·3 mm.

Culm surface

Epidermis: cells 4–6-sided; as long as or up to twice as long as wide; walls v. thick, wavy at outer surface, straight at lower focus. **Stomata:** subsidiary cells with acute end walls, long anticlinal wall sometimes wavy. **Cuticular marks** coarsely granular and warty.

Culm T.S.

Epidermis: cells all of same height, nearly 3 times higher than wide; outer walls v. thick, outer surface irregular, granular; anticlinal walls with tapering thickening, slightly wavy. **Stomata:** guard cells with lip at outer aperture.

Chlorenchyma. Cells in 2 layers. Protective cells extending in 2 layers from epidermis to parenchyma sheath; cells of outer layer frequently extending about half-way into inner chlorenchyma layer, cells of inner layer little-modified peg-cells. **Parenchyma sheath** 1–2-layered, cells slightly to 1½ times wider than high. **Silica:** none seen.

***R. tenuissimus* Kunth.** Culm diameter 0·7 mm.

Culm surface

Epidermis: cells mostly 4-sided, those opposite ends of stomata frequently shorter than wide, others as long as or *c.* 1½ times longer than wide; walls moderately thickened, wavy. **Stomata:** subsidiary cells with rounded ends and slightly wavy walls; guard cell pair with circular outline but individual cells with rounded ends.

Culm T.S.

Outline circular, with low ridges. **Epidermis:** cells as high as to twice as high as wide, those next to stomata being highest; outer walls v. thick. **Stomata:** guard cells with lip at outer aperture and slight ridge on inner wall. **Chlorenchyma.** Cells in 2 layers. Protective cells present in 2–3 layers extending from epidermis to parenchyma sheath; cells of outer layer extending to junction of chlorenchyma layers, or *c.* half-way into inner layer. **Parenchyma sheath** 1-layered, cells mostly nearly twice as high as wide. **Silica:** bodies present in some cells of parenchyma sheath.

***R. triflorus* Rottb.** Culm diameter 0·7–0·8 mm.

Culm surface

Epidermis: cells 4-sided, those opposite ends of stoma as long as wide, others as long as to 4 times longer than wide; walls moderately thickened, wavy. **Stomata:** subsidiary cells variable; guard cell pair wide, with lemon-shaped outline.

Culm T.S.

Epidermis: cells *c.* 1½ times wider than high, as high as or 1½ times higher than wide, the tallest being next to stomata; outer walls moderately thickened to thick. **Stomata:** guard cells with slight lip at outer aperture. **Chlorenchyma:** cells in 2 layers; protective cells represented by slightly thickened peg-cells. **Parenchyma sheath** 1–2–several-layered; cells as high as wide or slightly to twice as wide as high. **Silica:** bodies present in some cells of parenchyma sheath.

***R. vilis* Kunth.** Culm diameter 1·0 mm.

Culm surface

Epidermis: cells 4–6-sided, as long as wide; walls v. thick, wavy. **Stomata:** subsidiary cells with acute end walls; guard cell pair sometimes with circular outline.

Culm T.S.

Outline circular, with irregular, shallow depressions. **Cuticle** thick. **Epidermis:** cells mostly *c.* 4–5 times higher than wide, but some between stomata only about twice as high as wide; outer walls v. thick, straight or sloping towards shorter cells. **Stomata:** guard cells with lip at outer aperture. **Chlorenchyma:** cells in 1 or 2 layers; protective cells extending in 1 or 2 layers from epidermis to part way into inner chlorenchyma layer. **Parenchyma sheath** 1-layered, cells slightly to 1½ times higher than wide. **Silica** present as (i) spheroidal-nodular bodies in some cells of parenchyma sheath; (ii) granular material in some cells of central ground tissue.

***R. virgeus* Mast.** Culm diameter 1·0 mm.

Culm surface

Epidermis: cells 4–6-sided; as wide as long or slightly to 1½ times wider than long; walls heavily thickened, wavy at surface. **Stomata:** subsidiary cells variable.

Culm T.S.

Epidermis: cells all of same height, about twice as high as wide; outer walls v. thick, straight; anticlinal walls with tapering thickening, v. wavy. **Stomata:** subsidiary cells with pronounced outer ridge; guard cells with lip at outer aperture. **Chlorenchyma:** cells in 2 layers; protective cells extending in 1 or 2 layers from epidermis to inner chlorenchyma layer or parenchyma sheath. **Parenchyma sheath** 1-layered, cells as wide as or 1½–2 times wider than high. **Silica:** bodies present in some cells of parenchyma sheath.

RESTIO GROUP 4. South Africa

Leaf surface not seen.

Leaf T.S.

Sheathing base, seen in *R. scaber* only.

Hairs absent. **Papillae** small, present on some shorter abaxial epidermal cells. **Epidermis:** (i) abaxial cells similar to those in culm; (ii) adaxial cells 5-sided, mostly twice as wide as high and *c.* ¼ height of smaller abaxial cells; outer walls thick, other walls moderately thickened. **Stomata** superficial, raised on taller cells of abaxial surface; similar to those of culm. **Chlorenchyma:** (i) peg-cells, 2–3 times higher than wide with few, medium-sized to long pegs; arranged in 1 layer to inside of abaxial epidermis; (ii) more or less isodiametric, lobed cells in 1–2 layers to inside of peg-cells in regions between vb's. Protective cells similar to those in culm.

Vascular bundles each with a group of narrow to medium-sized, angular tracheids and a small phloem pole; some bundles slightly larger than others. **Bundle sheaths:** O.S. parenchymatous, cells present in 1 layer at phloem pole only; I.S. sclerenchymatous, fibres narrow, with thick walls, cells arranged in 1 layer at phloem pole and 1–2 layers on flanks and at xylem pole, continuous on flanks with fibres of ground tissue. **Sclerenchyma** represented by that of bundle sheaths and wider, thick-walled fibres in 2–3 layers to adaxial side of bundles, and on their flanks, between chlorenchyma and adaxial epidermis. **Air-cavities** absent. **Ground tissue** sclerenchymatous. **Silica:** spheroidal-nodular bodies present in some cells of outer bundle sheath. **Tannin:** none seen.

Culm surface

Hairs absent. **Papillae** present in 3 spp. only (*R. macer*, *R. patens*, and *R. scaberulus*). **Epidermis.** Cells usually 4-sided, sometimes 5- or 6-sided; generally as long as to 1½ times longer than wide, sometimes 2–3 or 4–6 times longer than wide, rarely shorter than wide. Walls frequently v. thick, often moderately thickened, sometimes slightly to moderately thickened, straight or wavy. **Stomata** normally raised on tall cells; subsidiary cells variable; guard cells mostly with rounded ends, guard cell pair sometimes wide, oval. **Silica** absent. **Tannin** present in epidermal cells of some spp. **Cuticular marks** granular, sometimes with superimposed longitudinal striations.

Culm T.S.

Outline normally circular, occasionally oval, with low, irregular mounds. **Cuticle** slightly thickened to v. thick. **Epidermis.** Cells arranged in 1 layer, 4-sided, those next to stomata normally higher than others, produced into low mounds, raising stomata above level of rest of culm surface; individual cells ranging from as high as wide to 4 times higher than wide. Cells not next to stomata normally as high as wide to 1½ or 2½ times higher than wide, shorter than cells surrounding stomata. Outer walls often thick or v. thick; anticlinal walls straight or wavy, frequently moderately thickened, sometimes with tapering thickening; inner walls slightly or moderately thickened. **Stomata** usually raised on tall epidermal cells; subsidiary cells variable. Guard cells frequently with lip at outer aperture only, occasionally with inner lip or with ridge on inner wall. **Chlorenchyma** composed of 2 or rarely 3 layers of peg-cells; protective cells present. **Parenchyma sheath** normally 1-layered, sometimes 1–3-layered, cells as wide as high, up to twice as wide as high, or 2 times higher than wide; cell walls slightly thickened. **Sclerenchyma:** sheath well developed, with slight, dome-shaped ribs opposite to peripheral vb's or, occasionally, with isolated fibres or small groups of fibres penetrating parenchyma sheath; fibres of outer layers normally thick- or v. thick-walled, those of innermost layers wider, with moderately thickened or thick walls.

Vascular bundles. (i) Peripheral vb's each composed of arc of tracheids and a circular or oval phloem pole; tracheids all narrow, or with wider ones in centre of arc, or with 1 wider one on either flank. (ii) Medullary vb's present in all spp., outer bundles small, frequently with 1 central, narrow or medium-sized, circular or angular mxv, other outer bundles with 1 mxv on either flank; inner bundles each with 1 narrow, medium-sized or wide mxv on either flank, the 2 vessels separated from each other by 1–2 or 3–4 rows of narrow cells. Phloem poles circular or oval, ensheathed by 1 layer of narrow cells with thick walls next to phloem, and slightly or moderately thickened walls next to xylem. Px poles present in all but smallest, outer bundles; some bundles usually free from culm scl. sheath. **Bundle sheaths:** fibres at poles usually narrow, thick- or v. thick-walled, arranged in 1 or 2–(3) layers; those on flanks usually wider, arranged in 1 layer, with slightly or moderately thickened walls. **Central ground tissue** parenchymatous, cells of outer layers surrounding vb's usually with moderately thickened walls, central cells thin-walled, often breaking down to form central cavity. **Silica:** form and distribution varying in different spp.; (i) spheroidal-nodular bodies in cells of parenchyma sheath and/or inner chlorenchyma; (ii) granular material in certain cells of central ground tissue and protective cells. **Tannin** present in epidermal cells except where stated to the contrary in species descriptions.

Rhizome and *root* not seen.

SPECIES DESCRIPTIONS

***R. cincinnatus* Mast.** Culm diameter 0·5 mm.

Culm surface

Epidermis: cells 4-sided; mostly 4–6 times longer than wide, those next to stomata shorter, walls slightly to moderately thickened and not more than slightly wavy. **Stomata** raised on small mounds; subsidiary cells with obtuse or acute end walls; guard cell pair with wide, oval outline. **Tannin:** none seen. **Cuticular marks** finely granular.

Culm T.S.

Cuticle slightly thickened. **Epidermis:** cells bordering stomata twice as high as wide, others as high as wide; outer walls slightly convex. **Stomata:** guard cells with lip at outer aperture. **Chlorenchyma:** protective cells extending in 1 layer from half-way up substomatal cavity lined by epidermal cells, to *c.* $\frac{1}{3}$ of way into inner chlorenchyma layer. **Sclerenchyma:** sheath outline circular, with occasional ribs 1–2 cells wide and 1 cell thick, extending into parenchyma sheath. **Silica:** bodies present in some cells of parenchyma sheath. **Tannin:** none seen.

***R. duthieae* Pillans.** Culm diameter 0·9 by 1·1 mm.

Culm surface

Epidermis: cells 4-sided; mainly $1\frac{1}{2}$ times wider than long; walls v. thick, wavy. **Stomata** raised; subsidiary cells with acute or obtuse end walls.

Culm T.S.

Outline oval. **Cuticle** thick. **Epidermis:** cells next to or near to stomata as high as wide, others between stomata $1\frac{1}{2}$ times wider than high; outer walls v. thick; anticlinal walls wavy, with tapering thickening. **Stomata:** guard cells with outer lip. **Chlorenchyma:** cells of outer layer 9–10 times higher than wide, pegs small, numerous; those of inner layer *c.* 3 times wider than outer cells, $1\frac{1}{2}$ times higher than wide, each with constriction, reducing width of cell by $\frac{1}{2}$, just over half-way down height of cell; protective cells poorly differentiated, similar to cells of outer chlorenchyma layer. **Parenchyma sheath** 1-layered, cells v. large, twice as high as shortest epidermal cells and $\frac{2}{3}$ height of outer chlorenchyma cells; each $1\frac{1}{2}$ times higher than wide. **Silica:** bodies present in some cells of inner chlorenchyma layer.

***R. filiformis* Poir.** Culm diameter 0·8 mm.

Culm surface

Epidermis: cells mainly 6-sided; some as long as wide but most frequently twice as long as wide; walls moderately thickened, not wavy. **Stomata:** subsidiary cells with acute end walls.

Culm T.S.

Cuticle moderately thickened. **Epidermis.** Cells of 2 distinct sizes; those next to stomata $2\frac{1}{2}$ times higher than wide, most others $1\frac{1}{2}$ times higher than wide. Outer walls v. thick, most slightly convex; anticlinal walls lining substomatal cavities wavy, with tapering thickening; walls of other cells slightly thickened, not wavy. **Stomata:** subsidiary cells sometimes with ridge on outer wall; guard cells with lip at outer aperture. **Chlorenchyma.** Cells in 2 similar layers.

Protective cells extending in 1 layer from ⅓ of way up substomatal cavity lined by epidermal cells, to part way into inner chlorenchyma layer; apertures between cells wide, oval, present near inner end of substomatal tube. **Parenchyma sheath** 1-layered, interrupted at irregular intervals by single fibres from scl. sheath, cells slightly to 1½ times wider than high. **Silica:** none seen.

R. fruticosus **Thunb.** Culm diameter 1·5, 2·0, 2·2 mm.

Culm surface

Epidermis: cells 4–6-sided; irregular, frequently slightly wider than long, or as long as wide; walls thick to v. thick. **Stomata:** subsidiary cells with acute end walls; guard cells shorter and wider than in most spp. of this group. **Cuticular marks:** faint longitudinal striations.

Culm T.S.

Cuticle v. thick. **Epidermis.** Cells all of same height in (24585), and *c.* 4 times higher than wide; of irregular heights in other specimens, the tallest next to stomata, nearly 4 times higher than wide; the shortest 2½ times higher than wide. Outer walls v. thick, outer surface irregular; anticlinal walls wavy, with tapering thickening. **Stomata:** guard cells with lip at outer aperture and ridge on inner wall. **Chlorenchyma:** protective cells extending in 2 layers from a short distance up substomatal cavity lined by epidermal cells, to parenchyma sheath. **Parenchyma sheath** 1–3-layered, cells rounded or slightly wider than high. **Vascular bundles:** peripheral with unusual distribution, several small bundles free from culm scl. sheath and embedded in parenchyma sheath in (24585). **Central ground tissue** parenchymatous, outer cells moderately thick-walled, interspersed with areas of thin-walled cells; central cells few, thin-walled; intercellular spaces frequent. **Silica:** none seen.

R. macer **Kunth.** Culm diameter 0·8 mm.

Culm surface

Papillae either short and isolated or formed by recurved outer and anticlinal walls of cells bordering stomata (*R. patens* q.v.). **Epidermis:** cells 4-sided; as long as to 1½ times longer than wide; walls moderately thickened, wavy. **Stomata:** subsidiary cells with obtuse end walls and irregular long anticlinal walls; guard cells wide and short. **Cuticular marks** irregular, longitudinal striations.

Culm T.S.

Papillae: see above under Culm surface. **Epidermis.** Cells next to stomata tallest, 2½ times higher than wide; others slightly higher than wide. Outer walls v. thick, outer surface irregular; longer anticlinal walls wavy, others straight, with tapering thickening. **Stomata** raised on tall epidermal cells; guard cells with lip at outer aperture. **Chlorenchyma:** cells in 2 layers; protective cells present as slightly thickened peg-cells of both layers. **Parenchyma sheath** 1–2-layered, cells mostly slightly wider than high. **Silica:** medium-sized bodies present in certain cells of parenchyma sheath.

R. marlothii **Pillans.** Culm diameter 1·0 mm.

Culm surface

Epidermis: cells 4–6-sided; mostly about as long as wide; walls v. thick.

Stomata: subsidiary cells with acute end walls; guard cell pair circular in outline with slightly rounded ends. **Tannin:** none seen in epidermal cells. **Cuticular marks** coarsely granular.

Culm T.S.

Epidermis: cells next to stomata and most others 4 times higher than wide, remainder 2–2½ times higher than wide; outer walls v. thick; anticlinal walls wavy, with tapering thickening. **Stomata:** guard cells with beak-like lip at outer and small lip at inner apertures. **Chlorenchyma:** protective cells extending in 2 layers from half-way up substomatal cavity lined by epidermal cells, to parenchyma sheath. **Parenchyma sheath** 1-layered, cells rounded, mostly as high as wide. **Silica:** bodies present in some cells of parenchyma sheath. **Tannin:** none seen.

R. monanthus **Mast.** Culm diameter 1·0 mm.

Culm surface

Epidermis: cells 4-sided; mostly as long as wide; walls v. thick, wavy. **Stomata** raised on tall cells; subsidiary cells with obtuse end walls; guard cells shorter and wider than in most spp. of this group. **Cuticular marks** coarsely granular.

Culm T.S.

Epidermis: cells next to stomata 2½ times higher than wide, others slightly higher than wide; outer walls v. thick; anticlinal walls wavy, with tapering thickening. **Stomata:** guard cells with large lips at outer aperture; walls to pore concave. **Chlorenchyma:** protective cells present as slightly thickened cells of outer palisade layer, with wavy long walls, cells extending almost to parenchyma sheath. **Parenchyma sheath** 1-layered, cells twice as high as wide and about twice as high as inner chlorenchyma cells. **Silica:** bodies present in many cells of inner chlorenchyma layer.

R. patens **Mast.** Culm diameter 1·0 by 0·7 mm.

Culm surface

Papillae present on tall cells next to stomata, curving away from free edge of cell. **Epidermis:** cells 4-sided; as long as to 1½ times longer than wide; walls moderately thickened to thick, wavy. **Stomata** raised on tall cells; subsidiary cells with acute end walls. **Tannin:** none seen. **Cuticular marks** granular, with superimposed fine, longitudinal striations; short, highly refractive lines crossing most walls at 90°, each line slightly longer than wall width.

Culm T.S.

Outline sub-circular with 1 flattened sector and low mounds. **Papillae:** see above under Culm surface. **Epidermis.** Cells next to stomata tallest, 2½–3 times higher than wide; those between stomata twice as high as wide. Outer walls v. thick; anticlinal walls frequently wavy. **Stomata:** guard cells with lip at outer aperture. **Chlorenchyma:** cells in 2–(3) layers. **Parenchyma sheath** 1-layered, cells wide, circular or slightly higher than wide. **Central ground tissue** parenchymatous, all cells with slightly to moderately thickened walls. **Silica** and **tannin:** none seen.

***R. purpurascens* Nees ex Mast.** Culm diameter 1·0 mm.

Culm surface

Epidermis: cells 4–6-sided; mostly as long as wide, some slightly to 1½ times longer than wide; walls v. thick, not wavy. **Stomata:** subsidiary cells with acute end walls. **Tannin** in some epidermal cells.

Culm T.S.

Epidermis: cells next to stomata *c.* 2½ times higher than wide, those between stomata slightly to 1½ times higher than wide; outer walls v. thick; anticlinal walls wavy, with tapering thickening. **Stomata:** guard cells with lip at outer aperture. **Chlorenchyma:** protective cells with slightly to moderately thickened walls; extending in 1 or 2 layers from ½ or ¼ of way up substomatal cavity lined by epidermal cells, to parenchyma sheath, or well into inner chlorenchyma. **Parenchyma sheath** mainly 1-, occasionally 2-layered, cells rounded to *c.* 1½ times wider than high. **Silica** as granular material in most protective cells.

***R. pygmaeus* Pillans.** Culm diameter 0·7 mm.

Culm surface

Epidermis: cells 4-sided; mostly as long as wide, some slightly wider than long; walls v. thick. **Stomata:** subsidiary cells with wavy long walls; guard cell pair with circular outline. **Tannin** in some epidermal cells. **Cuticular marks** finely granular, with superimposed, fine, longitudinal striations.

Culm T.S.

Epidermis: cells next to stomata 2½–3 times higher than wide, others *c.* 1½ times higher than wide; outer walls v. thick, slightly convex; anticlinal walls with tapering thickening. **Stomata:** guard cells with lip at outer aperture. **Chlorenchyma:** protective cells poorly developed, consisting of slightly thickened cells of outer palisade layer. **Parenchyma sheath** 1-layered, cells 1½ times higher than wide. **Central ground tissue:** outer cells thick-walled, central cells with moderately thickened walls. **Silica** present as (i) spheroidal-nodular bodies in some cells of parenchyma sheath, (ii) granular material in many central cells of ground tissue.

***R. scaber* Mast.** Culm diameter 0·75 mm.

Leaf T.S.: see group 4 genus description, p. 273.

Culm surface

Epidermis: cells mostly 4-sided; as long as wide to 3 times longer than wide, those surrounding stomata all as long as wide; walls moderately thickened, wavy. **Stomata:** subsidiary cells with acute end walls.

Culm T.S.

Epidermis: cells between stomata slightly to 1½ times higher than wide, those next to stomata twice as high as wide; outer walls v. thick, outer surface irregular; anticlinal walls wavy, with tapering thickening. **Stomata:** subsidiary cells with small, cuticular ridge on outer wall; guard cells with lip at outer and inner apertures. **Chlorenchyma:** protective cells extending in 2 layers from epidermis to parenchyma sheath; cells of outer layer extending about halfway into inner chlorenchyma layer, and overlapping the shorter cells of the inner layer. **Parenchyma sheath** 1–2-layered, cells as wide as high to 1½ times wider than high. **Silica** infrequent, as granular material in some protective cells.

R. scaberulus **N. E. Br.** Culm diameter 0·5 and 1·0 mm.

Culm surface

Papillae slightly developed, on some cells next to stomata. **Epidermis:** cells 4–6-sided; as long as wide to 2½ or 3 times longer than wide; walls wavy at surface, straight at lower focus, moderately thickened. **Stomata:** subsidiary cells variable; guard cell pair with wide oval outline. **Cuticular marks** granular, with superimposed faint longitudinal banding.

Culm T.S.

Papillae slight, on some cells next to stomata. **Epidermis:** cells ranging from as high as wide to 1½–2 times higher than wide, taller cells next to stomata; outer walls thick to v. thick, often convex; some anticlinal walls wavy. **Stomata** with well-developed outer and slight inner lips. **Chlorenchyma:** protective cells present as slightly elongated peg-cells of outer layer, with slightly thickened walls. **Parenchyma sheath** 1–2-layered, cells as high as wide to 1½–2 times wider than high. **Sclerenchyma:** sheath 6–9-layered, outline circular, slight ribs opposite peripheral vb's; occasional fibres interrupting parenchyma sheath. **Silica** represented by (i) bodies in some cells of parenchyma sheath, and (ii) granular particles, in some protective cells.

R. triticeus **Rottb.** Culm diameter 1·2, 2·0 mm.

Culm surface

Epidermis: cells 4–6-sided; as long as or up to twice as long as wide; walls moderately thickened to thick, wavy at surface. **Stomata:** subsidiary cells frequently with wavy long walls.

Culm T.S.

Epidermis: cells next to stomata 2½–3 times higher than wide, those between stomata 1½–2 times higher than wide, with intergrading sizes between the two; outer walls v. thick, slightly irregular at outer surface; anticlinal walls sometimes wavy. **Stomata** raised on taller epidermal cells; guard cells with lip at outer aperture and slight ridge on inner wall. **Chlorenchyma:** cells in 2–(3) layers; protective cells extending in 2 layers from inner end of substomatal cavity lined by epidermal cells, to parenchyma sheath; cells of outer layer extending about half-way into inner chlorenchyma layer. **Parenchyma sheath** 1–3-layered, cells up to 2½ times wider than high, walls thick to slightly thickened. **Central ground tissue:** many cells thin-walled, some cells moderately thick-walled. **Silica** as granular particles, in some cells of central ground tissue.

R. zwartbergensis **Pillans.** Culm diameter 0·4 mm.

Culm surface

Epidermis: cells mostly 4-sided; as long as to 2½ times longer than wide, those around stomata shortest; walls slightly to moderately thickened, wavy. **Stomata:** subsidiary cells with acute end walls.

Culm T.S.

Epidermis: some cells as high as wide; others, next to stomata, twice as high as wide; outer walls v. thick, often concave in shorter cells; anticlinal walls with tapering thickening. **Stomata:** guard cells with outer lip and slight ridge on inner wall. **Parenchyma sheath** 1-layered, cells slightly wider than high. **Central ground tissue:** outer cells with slightly thickened walls, central cells few, thin-walled. **Silica** absent.

RESTIO GROUP 5. Madagascar

As for group 3, but with 1 papilla on each culm epidermal cell.

***R. madagascariensis* H. Chermez.** Culm diameter 0·9 mm.

Culm surface

Papillae present, 1 to a cell (Fig. 37. D, p. 254). **Epidermis:** cells 4–6-sided; as long as to 2½ times longer than wide; walls moderately thickened, wavy. **Stomata:** subsidiary cells with acute or obtuse end walls; guard cell pair with oval outline. **Silica** and **tannin:** none seen.

Culm T.S. (Fig. 37. D)

Outline circular. **Cuticle** thick. **Papillae** parallel-sided with rounded ends, v. thick-walled, 1½–2 times higher than wide, *c.* ½ width of epidermal cells from which they arise. **Epidermis:** cells 4-sided, each with a papilla; cells 1–1½ times higher than wide and about as high as the papillae; outer walls v. thick, anticlinal and inner walls moderately thickened. **Stomata** slightly sunken; subsidiary cells large; guard cells with pronounced lip at outer and slight lip at inner apertures. **Chlorenchyma:** cells in 1 layer (Fig. 37. D); protective cells present as 1 layer of modified peg-cells, extending from epidermis to parenchyma sheath. **Parenchyma sheath** 1–3-layered, cells as wide as high to *c.* 1½ times wider than high; walls slightly or moderately thickened; some cells with thickened inner and anticlinal walls. **Sclerenchyma:** sheath 7–9-layered; outline circular, with low, dome-shaped ribs opposite to and between peripheral vb's; fibres medium-sized, v. thick-walled.

Vascular bundles. (i) Peripheral vb's: tracheids few, medium-sized, arranged in 1–2-layered arc, partially enclosing small, circular phloem pole. (ii) Medullary vb's: many outer bundles with 1 wide, central, rounded-angular mxv; flanking mxv's of inner bundles wide, with slightly thickened walls, separated from one another by 1 or 2 rows of narrow, thin-walled cells. Phloem pole oval, ensheathed by scl. Px absent from smaller, outer bundles; many bundles free from culm scl. sheath. **Bundle sheaths:** fibres narrow, with moderately thick to thick walls, in 2–3 layers at phloem pole and 1–2 layers at xylem pole, and with moderately thickened walls, in 1 layer on flanks. **Central ground tissue** parenchymatous, all cells with moderately thick walls. **Silica** and **tannin:** none seen.

RESTIO GROUP 6. South Africa

As for group 3, but outline of culm quadrangular, sclerenchyma sheath with outline following that of culm, and with a rectangular rib extending to epidermis at each angle (Pl. VII. B).

Species Descriptions

***R. quadratus* Mast.** Culm diameter 1 by 2 mm.

Culm surface

Hairs and **papillae** absent. **Epidermis:** cells 4-sided, as long as wide or up to 2½–3 times longer than wide, walls moderately thickened, wavy. **Stomata:** subsidiary cells with perpendicular or acute end walls, fairly narrow; guard cell pair with oval outline. **Tannin** in many epidermal cells. **Cuticular marks** finely granular, with superimposed longitudinal striations.

Culm T.S.

Epidermis: cells at angles up to twice as high as wide, those on faces mostly as high as wide; outer walls v. thick, slightly convex; anticlinal and inner walls slightly thickened. **Stomata** superficial; guard cells with lip at outer aperture. **Chlorenchyma.** Peg-cells in 2 or 3 layers. Protective cells poorly developed, consisting of v. slightly thickened cells of outer chlorenchyma layer; apertures wide, rounded, occurring between inner ends of cells. **Parenchyma sheath** 1–2-layered, cells mostly slightly wider than high. **Sclerenchyma:** sheath 5–7-layered; outline following that of culm, but with 4 parallel-sided ribs, extending to epidermis, each 5–8 cells wide, situated one at each angle of culm. Outermost fibres and those of ribs narrow, v. thick-walled; inner fibres wider, thick- or v. thick-walled. Fibres to outer side of peripheral vb's in 1 layer, narrow, with moderately thickened walls.

Vascular bundles. (i) Peripheral vb's: tracheids narrow and medium-sized, arranged in 1–2-layered arc, partially enclosing oval phloem pole. (ii) Medullary vb's: outer, smaller bundles with 1 central, wide, many-sided mx vessel; all others with 1 mxv on either flank. Flanking mxv's wide, many-sided, with slightly thickened walls, separated from one another by 1–6 rows of narrow, compressed, thin-walled cells. Phloem pole large, with several wide sieve-tubes, situated over and between mxv's, ensheathed by scl. Px absent from some outer bundles; many bundles free from culm scl. sheath. **Bundle sheaths:** fibres mostly narrow, thick-walled, in 1–2 layers at phloem pole; with moderately thickened walls, in 1 layer on flanks and in 1–2 layers at xylem pole. **Central ground tissue** parenchymatous, outer cells with moderately thick walls, central cells thin-walled, breaking down to form central cavity. **Silica** absent. **Tannin** in many cells of epidermis and some cells of central ground tissue.

***R. tetragonus* Thunb.** Culm diameter 1·5 by 2 mm.

As for *R. quadratus* except:

Culm surface

Epidermis: cells 4–6-sided; walls moderately thickened, sometimes slightly wavy. **Stomata:** subsidiary cells with acute end walls; guard cell pair wide, with oval outline. **Cuticular marks** coarsely granular.

Culm T.S. (Fig. 37. F, L, p. 254; Pl. VII. B)

Cuticle moderately thickened, inner part granular, outer part clear. **Epidermis:** cells *c.* 1½ times higher than wide, except opposite ribs, where 2–2½ times higher than wide; anticlinal walls moderately thickened, sometimes wavy. **Stomata:** subsidiary cells with pronounced outer ridge slightly over-arching guard cells; guard cells with beak-like outer and small inner lips. **Silica:** bodies present in cells of parenchyma sheath, particularly those next to ribs.

RESTIO GROUP 7. Australia

Leaf surface not seen.

Leaf T.S.

Seen in *R. complanatus* only.

Hairs absent. **Papillae:** outer walls of some cells of abaxial epidermis forming dome-shaped, low papillae. **Cuticle** thick, ridged over abaxial surface, thin over adaxial surface. **Epidermis:** (i) adaxial cells 4-sided, compressed, *c.* 4 times wider than high; walls slightly thickened; (ii) abaxial cells 4-sided, as high as to 1½ times higher than wide, similar in width to adaxial cells; outer walls thick, frequently dome-shaped, other walls slightly thickened. **Stomata** present in abaxial epidermis only, as in culm. **Chlorenchyma** represented by 1 layer of thin-walled, lobed, or pegged cells; cells slightly wider than high; layer interrupted at intervals by short pillar cells arising from scl. ribs over adaxial side of vb's or by the ribs themselves.

Vascular bundles mostly small, some medium-sized; tracheids narrow, angular, several to many in each bundle; phloem poles small. **Bundle sheaths:** O.S. parenchymatous, represented by 1 layer of several thin-walled, rounded cells opposite to phloem pole; these occasionally replaced by short pillar cells with slightly thickened walls; I.S. sclerenchymatous, fibres narrow, thick-walled, in 1 layer at phloem pole, wider, in 1–2 layers at xylem pole; those on flanks difficult to distinguish from scl. ground tissue. **Sclerenchyma** represented by that in bundle sheaths just described, and by 3–4 layers of fibres between vb's; layers bounded abaxially by chlorenchyma and adaxially by 1 layer of wide, thin-walled parenchymatous cells. Fibres of outer adaxial layer wide, those of outer abaxial layer narrow; all thick-walled. **Ground tissue** composed of 1 layer of wide, thin-walled parenchyma cells, situated between adaxial epidermis and scl. layers. **Silica** present as spheroidal-nodular bodies in some cells of parenchymatous O.S. **Tannin:** none seen.

Culm surface

Hairs multicellular, present in *R. megalotheca* and *R. nitens*. **Papillae** absent. **Epidermis:** cells 4- or 6-sided; mostly 1½–3 times longer than wide, some as long as wide, and some reaching 8 times longer than wide in occasional spp.; walls mostly slightly or moderately thickened, wavy. **Stomata:** subsidiary cells variable; guard cell pair frequently with lemon-shaped outline. **Silica** absent. **Tannin** in epidermal cells in some spp. **Cuticular marks** usually granular, rarely with longitudinal striations.

Culm T.S.

Outline circular, oval, or circular with ridges. **Cuticle** slightly thickened to thick. **Epidermis.** Cells 4-sided, arranged in 1 layer; all of same height or tallest opposite pillar cells, ranging from as high as wide to 1½–2 times higher than wide between pillar cells, and up to 3–4 times higher than wide opposite pillar cells; those next to stomata projecting into chlorenchyma in *R. nitens*.

Outer walls moderately thickened to v. thick, often convex; anticlinal walls either evenly and slightly or moderately thickened, or with tapering thickening; inner walls usually slightly or moderately thickened, sometimes thin. **Stomata** usually superficial, but sunken in spp. *R. sphacelatus* and *R. ornatus*; guard cells without pronounced lips or ridges but ridge on inner wall in *R. megalotheca*; walls to pore concave. **Chlorenchyma** composed of (1)–3 layers of peg-cells, layers interrupted at intervals by ribs from scl. sheath and their associated pillar cells; protective cells absent; for details of peg-cells, see individual spp.

Parenchyma sheath composed of 2 types of cell. (i) Small, rounded cells with slightly thickened walls, mostly about as high as wide, or slightly to 1½ times higher than wide or wider than high; cells arranged in 1 layer on flanks of girders or ribs from scl. sheath, and between them. (ii) Pillar cells, represented by palisade cells, often with moderately thickened walls, each cell about 3–4 times higher than wide, some taller, some shorter, radiating from most or all ribs from scl. sheath to epidermis. Pillar cells single or in groups of 2–3–(4), forming longitudinal files as seen in T.L.S. **Sclerenchyma:** sheath well developed, with pronounced ribs opposite to peripheral vb's.

Vascular bundles. (i) Peripheral vb's each with arc of tracheids partially enclosing phloem pole; tracheids all of same size, or those on flanks wider than others. (ii) Medullary vb's each with 1 narrow, medium-sized or wide vessel on either flank. Outer bundles small, often without px pole, inner bundles larger, with px. Phloem poles circular or oval, or overarching flanking mxv's, separated from xylem by sheath of 1 layer of narrow cells with thin walls facing phloem, and slightly or moderately thickened walls facing xylem. Some or no bundles free from culm scl. sheath. **Bundle sheaths:** fibres narrow, in 1–2 layers at poles, wider, usually in 1 layer on flanks; walls ranging in thickness from moderately thick to v. thick.

Central ground tissue parenchymatous, usually composed of several outer layers of cells with moderately thickened walls and a central area of thin-walled cells, these frequently breaking down; occasionally with all cells moderately thick-walled. **Silica** present (i) as spheroidal-nodular bodies in outer cells of scl. sheath, or in some cells of parenchyma sheath; (ii) in some spp. as granular material, in certain cells of central ground tissue. **Tannin** present in some cells of epidermis, except where noted in species descriptions.

Rhizome T.S.

Seen in *R. complanatus* only.

Hairs: few, short bicellular hairs present. **Epidermis:** cells slightly wider than high, 4–6-sided; walls all slightly thickened. **Cortex:** (i) outer part consisting of 4–5 layers of hexagonal cells, slightly wider than high, with slightly thickened walls; (ii) inner part composed of 4–5 layers of loosely packed, rounded, more or less isodiametric, thin-walled cells. **Endodermoid sheath** 1-layered; cells narrow, hexagonal, somewhat wider than high, with slightly thickened walls. **Vascular bundles** amphivasal, each composed of a ring of medium-sized, angular mxv's enclosing a circular phloem pole. **Bundle sheaths** sclerenchymatous, indistinguishable from ground tissue. **Central ground tissue:** cells of outer layers surrounding vb's with slightly to moderately thickened walls; cells of inner layers thin-walled, more or less isodiametric. **Silica** and **tannin:** none seen.

Root T.S.

Seen in *R. complanatus* only.

Root-hairs arising from cells similar to those of epidermis, shaft of hair *c.* ¼ of width of basal portion. **Epidermis:** cells 4–5-sided, slightly higher than wide, thin-walled. **Cortex:** (i) outer part composed of 2–3 layers of 6-sided cells, each cell *c.* 1½ times wider than high, with slightly thickened walls; (ii) middle part composed of radial files of rounded, thin-walled cells, those of outermost layers *c.* 4 times size of innermost cells, intermediate cells graded in size; (iii) inner part consisting of 1 or 2 layers of rounded, narrow cells (difficult to distinguish from inner layers of middle cortex). **Endodermis** 1-layered, cells 4-sided, 1½–2 times higher than wide, walls thin. **Pericycle** 1-layered, cells thin-walled, *c.* ⅙ height of endodermal cells. **Vascular cylinder:** mxv's thin-walled, many-sided, medium-sized, arranged in 1 ring to inside of that formed by phloem and px strands. **Central ground tissue** composed of narrow, thin-walled (lignified), hexagonal cells.

Species Descriptions

***R. amblycoleus* F. v. Muell.** Culm diameter 2·0 mm.

Culm surface

Epidermis: cells 4-sided; 1½–2 times longer than wide; walls moderately thickened, wavy. **Stomata:** subsidiary cells with obtuse end walls.

Culm T.S.

Outline circular. **Epidermis:** cells all of same height, as high as wide; outer walls thick, slightly convex; anticlinal walls not wavy. **Stomata:** subsidiary cells frequently with ridge on outer wall. **Chlorenchyma:** cells in 2 similar layers. **Silica:** bodies present in some cells of parenchyma sheath.

***R. chaunocoleus* F. v. Muell.** Culm diameter 1·0 mm.

Culm surface

Epidermis: cells 4–6-sided; as long as wide to 4 times longer than wide; walls slightly to moderately thickened, wavy. **Stomata:** subsidiary cells with obtuse or rounded end walls; guard cells narrow, elongated. **Tannin:** none seen. **Cuticular marks** granular, with superimposed, faint longitudinal striations.

Culm T.S.

Outline circular. **Epidermis:** cells mostly as high as to 1½ times higher than wide, those opposite pillar cells 3–4 times higher than wide. Outer walls v. thick, slightly convex; anticlinal walls with tapering thickening. **Chlorenchyma:** cells in 2 or 3 layers. **Central ground tissue:** all cells with slightly thickened walls. **Silica** present as (i) bodies in stegmata in parenchyma sheath; (ii) granular material in lumina of some cells of central ground tissue. **Tannin:** none seen.

***R. complanatus* R. Br.** Culm diameter 2·5 by 1·5; 2·5 by 0·8; 2·1 by 1·0 mm.

Culm surface

Epidermis: cells 4-sided; 3–8 times longer than wide, walls slightly thickened, wavy. **Stomata:** subsidiary cells slightly longer than guard cells, with acute end walls; guard cells long, narrow. **Cuticular marks:** none obvious.

Culm T.S.

Outline elliptical. **Epidermis:** cells as high as wide in (CA99), mostly 1½–2 times higher than wide in other samples. Outer walls moderately thickened or thick. **Chlorenchyma:** cells in 1 or 2 layers. **Silica:** bodies present in stegmata in parenchyma sheath.

Leaf, *Rhizome*, and *Root*; see group 7 genus description.

***R. gracillior* F. v. Muell. ex Benth.** Culm diameter 1·1 mm.

Culm surface

Epidermis: cells 4- or 6-sided; 1½–5 times longer than wide, walls slightly to moderately thickened, straight or wavy. **Stomata:** subsidiary cells with obtuse end walls.

Culm T.S.

Outline circular. **Epidermis:** cells slightly higher than wide, those opposite pillar cells 1½–2 times higher than wide; outer walls v. thick, frequently slightly convex. **Stomata:** guard cells narrow and elongated. **Chlorenchyma:** cells in 1 or 2 layers. **Silica:** bodies present in stegmata in parenchyma sheath.

***R. grispatus* R. Br.** Culm diameter 0·7 mm.

Culm surface

Epidermis: cells 4-sided; as long as to 2½ times longer than wide; walls moderately thickened, wavy. **Stomata:** subsidiary cells with perpendicular or acute end walls.

Culm T.S.

Outline more or less circular, with ridges. **Epidermis:** cells opposite pillar cells about twice as high as wide, others as high as or slightly higher than wide; outer walls v. thick, slightly convex. **Stomata:** subsidiary cells with ridge on outer wall. **Chlorenchyma:** cells in 1 or 2 layers. **Sclerenchyma:** sheath 4–5-layered; outline circular, with pronounced ribs opposite to and enclosing each peripheral vb; fibres medium-sized, v. thick-walled. **Silica** (i) as bodies in some stegmata in parenchyma sheath; (ii) as granular particles in certain cells of central ground tissue.

***R. laxus* R. Br.** Culm diameter 1·1 mm.

Culm surface

Epidermis: cells 4-sided; as long as to 1½ times longer than wide; walls thick, straight or wavy. **Stomata:** subsidiary cells with acute end walls and irregular, sometimes wavy, long anticlinal walls; guard cells narrower and more elongated than usual in species of this group. **Tannin:** none seen.

Culm T.S.

Outline more or less circular, with slight ridges opposite pillar cells.

Epidermis: cells opposite pillar cells 2–2½ times higher than wide, 1½ times taller and wider than those next to stomata; outer walls v. thick, frequently convex, outer surface granular; anticlinal walls slightly wavy, with tapering thickening. **Stomata:** subsidiary cells with ridge on outer wall. **Chlorenchyma:** cells in 2 or 3 layers. **Silica:** large, spheroidal-nodular bodies in some cells of parenchyma sheath. **Tannin:** none seen.

***R. megalotheca* F. v. Muell.** Culm diameter 1·0 mm.

Culm surface

Hairs multicellular, each with a short 2–3-celled stalk and several elongated, thin-walled branches orientated axially and lying flat on stem. **Epidermis:** cells 4-sided; slightly longer than wide, walls moderately thickened to thick, wavy. **Stomata:** subsidiary cells with acute end walls and irregular long anticlinal walls; guard cell pair with oval outline.

Culm T.S.

Outline more or less circular with irregular, round-topped mounds. **Epidermis:** cells opposite to pillar cells 3–4 times higher than wide; those next to stomata 1½–2 times higher than wide; intermediate sizes present between the extremes; outer walls v. thick, anticlinal walls occasionally slightly wavy. **Stomata:** subsidiary cells frequently with ridge on outer wall; guard cells with ridge on inner wall. **Chlorenchyma:** cells in 3 layers. **Silica:** spheroidal-nodular bodies present in some cells on flanks of scl. ribs.

***R. nitens* Nees.** Culm diameter 0·8 mm.

Culm surface

Hairs present at base of culm; branched, multicellular. **Epidermis:** cells 4-sided; those over pillar cells 3–6 times longer than wide, remainder wider, 2–4 times longer than wide; walls moderately thickened, wavy. **Stomata:** subsidiary cells with perpendicular or acute end walls, long anticlinal walls slightly wavy.

Culm T.S.

Outline circular. **Epidermis:** cells next to stomata projecting into chlorenchyma for up to ½ of their height, others nearly twice as high as wide; outer walls v. thick, straight or irregularly wavy at surface. **Stomata:** subsidiary cells with ridge on outer wall. **Chlorenchyma:** cells in 2 layers. **Silica:** none seen.

***R. ornatus* Steud.** Culm diameter 0·8 mm.

Culm surface

Epidermis: cells 4-sided; ranging from as long as to 3 times longer than wide; walls wavy, moderately thickened; cells surrounding stomata overarching them. **Stomata:** subsidiary cells with acute end walls.

Culm T.S.

Outline circular. **Epidermis:** cells 1½–2 times higher than wide, the tallest being next to stomata or opposite pillar cells. Outer walls v. thick, frequently convex; anticlinal and inner walls slightly thickened. **Stomata** sunken; subsidiary cells with ridge on outer wall. **Silica:** spheroidal-nodular bodies present in some cells of parenchyma sheath, particularly those next to the ribs.

***R. sphacelatus* R. Br.** Culm diameter 1 by 0·3 mm.

Culm surface

Epidermis: cells 4–6-sided; those next to stomata frequently as long as wide, others up to twice as long as wide; walls thick. **Stomata:** subsidiary cells with acute end walls. **Tannin:** none seen. **Cuticular marks** finely granular; cuticle continuous, extending over edges of each stoma, with a transversely orientated, bone-shaped aperture above the guard cells.

Culm T.S.

Outline oval. **Epidermis:** cells 4-sided, all of same height, mostly about twice as high as wide; outer walls v. thick, outer surface irregular; anticlinal walls wavy, with tapering thickening. **Stomata** sunken, partially covered by cuticle (see Culm surface); subsidiary cells mounted low on epidermal cells. **Chlorenchyma:** cells in (1)–2 layers. **Silica:** large bodies present in stegmata in outer layer of culm scl. sheath and in some cells of parenchyma sheath. **Tannin:** none seen.

RESTIO GROUP 8. Australia

Differing from other groups in characters of epidermal cells and sclerenchyma sheath (Fig. 37. E, G, p. 254).

***R. leptocarpoides* Benth.** Culm diameter 1 mm.

Leaf not seen.

Culm surface (Fig. 37. G)

Hairs absent. **Papillae** present on overarching anticlinal walls of larger epidermal cells (see Culm T.S.). **Epidermis.** Cells 4-sided, of 2 sizes: (i) large, slightly to 1½–2 times wider than long; walls moderately thickened; cells in longitudinal files, separated by 1–3 files of smaller cells; raised above smaller cells and overarching them; (ii) smaller cells 1½–2½ times longer than wide, often longer than larger cells; walls slightly to moderately thickened. **Stomata** present in grooves among smaller epidermal cells; subsidiary cells with rounded or acute end walls; guard cells with rounded ends. **Silica** absent. **Tannin** in some larger epidermal cells. **Cuticular marks** finely granular.

Culm T.S. (Fig. 37. E)

Outline oval. **Cuticle** thick. **Epidermis.** Cells present in 1 layer, 4-sided, of 2 different sizes. (i) Larger cells as high as to twice as high as wide, extending both to outer and inner sides of small cells, sometimes joining onto ribs from scl. sheath. Outer walls and exposed parts of anticlinal walls v. thick, inner part of anticlinal walls slightly to moderately thickened, the inner part of the cell with parallel sides, or with converging sides and tapering; inner walls slightly to moderately thickened. (ii) Smaller cells, 4-sided, $\frac{1}{4}$–$\frac{1}{5}$ height of larger cells. Slightly higher than wide, mounted *c.* $\frac{1}{2}$–$\frac{1}{3}$ of way down anticlinal walls of larger cells, and overarched by them to the outer side. Smaller cells mainly in rows of 1 or 2, alternating with larger cells; outer walls v. thick, inner and anticlinal walls moderately thickened. **Stomata** sunken, occurring among smaller epidermal cells; guard cells without lips; walls to pore concave.

Chlorenchyma divided into sectors by ribs from scl. sheath; composed of peg-cells in 1 layer; cells 10 to 18 times higher than wide, shorter cells being opposite large epidermal cells; pegs medium-sized, numerous. Protective cells absent. **Parenchyma sheath:** cells in 1 layer, situated between ribs from scl. sheath, cells rounded, smallest on flanks of ribs, widest between them. **Sclerenchyma.** Sheath 3–7-layered; outline circular with v. pronounced, tapering ribs, 3–4 cells wide at inner end, 2 cells wide at outer end, each extending in 3 to 4 layers towards and reaching a large epidermal cell, each opposite to and enclosing a peripheral vb. Fibres of ribs and outer layers of sheath v. thick-walled; inner fibres wider, thick-walled.

Vascular bundles. (i) Peripheral vb's characterized by several medium-sized tracheids arranged in 1- to 2-layered arc, with wide, rounded-angular tracheids on either flank; arc partially enclosing oval phloem pole. (ii) Medullary vb's: outline oval or circular; flanking mxv's medium-sized, angular, with slightly thickened walls, separated by 2–4 rows of narrow cells. Phloem ensheathed, oval. Px present in all bundles. No bundles free from culm scl. sheath. **Bundle sheaths:** fibres v. thick-walled, in 1 or 2 layers at phloem pole; wider, with slightly to moderately thickened walls, in 1 layer on flanks and at xylem pole. **Central ground tissue** parenchymatous, outer cells with slightly or moderately thickened walls. Central cells thin-walled, breaking down to form central cavity. **Silica:** bodies present in many cells of parenchyma sheath. **Tannin** in some larger epidermal cells.

Rhizome and *Root* not seen.

***R. leptocarpoides* var. *monostachyus* F. v. Muell. ex Benth.** Culm diameter 1 by 1·1 mm.

Culm T.S. (only)

As for *R. leptocarpoides*, but larger epidermal cells overhanging smaller ones to greater extent.

RESTIO GROUP 9. Australia

Leaf surface not seen.

Leaf T.S.

(Blade, seen in *R. tremulus* only.)

Outline flattened; 7–8 times wider than thick, with rounded margins. **Epidermis:** cells of abaxial and adaxial surfaces similar, as high as or slightly higher than wide; outer walls thick, anticlinal and inner walls slightly to moderately thickened. **Stomata** superficial, present in both surfaces; similar to those in culm. **Chlorenchyma** composed of peg-cells, each as high as wide or up to twice as high as wide, filling all space between, and in 1–2 layers above and below O.S. of vb's. **Vascular bundles:** 11 in leaf examined, all in line; some bundles fusing in pairs, but with distinct phloem poles; smaller and larger bundles alternating; each bundle with many central, narrow and medium-sized tracheids, some wide, flanking tracheids, and a small, ensheathed phloem pole. **Bundle sheaths:** O.S. parenchymatous, 1-layered, completely encircling or interrupted by fibres at phloem pole; I.S. sclerenchymatous, fibres medium-sized or narrow, thick-walled, in 2–3 layers on flanks and at xylem pole; narrow, v. thick-walled and in up to 4 layers at phloem pole. **Sclerenchyma** represented by I.S. only. **Ground tissue** chlorenchymatous, as described above. **Silica:** (i) medium-sized bodies present in stegmata in outer scl. layer at phloem poles, and (ii) large bodies present in cells of parenchyma sheath at xylem poles. **Tannin:** none seen.

Culm surface

Hairs and **papillae** absent. **Epidermis:** cells 4-sided; 1½–4 or 3–9 times longer than wide; walls slightly to moderately thickened, wavy. **Stomata:** subsidiary cells variable; guard cell pair with lemon-shaped outline. **Silica** absent. **Tannin** in some epidermal cells of *R. applanatus*. **Cuticular marks** coarsely granular; with superimposed longitudinal striations in *R. tremulus*.

Culm T.S.

Outline oval. **Cuticle** thin in *R. applanatus*, thick, with small ridges in *R. tremulus*. **Epidermis:** cells present in 1 layer, 4-sided; all cells as high as wide, or those opposite scl. up to twice as high as wide (*R. tremulus* only); outer walls thick, anticlinal walls straight or slightly wavy, anticlinal and inner walls slightly to moderately or moderately thickened. **Stomata** superficial; subsidiary cells frequently with ridge on outer wall; guard cells without lips in *R. applanatus*, with slight lips at outer and inner apertures in *R. tremulus*. **Chlorenchyma** divided into sectors by ribs from scl. sheath; composed of palisade cells in (1)–2–3–(4) layers; for details of cells see individual spp. Protective cells absent. **Parenchyma sheath** 1-layered, interrupted by scl. ribs; present on flanks of ribs, cells as high as, or 1½ times higher than wide; walls slightly thickened. **Sclerenchyma.** Sheath well developed, 4–5-layered, outline following that of culm with rectangular or wedge-shaped ribs opposite to peripheral vb's, extending towards and often reaching epidermis. Ribs 2–5 layers wide. Fibres v. thick-walled, those of ribs and outer layers of sheath narrow, those of inner layers wider.

Vascular bundles. (i) Peripheral vb's: tracheids medium-sized, sometimes with 1 wider tracheid on either flank, arranged in arc partially enclosing circular phloem pole. (ii) Medullary vb's each with 1 medium-sized or wide, rounded-angular mxv on either flank; phloem pole ensheathed; px absent from outer, smaller bundles; some or no bundles free from culm scl. sheath. **Bundle sheaths:** fibres narrow, v. thick-walled, arranged in 2 or 3 layers at poles, and 1 layer on flanks. **Central ground tissue** parenchymatous, outer cells with moderately thickened walls, central cells thin- or v. thin-walled, breaking down to form central cavity. **Silica:** small bodies present in some cells of parenchyma sheath in *R. tremulus*, and in stegmata in outer layers of scl. ribs in both spp. **Tannin:** in some epidermal cells in *R. applanatus*.

Rhizome and *Root* not seen.

Species Descriptions

***R. applanatus* Spreng.** Culm diameter 1·5 by 2·0 mm.

Culm surface

Epidermis: cells 3–9 times longer than wide, walls moderately thickened, wavy. **Stomata:** subsidiary cells with obtuse end walls, long walls sometimes wavy. **Tannin** in some epidermal cells.

Culm T.S.

Stomata: subsidiary cells with ridge on outer wall. **Chlorenchyma** composed of wide, lobed palisade cells in 2–3 layers, cells irregular (basal material).

***R. tremulus* R. Br.** Culm diameter 1·2 by 3 and 1 by 2 mm.

Culm surface

Epidermis: cells mostly 1½–4 times longer than wide; walls slightly to moderately thickened, wavy. **Stomata**: subsidiary cells with perpendicular or acute end walls, long anticlinal walls slightly wavy. **Tannin** absent. **Cuticular marks** as longitudinal striations.

Culm T.S.

Stomata: guard cells with slight lips at outer and inner apertures. **Chlorenchyma**: cells in 3 or 4 layers (or, occasionally, 1 layer between scl. ribs and epidermis); cells as high as or up to 3 or 4 times higher than wide; pegs medium-sized, numerous.

RESTIO GROUP 10. Australia

As for group 1, but with papillae, and stomata arranged in longitudinal grooves.

***R. fastigiatus* R. Br.** Culm diameter 1–2 mm.

Culm surface (Fig. 37. I, p. 254)

Papillae present on epidermal cells bordering on stomatal grooves; individual papillae overarching grooves. **Epidermis**: cells 4–6-sided; as long as to 2½–3 times longer than wide; walls slightly to moderately thickened, wavy at low focus. **Stomata** in longitudinal files, in grooves; subsidiary cells with acute end walls; guard cell pair narrow, more or less lemon-shaped. **Tannin** in some epidermal cells. **Cuticular marks** granular, sometimes with faint longitudinal striations.

Culm T.S. (Fig. 37. K, p. 254)

Outline circular, with shallow grooves, these often overarched by papillae from adjacent epidermal cells. **Papillae**: see above under Culm surface. **Epidermis**: cells at base of grooves *c.* ¼ height of shortest outer cells, and up to twice as wide as high; those between stomatal grooves as high as to twice as high as wide, those bordering stomatal grooves 2½–3 times higher than wide, with lateral papillae; outer walls v. thick; anticlinal walls thick at outer ends, thickening tapering, and inner ends slightly to moderately thickened. **Stomata**: guard cells without lips. **Chlorenchyma**: peg-cells in 1 layer, except at culm base where locally 2-layered; cells usually 5–7 times higher than wide, those opposite stomata shorter, 2–4 times higher than wide; pegs medium-sized, frequent. **Parenchyma sheath** 1-layered, cells slightly to 1½–2 times wider than high, walls slightly thickened. Stegmata present, with moderately thickened inner and anticlinal walls and thin outer walls. **Sclerenchyma**: sheath 5–9-layered, outline circular, with slight or pronounced ribs opposite to peripheral vb's; fibres v. thick-walled.

Vascular bundles. (i) Peripheral vb's: tracheids medium-sized and narrow, arranged in 1–3-layered arc, partially enclosing oval phloem pole, or with 1 wider tracheid on either flank of arc of narrower tracheids. (ii) Medullary vb's: outline oval to circular, flanking mvx's angular, normally rounded, sometimes oval, medium-sized to wide, walls slightly thickened, separated from one another by 1–3 rows of narrow, thin-walled cells. Phloem pole oval or slightly overarching flanking mxv's, ensheathed by 1–2 layers of narrow cells with slightly thickened walls. Px present in most bundles. Many bundles free from culm scl. sheath. **Bundle sheaths:** fibres thick-walled, medium-sized and/or narrow, in 3–4 layers at phloem pole, 1–2 layers at xylem pole; wider, with moderately thickened walls, in 1 layer on flanks. **Central ground tissue** parenchymatous, outer cells with slightly to moderately thickened walls, central cells thin-walled, breaking down to form cavity. **Silica** present as: (i) bodies in stegmata in parenchyma sheath; (ii) granular material in some cells of central ground tissue.

Taxonomic Notes

Restio has been thought of at various times as containing all or most of the species of the following genera: *Thamnochortus* Berg., *Chondropetalum* Rottb., *Elegia* L., *Calorophus* Labill., *Leptocarpus* R. Br., *Cannomois* Beauv., *Calopsis* Beauv., *Lepidanthus* Nees, *Rhodocoma* Nees, *Craspedolepis* Steud., *Ischyrolepis* Steud., *Baloskion* Rafin., *Chondropetalon* Rafin., *Leieana* Rafin., and *Megalotheca* F. v. Muell. The last 7 of these were still included in *Restio* by Gilg-Benedict (1930).

It is clear from their synonymy that certain species of most genera of Restionaceae have, at one time or another, been included in *Restio* itself. Indeed, it is evident that since the genus was first described there have been many different views concerning its composition. Furthermore, since no comprehensive account of the anatomy of the genus had been published when the present investigation was started, it is not surprising that the anatomical evidence, based on a study of 84 species (out of about 125), has now revealed that *Restio* falls into several distinct groups.

Little or nothing is known about the possibility of interspecific hybridization between the South African species, but the probability of this happening is quite high, because hybridization is known in the *R. gracilis* complex of Australian species (Johnson and Evans 1963*a*).

At present the anatomical descriptions of most of the South African species should be regarded as a source of additional characters to be used in conjunction with morphology when attempting to identify a specimen. It may be possible in the future to find 'key' anatomical characters for individual species.

The clear anatomical distinction between species of *Restio* from Australia and South Africa respectively is of major importance and probably indicates a taxonomic discontinuity within the genus. Both *Hypolaena* and *Leptocarpus* have now been shown to be exclusively Australian, and South African species previously included in these 2 genera have since been transferred to *Mastersiella* and *Calopsis* respectively.

Although the present work has revealed a similar state of affairs for *Restio* in a broad sense, there is the added complication that the name *Restio* has priority of publication for a South African species (Bullock 1959). For this reason and also because their anatomical structure is so different, a new name (or names) will probably be needed for Australian species that can now be regarded only as misfits if they are allowed to remain in *Restio*.

GROUP 1. Australia

This group contains all the species described by Johnson and Evans (1963*a*) in their paper on the *R. gracilis* complex, together with *R. oligocephalus* F. v. Muell., *R. monocephalus* R. Br., and *R. dimorphus* R. Br. These species all show the simplest form of culm organization found in *Restio*. The chlorenchyma is composed of palisade-like peg-cells usually 1-layered in most, but locally 2–3-layered in a few species. It is 2–3-layered in the specimens examined of *R. monocephalus* R. Br., *R. oligocephalus* F. v. Muell., and *R. stenocoleus* L. Johnson and O. Evans.

In their study, Johnson and Evans found stomatal position relative to the culm surface to be diagnostic; stomata are sunken in *R. australis* R. Br. and *R. dimorphus* R. Br. but only slightly sunken in *R. longipes* L. Johnson and O. Evans and superficial in *R. monocephalus* R. Br.

The present author's descriptions of the appearance of the epidermis in surface view and T.S. differ slightly from those published by Johnson and Evans. For example, the anticlinal walls of epidermal cells as seen in surface view are wavy, and the cells longer than wide in the material of *R. pallens* R. Br. from the Kew Herbarium. The walls were stated to be 'not corrugated' and the cells predominantly about as long as wide by Johnson and Evans. Since Johnson himself checked the identity of specimens from the *R. gracilis* complex used for the present study, the possibility of incorrect naming is unlikely. The reason for the variation may be that the material described by Johnson and Evans came from 'the upper region of the culm some distance below the inflorescence' and the material described here came from as near the middle of the culm as possible. Slight variations are known to occur between material taken from different positions on the culm. If this is not the explanation it may be that epidermal characters as defined by Johnson and Evans are not of such diagnostic value in this group as these authors suggest. It is evident that more material of this species should be examined.

Johnson and Evans reported the presence of 'rays' of parenchymatous tissue in the chlorenchyma of *R. pallens* R. Br. and indicated that this feature may be of taxonomic value. The 'rays' could not be found in any of the material examined for this work, and the value of this character must therefore remain dubious.

Johnson and Evans indicated that the undulations of the scl. sheath opposite the peripheral vascular bundles are more prominent in some species than others. The specimens described here do not all conform with the descriptions given by Johnson and Evans. Variations in the sclerenchyma sheath are also known in other species and should not be regarded as being of diagnostic value in the *R. gracilis* complex.

The number of layers of cells in the parenchyma sheath could be diagnostic for individual species, as suggested by Johnson and Evans, but in the author's experience variations in thickness of the order of 1 or 2 layers do not normally appear to be significant.

Briggs's (1963) cytological information for certain *Restio* species is of interest in this group especially as Johnson and Evans correlated it with their observations. Chromosome numbers of $2n = 14$, 22, and 24 are recorded by Briggs for members of the *R. gracilis* complex. *R. monocephalus* R. Br. has $2n = 32$ and *R. dimorphus* R. Br. has $2n = 14$; neither of these is a member of the complex according to Johnson and Evans, but they both share a similar culm anatomy with members of the complex. Briggs found the following range of numbers for *Restio* species outside the complex: $2n = 14$, 22, 24, and *c.* 44.

Since some species outside the *R. gracilis* complex have the same chromosome numbers as those within the complex, the existing cytological evidence does not appear to help in the classification of the *Restio* species of group 1.

It may be that the simple form of anatomy has been retained by several groups of species which have diverged with respect to other characters, including chromosome numbers. If this is so, the present taxonomic status of the species concerned would not be indicated by their anatomy. The simple anatomical organization of the species of this group has already been referred to under *Calorophus* (p. 138). *Calorophus minor* Hk. f. and *C. gracillimus* Hk. f. share many of the anatomical details exhibited by species of *Restio* belonging to group 1. This similarity seems likely to be reliable as evidence of close relationship.

GROUP 2. Australia

One species only, *R. confertospicatus* Steud., is included in this group. The basic anatomical structure of its culm is similar to that of species belonging to group 1, but it differs from them in having hairs and the elongated epidermal cells surrounding the stomata extend into the chlorenchyma. In this last respect it resembles species of *Lepidobolus* Nees, *Loxocarya* spp. excl. *L. pubescens* Benth., *Harperia* Fitz., *Coleocarya* S. T. Blake, and *Onychosepalum* Steud. There is a discussion of the possible interrelationships between these species on p. 159, in the *Coleocarya* section.

GROUPS 3 AND 4. South Africa

Species in these groups differ from those of group 1 in having protective cells. Species belonging to group 4 have culm stomata raised on mounds, while those of group 3 do not. The basic uniformity in anatomical organization indicates that Pillans's delimitation of *Restio* in South Africa is probably accurate. In some instances where 2 or more species are very similar to one another, it is possible that they may be conspecific, but further sampling is necessary before finer points like this can be established from anatomical data.

Members of groups 3 and 4 appear on anatomical grounds to be closely related to '*Leptocarpus*' species from South Africa. There is some evidence that the similarity may be more than superficial. For example, protective cells occur in both genera. However, it is not clear at present whether protective cells have developed independently in different genera, or whether their occurrence affords reliable evidence of the taxonomic affinity between the genera in which they are found.

There are some genera (see p. 291) that are still included in *Restio* as synonyms; among these are the following:

Rhodocoma capense Steud. is now called *Restio rhodocoma* Mast. It shows most of the anatomical characters of *Restio* species belonging to group 3. The islands of thin-walled parenchymatous tissue amongst the ground tissue, which is composed of cells with slightly or moderately thickened walls, are similar to those of *Thamnochortus* species. These by themselves do not provide enough evidence to justify removing the species from *Restio*.

Restio filiformis Poir. includes *Craspedolepis verrauxii* Steud.; it fits well into group 4.

Ischyrolepis Steud., now *R. subverticillatus* Mast., has not been examined.

Individual species in these groups are normally difficult to distinguish on anatomical characters alone, but some of the observations which follow may help in the identification of some specimens.

Culm rounded in all but *R. bolusii* Pillans, *R. callistachyus* Kunth, *R. compressus* Rottb., *R. praeacutus* Mast. and *R. stereocaulis* Mast., where it is oval.

Species with sunken stomata (all from group 3) include *R. callistachyus* Kunth, *R. compressus* Rottb., *R. foliosus* N. E. Br., *R. galpinii* Pillans, and *R. major* (Mast.) Pillans. *R. foliosus* has many diagnostic characters of a *Thamnochortus* species, to which genus it may possibly belong.

R. miser Kunth is the only species which has 1 layer of chlorenchyma (this character may be variable and should be used with care).

Among species in which no spheroidal-nodular silica-bodies were seen are: *R. bifarius* Mast., *R. bifurcus* Nees ex Mast., *R. bolusii* Pillans, *R. brunneus* Pillans, *R. callistachyus* Kunth, *R. egregius* Hochst., *R. filiformis* Poir., *R. fruticosus* Thunb., *R. giganteus* (Kunth) N. E. Br., *R. multiflorus* Spreng., *R. nodus* Pillans, *R. pachystachyus* Kunth, *R. patens* Mast., *R. quinquefarius* Nees, *R. rhodocoma* Mast., *R. saroclados* Mast., *R. stereocaulis* Mast., *R. strobilifer* Kunth, *R. triticeus* Rottb., and *R. zwartbergensis* Pillans.

The following species have epidermal cells with wavy as opposed to straight anticlinal walls as seen in surface view: *R. callistachyus* Kunth, *R. cincinnatus* Mast., *R. curviramis* Kunth, *R. cuspidatus* Thunb., *R. dodii* Pillans, *R. duthieae* Pillans, *R. foliosus* N. E. Br., *R. giganteus* (Kunth) N. E. Br., *R. gossypinus* Mast., *R. laniger* Kunth, *R. leptocladus* Mast., *R. macer* Kunth, *R. micans* Nees, *R. miser* Kunth, *R. monanthus* Mast., *R. nodus* Pillans, *R. patens* Mast., *R. quinquefarius* Nees, *R. rhodocoma* Mast., *R. rottboellioides* Kunth, *R. scaber* Mast., *R. scaberulus* N. E. Br., *R. setiger* Kunth, *R. sieberi* Kunth var. *venustulus* (Kunth) Pillans, *R. stereocaulis* Mast., *R. tenuissimus* Kunth, *R. triflorus* Rottb., *R. vilis* Kunth, and *R. zwartbergensis* Pillans.

GROUP 5. Madagascar

This group contains the only member of the Restionaceae known to occur in Madagascar, *R. madagascariensis* H. Chermez. It is anatomically similar to species belonging to group 3, except that each epidermal cell bears a single papilla. Affinities are therefore apparently with South African rather than Australian species.

GROUP 6. South Africa

There are 2 species in this group, *R. quadratus* Mast. and *R. tetragonus* Thunb. As their names suggest, the culms are 4-sided in cross-section.

They are otherwise anatomically similar to South African species of *Restio* belonging to group 3, except that they have a scl. rib extending from the scl. sheath to the epidermis at each of the 4 angles.

GROUP 7. Australia

Species belonging to this group differ from all others in having pillar cells extending from the ridges of the scl. sheath to the epidermis thereby dividing the chlorenchyma into sectors. Certain species can be recognized anatomically; for example, the culm is oval in T.S. in *R. sphacelatus* R. Br. and *R. complanatus* R. Br., circular in other species; the stomata are sunken in *R. sphacelatus* R. Br. and *R. ornatus* Steud., superficial in other species. Hairs were seen in *R. megalotheca* F. v. Muell. and in *R. nitens* Nees, but in this last species they were seen only at the base of the culm. Pillar cells are present opposite to most ridges of the sclerenchyma sheath in all species except *R. complanatus* R. Br. and *R. gracillior* F. v. Muell. ex Benth. *R. nitens* Nees is unique in group 7 in having the elongated cells next to stomata extending into the chlorenchyma. The culm outline is slightly ridged in *R. grispatus* R. Br., *R. laxus* R. Br., *R. megalotheca* F. v. Muell., and *R. ornatus* Steud., in which the epidermal cells opposite the pillar cells are taller than the others. Because these species with pillar cells exhibit unusual characters for the genus *Restio*, type material of several species has been examined to ensure correct identification. They show a marked anatomical similarity to Australian species of *Leptocarpus* and *Hypolaena*, *Chaetanthus leptocarpoides* R. Br., *Loxocarya pubescens* Benth., and *Meeboldina denmarkica* Suess., the interrelationships of which are discussed on p. 147. It is possible that further investigations may lead to a reduction in the number of genera in this group.

Possible affinities are indicated with species of *Restio* group 9 (q.v.).

GROUP 8. Australia

This group consists of 1 species, *R. leptocarpoides* Benth. and its variety *R. leptocarpoides* var. *monostachyus* F. v. Muell. ex Benth. These plants have certain anatomical features not found in other members of the Restionaceae so far examined, but they are sufficiently similar to warrant their retention in the family. The alternating rows of 1 large and (1)–2–3 small epidermal cells, and the scl. ribs extending to many of the files of large epidermal cells are diagnostic (see p. 287). *R. leptocarpoides* is discussed in connection with *Dielsia* on p. 161. It is suggested that on anatomical grounds the species and its variety should be placed in a new genus.

GROUP 9. Australia

Two species are included in this group, *R. applanatus* Spreng. and *R. tremulus* R. Br. Both have an oval culm outline in T.S. They are distinguished from other members of the genus by having sclerenchyma girders opposite to the peripheral vascular bundles and extending from the sclerenchyma sheath to the epidermis, and they differ from *R. leptocarpoides* Benth. in not having alternating rows of epidermal cells of 2 distinct sizes. There are no clear affinities that can be deduced from the anatomical characters on their own. It is possible that these species may be related to those of *Restio* group 7 if it can be envisaged that they have either lost or have not developed pillar cells.

It is suggested that if no obvious affinities can be found with other members of the family on the basis of other evidence, the 2 species should be made to constitute a new genus.

GROUP 10. Australia

R. fastigiatus R. Br. differs from all other *Restio* species in having its stomata overhung by papillae, and arranged in longitudinal grooves. It differs from *R. leptocarpoides* Benth. in not having epidermal cells of 2 very unequal sizes, and by lacking sclerenchyma girders. Apart from the characters just mentioned, the species is anatomically similar to species of *Restio* belonging to group 1.

There are no obvious anatomical affinities with other genera. If no other evidence can be found to link this species with *Restio* it is suggested that it should be raised to generic rank.

***Restio* : general conclusions**

The discontinuity between species of *Restio* in South Africa and Australia is such that it may be sensible to regard the 2 continental groups as separate at the generic level.

Apart from some minor sub-groupings for which there is anatomical evidence, the South African species appear to form a reasonably natural genus. The Australian species, on the other hand, are so varied in their anatomical structure that they do not seem to constitute a single group. Some of them may eventually be transferred to genera that are already recognized, but it may be necessary to use new generic names for others. For example, the species here included in group 1 (see p. 247) may be found to require a new generic name, since they are very different from all other Australian genera except *Calorophus* in part. Indeed, since, as we have already seen (p. 291),[1] the name *Restio* has priority for a South African species, it is doubtful whether there is justification for retaining any of the Australian taxa in *Restio* itself.

SPORADANTHUS F. v. Muell. ex J. Buchanan

No. of spp. 1; examined 1.

GENUS AND SPECIES DESCRIPTION

Leaf surface (abaxial, scale leaf)

Epidermis: cells hexagonal, *c.* 3 times wider than long; walls v. thick. **Stomata :** none seen.

Leaf T.S. (scale leaf)

Epidermis: (i) adaxial cells 5-sided, 1½ times wider than high; outer walls flat, all walls thick; (ii) abaxial cells *c.* twice as large as those of adaxial surface; twice as wide as high; outer walls v. thick, other walls thick. **Stomata :** none seen. **Chlorenchyma** restricted to groups of several lobed cells, between abaxial epidermis and scl. **Vascular bundles** all small, some with 2–4 narrow tracheids and several phloem cells, others with up to 10 narrow tracheids and 4–6 phloem cells. **Bundle sheaths** sclerenchymatous; fibres thick-walled, in 1–2 layers at xylem pole, wider and in 3–4 layers at phloem pole, continuous

[1] See also Johnson and Evans (1966).

on flanks with sclerenchymatous ground tissue. **Sclerenchyma** in bundle sheaths described above. **Ground tissue** sclerenchymatous. **Silica** and **tannin:** none seen.

Culm surface

Hairs absent. **Epidermis:** cells square or slightly wider than long or up to 1½ times longer than wide; walls thick, occasionally wavy at surface, straight at lower focus; most cells surrounding stomata produced into short, over-arching papillae. **Stomata:** subsidiary cells mostly with rounded or acute end walls, partly obscured by papillae; guard cells with rounded ends. **Silica** and **tannin:** none seen. **Cuticular marks:** longitudinal striations.

Culm T.S. (Fig. 38. A, D)

Outline circular. **Cuticle** thick, with small ridges. **Papillae** present, 1 to a cell, on cells next to stomata, each papilla overarching a stoma. **Epidermis:** cells slightly higher than wide, outer walls thick, straight to slightly convex; anticlinal walls moderately thickened, slightly or not wavy; inner walls moderately thickened. **Stomata** sunken; subsidiary cells with ridge on outer wall; guard cells with pronounced beak-like lip at outer aperture and slight ridge on inner wall. **Chlorenchyma** composed of palisade cells in 2 and occasionally 3 layers; cells interdigitating shortly at junction between layers and separated from one another along their anticlinal walls except for occasional pegs (in T.L.S. cells seen to be arranged in single-layered, longitudinally orientated plates). Protective cells present as modified palisade cells, with moderately thickened walls, extending from epidermis to parenchyma sheath; inner ends of individual cells curving towards centre of substomatal tube, and closing its inner end; walls of tube 2–3 cells thick. (Substomatal cavities bounded by tube of protective cells oval as seen in T.L.S.; several tubes frequently occurring close together or touching one another. Apertures between protective cells leading from substomatal cavity to chlorenchyma small, rounded, and sparsely distributed. **Parenchyma sheath** 2–3-layered; cells mostly rounded-hexagonal, slightly wider than high, with slightly to moderately thickened walls. **Sclerenchyma:** sheath 7–8-layered, outline circular, with slight ribs opposite peripheral vb's; fibres of outer 5–6 layers narrow, v. thick-walled, those of inner layers wider, with thick walls and wider lumina.

Vascular bundles. (i) Peripheral vb's: some v. small, with 1–2 narrow tracheids and several phloem cells; majority with 4–6 angular, medium-sized, thin-walled tracheids arranged in arc partially enclosing phloem pole. (ii) Medullary vb's: some outer bundles with 1 central, medium-sized, angular, mxv; outline of central bundles oval, with 1 wide, angular, mxv with slightly thickened walls on either flank, these separated by 1–4 rows of narrow cells. Lateral wall-pitting of mxv alternate-scalariform; perforation plates simple, oblique (those in outer bundles scalariform, oblique). **Phloem** pole capping and abutting directly onto xylem. Px pole present in all but smaller, outer bundles (Fig. 38. D). **Bundle sheaths:** fibres at poles narrow, moderately thick-walled, in 1 or 2 layers, those on flanks wider, in 1 layer. **Central ground tissue:** cells of outer layers wide, with moderately thickened, lignified walls (some areas of thin-walled cells present), those of inner layers with moderately thickened, cellulose walls. **Silica** and **tannin:** none seen.

Culm T.S. (basal region)

Unusual structural features present in this genus.

Epidermis: cells short and wide, thick-walled. **Hypodermis** several-layered, cells thick-walled. **Chlorenchyma:** cells short, spongy, thin-walled, in many layers, making up $\frac{1}{4}$–$\frac{1}{3}$ of culm diameter. **Parenchyma sheath** 4–5-layered, cells narrow, rounded, walls moderately thickened. **Sclerenchyma:** fibres of sheath present in *c.* 30 layers; those of outer 5–6 layers v. thick-walled, with v. small or no lumina, those of inner 25 or so layers slightly wider, with thick or moderately thick walls. **Vascular bundles.** (i) Peripheral vb's: as for higher culm section (see p. 297). (ii) Medullary vb's: many of outer bundles with 1 central, wide, mxv, others as for higher culm section. **Central ground tissue** parenchymatous; all bundles confined to outer region of ground tissue.

Rhizome T.S. (Fig. 38. C)

Epidermis: cells 4-sided, as high as to 1$\frac{1}{2}$ times higher than wide, outer walls slightly thickened, other walls thin. **Cortex:** (i) outer part composed of 12–14 layers of hexagonal cells; cells mostly 1$\frac{1}{2}$–3 times wider than high, those of outer 4–5 layers thin-walled, those of inner layers moderately thick- to thick-walled; (ii) inner part composed of 10 or 12 layers of rounded, thin-walled cells. **Endodermoid sheath** 1–2-layered, cells irregular, with thick to v. thick inner and anticlinal walls and thin or slightly thickened outer walls. **Vascular bundles** mostly amphivasal, each composed of ring of wide angular vessels with slightly to moderately thickened walls, enclosing rounded phloem pole (Fig. 38. C). **Bundle sheaths** difficult to distinguish from surrounding ground tissue cells. **Central ground tissue** composed of matrix of narrow to medium-sized, hexagonal cells with slightly thickened walls, interspersed with areas of thin-walled cells. **Silica** and **tannin:** none seen.

Root T.S. (Fig. 38. B)

Epidermis and outer **cortex** not seen. **Cortex:** (i) middle part composed of radiating plates of rounded, thin-walled cells; plates separated from one another by air-spaces; each plate 1 cell wide and *c.* 20–30 cells long; (ii) inner part 1–2-layered, cells rounded-hexagonal, intercellular spaces present. **Endodermis** 1-layered; cells slightly higher to 1$\frac{1}{2}$ times higher than wide, inner and anticlinal walls heavily thickened, outer walls thin. **Pericycle** 1–2-layered, cells hexagonal, mostly slightly wider than high, $\frac{1}{3}$–$\frac{1}{4}$ of height of endodermal cells; walls thick, with conspicuous pitting. **Vascular system:** mxv's oval, wide, 21 (in material examined) arranged in 1 ring to inner side of that formed by phloem and px strands. **Central ground tissue** sclerenchymatous; cells hexagonal, with moderately thick to thick walls and conspicuous pitting. **Tannin:** none seen.

Material Examined

Sporadanthus traversii F. v. Muell. ex J. Buchanan: (i) Cheeseman, T. F. (as *Lepyrodia traversii* F. v. Muell. in Kew Herb.); New Zealand (K). (ii) Cheeseman, T. F. 7/79; New Zealand (K). (iii) Cheeseman, T. F. 208; Chanpo Waikato, New Zealand (K). (iv) Butcher, E. W. E., An. 1963; Moana Tua Tua Swamp, nr. Hamilton, New Zealand (S).

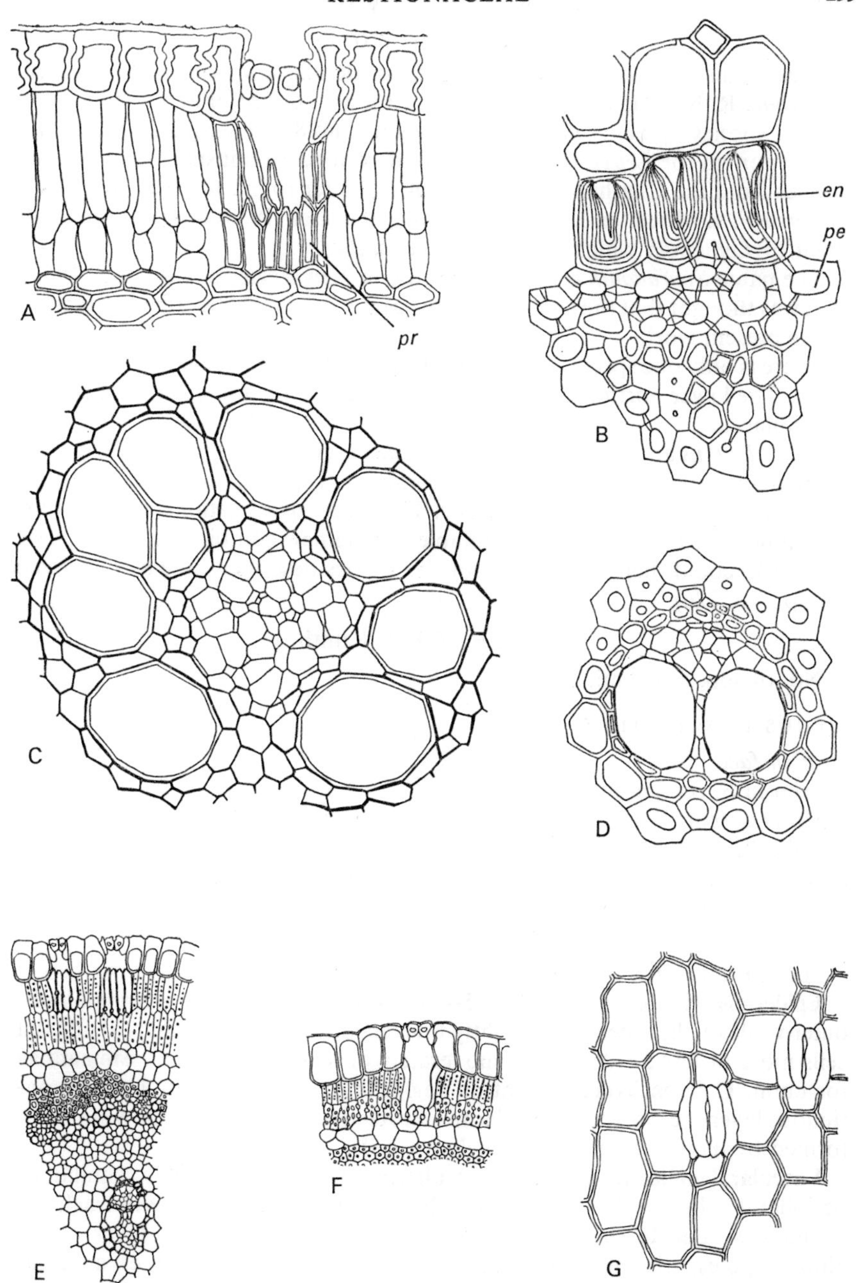

FIG. 38. Restionaceae. A–D, *Sporadanthus traversii*: A, detail of culm T.S.; pr = protective cell (×235); B, detail of root T.S.; en = endodermis, pe = pericycle (×350); C, vascular bundle from rhizome (×235); D, vascular bundle from culm (×225). E–G, *Staberoha*. E, *S. cernua*, detail of culm T.S. (×100). F, G, *S. banksii*: F, detail of culm T.S. (×100); G, culm surface (×200).

Taxonomic Notes

This genus was first erected in 1878, to include a species removed from *Lepyrodia* R. Br.; Masters (1878) described it as a doubtful and insufficiently known species of *Lepyrodia* and Hieronymus (1888) regarded *Sporadanthus* as a synonym for *Lepyrodia.* Gilg (1890) described an unusual '*Leptocarpus*' species from the Chatham Islands which was later found to be *Sporadanthus traversii* F. v. Muell. ex J. Buchanan. Gilg-Benedict (1930) retained the treatment of *Lepyrodia* which includes *Sporadanthus.*

The name *Sporadanthus* is in current use in New Zealand (Campbell 1964), so it was thought that the anatomy of this genus should be described separately in the present work.

There is only one species, *S. traversii* F. v. Muell. ex J. Buchanan. It is a bog plant. The lower portion of the culm is swollen in a manner unusual for the Restionaceae.

Sporadanthus and the species of *Lepyrodia* described here as group 1 (p. 225) share many anatomical characters in common. They are so similar that they could well constitute a natural genus. If further investigations confirm this, the name *Sporadanthus* may be suitable for the new combination. The name *Lepyrodia* could then be retained for the species described here as *Lepyrodia* group 2 (p. 226).

STABEROHA Kunth

No. of spp. 8; examined 5.

Genus Description

Leaf surface

Seen in *S. cernua* only.

Hairs absent. **Epidermis:** (i) abaxial cells oblong, costal longer than intercostal cells; anticlinal walls thin, wavy; (ii) adaxial cells longer and narrower; anticlinal walls straight, end walls oblique. **Stomata** as in culm; present on abaxial surface only.

Leaf T.S.

Seen in *S. cernua*, *S. distachya*, and *S. vaginata.*

Epidermis: (i) abaxial cells 4- or 5-sided, square or slightly wider than high; outer walls thick, other walls slightly thickened; all walls straight; (ii) adaxial cells more or less 5-sided, walls moderately or heavily thickened. **Stomata** as for culm. **Chlorenchyma** present as peg-cells or lobed cells in 1–2 discontinuous layers below abaxial epidermis. Protective cells present below stomata, formed by lignification of chlorenchyma cells.

Vascular bundles of 2 sizes, these alternating. (i) Larger vb's, similar to peripheral culm bundles of corresponding sp. (ii) Smaller vb's, with tracheids all narrow. **Bundle sheaths:** O.S. parenchymatous, cells in 1 layer above phloem poles, discontinuous round bundle flanks, but those of adjacent bundles joined laterally (sheath absent from *S. cernua*); I.S. sclerenchymatous, fibres arranged in up to 7 layers, thick-walled. **Sclerenchyma** present as: (i) fibres of bundle sheaths, just described, and (ii) (in *S. cernua* and *distachya*) a double layer of v. thick-walled fibres lying immediately subjacent to heavily thickened, adaxial epidermis, and joining up bundle sheaths. **Ground tissue**

parenchymatous, present as several layers between parenchyma bundle sheaths and adaxial epidermis in *S. vaginata*, and also between vb's except in *S. cernua*. **Air-spaces:** none present. **Silica** and **tannin:** none observed.

Culm surface (Fig. 38. G)

Epidermis: cells rectangular to elongated hexagonal, elongated axially, walls slightly to moderately thickened, frequently wavy, occasionally straight. **Stomata:** subsidiary cell walls thin to slightly thickened, long anticlinal walls straight or slightly wavy; guard cells thick-walled, apertures sometimes wider at either end but ends usually rounded (Fig. 38. G). **Tannin** in some epidermal cells in certain spp. **Cuticular marks** granular.

Culm T.S.

Outline circular. **Hairs:** none seen. **Cuticle** moderately thick. **Epidermis:** cells square or oblong, all of equal height or some shorter and forming depressions; outer walls thick or v. thick, anticlinal walls straight, slightly or moderately thickened. **Stomata** superficial; subsidiary cell walls extending inwards below guard cells, outer wall protruding as a ridge; guard cells with cuticular lip at outer aperture and sometimes with lip at inner aperture. **Hypodermis** absent. **Chlorenchyma** consisting of peg-cells arranged in 2 layers, of equal or unequal thickness; individual cells mostly 5–6 times higher than wide; protective cells present as: (i) slightly lignified peg-cells, or (ii) more fully differentiated, thick-walled, elongated cells. **Parenchyma sheath** 1, 2, or several cells thick, cell walls slightly or moderately thickened, cells polyhedral in T.S. and oblong in L.S. **Sclerenchyma:** sheath well developed, outline circular, usually clearly distinct from central ground tissue, enclosing all peripheral vb's and some medullary vb's. Individual fibres thick- or moderately thick-walled.

Vascular bundles. (i) Peripheral vb's with 1 medium-sized tracheid on either flank, or with several arranged in an arc and partially enclosing the well-developed phloem pole; phloem abutting directly on xylem. (ii) Medullary vb's of 2 types: type (*a*) characterized by circular phloem pole separated from xylem by 1 layer of parenchyma cells with slightly lignified walls; wall adjacent to phloem frequently unlignified; xylem either with 1 many-sided to circular mxv on either flank, mxv's separated from one another by several rows of narrow vessels, tracheids and parenchyma cells, or xylem with 2–3 medium sized mxv's on either flank (*S. cernua* only); type (*b*) with 1 central, large mxv and a few px elements opposite to the phloem pole; vb's of type (*b*) usually occurring in a ring to inner side of and close to ring of pxv's. **Bundle sheaths** weak, present as caps of thick-walled fibres at phloem poles and, infrequently, also as 1 layer of moderately thick-walled fibres on either flank. **Central ground tissue** parenchymatous, occasionally breaking down in centre of culm. Cell walls slightly thickened. **Silica:** none seen. **Tannin** in epidermis of some spp.

Rhizome T.S.

Seen in *S. cernua* only.

Epidermis: cells more or less square, outer walls slightly to moderately thickened, other walls thin. **Hypodermis:** 1 layer of thin-walled, parenchymatous cells. **Cortex** composed of up to 6 layers of lobed, parenchymatous cells interspersed with larger, isodiametric cells. **Endodermoid sheath** 2–3-layered, cells 5–6-sided, inner walls and inner half of anticlinal walls thickened.

Vascular bundles amphivasal or collateral, phloem poles separated from xylem by single layer of narrow, parenchymatous cells with all walls lignified except those adjacent to phloem cells. Mxv's angular, of 2 sizes: (i) wide, thick-walled, and (ii) medium-sized, with thinner walls. Lateral wall-pitting opposite, scalariform; perforation plates simple, slightly oblique. **Bundle sheaths** completely encircling, 2–3 layers thick; individual fibres thick-walled. **Central ground tissue:** outer ring 4–8 cells wide, sclerenchymatous, forming mechanical cylinder enclosing outer vb's; inner tissue loosely packed, composed of more or less isodiametric, slightly lobed, parenchymatous cells. **Silica** and **tannin** absent.

Root T.S.

Seen in *S. cernua* only.

Root-hairs arising from cells similar in dimensions to those of epidermis; hair shaft narrower than base. **Hypodermis** 1-layered, cells rectangular, slightly wider than high, walls thin. **Exodermis:** 1–3 layers of cells with slightly thickened, lignified walls. **Cortex.** (i) Outer part composed of more or less isodiametric, parenchymatous cells in 1–2 layers. (ii) Middle part with more or less isodiametric cells forming radial plates 1 cell thick, with either much larger, thin-walled cells or air-spaces flanking them. (iii) Inner part, consisting of 1–2 layers of narrow, lobed, parenchymatous cells with intercellular spaces between them. **Endodermis:** 1-layered, cells rectangular, slightly wider than high; outer and anticlinal walls slightly thickened, inner walls moderately but evenly thickened. **Pericycle** 8–10 cells wide; cells parenchymatous, thick-walled, hexagonal (end walls transverse as seen in L.S.); wall-pits simple, circular.

Vascular system in 2 concentric polyarch rings. Phloem in narrow strands, each composed of *c.* 2 sieve-tubes and companion cells and some parenchyma. Strands occurring: (i) in outer ring immediately inside pericycle, (ii) in an inner ring separated from the outer mxv ring by *c.* 4 layers of ground tissue cells, and (iii) 1 in central position. Xylem composed of px groups near to phloem strands of outer ring, and wide, polygonal to rounded mxv's arranged in 2 concentric rings. Lateral wall pitting of vessels scalariform; perforation plates simple, transverse. Individual mxv's surrounded by 1 layer of narrow, flattened, parenchyma cells having slightly lignified walls. **Central ground tissue** parenchymatous, cells thick-walled, mainly starch-filled. **Silica:** small, silica-like particles present in many ground tissue cells. **Tannin:** none seen.

MATERIAL EXAMINED

Staberoha aemula (Kunth) Pillans: mixed sheet 36 (K).

S. banksii Pillans: Burke; Cape Flats (K).

S. cernua (L.f.) Dur. and Schinz.: (i) Burchell, Museum material (M). (ii) Cheadle CA817; Cape (S).

S. distachya (Rottb.) Kunth: Bolus 4239 (K).

S. vaginata (Thunb.) Pillans: Parker, R. N. 4862 (K).

Species Descriptions

***S. aemula* (Kunth) Pillans.** Culm diameter 1·5 mm.

Culm surface not seen.

Culm T.S.

Epidermis: cells more or less square, cell heights varying slightly, outer walls thick. **Stomata:** subsidiary cells as for genus; guard cells without lips. **Chlorenchyma:** protective cells consisting of thickened, slightly elongated peg-cells. **Central ground tissue** parenchymatous, with central cavity.

***S. banksii* Pillans.** Culm diameter 1·3 mm.

Culm surface (Fig. 38. G)

Epidermis: cells rectangular or 6-sided, 1½–2 times longer than wide, anticlinal walls slightly thickened, straight. **Stomata:** subsidiary cells with acute end walls. **Tannin:** not seen.

Culm T.S. (Fig. 38. F)

Epidermis: cells twice as high as wide, outer walls thick, other walls slightly thickened. **Stomata** as in *S. cernua.* **Chlorenchyma:** protective cells v. elongated. **Bundle sheaths** reduced to strands of 4–8 fibres at either pole. **Central ground tissue** parenchymatous, without central cavity.

***S. cernua* (L. f.) Dur. and Schinz.** Culm diameter 1·25 mm.

Culm surface

Epidermis: cells oblong, slightly longer than wide; epidermal cells at either end of stomatal apparatus shortened, giving a pseudo-tetracytic appearance; walls slightly thickened, wavy. **Stomata:** subsidiary cells with wavy long walls. **Tannin** in some epidermal cells.

Culm T.S. (Fig. 38. E)

Epidermis: cells twice as high as wide, some shorter than others; outer walls v. thick, other walls slightly thickened; anticlinal walls straight in (M), some slightly wavy in (CA817). **Stomata:** guard cells with lips at both apertures. **Chlorenchyma:** protective cells formed from slightly elongated peg-cells from outer layer; walls lining substomatal cavity slightly thickened. **Parenchyma sheath** 2–3-layered, cells of 2 sizes, the smaller *c.* ½ size of the larger.

Leaf, *rhizome*, and *root*: as described for the genus.

***S. distachya* (Rottb.) Kunth.** Culm diameter 1·5 mm.

Note: material taken from part of culm enclosed by leaf sheath and therefore probably unrepresentative.

Leaf T.S.: see genus description.

Culm surface

Epidermis: cells oblong; anticlinal walls moderately thickened, wavy. **Stomata:** subsidiary cells slightly longer than guard cells, with rounded end walls. **Tannin:** none seen.

Culm T.S.

Epidermis: cells more or less square, outer walls thick, other walls moderately thickened. **Stomata** as for genus. **Chlorenchyma:** cells mainly in 2 layers, but v. large air-spaces present (see note above); protective cells short. **Parenchyma sheath** composed of 1–several layers of cells; walls slightly thickened. **Bundle sheaths** present as sclerenchymatous caps, 2–3 cells thick. **Central ground tissue** parenchymatous; islands of thin-walled cells present in matrix of cells with slightly thickened walls.

***S. vaginata* (Thunb.) Pillans.** Culm diameter 0·9 mm.

Note: culm sections taken from part of culm enclosed by leaf sheath and therefore probably unrepresentative.

Leaf T.S.

Vascular bundles all small; some with only 1 narrow tracheid on either flank; xylem otherwise composed only of v. narrow tracheids.

Culm surface

Epidermis: cells rectangular, anticlinal walls moderately thickened, wavy. **Stomata:** long anticlinal walls of subsidiary cells wavy.

Culm T.S.

Epidermis: cells slightly wider than high; outer walls v. thick. **Stomata:** subsidiary cells with ridge on outer wall, walls slightly thickened. **Chlorenchyma:** protective cells represented by thickened peg-cells. **Parenchyma sheath** composed of 1 layer of tangentially compressed cells. **Bundle sheaths** represented by scl. bundle caps only.

Taxonomic Notes

Staberoha was accepted by all authors from the date of its description (1841) until 1878, when Masters included it in *Thamnochortus* Berg. Bentham and Hooker (1883) did not agree with this treatment of the genus, and separated the two once more. The 2 genera have been regarded as being distinct from one another since that time. Pillans (1928) gave synonyms in *Thamnochortus* for 3 and in *Restio* for 4 of the 5 *Staberoha* species he described.

Staberoha is anatomically distinct from *Thamnochortus*; it lacks the areas of thin-walled, ground tissue cells scattered amongst the thicker-walled central cells, and also lacks tannin in cells of the central ground tissue which is present in most *Thamnochortus* species. The perforation plates of the conspicuous flanking metaxylem vessels of medullary vascular bundles are oblique and scalariform in *Staberoha*. In *Thamnochortus* they are simple and nearly transverse or transverse. Hairs are unknown in *Staberoha*, but present in 3 species of *Thamnochortus*.

The anatomical structure and organization of the culm in *Staberoha* does not suggest any particular affinities within the family. It is basically similar to that of many South African species of *Restio* (groups 3 and 4) and *Leptocarpus*, and is clearly distinct from *Thamnochortus*.

THAMNOCHORTUS Berg.

No. of spp. *c.* 37; examined 20.

Genus Description

Leaf surface

Seen in *T. dichotomus* (Rottb.) R. Br. only.

Epidermis: cells (abaxial only seen) rectangular, 2–3 times longer than wide, walls wavy. **Stomata:** none seen. **Silica** and **tannin:** none seen.

Leaf T.S.

Seen in *T. dichotomus* (Rottb.) R. Br., *T. erectus* (Thunb.) Mast., and *T. muirii* Pillans.

T.S. middle of leaf

Epidermis: (i) adaxial cells 5-sided, outer walls straight, thick, other walls moderately thickened; (ii) abaxial cells 5-sided, outer walls thick, slightly convex, other walls moderately thickened. **Stomata** present in abaxial epidermis only, superficial; subsidiary cells with ridge on outer wall and elongated inwards; guard cells thick-walled, outer wall sloping towards subsidiary cell, without conspicuous lips or ridges. **Chlorenchyma:** cells lobed, present in 2–3 layers, confined to longitudinal channels between veins and next to abaxial epidermis; cells with slightly thickened walls occurring around substomatal cavities. **Vascular bundles** without wide vessels or tracheids; xylem and phloem poles present, but restricted to few, narrow cells. **Bundle sheaths** sclerenchymatous, usually 1-layered but 3–4-layered in *T. muirii.* **Ground tissue** parenchymatous, cells wide, walls moderately thickened. **Air-cavities** absent. **Silica** and **tannin:** none seen.

T.S. base of leaf

Epidermis: (i) adaxial cells 5-sided, as high as wide, thick-walled, lumina v. narrow; (ii) abaxial cells *c.* 3 times larger than those of adaxial surface, about twice as wide as high, walls thick. **Stomata:** none seen. **Chlorenchyma** as in section through the middle of the leaf but channels narrower. **Vascular bundles** as for middle section. **Bundle sheaths** sclerenchymatous, fibres indistinguishable from ground tissue. **Ground tissue** composed of fibres.

Culm surface

Hairs present in *T. argenteus*, *T. dichotomus* var. *gracilus* and *T. fruticosus*, simple, to *c.* 300 μm long and *c.* 10 μm wide; sides parallel or slightly tapered, ends rounded. **Papillae** short, present in *T. muirii.* **Epidermis:** cells (i) square or hexagonal and as wide as long; walls v. thick, wavy in most spp.; or (ii) rectangular or elongated hexagonal, with transverse end walls; walls moderately thickened, straight or wavy. **Stomata:** subsidiary cells as long as, or slightly longer than the variable guard cells. **Silica:** irregularly shaped inclusions present in some epidermal cells of *T. bachmannii.* **Tannin** frequent in epidermal cells. **Cuticular marks** granular.

Culm T.S. (Fig. 39. A)

Outline circular. **Hairs:** see culm surface. **Cuticle** thick or v. thick. **Epidermis** 1-layered; cells normally 2–4 times higher than wide, 4-sided; outer walls v. thick, often flattened dome-shaped, with slight depressions over anticlinal walls (Fig. 39. A); anticlinal walls wavy, v. thick at outer ends, but thickening tapering; inner walls slightly or moderately thickened; lumina frequently flask-shaped. Cells of uneven heights in some spp. **Stomata** superficial in most spp., slightly sunken in several; subsidiary cells variable; guard cells thick-walled, with cuticular lip at outer aperture (v. pronounced in some spp.), occasionally with lip at inner aperture or ridge on inner wall in some spp. **Chlorenchyma** composed of 2 layers of peg-cells. Pegs small and close together on cells of outer layer, larger and more widely spaced on cells of inner layer. Cells 4–16 times longer than wide. Protective cells extending from epidermal cells, in 2 layers, to parenchyma sheath; cell walls slightly or moderately thickened, cells retaining many characters of adjacent peg-cells. Apertures between protective cells narrowly rounded, in files (as between peg-cells); often confined to inner layer.

Parenchyma sheath frequently 1–2-layered, occasionally 2–3- or 4–6-layered in some spp.; cells wide, often narrowest opposite peripheral vb's and widest between them; walls slightly thickened (cellulose). **Sclerenchyma.** Sheath well developed in all spp., frequently 7–8-layered, sometimes 10–15-layered; sheath with low, dome-shaped ribs, each accommodating 1 peripheral vb in most spp. Outer medullary vb's also embedded in scl. sheath in some spp. Fibres thick-walled; narrow in outer layers, wider in inner layers.

Vascular bundles. (i) Peripheral vb's frequently with several medium-sized to wide, angular tracheids arranged in arc; terminal tracheids on either side of arc sometimes wider than remainder; phloem pole situated in concavity of arc, to outer side of xylem, usually circular or oval in outline. (ii) Medullary vb's with 1 medium-sized or wide, mxv on either flank, these mxv's normally rounded, infrequently angular, separated from one another by 2–3 files of narrow, thin- or thick-walled cells. Lateral wall-pitting of mxv's scalariform; perforation plates simple and more or less transverse. **Bundle sheaths** well developed in all spp.; fibres frequently 2-layered at xylem and phloem poles and 1-layered on bundle flanks, occasionally 4-layered at poles and 1-layered on flanks; walls frequently v. thick. **Central ground tissue** parenchymatous, cell walls slightly to moderately thickened; areas of 2–7 thin-walled cells scattered between medullary vb's. **Silica :** none seen. **Tannin** in some epidermal cells, some cells of parenchyma sheath, and also isolated cells of central ground tissue. (See Fig. 39. G, *Thamnochortus* sp.).

Rhizome T.S.

Seen only in *T. argenteus* (Thunb.) Kunth and *T. fruticosus* Berg.

Epidermis : cells 4-sided, as high as wide or *c.* 1½ times higher than wide; outer walls thick, other walls moderately thickened. **Hypodermis** 1–3-layered, cells of outer 2 layers with v. thick walls, those of inner layer thick (if 1-layered, then walls moderately thickened, e.g. *T. fruticosus*). **Cortex** composed of *c.* 6 layers of parenchymatous, narrow, regularly hexagonal cells, walls thin (sometimes lobed). **Endodermoid sheath :** cells in 1–2 layers, outwardly directed walls thin, other walls thick. **Pericycle :** sclerenchymatous, *c.* 8-layered. **Vascular bundles** scattered throughout ground tissue; some amphivasal, others with 1–2-layered arc of narrow, thick-walled, rounded-angular vessels; phloem abutting directly on xylem. Lumina of many vessels occluded by tannin. **Bundle sheaths** *c.* 4-layered, walls of fibres moderately thickened. **Central ground tissue:** cells narrow, parenchymatous, with moderately thickened walls. **Silica:** none seen.

Root T.S.

Seen in *T. bachmannii* Mast., *T. glaber* (Mast.) Pillans, *T. gracilis* Mast., *T. muirii* Pillans, *T. paniculatus* Mast., and *T. spicigerus* (Thunb.) R. Br.

Root-hairs numerous, arising from cells similar in dimensions to remainder in epidermis (*c.* 1½ times wider than high), hair shaft slightly narrower than basal portion. **Cortex:** (i) outer part composed of 1 layer of more or less square, thin-walled cells; (ii) middle part consisting of 4–6 layers of flattened, thick-walled, parenchymatous cells (outer layer fibrous in *T. glaber*, and with thin outer and thick inner and side walls in *T. paniculatus*); (iii) inner part consisting of 1–2 layers of compressed, thin-walled cells (4 layers of tangentially compressed, thick-walled cells in *T. gracilis*). **Endodermis** 1-layered, outer walls thin, other walls thick. **Pericycle:** fibres in 4–6 layers, walls thick, lumina wide. **Vascular system:** mxv's arranged in 2 rings or scattered (*T. bachmannii*, *T. glaber*, and *T. spicigerus*), rounded or oval, solitary, paired, or in tangential groups. **Central ground tissue** sclerenchymatous; cells thick-walled, lumina wide.

MATERIAL EXAMINED

South African endemic

Thamnochortus argenteus (Thunb.) Kunth: (i) Acocks 18267 ♀, Sheet 2/2, An. 1955; Prince Albert Div. (K). (ii) Gill; Oliphants R. (K). (iii) Cheadle, V. I. CA769, An. 1961; Cape (S).

T. bachmannii Mast.: Esterhuysen 23739 (K).

T. comptonii Pillans: Esterhuysen 3235; Clanwilliam (K).

T. dichotomus (Rottb.) R. Br.: (i) Bolus 4758 (K). (ii) MacOwan 1678 (K). (iii) Pickstone 15 (K).

var. *β consanguineus* Pillans: (i) Burchell 525 (K). (ii) Bolus 2883, An. 1873 (K).

var. *γ gracilis* Pillans: Schlechter 10349 (K).

T. dumosus Mast.: Pickstone (K).

T. erectus (Thunb.) Mast.: (i) 1569, An. 1921 (K). (ii) Bolus 4449 ♂ (as *T. floribundus* in Kew Herb.); Cape Flats (K). (iii) Bolus 12897 ♂ (K).

T. fruticosus Berg.: (i) Garside 1709, An. 1920; hills at foot of Victoria Peak, Jonkershoek, Stellenbosch (K). (ii) Niekerk, G. van, 712, An. 1956; stony hillside, Touws R. (K).

T. glaber (Mast.) Pillans: (i) Pillans 4363 (K). (ii) Burchell 5462; Melville, Knysna Div. (K). (iii) 410 ♂, An. 1954; Bathurst (K).

T. gracilis Mast.: Parker, R. N. 4866 ♀, An. 1953 (K).

T. insignis Mast.: Esterhuysen 24203 (K).

T. levynsae Pillans: Pillans 4925; mts. above Kalk Bay, Cape Peninsula (K).

T. muirii Pillans: 628A; Mossel Bay (K).

T. nutans (Thunb.) Pillans: Bolus 2883; summit of Table Mt. (K).

T. paniculatus Mast.: Esterhuysen 23165 (K).

T. platypteris Kunth: (i) 2502 a (K). (ii) 2502 b ♂ (K).
T. pluristachyus Mast.: no number; on sheet with *T. floribundus* (K).
T. punctatus Pillans: Wolley Dod 2595 (K).
T. scabridus Pillans: Compton 3145, An. 1926; Tweedside, Laingsburg Div. (K).
T. spicigerus (Thunb.) R. Br.: Parker, R. N. 3511 (K).
T. stokoei Pillans: Esterhuysen 11302; summit of Wemmershoek Peak (K).

Species Descriptions

***T. argenteus* (Thunb.) Kunth.** Culm diameter 2·0 to 2·5 mm.

Culm surface

Hairs unicellular, 1 arising from most epidermal cells, up to *c.* 220 μm long and 8–10 μm wide, but tapering to rounded apex; walls slightly thickened, not staining with safranin/haematoxylin mixture, lumina containing substance staining with safranin (Fig. 39. C). **Epidermis:** cells 4- or 6-sided, as long as wide, walls thick, wavy. **Stomata:** subsidiary cells with acute end walls, slightly longer than guard cells. **Tannin** in some cells.

Culm T.S.

Hairs: see Culm surface. **Epidermis:** cells 1½–2 times higher than wide. **Stomata** superficial; subsidiary cells with ridge on outer wall. **Parenchyma sheath** 1–2-layered. **Tannin** in some epidermal cells.

Rhizome T.S.: see genus description.

***T. bachmannii* Mast.** Culm diameter 1 mm.

Culm surface

Epidermis: cells 4–6-sided, mostly as long as wide; walls v. thick, wavy; lumina v. narrow. **Stomata:** subsidiary cells with acute end walls, long anticlinal walls slightly wavy. **Silica:** small bodies present in a few cells. **Tannin:** none seen.

Culm T.S. (Fig. 39. B)

Epidermis: cells 2–3 times higher than wide. **Stomata** slightly sunken. **Parenchyma sheath** 1–2-layered, cells slightly higher than wide. **Tannin** in some cells of central ground tissue.

Root T.S.: see genus description.

***T. comptonii* Pillans.** Culm diameter 1·5 mm.

Culm surface

Epidermis: cells 4–6-sided, mostly as long as wide; walls v. thick, wavy; lumina narrow. **Stomata:** subsidiary cells with acute end walls, long walls slightly wavy. **Silica** and **tannin:** none seen.

Culm T.S.

Epidermis: cells twice as high as wide; outer walls v. thick, anticlinal walls wavy, with tapering thickening. **Stomata** superficial; guard cells with outer lip and ridge on inner wall. **Parenchyma sheath** 1–2-layered. **Tannin** in some cells of central ground tissue.

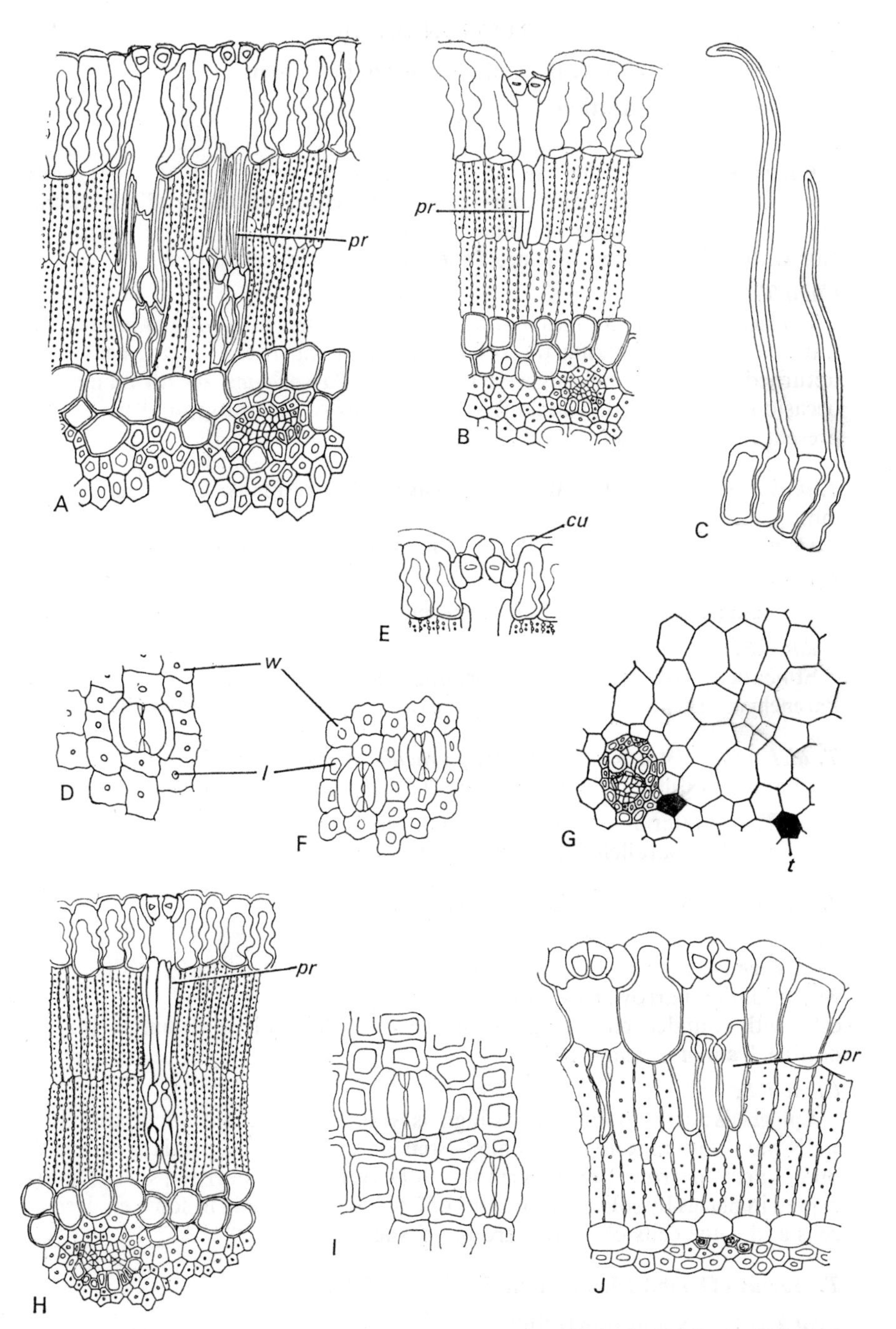

FIG. 39. Restionaceae. A–H, *Thamnochortus*. A, *T. paniculatus*, detail of culm T.S. B, *T. bachmannii*, detail of culm T.S. C, *T, argenteus*, hairs on culm. D, *T. platypteris*, culm surface. E, *T. spicigerus*, stoma; note cuticular lips. F, H, *T. insignis*: F, culm surface; H, detail of culm T.S. G, *Thamnochortus* sp., ground tissue and vascular bundle; note tannin (black) and small area of thin-walled cells. I, J, *Willdenowia teres*: I, culm surface; J, detail of culm T.S. (All × 200.)
cu = cuticle; l = lumen; pr = protective cells; t = tannin; w = cell wall.

***T. dichotomus* (Rottb.) R. Br.** Culm diameter 1–1·5 mm.

Leaf: see genus description.

Culm surface

Epidermis: cells 4-sided, as long as or up to twice as long as wide; walls thick, wavy. **Stomata:** subsidiary cells with acute end walls, all extending beyond guard cells more at 1 end than the other in 1 specimen; guard cell pair with lemon-shaped outline. **Tannin** in some epidermal cells.

Culm T.S.

Epidermis: cells *c.* 3 times higher than wide; outer walls v. thick, anticlinal walls wavy, with tapering thickening. **Stomata** superficial; guard cells with upturned outer lip and ridge on inner wall. **Parenchyma sheath** mainly 1-, occasionally 2-layered. **Tannin** frequent in cells of central ground tissue and present in some epidermal cells.

***T. dichotomus* (Rottb.) R. Br. var. *β consanguineus* Pillans.** Culm diameter 2 mm.

Characters differing from those in *T. dichotomus* (Rottb.) R. Br.:

Culm surface

Stomata: guard cells with rounded ends.

Culm T.S.

Stomata: guard cells with less pronounced outer lip than in *T. dichotomus*. **Parenchyma sheath** 2–3 cells wide, narrow opposite peripheral vb's.

***T. dichotomus* (Rottb.) R. Br. var. *γ gracilis* Pillans.** Culm diameter 1·5 mm.

Characters differing from those in *T. dichotomus* (Rottb.) R. Br.:

Culm T.S. (Material examined from culm base)

Hairs short, parallel-sided, with rounded ends.

***T. dumosus* Mast.** Culm diameter 0·8 mm.

Culm surface

Epidermis: cells as long as to 1½–2 times longer than wide; walls thick, wavy; lumina narrow. **Stomata:** subsidiary cells with acute end walls; guard cells with rounded ends or guard cell pair occasionally with lemon-shaped outline. **Tannin** in many cells.

Culm T.S.

Epidermis: cells about twice as high as wide; outer walls v. thick; anticlinal walls wavy with tapering thickening. **Stomata** superficial; subsidiary cells with outer part v. narrow. **Parenchyma sheath** 1-layered, cells narrow; sheath interrupted in places by solitary or groups of 2 or 3 fibres. **Tannin** in some epidermal cells and some cells of central ground tissue.

***T. erectus* (Thunb.) Mast.** Culm diameter 2–2·5 mm.

Leaf T.S.: see genus description.

Culm surface

Epidermis: cells 4-sided, about as long as wide, except in (1569), where up to 4 times longer than wide; walls thick to v. thick, wavy. **Stomata:** subsidiary cells slightly longer than guard cells, with acute end walls. **Silica:** silica-like deposits present in occasional depressions in outer walls of epidermal cells. **Tannin:** none seen.

Culm T.S.

Epidermis: cells twice as high as wide, some isolated cells shorter than others (slightly wider than high, some with short papillae, in (1569)); outer walls thick; anticlinal walls v. wavy. **Stomata** superficial. **Parenchyma sheath** 1–3-layered, widest between peripheral vb's; cells wide (outer cells slightly lignified in (1569)). **Tannin** in some cells of parenchyma sheath and central ground tissue.

***T. fruticosus* Berg.** Culm diameter 2 and 2·5 mm.

Culm surface

Hairs unicellular, straight or curved, sides parallel for most of length, tapering to apex; up to 300 μm long and *c.* 10 μm wide; walls thick, not staining; lumina narrow, contents staining red with safranin. **Epidermis:** cells 4-sided, as long as, or up to twice as long as wide, walls moderately thickened to thick, slightly wavy. **Stomata:** subsidiary cells variable. **Tannin** in some epidermal cells. **Cuticular marks** granular.

Culm T.S.

Hairs: see Culm surface. **Epidermis:** cells 2–3 times higher than wide; outer walls thick, anticlinal walls wavy. **Stomata** superficial; subsidiary cells with ridge on outer wall; guard cells with outer and inner lips. **Parenchyma sheath** 1-layered opposite to peripheral vb's, 2-layered between them; cells wide. **Tannin** in some cells of epidermis, parenchyma sheath and central ground tissue.

Rhizome T.S.: see genus description.

***T. glaber* (Mast.) Pillans.** Culm diameter *c.* 2 mm.

Culm surface

Epidermis: cells 4- or 6-sided, as long as wide; walls v. thick, wavy. **Stomata:** subsidiary cells with acute end walls. **Tannin** in some epidermal cells.

Culm T.S.

Outline circular in (410) and (5462), semicircular in (4363) (a lateral branch). **Epidermis:** cells as high as wide (410, 4363) or 2–3 times higher than wide (5462); outer walls v. thick or thick, curving inwards slightly at cell margins; anticlinal walls wavy. **Stomata** superficial; guard cells with lips at outer and inner apertures. **Parenchyma sheath** 2-layered opposite to peripheral vb's, 3-layered between them; cells wide. **Tannin** in some epidermal and ground tissue cells.

Root T.S.: see genus description.

***T. gracilis* Mast.** Culm diameter 1·5 mm.

Culm surface

Epidermis: cells hexagonal or slightly irregular, mostly as long as wide; walls v. thick, wavy. **Stomata:** subsidiary cells slightly longer than guard cells, with acute end walls. **Tannin** in some cells.

Culm T.S.

Epidermis: cells mostly 3 times higher than wide; outer walls v. thick, slightly convex; anticlinal walls wavy, with rapidly tapering thickening. **Stomata** superficial; guard cells with lips at outer and inner apertures. **Parenchyma sheath** 1–2-layered; cells wide. **Tannin** in some cells of epidermis and central ground tissue.

Root T.S.: see genus description.

***T. insignis* Mast.** Culm diameter 2 mm.

Culm surface (Fig. 39. F)

Epidermis: cells 4–6-sided, mostly about as long as wide; walls v. thick, wavy. **Stomata:** subsidiary cells with acute end walls. **Tannin** in some epidermal cells.

Culm T.S. (Fig. 39. H)

Epidermis: cells twice as high as wide; outer walls thick, curving inwards slightly at margins; anticlinal walls wavy, with tapering thickening. **Stomata** superficial; guard cells with pronounced cuticular lip at outer aperture and slight ridge on inner wall. **Chlorenchyma:** peg-cells longer and narrower than in most spp. **Parenchyma sheath** 1- or 2-layered opposite to peripheral vb's, 2-layered between them; cells wide. **Tannin** in some cells of epidermis, parenchyma sheath, and central ground tissue.

***T. levynsae* Pillans.** Culm diameter 2 mm.

Culm surface

Epidermis: cells 6-sided, slightly longer than wide; walls v. thick, wavy. **Stomata:** subsidiary cells with acute end walls, but with narrow central part. **Tannin:** none seen.

Culm T.S.

Epidermis: cells twice as high as wide; outer walls thick, slightly convex, curving inwards at cell margins; anticlinal walls wavy, with tapering thickening. **Stomata** superficial; guard cells with outer and inner lips. **Parenchyma sheath** 2–3-layered, cells up to twice as high as wide. **Sclerenchyma.** Sheath composed of fibres of 2 types: (i) majority, thick-walled (innermost layers gelatinous) with narrow lumina; (ii) scattered, with moderately thickened walls and wide lumina. **Tannin:** none seen.

***T. muirii* Pillans.** Culm diameter 2 mm.

Culm surface

Papillae short, present on some epidermal cells. **Epidermis:** cells hexagonal, mostly as long as wide, walls v. thick, wavy. **Stomata:** subsidiary cells with perpendicular end walls. **Tannin** in some epidermal cells.

Culm T.S.

Epidermis: cells 3 times higher than wide; outer walls v. thick, curving inwards slightly at margins; anticlinal walls wavy, with tapering thickening. **Stomata** superficial; guard cells with pronounced lip at outer aperture and ridge on inner wall. **Parenchyma sheath** 1-layered opposite peripheral vb's, 2-layered between them. **Tannin** in some epidermal cells.

Leaf and *root T.S.*: see genus description.

***T. nutans* (Thunb.) Pillans.** Culm diameter 2 mm.

Culm surface

Epidermis: cells hexagonal, as long as wide; walls v. thick, wavy. **Stomata:** subsidiary cells with acute end walls, long wall slightly curved towards guard cell. **Tannin:** none seen.

Culm T.S.

Epidermis: cells *c.* 3 times higher than wide; outer walls v. thick, curving inwards at cell margins; anticlinal walls wavy, with tapering thickening. **Stomata** superficial; guard cells with lips at outer and inner apertures. **Parenchyma sheath** 1–2-layered opposite peripheral vb's, 3-layered between them, walls slightly thickened, those of outer layers mainly cellulose, those of inner layer slightly lignified. **Tannin:** none seen.

***T. paniculatus* Mast.** Culm diameter 2 mm.

Culm surface

Epidermis: cells 4- or 6-sided, as wide as long; walls v. thick, wavy at surface only. **Stomata:** subsidiary cells with acute end walls. **Tannin** in some epidermal cells.

Culm T.S. (Fig. 39. A)

Epidermis: cells 3–4 times higher than wide; outer walls v. thick, curving inwards at margins; anticlinal walls v. wavy, v. thick, with tapering thickening. **Stomata** superficial; guard cells with lips at outer and inner apertures. **Parenchyma sheath** 1-layered opposite to peripheral vb's, 2-layered between them. **Tannin** in some cells of epidermis, parenchyma sheath, and central ground tissue.

Root T.S.: see genus description.

***T. platypteris* Kunth.** Culm diameter 1·5 and 1·0 mm.

Culm surface (Fig. 39. D)

Epidermis: cells 4- or 6-sided, as long as wide; walls v. thick, wavy. **Stomata:** subsidiary cells slightly irregular, sometimes longer than guard cells, long anticlinal wall wavy, end walls acute. **Tannin:** none seen.

Culm T.S.

Epidermis. Cells of 2 main types: (i) those next to stomata, about twice as high as wide; (ii) those between stomata, 1½ times higher than wide. Outer walls thick to v. thick; anticlinal walls wavy, with tapering thickening. **Stomata** superficial, raised on taller epidermal cells; guard cells with lip at outer aperture. **Parenchyma sheath:** 1-layered opposite peripheral vb's, 1–2-layered between them. **Tannin** in some cells of central ground tissue.

***T. pluristachyus* Mast.** Culm diameter 2 mm.

Culm surface

Epidermis: cells 4–6-sided, as long as wide; walls v. thick, wavy. **Stomata:** subsidiary cells slightly irregular, mostly with acute end walls. **Tannin:** none seen.

Culm T.S.

Epidermis. Cells of 2 sizes: (i) more frequent type 4–5 times higher than wide; (ii) less frequent type not situated next to stomata; 3 times higher than wide. Outer walls v. thick, curving inwards slightly at margins; anticlinal walls wavy, v. thick at outer ends, tapering, and slightly thickened at inner ends; inner walls slightly thickened. **Stomata** superficial; guard cells with outer and inner lips. **Chlorenchyma:** pegs of inner cells larger and less frequent than those on outer cells. **Parenchyma sheath:** cells mainly 1-, occasionally 2-layered, up to twice as high as wide. **Tannin** scattered in cells of central ground tissue.

***T. punctatus* Pillans.** Culm diameter 1·3 mm.

Culm surface

Epidermis: cells 4–6-sided, as long as wide; walls v. thick, wavy. **Stomata:** subsidiary cells mostly with acute end walls. **Tannin:** none seen.

Culm T.S.

Epidermis: cells mostly 3–4 times higher than wide, some, in small areas, slightly shorter; outer walls v. thick, straight; anticlinal walls wavy, with tapering thickening. **Stomata** slightly sunken; guard cells with pronounced outer lips. **Parenchyma sheath** mostly 1-layered opposite peripheral vb's and 2-layered between them. **Tannin:** none seen.

***T. scabridus* Pillans.** Culm diameter 1 mm.

Culm surface

Epidermis. Cells of 2 sizes: (i) 4-sided, as long as wide, situated next to stomata; (ii) 4–6-sided, up to 1½ times longer than wide; walls v. thick, wavy. **Stomata:** subsidiary cells with wavy long wall. **Tannin** in some epidermal cells.

Culm T.S.

Epidermis. Cells of 2 types: (i) those next to stomata, *c.* 3 times higher than wide; (ii) those between stomata, about twice as high as wide; cells of intermediate height also present. Outer walls v. thick; anticlinal walls with tapering thickening. **Stomata** superficial; subsidiary cells with ridge on outer wall; guard cells with lip at outer aperture. **Parenchyma sheath** mainly 1-layered, occasionally 2-layered; cells 1½–2 times higher than wide. **Tannin** in some cells of epidermis and central ground tissue.

***T. spicigerus* (Thunb.) R. Br.** Culm diameter 2·5 mm.

Culm surface

Epidermis: cells 4- or 6-sided, as long as wide; walls v. thick, wavy. **Stomata:** subsidiary cells with slightly or v. wavy long wall. **Tannin:** none seen.

Culm T.S. (Fig. 39. E)

Cuticle v. thick. **Epidermis:** cells mostly 3 times higher than wide; outer walls v. thick, anticlinal walls v. wavy, thickening tapering. **Stomata** slightly sunken; guard cells with exceptionally well-developed, hooked, cuticular lip at outer aperture and slight lip at inner aperture (Fig. 39. E). **Chlorenchyma:** peg-cells longer than in most other species. **Parenchyma sheath** 2–3-layered opposite peripheral vb's, 5–6-layered between them; some cells several times larger than others. **Tannin** in some cells of central ground tissue.

Root T.S.: see genus description.

***T. stokoei* Pillans.** Culm diameter 1 mm.

Culm surface

Epidermis: cells 4–6-sided, as long as wide; walls thick, wavy. **Stomata:** subsidiary cells with slightly or v. wavy long walls. **Tannin:** none seen.

Culm T.S.

Epidermis: cells 3½–4 times higher than wide; outer walls v. thick, curving inwards at margins; anticlinal walls v. wavy, v. thick at outer ends, thickening tapering abruptly, and inner $\frac{1}{6}$ only slightly thickened; inner walls slightly thickened. **Stomata** superficial; guard cells with outer and inner lips. **Parenchyma sheath** 2-layered opposite peripheral vb's, 3-layered between them, cells slightly to 1½ times higher than wide. **Tannin** in some cells of central ground tissue.

TAXONOMIC NOTES

Thamnochortus has been included in *Restio* on 2 occasions since its description in 1767. It has been regarded as a genus in its own right since 1836, but Masters included *Staberoha* in it. Gilg-Benedict (1930) found that the two groups into which the genus was previously divided on gross morphological characters by Pillans can also be distinguished from one another anatomically.

The species of *Thamnochortus* taken collectively have a distinctive anatomy and the individual species differ from one another in minor details only (see below). Although sharing some features with South African species of *Restio* belonging to groups 3 and 4 (notably protective cells, 1-layered epidermis), *Thamnochortus* differs from them on several points. The individual epidermal cells in *Thamnochortus* have wavy walls and frequently the ratio of height to width in these cells is higher than in *Restio*. Occasional *Restio* species, such as *R. foliosus* N. E. Br., have scattered tannin-containing cells and areas of thin-walled cells in the central ground tissue. Similar thin-walled cells are found in all, and the scattered tannin cells in many *Thamnochortus* species. Although the occurrence of tannin in species of other families and other genera of Restionaceae is often very variable, it does seem to be a constant character of many *Thamnochortus* species. The anatomical evidence is not strong for close affinity between *Restio* species belonging to groups 3 and 4 and *Thamnochortus*. It is possible, however, that *Restio* species exhibiting characters similar to those of *Thamnochortus* may be incorrectly classified.

There are several anatomical characters that are diagnostic for certain species of *Thamnochortus*. Unicellular hairs (unlike any found in other restionaceous plants) are present in *T. argenteus* (Thunb.) Kunth and *T. fruticosus* Berg. and in material from the base of the culm in *T. dichotomus* (Rottb.) R. Br. var. *gracilis* Pillans.

T. bachmannii Mast. is unique in the genus in having spheroidal-nodular silica-bodies in some epidermal cells.

Epidermal cells are of uniform height as seen in T.S. in all except the following species, which have some areas of shorter cells between the stomata: *T. erectus* (Thunb.) Mast., *T. platypteris* Kunth, *T. pluristachyus* Mast., *T. punctatus* Pillans, and *T. scabridus* Pillans.

Stomata are superficial in all species except *T. spicigerus* (Thunb.) R. Br.

T. bachmannii Mast., *T. punctatus* Pillans, where slightly sunken. *T. spicigerus* has pronounced beak-like cuticular lips at the outer stomatal aperture; the lips are short in *T. bachmannii*, so these 2 species can be readily separated.

Now that more anatomical information is available about *Thamnochortus* it appears that the two groups recognized by Gilg-Benedict (1930) are not very well defined. It is possible to separate species with hairs, or with straight anticlinal walls to the epidermal cells as seen in T.S., from the remaining species as Gilg-Benedict did. However, other distinctive anatomical characters shared by all species seem to indicate that *Thamnochortus* should be regarded as a single taxon.

WILLDENOWIA Thunb.

No. of spp. 15; examined 8.

Genus Description

Leaf surface

Scale leaf, seen in *W. argentea* only.

Epidermis: cells (adaxial) hexagonal, 2–3 times wider than long; walls v. thick. **Stomata:** none seen.

Leaf T.S.

Seen in *W. arescens*, *W. argentea*, *W. lucaeana*, and *W. teres*.

Epidermis: (i) adaxial cells 4-sided, flattened, *c.* 3 times wider than high; walls thick, lumina v. narrow; (ii) abaxial cells 4-sided, as high as wide (twice as high as wide in *W. arescens*); outer walls thick or v. thick, slightly dome-shaped in *W. lucaeana*; anticlinal walls straight or wavy, moderately thickened; inner walls moderately thickened. **Stomata** superficial (sunken in *W. arescens*), present in abaxial epidermis only, except in *W. lucaeana*; guard cells with lips at outer and inner apertures in *W. arescens*, v. flattened and elongated in *W. lucaeana* and *W. teres*. **Chlorenchyma** composed of 1 or 2 layers of peg-cells with few pegs; layers continuous (e.g. *W. arescens*) or interrupted by scl. bundle sheaths. Protective cells present as 1 layer of shorter cells, with moderately thickened walls, encircling substomatal cavity; apertures present at outer end of substomatal tube. **Parenchyma** present as 1 interrupted layer of rounded cells separating chlorenchyma from scl. or ground tissue.

Vascular bundles similar to peripheral vb's of culm of corresponding sp. **Bundle sheaths** sclerenchymatous; fibres in 1–2 layers at phloem pole and xylem pole, continuous on flanks with those of ground tissue (except in *W. arescens*, where 1–2-layered). **Sclerenchyma** present in bundle sheaths and as (1)–3–5 layers of wide, thick-walled fibres between bundles in all spp. except *W. arescens*. **Ground tissue** parenchymatous, cells present in 2–3 layers between scl. and adaxial epidermis in *W. lucaeana*, and between bundle sheaths in *W. arescens*. **Air-cavities** absent. **Silica:** spheroidal-nodular bodies present in some parenchyma cells. **Tannin:** none seen.

Culm surface

Hairs and **papillae** absent. **Epidermis:** cells square, or rectangular and up to 3 times longer than wide, or hexagonal and 2–3 times wider than long, or as long as wide; walls wavy at surface, often straight at lower focus; all walls moderately thickened or all thick or v. thick. **Stomata:** subsidiary cells variable, as long as, or slightly shorter or longer than guard cells; guard cell pair with oval or widely oval outline, individual guard cells sometimes with rounded ends. **Silica:** none seen. **Tannin** in epidermal cells of some spp. **Cuticular marks** granular.

Culm T.S.

Outline circular. **Cuticle** thick. **Epidermis:** cells 1-layered; as high as to 2–3 times higher than wide; outer walls frequently v. thick; anticlinal walls wavy or straight, slightly to moderately thickened or thick; inner walls slightly to moderately thickened. **Stomata** superficial; subsidiary cells variable; guard cells sometimes with lip at outer aperture, sometimes with lip at inner aperture or with ridge on inner wall. **Hypodermis** absent. **Chlorenchyma** composed of peg-cells in (1)–2–3 layers, individual cells normally 4–6 times higher than wide; pegs frequently sparse except on radial walls. Chlorenchyma layers in some spp. divided into sectors by up to 30 radiating scl. ribs and their associated pillar cells. Protective cells represented by (i) small, moderately thick-walled cells, with rounded apertures, surrounding substomatal cavity and extending inwards through 1 chlorenchyma layer only or (ii) longer, moderately thick-walled cells extending from epidermis, in 1 layer, to parenchyma sheath, or only part way into inner chlorenchyma layer. Elongated apertures present between cells, mainly in outer half of substomatal tube (present only in spp. without pillar cells).

Parenchyma sheath 1-layered (occasionally 2-layered in places), cells medium-sized, with slightly thickened walls; pillar cells present in some spp., extending from ribs radiating from scl. sheath, to epidermis. **Sclerenchyma.** Sheath of 3 distinct types. (i) Outline circular; 3–6-layered, enclosing all peripheral vb's and some or no medullary vb's; all fibres thick- or v. thick-walled. (ii) Outline circular; 4–10-layered, with all peripheral vb's and some medullary vb's embedded in it. Fibres in the 2–3 layers on the outer side of the peripheral vb's and sometimes those on flanks of and to inner side of peripheral vb's with slightly to moderately thickened walls, other fibres thick- or v. thick-walled. (iii) Outline circular, with dome-shaped or rectangular ribs 2–4 fibres wide and 4–14 fibres high, extending from scl. sheath to epidermis, or to pillar cells, and dividing chlorenchyma into up to *c.* 30 sectors. Portions of sheath between ribs 3–7 fibres thick. All peripheral vb's embedded in sheath but alternating with ribs, some medullary vb's also embedded. Fibres thick- or v. thick-walled; secondary thickening of some inner fibres gelatinous.

Vascular bundles. (i) Peripheral vb's: tracheids (narrow) medium-sized or wide, arranged in arc partially enclosing circular or oval phloem pole; occasionally with 1 or 2 wider tracheids on either flank (*W. sulcata*); alternating with ribs from scl. sheath (if present). (ii) Medullary vb's: some outer bundles with 1 wide, central mxv, most inner bundles with 1 medium-sized to wide mxv on either flank; mxv's with scalariform or alternate lateral wall pitting and oblique, scalariform perforation plates (simple perforation plates in *W. striata* and *W. sulcata*). Phloem frequently abutting directly on and over-arching flanking mxv's, ensheathed by 1 layer of narrow, thin-walled cells in

W. striata. Px present in all but smallest, outer bundles. All spp. with some medullary vb's scattered in central ground tissue.

Bundle sheaths: fibres narrow, with moderately thickened or thick walls, in 2–3 layers at phloem pole; those on flanks wider, with slightly or moderately thickened walls, in 1–2 layers, or absent; those at xylem pole narrow, with slightly to moderately thickened walls, in 1 or 2 layers. **Central ground tissue** parenchymatous; outer cells, surrounding scattered medullary vb's, with slightly or moderately thickened walls; occasionally with areas of thin-walled cells, scattered amongst vb's; central cells thin-walled, often breaking down to form central cavity. **Silica** present in some spp. as: (i) spheroidal-nodular bodies in some cells of parenchyma sheath, or in some outer cells of scl. sheath; (ii) granular material, filling lumina of some cells of central ground tissue. **Tannin** in some epidermal cells of certain spp.

Rhizome T.S. not seen.

Root T.S.

Seen in *W. arescens*.

Cortex not seen. **Endodermis:** 1-layered; inner walls and inner half of anticlinal walls thick, other walls thin. **Pericycle:** cells narrow, thick-walled, arranged in *c.* 8 layers. **Vascular system:** mxv's wide, oval, solitary, arranged in 1 ring. **Central ground tissue:** most cells thick-walled; central cells thin-walled.

MATERIAL EXAMINED

Willdenowia arescens Kunth: Acocks, J. P. H. 17410, An. 1953; Calvinia, Cape Prov. (K).

W. argentea (Nees) Hieron.: Burchell, W. J. ?7659 ♀ '*Ceratocaryum argenteum* Kunth' (M).

W. fimbriata Kunth: Schlechter 7447, An. 1897, Pillans, N. S. 7905; Baviaans Kloof, Caledon Div. (K).

W. humilis Mast.: (i) Schlechter 489 ♀ '*Hypodiscus dodii* Mast.'; Cape Flats (K). (ii) Wolley Dod, A. H. 2578 ♂ (K).

W. lucaeana Kunth: Esterhuysen 1968, An. 1940; Bain's Kloof (K).

W. striata Thunb.: (i) Howes, F. N. 179; Piquetberg, Cape Prov. (K). (ii) '*W. ecklonii* Endl.', '*Nematanthus ecklonii* Nees ab Esenb.' (K).

W. sulcata Mast.: (i) Esterhuysen 7472, An. 1941; Cederberg Mts. (K). (ii) Burchell, W. J. Herb. 787 ♂, An. 1811 (K).

W. teres Thunb.: (i) Esterhuysen 23143; Mowbray-Faure Rd., Cape Prov. (K). (ii) Parker, R. N. 3527, An. 1940; nr. Faure (K).

SPECIES DESCRIPTIONS

***W. arescens* Kunth.** Culm diameter 1 mm.

Culm surface not seen.

Culm T.S.

Epidermis: cells of varying heights; shortest about twice as high as wide, tallest *c.* 3 times higher than wide, with intermediate sizes; anticlinal walls wavy or straight. **Stomata:** guard cells with outer and inner lips. **Chlorenchyma:** cells in 2 layers completely encircling culm; pegs not seen, cells arranged in alternating, transverse plates as seen in L.S.; protective cells present in single outer layer. **Parenchyma sheath:** cells of 2 main sizes, with intermediates; widest opposite peripheral vb's, narrowest between them. **Sclerenchyma:** sheath of type (ii) as described for the genus. **Central ground tissue** devoid of outer areas of thin-walled cells. **Silica:** bodies present in some smaller cells of parenchyma sheath, and in some outer cells of scl. sheath. **Tannin:** none present.

Leaf and *root*: see genus description.

W. argentea **(Nees) Hieron.** Culm diameter 3 mm.

Leaf surface and *T.S.*: see genus description.

Culm surface (Fig. 40. B)

Epidermis: cells 4–6-sided, up to 3 times wider than long; walls thick. **Stomata:** subsidiary cells with rounded end walls; guard cells with rounded ends. **Tannin:** none present.

Culm T.S. (Fig. 40. D)

Epidermis: cells mostly 2 or 3 times higher than wide, occasional cells shorter; outer walls thick; anticlinal walls wavy with tapering thickening. **Stomata:** guard cells with ridges on outer and inner walls. **Chlorenchyma:** cells in 2 layers; protective cells elongated, extending in 1 or 2 layers from *c.* $\frac{1}{4}$ of way up substomatal cavity lined by epidermal cells, almost to parenchyma sheath. **Parenchyma sheath** 1-layered, cells mainly as high as wide and of uniform size. **Sclerenchyma:** sheath 5–6-layered, of type (ii) as described for the genus. **Silica:** bodies present in some cells of parenchyma sheath. **Tannin:** none present.

W. fimbriata **Kunth.** Culm diameter 3 mm.

Culm surface

Epidermis: cells hexagonal, mostly twice as wide as long; walls v. thick, lumina narrow. **Stomata:** subsidiary cells mainly with obtuse end walls; guard cell pair with oval outline. **Tannin:** none present.

Culm T.S.

Epidermis: cells all *c.* 3 times higher than wide; anticlinal walls wavy. **Stomata:** subsidiary cells with ridge on outer wall; guard cells with outer and inner lips. **Chlorenchyma** consisting of 2 layers of cells. Protective cells well developed, extending in 1 layer from *c.* $\frac{1}{4}$ of way up substomatal cavity lined by epidermal cells, almost to parenchyma sheath. **Parenchyma sheath:** cells in 1 layer, mostly as high as wide, largest opposite peripheral vb's. **Sclerenchyma:** sheath 5–8-layered, of type (ii) described for the genus. **Bundle sheaths** poorly developed, composed of cap of 1 layer of narrow, thick-walled fibres at phloem pole; cells on flanks and at xylem pole similar to those of ground tissue. **Silica:** bodies present in some cells of outer layer of scl. sheath. **Tannin:** none seen.

W. humilis **Mast.** Culm diameter 1 mm.

Culm surface (Fig. 40. C)

Epidermis: cells 4–6-sided, as long as to twice as long as wide; walls thick,

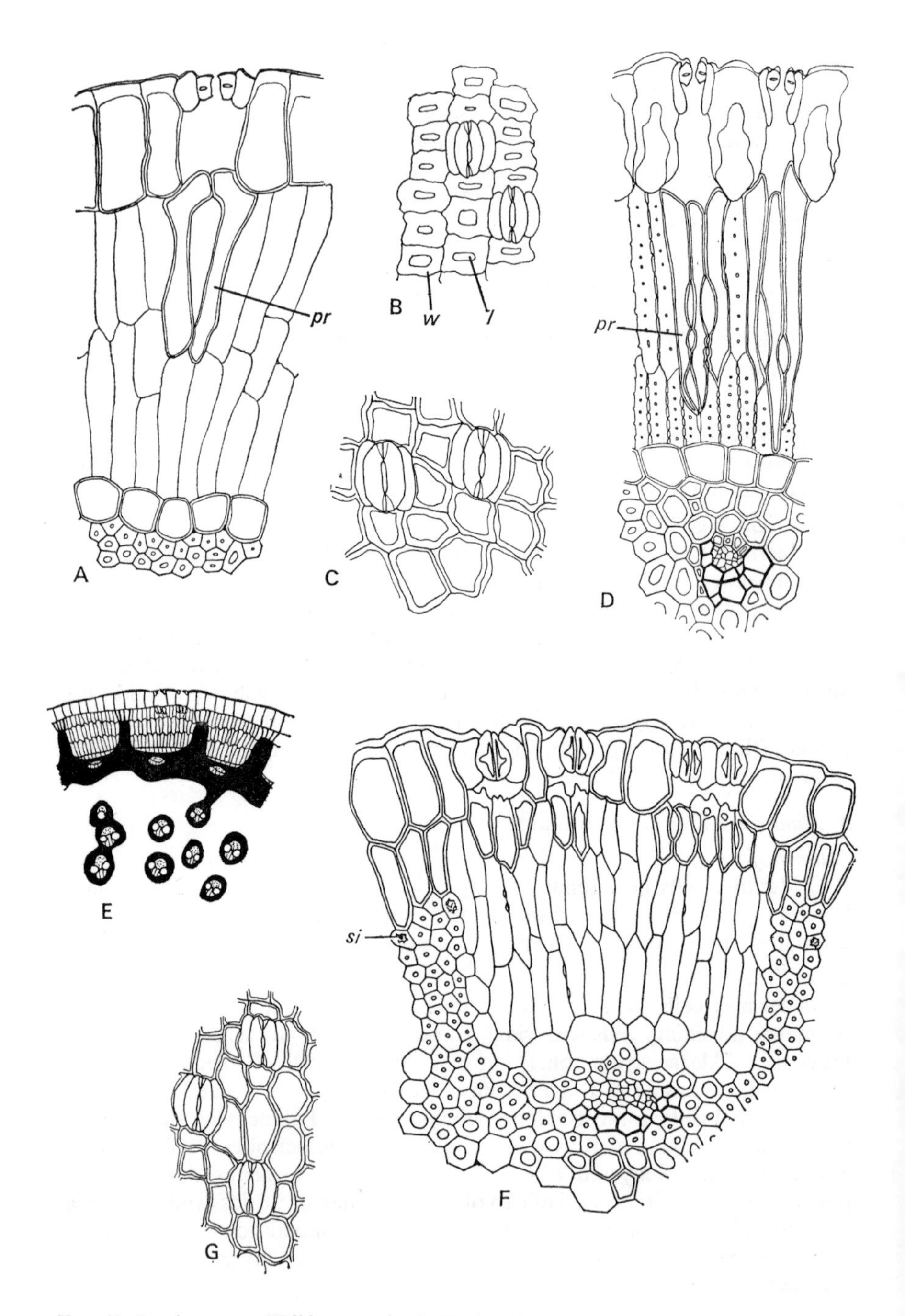

FIG. 40. Restionaceae. *Willdenowia*. A, C, *W. humilis*: A, detail of culm T.S. (×200); C, culm surface (×200). B, D, *W. argentea*: B, culm surface (×200); D, detail of culm T.S. (×200). E, G, *W. striata*: E, diagram of sector of culm T.S. (×45); G, culm surface (×200). F, *W. striata*, forma *monstrosa*: detail of culm T.S. (×200).
l = lumen; pr = protective cells; si = silica-body; w = cell wall.

wavy at surface, straight at lower focus. **Stomata:** subsidiary cells mainly with obtuse end walls and long anticlinal walls occasionally slightly wavy; guard cell pair with widely oval outline. **Tannin:** none present.

Culm T.S. (Fig. 40. A)

Epidermis: cells 2–3 times higher than wide; some anticlinal walls wavy. **Stomata:** guard cells with outer and inner lips. **Chlorenchyma:** cells in 2 layers; protective cells short, consisting of little-modified outer palisade cells. **Parenchyma sheath** 1-layered, cells mostly as high as wide and of uniform size. **Sclerenchyma:** sheath 4–5-layered, of type (i) described for the genus. **Central ground tissue** devoid of areas of thin-walled cells in outer region. **Silica:** bodies present in some cells of outer layer of scl. sheath. **Tannin** in some vessels and tracheids (pathological).

W. lucaeana **Kunth.** Culm diameter 1 mm.

Leaf T.S.: see genus description.

Culm surface

Epidermis: cells 4-sided, as long as to 2, or occasionally 3, times longer than wide, smallest cells next to stomata; walls thick, not wavy. **Stomata:** subsidiary cells with acute end walls, guard cell pair with oval outline. **Tannin:** in some cells.

Culm T.S.

Epidermis: cells mostly 1½ times higher than wide; outer walls slightly concave; anticlinal walls occasionally wavy with rapidly tapering thickening. **Stomata:** guard cells with outer and inner lips. **Chlorenchyma:** cells in 2–3 layers. **Parenchyma sheath** 1-layered, cells as wide as high, circular; cells opposite peripheral vb's wider than those opposite low, dome-shaped ribs from scl. sheath. **Central ground tissue** devoid of areas of thin-walled cells in outer regions. **Silica:** none present. **Tannin** in some epidermal cells.

W. striata **Thunb.** Culm diameter 2 mm.

Culm surface (Fig. 40. G)

Epidermis: cells 4–6-sided, as long as or up to twice as long as wide, with shorter cells next to stomata; walls moderately thickened to thick, wavy at surface, straight at lower focus. **Stomata:** subsidiary cells with acute or obtuse end walls, long wall sometimes slightly wavy; guard cells with rounded ends, or guard cell pair with lemon-shaped outline. **Tannin** in some epidermal cells.

Culm T.S. (Figs. 40. E, F; Pl. VI. B)

Epidermis: cells 1½–3 times higher than wide, those next to stomata slightly shorter than others, longest cells opposite to scl. ribs; anticlinal walls straight or slightly wavy. **Stomata:** guard cells with lip at outer aperture and ridge on inner wall. **Chlorenchyma** composed of (1)–3 layers of cells, divided into *c.* 30 sectors by ribs from scl. sheath and their associated pillar cells. **Parenchyma sheath** composed of 2 different types of cell: (i) pillar cells, in longitudinal files of 2–4, connecting ribs from scl. sheath to epidermis; individual cells 2–4 times higher than wide, with slightly to moderately thickened walls; (ii) rounded-hexagonal, thin-walled cells, arranged in 1–(2) layer(s), cells sometimes narrow on inner ends of flanks of scl. ribs and wider next to cylindrical part of scl. sheath. **Sclerenchyma:** sheath 4–7-layered, with pronounced ribs of type (iii) as described for the genus; ribs rectangular, 2–4 cells wide and 8–14 cells high, radiating at more or less regular intervals from cylindrical part of sheath. **Central ground tissue** devoid of areas of thin-walled cells in outer region. **Silica:** bodies present in some cells flanking scl. ribs. **Tannin** in some epidermal cells.

***W. sulcata* Mast.** Culm diameter 1, 2 mm.

Culm surface

Epidermis: cells 4–6-sided, as long as to 2–3 times longer than wide, with shorter cells next to stomata; walls moderately thickened, wavy at surface, straight at lower focus. **Stomata:** subsidiary cells with acute or obtuse end walls; guard cells with rounded ends. **Tannin** in some epidermal cells.

Culm T.S.

Epidermis: cells as high as wide (787) or 2–3 times higher than wide (7472), tallest opposite scl. ribs; anticlinal walls more wavy in (7472) than (787). **Stomata:** guard cells with lip at outer aperture and ridge on inner wall. **Chlorenchyma** and **parenchyma sheath** as in *W. striata.* **Sclerenchyma:** sheath as in *W. striata,* but ribs in (787) slightly shorter, and only 4–6 cells high. **Central ground tissue, silica** and **tannin**; as in *W. striata.*

***W. teres* Thunb.** Culm diameter 1 mm.

Leaf T.S.: see genus description.

Culm surface (Fig. 39. I, p. 309)

Epidermis: cells 4-sided, as long as or up to twice as long as wide, with the smaller cells next to stomata; walls moderately thickened to thick, wavy. **Stomata:** subsidiary cells with perpendicular to acute end walls and often with slightly wavy long anticlinal walls; guard cell pair with lemon-shaped outline. **Tannin:** none present.

Culm T.S. (Fig. 39. J)

Epidermis: cells 2–3 times higher than wide; anticlinal walls v. wavy in (3527), not so wavy in (23143). **Stomata:** guard cells with pronounced beak-like lip at outer aperture and ridge on inner wall. **Chlorenchyma** composed of 2 uninterrupted layers of cells with some pegs on lateral walls. Protective cells present as modified palisade cells of outer layer. **Parenchyma sheath** mainly 1-, occasionally 2-layered, cells slightly wider than high. **Sclerenchyma:** sheath 3–4-layered, of type (i) as described for the genus. **Central ground tissue** devoid of areas of thin-walled cells in outer region. **Silica:** bodies present in some cells of outer layer of scl. sheath. **Tannin:** none present.

Taxonomic Notes

Willdenowia has been recognized by all authorities since it was described in 1790. Several genera have been included with it as synonyms from time to time. For example, Pillans (1928) had *Nematanthus* Nees and *Ceratocaryum* Nees in the synonymy for the genus, and stated that several *Willdenowia* species have synonyms in *Restio,* and one species is synonymous with a *Hypodiscus* species. Gilg-Benedict (1930) regarded *Ceratocaryum* Nees as a good genus, but retained *Nematanthus* Nees in *Willdenowia* as did Pillans, and she also included *Spirostylis* Mast. in *Willdenowia*. She stated that 2 anatomical groups exist in *Willdenowia*; one has sclerenchyma ribs in the culm, the other has not.

The present investigation has confirmed Gilg-Benedict's opinion that the species of *Willdenowia* fall into 2 groups. Group 1 contains *W. striata* Thunb. and *W. sulcata* Mast., species characterized by sclerenchyma ribs extending to pillar cells, as mentioned on p. 317. These species are very similar to *Hypodiscus* species here included in group 1. Species belonging to group 2 of *Willdenowia* have a sclerenchyma sheath with low, dome-shaped ridges or a rounded outline and they lack pillar cells. These species are *W. arescens* Kunth, *W. argentea* (Nees) Hieron., *W. fimbriata* Kunth, *W. humilis* Mast., *W. lucaeana* Kunth, and *W. teres* Thunb. Of group 2, *W. arescens* and *W. humilis* have no peg-cells and the chlorenchyma consists of palisade cells, but in the remaining species belonging to this group, peg-cells are well developed.

W. striata Thunb., under which Pillans includes *Nematanthus ecklonii* Nees, has a very striking anatomy. As just noted it is similar to that of *W. sulcata* Mast., and almost identical with that of *Hypodiscus argenteus* (Thunb.) Mast. as indicated on p. 187, and it is also very similar to several other species of *Hypodiscus* belonging to group 1. This complex will need to be examined further in order to work out a more natural classification. When this is undertaken *Leucoploeus* Nees, *Lepidanthus* Nees, and *Nematanthus* Nees will all have to be considered because one or more of these names may need to be used for new combinations.

The species of *Willdenowia* that have synonyms in *Ceratocaryum* Nees are *W. argentea* (Nees) Hieron. and *W. fistulosa* Pillans. Of these, only the first has been examined anatomically. It seems to fit well with other species of *Willdenowia* belonging to group 2. There is insufficient anatomical evidence to justify following Gilg-Benedict by treating *Ceratocaryum* as a distinct genus. Pillans's treatment of the species seems more worthy of support.

It is clear that *Willdenowia* and *Hypodiscus* need extensive revision.

Restionaceae. Lists of genera and species in which certain diagnostic characters occur

Hairs present on culm

Chaetanthus leptocarpoides
Hypolaena exsulca
H. fasciculata
H. fastigiata
Lepidobolus chaetocephalus
L. preissianus
Leptocarpus aristatus
L. brownii
L. canus
L. chilensis
L. coangustatus
L. disjunctus
L. erianthus
L. simplex
L. spathaceus
L. tenax
Loxocarya cinerea
L. fasciculata
L. flexuosa
L. pubescens
L. vestita
L. virgata
Meeboldina denmarkica
Restio confertospicatus
R. megalotheca
R. nitens
Thamnochortus argenteus
T. dichotomus var. *gracilis* (basal)
T. fruticosus

Anticlinal walls of epidermal cells markedly wavy as seen in T.S. culm

Calorophus elongatus
Cannomois parviflora
C. scirpoides
Coleocarya gracilis
Elegia cuspidata
E. glauca
Harperia lateriflora
Hypodiscus argenteus
H. synchroolepis
Lepidobolus chaetocephalus
L. drapetocoleus
L. preissianus
Leptocarpus burchellii
L. esterhuysianae
L. membranaceus
Lepyrodia anarthria
L. caudata
L. drummondiana
L. flexuosa
L. glauca
L. gracilis
L. heleocharoides
L. hermaphrodita
L. leptocaulis
L. macra
L. monoica
L. monoica var. *foliosa*
L. muelleri
L. muirii
L. scariosa
L. tasmanica
L. valliculae
Loxocarya cinerea
L. flexuosa
L. virgata
Lyginia barbata
L. tenax
Mastersiella browniana
M. hyalina
M. laxiflora
Onychosepalum laxiflorum
Restio bifarius
R. bifidus
R. bifurcus
R. bolusii
R. brunneus
R. curviramis
R. dispar
R. dodii
R. duthieae
R. filiformis
R. foliosus
R. fraternus
R. fruticosus
R. laniger
R. macer
R. marlothii
R. monanthus
R. multiflorus
R. ocreatus
R. pachystachyus
R. patens
R. purpurascens
R. quinquefarius
R. rhodocoma
R. rottboellioides
R. scaber
R. scaberulus
R. setiger
R. sieberi
R. sieberi var. *venustulus*
R. sphacelatus
R. stenocoleus
R. tetragonus
R. triticeus
R. vilis
R. virgeus
Thamnochortus argenteus
T. bachmannii
T. comptonii
T. dichotomus and vars.
T. dumosus
T. erectus
T. fruticosus
T. glaber
T. gracilis
T. insignis
T. levynsae
T. muirii
T. nutans
T. paniculatus
T. platypteris
T. pluristachyus
T. punctatus
T. scabridus
T. spicigerus
T. stokoei
Willdenowia arescens
W. argentea
W. fimbriata
W. humilis
W. teres

Culm epidermis 2- or more-layered

Chondropetalum aggregatum [(2)
C. deustum (2)
C. ebracteatum (2)
C. hookerianum (2)
C. macrocarpum (2–3)
C. marlothii (2)
C. microcarpum (2)
C. mucronatum (2)
C. nudum (1–2)
C. rectum (2–3(–4))
C. tectorum (1–2)
Elegia asperiflora (2)
E. coleura (2(–3))
E. cuspidata (2)
E. equisetacea (2)
E. fistulosa (2–3)
E. galpinii (2)
E. glauca (2)
E. grandis (2(–3))
E. intermedia (2)
E. juncea (2)
E. muirii (2)
E. neesii (2)
E. obtusiflora (2)
E. racemosa (2)
E. spathacea (2)
E. squamosa (2)
E. stipularis (2)
E. thyrsoidea (2)
E. vaginulata (2)
E. verreauxii (2(–3–4))
E. verticillaris (1–2)
(*Restio callistachyus*)

Epidermal cells next to stomata of culm extending inwards

Coleocarya gracilis
Harperia lateriflora
Lepidobolus chaetocephalus
L. drapetocoleus
L. preissianus
Leptocarpus tenax
Lepyrodia caudata
L. gracilis
L. tasmanica
Loxocarya cinerea
L. fasciculata
L. flexuosa
L. vestita
L. virgata
Onychosepalum laxiflorum
Restio confertospicatus
R. nitens
R. stenocoleus

Stomata of culm raised on tall epidermal cells

Leptocarpus asper
L. impolitus
Loxocarya fasciculata
Mastersiella diffusa
M. hyalina
Restio cincinnatus
R. duthieae
R. festuciformis
R. filiformis
R. macer
R. monanthus
R. patens
R. purpurascens
R. pygmaeus
R. scaber
R. scaberulus
R. triticeus
R. zwartbergensis
Thamnochortus platypteris
T. scabridus

Culm stomata sunken

( ) = slightly sunken

Cannomois nitida
C. parviflora
C. scirpoides
(*Chaetanthus leptocarpoides*)
(*Chondropetalum aggregatum*)
C. macrocarpum
(*C. marlothii*)
C. microcarpum
C. rectum
Elegia fistulosa
(*E. galpinii*)
E. glauca
E. grandis
E. muirii
E. neesii
E. obtusiflora (or superficial)
(*E. vaginulata*)
Hopkinsia anoectocolea
Hypodiscus neesii
(*Hypolaena fasciculata*)
(*Leptocarpus aristatus*)
L. brownii
(*L. canus*)
L. chilensis
L. coangustatus
L. simplex
L. tenax
*Lepyrodia glauca**
(*L. muirii*)
Lyginia barbata
L. tenax
Restio australis
R. compressus
R. dimorphus
R. foliosus
R. galpinii
R. leptocarpoides
R. leptocarpoides var. *monostachyus*
(*R. longipes*)
(*R. madagascariensis*)
R. major
R. ornatus
R. sphacelatus
Sporadanthus traversii
(*Thamnochortus bachmannii*)
(*T. punctatus*)
(*T. spicigerus*)

* Subsidiary cells superficial but guard cells mounted low on them.

Chlorenchyma cells in 1 layer as seen in culm T.S.

Calorophus gracillimus
C. minor (1–2)
Coleocarya gracilis
Harperia lateriflora
Hypodiscus neesii (1, 2, or 3)
Hypolaena crinalis (1–2)
Lepidobolus chaetocephalus
L. drapetocoleus
L. preissianus
Lepyrodia glauca
L. heleocharoides
Loxocarya cinerea
L. fasciculata (1–2)
L. flexuosa (1–2)
L. virgata
Onychosepalum laxiflorum
Restio complanatus (1–2)
R. compressus
R. confertospicatus
R. dimorphus
R. fastigiatus (occ. 2)
R. festuciformis (1–2)
R. gracilis
R. gracillior (1–2)
R. grispatus (1–2)
R. leptocarpoides
R. leptocarpoides var. *monostachyus*
R. madagascariensis
R. major (1–2)
R. miser
R. pallens (1–2)
R. stereocaulis
R. vilis (1–2)

Protective cells present in culm

Calorophus elongatus
Cannomois acuminata
C. drègei
C. nitida
C. parviflora
C. scirpoides (rachis)
C. virgata
Chondropetalum aggregatum
C. andreaeanum
C. capitatum
C. chartaceum
C. deustum
C. ebracteatum
C. hookerianum
C. macrocarpum
C. marlothii
C. microcarpum
C. mucronatum
C. nitidum
C. nudum
C. paniculatum
C. rectum
C. tectorum
Elegia asperiflora

E. coleura
E. cuspidata
E. equisetacea
E. fistulosa
E. galpinii
E. glauca
E. grandis
E. intermedia
E. juncea
E. muirii
E. neesii
E. obtusiflora
E. racemosa
E. spathacea
E. squamosa
E. stipularis
E. thyrsoidea
E. vaginulata
E. verreauxii
E. verticillaris
Hypodiscus albo-aristatus
H. alternans
H. argenteus
H. aristatus
H. binatus
H. neesii
H. striatus
H. synchroolepis
H. willdenowia
Hypolaena crinalis
H. graminifolia
H. purpurea
H. spathulata
Leptocarpus asper
L. burchellii
L. esterhuysianae
L. impolitus
L. membranaceus
L. parkeri
L. rigoratus
L. vimineus
Lepyrodia anarthria
L. caudata
L. flexuosa
L. glauca
L. gracilis
L. hermaphrodita
L. interrupta
L. leptocaulis
L. macra
L. muirii
L. scariosa
L. tasmanica
Lyginia barbata
L. tenax
Mastersiella browniana
M. diffusa
M. digitata
M. hyalina
M. laxiflora
Restio aridus
R. bifarius
R. bifidus
R. bifurcus
R. bolusii
R. brunneus
R. callistachyus
R. cincinnatus
R. compressus
R. curviramis
R. cuspidatus
R. dispar
R. dodii
R. duthieae
R. egregius
R. festuciformis
R. filiformis
R. foliosus
R. fourcadei
R. fraternus
R. fruticosus
R. galpinii
R. giganteus
R. gossypinus
R. laniger
R. leptocladus
R. macer
R. madagascariensis
R. major
R. marlothii
R. micans
R. miser
R. monanthus
R. multiflorus
R. nodus
R. ocreatus
R. pachystachyus
R. patens
R. pedicellatus
R. praeacutus
R. purpurascens
R. pygmaeus
R. quadratus
R. quinquefarius
R. rhodocoma
R. rottboellioides
R. sarocladus
R. scaber
R. scaberulus
R. setiger
R. sieberi
R. sieberi var. *venustulus*
R. stereocaulis
R. strobilifer
R. tenuissimus
R. tetragonus
R. triflorus
R. triticeus
R. vilis
R. virgeus
R. zwartbergensis
Sporadanthus traversii
Staberoha aemula
S. banksii
S. cernua
S. distachya
S. vaginata
Thamnochortus argenteus
T. bachmannii
T. comptonii
T. dichotomus
T. dichotomus var. *consanguineus*
T. dichotomus var. *gracilis*
T. dumosus
T. erectus
T. fruticosus
T. glaber
T. gracilis
T. insignis
T. levynsae
T. muirii
T. nutans
T. paniculatus
T. platypteris
T. pluristachyus
T. punctatus
T. scabridus
T. spicigerus
T. stokoei
Willdenowia arescens
W. argentea
W. fimbriata
W. humilis
W. lucaeana
W. striata
W. sulcata
W. teres

Pillar cells present in culm

Chaetanthus leptocarpoides
Hypodiscus argenteus
H. striatus
Hypolaena exsulca
H. fasciculata
H. fastigiata

Leptocarpus aristatus	*L. spathaceus*	*R. laxus*
L. brownii	*L. tenax*	*R. megalotheca*
L. canus	*Loxocarya pubescens*	*R. nitens*
L. chilensis	*Meeboldina denmarkica*	*R. ornatus*
L. coangustatus	*Restio amblycoleus*	*R. sphacelatus*
L. disjunctus	*R. chaunocoleus*	*Willdenowia striata*
L. erianthus	*R. complanatus*	*W. sulcata*
L. ramosus	*R. gracillior*	
L. simplex	*R. grispatus*	

Species in which the culm sclerenchyma cylinder has girders (G), well-developed partial girders (P) or ribs (R—extending to epidermis; Q—not reaching epidermis)

Anthochortus ecklonii	R	*Mastersiella laxiflora*	Q (very low)
Cannomois acuminata	Q	*Phyllocomos insignis*	R
C. virgata	Q	*Restio applanatus*	G, P
Dielsia cygnorum	G	*R. leptocarpoides*	G, P
Hypodiscus albo-aristatus	Q	*R. leptocarpoides* var. *monostachyus*	G, P
H. alternans	Q	*R. quadratus*	R
H. argenteus	Q	*R. tetragonus*	R
H. aristatus	Q	*R. tremulus*	G, P
H. binatus	Q	*Willdenowia striata*	Q
H. neesii	Q	*W. sulcata*	Q
H. striatus	Q, R		
H. synchroolepis	Q		
H. willdenowia	Q, R		

Note: Girders and partial girders are, by definition, opposite to vascular bundles, in this case peripheral vascular bundles. Ribs alternate with vascular bundles. All must penetrate the parenchyma sheath to qualify for inclusion in the above list. In many spp. 'ribs' or 'ridges' are present on the sclerenchyma cylinder opposite to the peripheral vascular bundles, but these do not penetrate the parenchyma sheath. Species with pillar cells have what might be termed partial girders; these are in the preceding list.

Silica-bodies present in culm epidermis

Harperia lateriflora	*L. hermaphrodita*	*L. muirii*
Lepyrodia anarthria	*L. leptocaulis*	*L. scariosa*
L. drummondiana	*L. macra*	*L. valliculae*
L. flexuosa	*L. monoica*	*Thamnochortus bachmannii*
L. glauca	*L. monoica* var. *foliosa*	
L. heleocharoides	*L. muelleri*	

Granular silica present in culm epidermis

Lepyrodia caudata	*L. stricta*	*L. tasmanica*

Silica-bodies present in culm (excluding epidermis)

C = in inner chlorenchyma layer

P = in parenchyma sheath; S = outer layer of sclerenchyma cylinder.

Anthochortus ecklonii	P	*Cannomois acuminata*	S
Calorophus elongatus	P	*C. drègei*	S
C. gracillimus	P	*C. nitida*	S
C. minor	P	*C. parviflora*	S

C. scirpoides	S
C. virgata	S
Chaetanthus leptocarpoides	P
Coleocarya gracilis	P
Dielsia cygnorum	P
Harperia lateriflora	S
Hypodiscus alternans	S
H. argenteus	S
H. binatus	S
H. neesii	S
H. striatus	S
H. synchroolepis	P
H. willdenowia	S
Hypolaena crinalis	S
H. exsulca	S
H. fastigiata	S
H. graminifolia	S
H. purpurea	S
H. spathulata	S
Lepidobolus chaetocephalus	P
L. drapetocoleus	P
L. preissianus	P
Leptocarpus aristatus	S
L. brownii	S
L. canus	S and pillar cells
L. chilensis	S
L. coangustatus	S and pillar cells
L. disjunctus	S
L. erianthus	S
L. impolitus	protective cells
L. membranaceus	C
L. parkeri	P
L. ramosus	S
L. simplex	S
L. spathaceus	S
L. tenax	P
Lepyrodia glauca	P
Loxocarya cinerea	P
L. fasciculata	S
L. flexuosa	S
L. pubescens	S
L. vestita	P
L. virgata	P
Mastersiella digitata	S
Meeboldina denmarkica	S
Onychosepalum laxiflorum	P
Phyllocomos insignis	P
Restio amblycoleus	P
R. applanatus	S
R. aridus	P
R. australis	S
R. chaunocoleus	P
R. cincinnatus	P
R. complanatus	P
R. confertospicatus	P
R. curviramis	P
R. cuspidatus	P
R. dimorphus	P
R. dodii	P and C
R. duthieae	C
R. fastigiatus	P
R. festuciformis	P and C
R. fimbriatus	P
R. fourcadei	P
R. fraternus	P
R. gossypinus	P
R. gracilis	S
R. gracillior	P
R. grispatus	P
R. laniger	P
R. laxus	P
R. leptocarpoides	P (and var. *monostachyus*)
R. leptocladus	P
R. longipes	S
R. macer	P
R. marlothii	P
R. megalotheca	P
R. micans	P
R. monanthus	C
R. monocephalus	P
R. oligocephalus	P
R. ornatus	P
R. pallens	P and S
R. pygmaeus	P
R. rottboellioides	P and C
R. scaberulus	P
R. setiger	P
R. sieberi	P and central ground tissue
R. sieberi var. *venustulus*	P
R. sphacelatus	P and S
R. stenocoleus	P
R. tenuissimus	P
R. tetragonus	P
R. tremulus	P and S
R. triflorus	P
R. vilis	P
R. virgeus	P
Willdenowia arescens	P and S
W. argentea	P
W. fimbriata	S
W. humilis	S
W. striata	S
W. sulcata	S
W. teres	S

BIBLIOGRAPHY FOR RESTIONACEAE

ARBER, A. (1922) Leaves of the Farinosae. *Bot. Gaz.* **74**, 80–94.
—— (1925) *Monocotyledons. A morphological study*. Cambridge University Press.
BACKER, K. (1957) Restionaceae in *Flora Malesiana* ser. 1, **5**, 416–20.
BAILEY, F. M. (1902) *The Queensland flora*, Vol. 6, pp. 1720–6.
BAILLON, H. E. (1878) *Hist. Pl., Genève* **12**, 6–7.
BEAUVOIS, P. (1828)—see DESVAUX, M.
BENTHAM, G. (1878) *Flora Australiensis*, Vol. 7, pp. i–xii, 1–806.
—— and HOOKER, J. D. (1883) *Genera plantarum*, Vol. 3.
BERGIUS, P. J. (1767) *Descr. plant. ex Capite Bonae Spei*, p. 353.
BLAKE, S. T. (1943) *Coleocarya*, a new genus of the Restionaceae from S.E. Queensland. *Proc. R. Soc. Qd* **54**, 75–7.
BORTENSCHLAGER, S., ERDTMAN, G., and PRAGLOWSKI, J. (1966) Pollenmorphologische Notizen über einige Blütenpflanzen incertae sedis. *Bot. Notiser* **119**, 160–8.
BRIGGS, B. G. (1963) Chromosome numbers in *Lepyrodia* and *Restio* in Australia. *Contr. N.S.W. natn. Herb.* **3**, 228–32.
BROWN, R. (1810) *Prodromus Florae Novae Hollandiae*, pp. 243, 247, 248, 249, 251.
BULLOCK, A. A. (1959) The generic name *Restio*. *Taxon* **8**, 107.
CAMPBELL, E. O. (1964) The restiad peat bogs at Motumaoho and Moanatuatua. *Trans. R. Soc. N.Z.* Bot. **2**, 219–27.
CARLQUIST, S. (1961) *Comparative plant anatomy*. New York.
CHEADLE, V. I. (1955) The taxonomic use of specialisation of vessels in the metaxylem of Gramineae, Cyperaceae, Juncaceae and Restionaceae. *J. Arnold Arbor.* **36**, 147–57.
CUTLER, D. F. (1963) Inverted vascular bundles in the leaf of the Thurniaceae. *Nature, Lond.* **198**, 1111–12.
—— (1964) Three cell types occurring in the cortex of the culm of various species of Restionaceae. *Notes Jodrell Lab.* **1**, 11–13.
—— (1965*a*) The taxonomic significance of the anatomy of the Restionaceae. Thesis, London. 778 pp. (unpublished).
—— (1965*b*) Vegetative anatomy of the Thurniaceae. *Kew Bull.* **19**, 431–41.
—— (1966) Anatomy and taxonomy of the Restionaceae. *Notes Jodrell Lab.* **4**, 1–25.
—— (1967) The correct name of *Hopkinsia scabrida* Fitzgerald (Restionaceae). *Kew Bull.* **21**, 67.
—— and SHAW, H. K. A. (1965) Anarthriaceae and Ecdeiocoleaceae: two new monocotyledonous families, separated from the Restionaceae. Ibid. **19**, 489–99.
DESVAUX, M. (1828) Observations sur quelques familles des plantes monocotylédones, d'après les manuscrits de feu le Baron Palisot de Beauvois. *Annls. Sci. nat.* ser. 1, **13**, 37–52.
DIELS, L. and PRITZEL, E. (1904) Fragmenta Phytographiae Occidentalis. *Bot. Jb.* **35**, 83–91.
ENDLICHER, S. L. (1836) *Genera plantarum*, p. 120.
FITZGERALD, W. (1904) Additions to the West Australian flora. *J. W. Aust. nat. Hist. Soc.* **1**, 33, 34.
GARDNER, C. A. (1941–2) The Vegetation of Western Australia. *J.R. Soc. W. Aust.* **28**, pp. xi–lxxxvii.
GAY, C. (1853) *Flora de Chile*, Vol. 6, p. 153. Paris.
GILG, E. (1890) Beiträge zur vergleichenden Anatomie der xerophilen Familie der Restionaceae. *Bot. Jb.* **13**, 541–606.
GILG-BENEDICT, C. (1930) Restionaceae in: Engler and Prantl, *Die natürlichen Pflanzenfamilien*, 2nd edn, Vol. 15a, pp. 8–27.
HAMANN, U. (1961) Merkmalsbestand und Verwandtschaftsbeziehungen der Farinosae. *Willdenowia* **2**, 639–768.
—— (1962) Weiteres über Merkmalsbestand und Verwandtschaftsbeziehungen der Farinosae. Ibid. **3**, 169–207.
HIERONYMUS, G. (1888) Restionaceae in: Engler and Prantl, *Die natürlichen Pflanzenfamilien*, Vol. II. 4, pp. 3–10.
HOOKER, J. D. (1864) *Handbook of the New Zealand flora*, Pt. 1, pp. 293–6.

HUTCHINSON, J. (1959) *Families of flowering plants. II. Monocotyledons*, 2nd edn, pp. 700–2. Clarendon Press, Oxford.

JOHNSON, L. A. S. and EVANS, O. D. (1963*a*) A revision of the *Restio gracilis* complex. *Contr. N.S.W. natn. Herb.* **3,** 200–17.

—— —— (1963*b*) Geographic races in *Restio tetraphyllus* Labill. Ibid. 218–22.

—— —— (1963*c*) Intrageneric groups and new species in *Lepyrodia*. Ibid. 223–7.

—— —— (1966) Flora of New South Wales. Restionaceae. *Contr. N.S.W. natn. Herb.* Flora ser. No. 25, pp. 2–28.

JORDAAN, P. G. (1946) Die Anatomie van die Wortel van *Restio paludosus* en *Leptocarpus vimineus*. *Jl. S. Afr. Bot.* **12,** 115–20.

JUSSIEU, A. L. DE (1789) *Genera plantarum*. Paris.

KIMURA, Y. (1956) Système et phylogénie des monocotylédones. *Phanérogamie* **15,** 137–59.

KUNTH, C. S. (1841) *Enum. Pl.* **3,** 381–486.

LABILLARDIÈRE, J. J. H. DE (1804–7) *Novae Hollandiae plantarum specimen*, Vol. 2, pp. 77–8, Pl. 226 and 228. Paris.

LINDLEY, J. (1846) *The vegetable kingdom* (2nd edn, 1847; 3rd edn 1853), p. 121.

LINNAEUS, C. (1766) *Syst. Nat.* 12th edn, Addenda, p. 735.

—— (1771) *Mant.* II (altera), pp. 162, 297.

LOWE, J. (1961) The phylogeny of monocotyledons. *New Phytol.* **60,** 355–71.

MALMANCHE, L. A. (1919) *Contribution à l'étude anatomique des Eriocaulonacées et des familles voisines: Restiacées, Centrolépidacées, Xyridacées, Philydracées, Mayacacées.* Thesis, St. Cloud. 165 pp.

MASTERS, M. T. (1865) Observations on the morphology and anatomy of the genus *Restio* L. together with an enumeration of the South African species. *J. Linn. Soc.* (Bot.) **8,** 211–55.

—— (1867) Synopsis of the S. African Restionaceae. Ibid. **10,** 210–79.

—— (1878) Restionaceae in: De Candolle, *Monographicae* **1,** 218–398. Paris.

—— (1900) Restionaceae novae capenses herbarii Berolinensis imprimis schlechterianae. *Bot. Jb.* **29,** Beibl. 66.

METCALFE, C. R. (1959) A vista in plant anatomy. *Vistas in botany*, pp. 91–4. Pergamon Press, London.

—— (1960) *Anatomy of the monocotyledons. I. Gramineae.* Clarendon Press, Oxford.

—— (1961) The anatomical approach to systematics. *Recent Advances in Botany*, pp. 146–50. Toronto.

—— (1963) Comparative anatomy as a modern botanical discipline. *Adv. bot. Res.* **1,** 101–47.

MEYER, F. J.—see SOLEREDER, H., and MEYER, F. J.

MUELLER, F. VON (1873) *Fragm. Phytogr. Austr.* **8,** 100.

—— (1874) ex J. Buch. in *Trans. N.Z. Inst.* **6,** 340.

NEES VON ESENBECK, C. G. (1830) Beitrag zur Kenntniss der Familien der Restiaceen in Rücksicht auf Gattungen und Arten. *Linnaea* **5,** 627–66.

—— (1832) In Lindley's *Intr. Nat. Syst.* 2nd edn, p. 614.

—— (1836) In Lindley's *Nat. Syst.* 2nd edn, p. 450.

—— (1846) In *Lehm. Pl. Preiss* **2,** 66.

NOEL, A. R. A. (1959) The stem anatomy of *Restio triticeus* Rottb., etc. *Jl. S. Afr. Bot.* **25,** 357–70.

PEISL, P. (1957) Die Binsenform. *Ber. schweiz. bot. Ges.* **67,** 99–212.

PFITZER, E. (1869–70) Beiträge zur Kenntnis der Hautgewebe der Pflanzen. II. Über das Hautgewebe einiger Restionaceen. *Jb. wiss. Bot.* **7,** 561–84.

PILLANS, N. S. (1928) The African genera and species of Restionaceae. *Trans. R. Soc. S. Afr.* **16,** 207–440.

—— (1945) New and hitherto imperfectly known species of African Restionaceae. Ibid. **30,** 245–66.

—— (1950) Restionaceae in: R. S. Adamson and T. M. Salters, *Flora of the Cape Peninsula.* Cape Town and Johannesburg.

RAFINESQUE, C. S. (1838) *Flora Tellur.* **4,** 32–3.

RIDLEY, H. N. (1925) *The Flora of the Malay Peninsula*, Vol. 5, p. 136.

RODWAY, L. (1903) *The Tasmanian flora*, pp. 233–7. Hobart.

SCHWENDENER, S. (1874) *Das mechanische Princip im anatomischen Bau der Monokotylen*, Leipzig.

SMITHSON, R. (1956) [1957] The comparative anatomy of the Flagellariaceae. *Kew Bull.* **11,** 491–501.

SOLEREDER, H. and MEYER, F. J. (1929) *Systematische Anatomie der Monokotyledonen*, Vol. 4, Farinosae, pp. 6–30. Berlin.

SPRAGUE, T. A. (1929) Bergius, Descriptiones Plantarum and Linné, Mantissa prima. *Kew Bull.* 88–9.

STEUDEL, E. G. (1855) *Synopsis plantarum Glumacearum*, Vol. 2, p. 246.

SUESSENGUTH, K. (1943) Über eine neue Gattung der Restionaceeen. *Boissiera* **7,** 20–6.

THISELTON-DYER, W. T. (ed.) (1900) M. T. Masters Restionaceae. *Flora Capensis*, Vol. 8, pp. 59–149.

THUNBERG, C. P. (1788) *Diss. de Restione.*

—— (1790) *Act. Holm.* **11,** 26t. 2f, 2.

TOMLINSON, P. B. (1964) Notes on the anatomy of *Triceratella* (Commelinaceae). *Kirkia*, Vol. 4, p. 209.

TSCHIRCH, A. (1880) Ueber einige Beziehungen des anatomischen Baues der Assimilationsorgane zu Klima und Standort. *Linnaea* **43,** 143, 223, 227, 238, and 241.

UBERFELD, M. (1926) Beiträge zur Kenntnis des sexuellen Dimorphismus der Restionaceen. *Bot. Jb.* **60,** 176–206.

VAN TIEGHEM, P. (1887) Sur l'exoderme de la racine des Restionacées. *Bull. Soc. bot. fr.* **34,** 448–51.

—— (1887) Racine des Centrolépidacées, etc. *J. Bot., Paris* **1,** 305–15.

—— and DOULIOT, H. (1888) Recherches comparatives sur l'origine des membres endogènes dans les plantes vasculaires. *Annls Sci. nat.* b, ser. 7, **8,** 1–656.

ANARTHRIACEAE

(Figs. 41, 42)

INTRODUCTION

ANARTHRIACEAE are a monogeneric family of perennial rush-like herbs, separated from the Restionaceae in 1965 (Cutler and Shaw). The genus *Anarthria* contains 5 species. The family differs from the Restionaceae in four main characters of gross morphology. It has basal leaves, the leaves are sometimes laterally flattened and equitant, the anthers are bilocular, free from one another and laterally dehiscent, and the flowers are bracteate. The Restionaceae in contrast never develop basal leaves, the leaves are dorsiventrally flattened, never equitant; the anthers are unilocular (rarely bilocular except in *Lyginia*, but then laterally connate for middle third and introrse) and the flowers are ebracteate (v. rarely bracteate and bracteolate, in *Lepyrodia*).

The family is endemic in south-western Australia.

ANARTHRIA R. Br.

No. of spp. 5; examined 5.

GENUS DESCRIPTION

Leaf surface

Epidermis: cells axially elongated, oblong or hexagonal, anticlinal walls straight to very wavy; cells over scl. girders longer and narrower. **Papillae** pronounced on abaxial and adaxial angles of lamina. **Hairs** absent. **Stomata** paracytic, intercostal, axially orientated.

Leaf T.S.

Outline elliptical, becoming oblong truncated near transition with sheathing base, sometimes with longitudinal grooves (Fig. 41. E, F). **Cuticle** frequently thick. **Epidermis.** Cells in 1 layer, square to upright oblong, with or without well-developed papillae. Outer and anticlinal cell walls frequently thickened, lumen square to oval. Larger, adaxial cells present only at sheathing base. **Stomata** sunken or superficial. **Vascular bundles** restricted to 1 ring, lying immediately inside chlorenchyma layer(s). Bundles of 2 main sizes: (i) large, with 1 conspicuous angular mx vessel on either flank and a well-developed phloem pole; numerous narrower mx and px vessels (Fig. 42. E); (ii) small, with scarcely differentiated xylem and phloem poles. Large and small bundles alternating. Flattened leaves having 1 large vb at abaxial angle and 2 small vb's at adaxial angle corresponding to marginal vb's of leaf sheath (Fig. 41. F). **Bundle sheaths:** I.S. sclerenchymatous, frequently connected to cortical fibre strands or girders; O.S. represented by 1 layer of narrow parenchymatous cells on flanks and around xylem pole. **Chlorenchyma** composed of several layers of elongated peg-cells orientated axially and in long chains, or 1 layer of peg-cells orientated radially. **Sclerenchyma** present (i) as tabular or wedge-shaped girders or partial girders, 1–2 or 3–8 cells wide, situated opposite

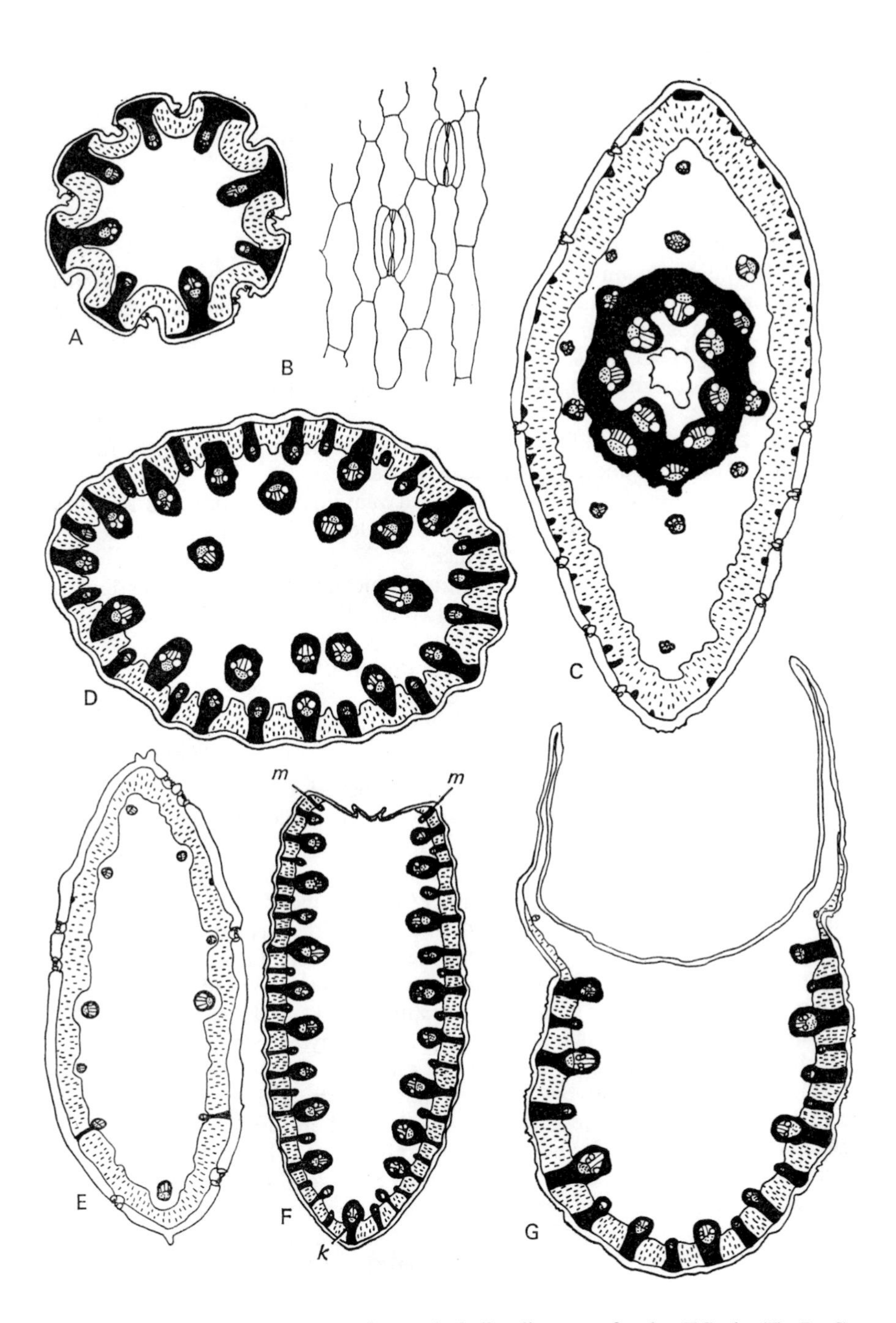

FIG. 41. Anarthriaceae. A, *Anarthria polyphylla*, diagram of culm T.S. (×45). B, C, *A. prolifera*: B, culm epidermis, surface view (×180); C, diagram of culm T.S. (×45). D, *A. gracilis*, diagram of culm T.S. (×45). E, *A. prolifera*, diagram of leaf T.S. (×45). F, *A. laevis*, diagram of leaf T.S.; k = keel bundle; m = marginal bundles (×20); G, *A. scabra*, diagram of leaf base T.S. (×45).

some or all vb's and dividing chlorenchyma into sectors; (ii) as cortical strands of 2–6 fibres in T.S. **Sclereids:** unbranched osteosclereids occurring infrequently in chlorenchyma of some spp. **Central ground tissue** parenchymatous, cells wide. **Crystals** present in some spp., rhombic. **Silica-bodies:** none seen. **Tannin** occasional (pathological ?).

Leaf base–blade transition region

The transition region between the sheathing leaf base and laterally flattened leaf blade is unusual. The characteristic appearance of transverse sections at different levels is as follows, starting from the base, working upwards (Figs. 42. A–D, also 41. G and 42. F):

(1) deeply crescentiform, with two halves of adaxial epidermis facing one another;
(2) v. thickly crescentiform with adaxial epidermis consequently further from midrib bundle and reduced in area;
(3) oblong truncated, with original leaf margins fused inwards and new secondary margins (*a*, Fig. 42. C) developed from abaxial epidermis. Adaxial epidermis further reduced;
(4) narrowly oval owing to continued thickening of midrib leading to fusion of secondary margins and elimination of adaxial epidermis and primary margins.

(Primary margins cut off as a ligule in one sp.).

Culm surface (Fig. 41. B)

Epidermis and **stomata** as in leaf.

Culm T.S. (Fig. 41. A, C, D).

Outline elliptical or ovoid to circular and grooved. **Cuticle** frequently thick. **Epidermis** and **stomata** as in leaf. **Hairs** absent. **Chlorenchyma** as in leaf. **Vascular bundles,** large and small, as in leaf; arranged (i) in 1 ring, or (ii) in outer ring and scattered in central ground tissue, or (iii) in central ring and scattered in cortex. Individual vb's either (*a*) surrounded by scl. sheath alone, and all free, or (*b*) having sheath continuous with cortical fibre girder or partial girder, or (*c*) embedded in central scl. cylinder. Large and small vb's alternating in most spp. Occasional transverse connections observed. Vessel lateral wall pitting either scalariform or annular to spiral; perforation plates oblique, scalariform. **Bundle sheaths:** I.S. sclerenchymatous, up to 10-layered at phloem pole, reducing to 2–3-layered on flanks and up to 5–6-layered at xylem pole; O.S. cells narrow, parenchymatous, arranged in 1 layer and completely ensheathing or restricted to flanks and around the xylem pole. **Sclerenchyma** as in leaf, and in one sp. also as 2–5-layered cylinder connecting inner vb's to one another. **Sclereids** as in leaf. **Central ground tissue** parenchymatous, cells thin-walled, more or less hexagonal in T.S., elongated axially, with square ends; intercellular spaces present. **Crystals** present in some spp., rhombic. **Silica-bodies** and **tannin:** none seen.

Root T.S.: see *A.prolifera*, p. 337.

MATERIAL EXAMINED

Anarthria gracilis R. Br.: (i) Garland, L. s.n.; Albany, W. Australia, Nov. 1921 (K). (ii) Andrews, C. 1077, 1078; Torbay Junction, nr. Albany, W. Australia (K).

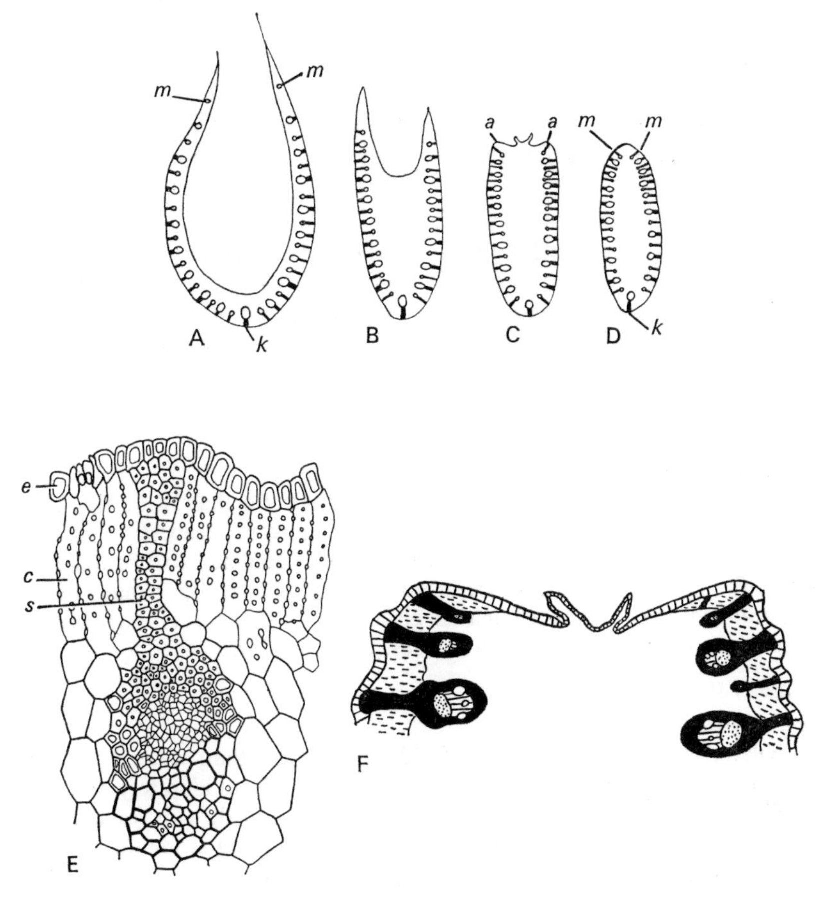

FIG. 42. Anarthriaceae. A–F, *Anarthria laevis*: A–D, diagrammatic T.S. from different parts of leaf (×8); a = secondary margin; k = keel bundle; m = marginal bundles; A, sheathing base; B, *c.* 2·5 cm from base; C, *c.* 6 cm from base; D, blade, *c.* 10 cm from base. E, detail of leaf T.S. (×180); c = chlorenchyma; e = epidermis; s = sclerenchyma girder. F, adaxial part of leaf T.S. (detail from C) (×45).

A. laevis R. Br.: (i) Andrews, C. 1075; nr. Albany, W. Australia, 22 Dec. 1902 (K). (ii) Cheadle, V. I. CA340; Perth, W. Australia, 1960 (S).

A. polyphylla Nees: Drummond s.n.; without locality, in Herb. Mueller, W. Australia (K).

A. prolifera R. Br.: (i) Garland, L. s.n.; Albany, W. Australia, Nov. 1921 (K). (ii) Eames and Hotchkiss s.n.; Busselton, W. Australia, Sept. 1953 (K). (iii) Cheadle, V. I. CA341; Perth, W. Australia 1960 (S).

A. scabra R. Br.: (i) Drummond 271; Swan River, W. Australia (K). (ii) Elder 88; 25 ml. E. of Albany, W. Australia, Oct. 1932 (K). (iii) Oldfield 638; King George's Sound, W. Australia (K). (iv) Williamson, s.n.; without locality, Oct. 1924 (K). (v) Cheadle, V. I. CA463; without locality, 1960 (S).

SPECIES DESCRIPTIONS

***A. gracilis* R. Br.**

Leaf surface

Epidermis: cells 4-sided, 2–6 times longer than wide, walls slightly wavy. **Papillae** numerous, small, several or many to each cell, most pronounced on intercostal cells.

Leaf T.S. Diameter 2×1·5 mm.

Outline oval, with slight longitudinal striations. **Stomata** sunken; overarched by papillae from surrounding epidermal cells. **Chlorenchyma:** 3–4 layers of axially elongated cells. **Sclerenchyma:** girders opposite all vb's; largest girders widest at epidermis and next to vb, with concave sides; narrowest girders more or less rectangular.

Culm surface

As leaf.

Culm T.S. (Fig. 41. D). Diameter 1×1·5 mm.

Outline oval, with slight longitudinal grooves. **Papillae** more frequent in Andrews than Garland material. **Stomata** sunken, deepest in Andrews material. **Chlorenchyma** as leaf. **Vascular bundles** confined to 1, outer ring in Andrews material; besides those in outer ring some free in ground tissue in Garland material. **Sclerenchyma** as girders or partial girders; groups rectangular, rectangular with concave sides, or triangular.

***A. laevis* R. Br.**

Leaf surface

Epidermis: costal cells mostly 15–20 times longer than wide, walls thick; intercostal cells 2–8 times longer than wide, walls moderately thickened. **Papillae** large, usually 1–2 per cell, frequently 2–3-lobed, with rounded ends, thick-walled, occurring on intercostal cells, overarching stomata.

Leaf T.S. (Figs. 41. F; 42. A–F). Diameter 3×1 mm.

Outline narrowly oval, with rounded abaxial and adaxial angles and shallow grooves. **Stomata** superficial. **Chlorenchyma** consisting of radially elongated peg-cells. **Sclerenchyma:** as girders opposite all vb's, widest at epidermis, with slightly concave sides, and frequently 4–5 times taller than wide. **Crystals** conspicuous, in outer bundle sheaths.

Culm surface

As leaf, but most **papillae** shorter and less frequent.

Culm T.S. Diameter 4·5×1 mm.

Outline elliptical to oval. **Chlorenchyma** as leaf. **Vascular bundles** arranged in 1 ring, each opposite a scl. girder. **Sclerenchyma:** girders as in leaf, but narrower; occasionally present between vb's.

***A. polyphylla* Nees**

Leaf not seen.

Culm surface

Epidermis: costal cells 15–20 times longer than wide, intercostal cells 2–10 times longer than wide. **Cuticle** with pronounced, widely spaced longitudinal striations. **Papillae** tall, 2–3 times higher than wide, gradually tapering to rounded apex; 1-several per cell; overarching stomata, otherwise more or less perpendicular to cell wall.

Culm T.S. (Fig. 41. A). Diameter 0·6–0·8 mm.

Outline circular, with deep grooves (8 in specimen examined). **Epidermis:** costal cells small, about $\frac{1}{4}$ height and width of intercostal cells. **Stomata** superficial, but present only amongst those intercostal cells lining grooves. **Chlorenchyma:** cells axially elongated; in 3–4 layers. **Vascular bundles** in 1, outer ring. **Sclerenchyma:** girders with concave sides, broadening rapidly to epidermis, becoming more or less T-shaped, outer end extending for width of ridge between culm grooves.

A. prolifera R. Br.

Leaf surface

Epidermis: most cells between 2 and 6 times longer than wide; walls v. wavy. **Cuticle** v. thick. **Papillae** tall, wide, ends obtuse or slightly spathulate; present on some epidermal cells, on abaxial and adaxial edges of leaf. **Tannin** present in most cells.

Leaf T.S. (Fig. 41. E). Diameter 1·2×0·5 mm.

Outline elliptical. **Papillae** confined to adaxial and abaxial edges of leaf, see leaf surface. **Stomata** superficial. **Chlorenchyma** cells axially elongated, in 3–4 layers. **Sclerenchyma:** (i) girders, rectangular, opposite 2 vb's only; (ii) subepidermal strands each consisting of 2 fibres.

Culm surface

As leaf (Fig. 41. B).

Culm T.S. (Fig. 41. C). Diameter 2×1 mm.

Outline elliptical. **Papillae** shorter than in leaf, situated at angular edges of culm and on occasional other cells. **Chlorenchyma** as leaf. **Vascular bundles:** (i) outer vb's free from scl. and embedded in cortical ground tissue; (ii) inner vb's in 2–3 rings and embedded in scl. cylinder. **Sclerenchyma:** (i) subepidermal strands of 2–10 fibres arranged at irregular intervals around the culm; (ii) 3–5-layered cylinder. **Ground tissue** parenchymatous, present between chlorenchyma and scl. cylinder and also in culm centre. **Air-canal** central.

Root T.S.

Root-hairs arising from cells similar in size to those of remainder of epidermis; base of hair as wide as epidermal cells, tapering gradually and becoming cylindrical. **Exodermis** 2-layered, cells thin-walled, lignified, similar in size to epidermal cells. **Cortex** 6-layered, cells thin-walled parenchymatous, with cellulose walls. **Endodermis:** cells mostly in 1 layer but locally 2-layered; mostly 3–4 times higher than wide, narrow; walls all thick. **Pericycle** 1-layered, cells narrow, thin-walled. **Vascular tissue:** phloem in strands 8–10 cells wide, separated from one another by protoxylem and situated next to pericycle; 10–12 mx vessels arranged in 1 ring. **Central ground tissue** parenchymatous.

***A. scabra* R. Br.**

Leaf surface

Epidermis: costal cells 6–12 times longer than wide, intercostal cells 1–6 times longer than wide; walls moderately thickened, slightly wavy.

Leaf T.S. Diameter 4–13 × 0·8–1·5 mm.

Outline elliptical. **Epidermis:** cells mostly about as high as wide; those on abaxial and adaxial margins up to twice as wide and 3–4 times higher than the others, and in 1–2 layers. **Stomata** superficial. **Papillae:** none seen. **Chlorenchyma** consisting of radially elongated peg-cells. **Sclerenchyma** present (i) as girders, widest at either end, 3–4 times as high as wide; (ii) as strands similar in size and shape to girders, extending through chlorenchyma but not opposite to vb's; (iii) as small strands composed of 1–5 fibres, to inner side of chlorenchyma. Fig. 41. G, leaf base T.S.

Culm surface

As leaf.

Culm T.S. Diameter 1·5 × 7 mm.

Outline elliptical. **Epidermis** and **chlorenchyma** as in leaf. **Vascular bundles:** (i) outer, mostly with girders or partial girders; (ii) inner, scattered in central ground tissue. **Sclerenchyma:** girders as in leaf; no large strands but occasional small strands present.

Taxonomic Notes

The 5 species can be divided into 2 groups on chlorenchyma characters. (i) Peg-cells elongated axially: *A. gracilis*, *A. polyphylla*, and *A. prolifera*. (ii) Peg-cells elongated radially: *A. laevis* and *A. scabra*. The peg-cell arrangement appears to be related to the habit of the plants; those species with only radical leaves have radially elongated peg-cells.

The Anarthriaceae differ anatomically from the Restionaceae mainly in leaf structure and in the distribution of sclerenchyma in leaf and culm. No culm vb's occur outside the sclerenchyma cylinder in the Restionaceae (except *Hypolaena graminifolia*, p. 194). In contrast, only *A. prolifera* has a sclerenchyma cylinder and there are free bundles in the cortex of this species. The condition found in most members of the Anarthriaceae, with girders extending from individual 'free' vb's to the culm surface, is unknown in Restionaceae. The presence of vestigial girders (strands) in the cortical region of the culm of *A. prolifera*, a species with few or no subtended vb's, may indicate that its mechanical cylinder is a secondary character.

The dorsiventrally flattened leaves of the Restionaceae are very reduced, with one row of vb's, each with its own sheath, and occasionally an arc of sclerenchyma joins most of the vb's together, but there is nothing reminiscent of the girders of Anarthriaceae.

The apparent similarity of the leaf structure of *Johnsonia lupulina* R. Br. (Liliaceae) to *Anarthria* was noted by Arber (1925). The similarity is, however, only superficial. *Johnsonia* has anomocytic stomata, and lacks large mx vessels in the vb's. The lack in this species of a sclerenchyma sheath to the vb's and the presence of a micro-papillate cuticle show the real discontinuity.

Both *Johnsonia* and *Anarthria* grow in the same area, and their similarity in gross morphology may be related to their responses to the same environment.

Leptosperma laterale R. Br. (Cyperaceae) shows similar leaf form, but it has conical silica-bodies in the epidermal cells of the leaf, and is thus clearly distinct. This also comes from the same region of Australia.

Although the anatomical evidence indicates that Anarthriaceae are distantly related to other members of the Juncales, the possession of peg-cells, the nature of the vb's and stomata seem to indicate an affinity with that group.

Bibliography for Anarthriaceae

ARBER, A. (1922) Leaves of the Farinosae. *Bot. Gaz.* **74**, 80–94.

—— (1925) *Monocotyledons. A morphological study*. Cambridge University Press.

BORTENSCHLAGER, S., ERDTMAN, G., and PRAGLOWSKI, J. (1966) Pollenmorphologische Notizen über einige Blütenpflanzen incertae sedis. *Bot. Notiser* **119**, 160–8.

CUTLER, D. F., and SHAW, H. K. A. (1965) Anarthriaceae and Ecdeiocoleaceae: two new monocotyledonous families, separated from the Restionaceae. *Kew Bull.* **19**, 489–99.

GILG-BENEDICT, C. (1930) Restionaceae in Engler and Prantl, *Die natürlichen Pflanzenfamilien*, 2nd edn, Vol. 15a, p. 17.

MALMANCHE, L.-A. (1919) *Contribution à l'étude anatomique des Eriocaulonacées et des familles voisines: Restiacées, Centrolépidacées, Xyridacées, Philydracées, Mayacacées*. Thesis, St. Cloud. 165 pp.

ECDEIOCOLEACEAE

(Fig. 43)

INTRODUCTION

THE single representative of this family, *Ecdeiocolea monostachya*, is a rush-like, perennial plant from Western Australia. The family was separated from the Restionaceae in 1965 (Cutler and Shaw) on the basis of characters of anatomy and gross morphology.

ECDEIOCOLEA F. v. Muell.

GENUS AND SPECIES DESCRIPTION

Leaf surface not seen.

Leaf blade T.S. (lamina only 0·5 mm long, 84 μm thick).

Epidermis 1-layered; (i) adaxial cells thin-walled, flattened, 4 times wider than high; (ii) abaxial cells with outer walls thickened, cells half width of adaxial cells but equally high. **Stomata** superficial, paracytic, present only in abaxial epidermis. **Vascular bundles:** 7 present in specimen examined. Phloem and xylem poles small; tracheids angular, thick-walled; wall-pitting scalariform. **Bundle sheaths** sclerenchymatous, weak. **Sclerenchyma:** poorly developed girders to abaxial epidermis. **Ground tissue** indistinguishable from chlorenchyma; cells parenchymatous, thin-walled. **Air-canals** absent. **Tannin** present in certain tracheids, ground tissue, and abaxial epidermal cells. **Crystals** and **silica:** none observed.

Leaf sheath T.S.

Encircling culm with overlap of $\frac{1}{2}$ turn; 140 μm thick, tapering to margins.

Epidermis: adaxial cells 3 times wider than high, thin-walled; abaxial cells twice as wide as high, with thick outer walls. **Stomata** superficial, paracytic. **Hypodermis** present to inner side of abaxial epidermis as 2 layers of fibres. **Vascular bundles:** xylem pole with 1 large tracheid on either flank, wall-pitting scalariform. Phloem pole well developed. **Bundle sheaths** sclerenchymatous, 3–4 times thicker at xylem pole than phloem pole. **Sclerenchyma** represented by hypodermal fibres. **Ground tissue** parenchymatous, surrounding vb's and in 1–2 layers next to hypodermis and adaxial epidermis. **Air-canals** present between vb's.

Culm surface

Epidermis. Cells of 3 types: (i) more or less square to slightly rectangular, on outer face of culm; (ii) rectangular, 2–3 times longer than wide, on sides of grooves; (iii) up to 12 times longer than wide, at bases of grooves. **Stomata** paracytic, situated on sides of grooves; epidermal cells encircling stomata losing their strictly axial arrangement.

Culm T.S. (Fig. 43, A, B). Diameter 0·8 to 1·2 mm.

Outline circular, with evenly and widely spaced, narrow, longitudinal grooves; each extending inwards to a depth of about $\frac{1}{4}$ of radius of culm. Grooves wider at blind end than opening. **Cuticle** thick. **Epidermis** 1-layered; exhibiting 3 cell types as above: (i) height equal to width, 5-sided (inner wall having 2 faces); (ii) 1–2 times wider than high, mostly papillate; (iii) 2–3 times wider than high, without papillae. All walls thin. **Stomata** situated on sides, towards inner end of grooves, superficial. **Chlorenchyma:** peg-cells in 2 layers, arranged in longitudinal strands, between deep grooves, each bounded on outer side and flanks by hypodermal scl. and on inner side by parenchymatous

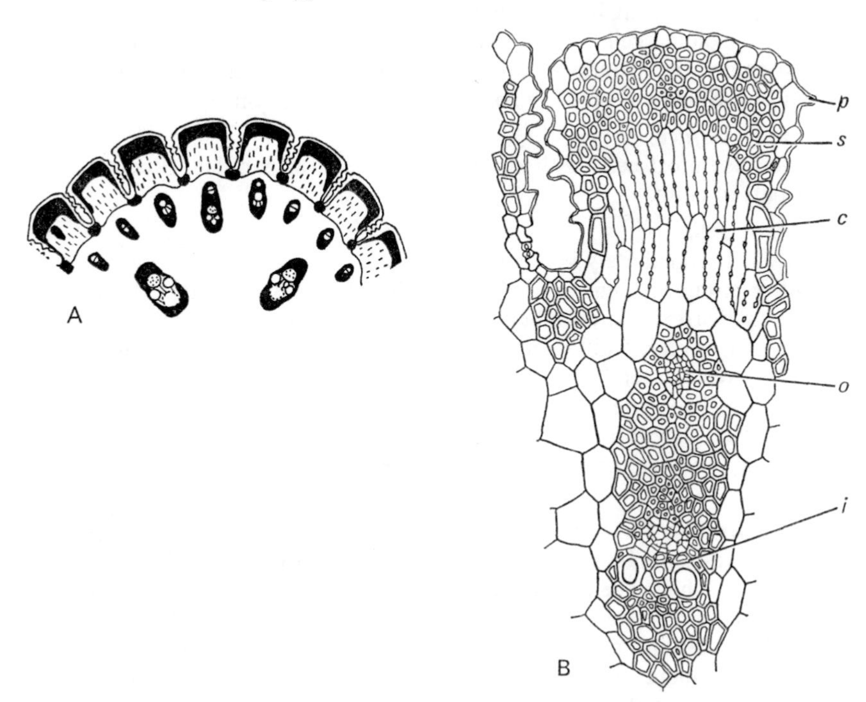

FIG. 43. Ecdeiocoleaceae. A, B, *Ecdeiocolea monostachya*: A, diagram of sector of culm T.S. (×45); B, detail of part of A (×220).
c = chlorenchyma; i = inner vascular bundle; o = peripheral vascular bundle joined to i; p = papillate epidermis; s = sclerenchyma layers immediately inside epidermis.

cells of ground tissue. **Vascular bundles** scattered, of 2 main sizes: (i) peripheral, small, with well-developed phloem, and xylem composed of tracheids and narrow vessels; (ii) medullary, medium-sized to large, having well-developed phloem; xylem exhibiting a large mx vessel on either flank; central medullary bundles with cavity in place of px pole. Vessels moderately thick-walled, lateral wall-pitting scalariform, pits with narrow borders or none. Perforation plates oblique, simple or scalariform. Phloem with much elongated sieve-tube elements and more or less transverse sieve-plates. Peripheral vb's on same radii as chlorenchyma strands and frequently opposite to outer, medullary bundles, the scl. sheaths of such a bundle pair attached to one another by a radial bridge of scl. Remaining peripheral and medullary bundles separate from one another. Transverse connections between vb's fairly frequent, short. **Bundle sheaths:** I.S. sclerenchymatous, thicker at poles than on flanks, frequently up to 10 cells thick at phloem pole; O.S. parenchymatous, 1-layered, frequently difficult to distinguish from ground parenchyma. **Sclerenchyma** represented by fibres and sclereids. Fibres much elongated, with pointed ends and simple, oblique pits; occurring as: (i) 4–6-layered hypodermis to inside of outwardly facing epidermal cells and 1–2-layered along flanks of grooves, except beneath stomata; (ii) longitudinal strands of 10–15 cells in T.S., at base of each groove. Sclereids of 2 types: (i) radially elongated, isolated cells in chlorenchyma; (ii) more or less isodiametric cells next to epidermis, around substomatal cavities. **Central ground tissue** parenchymatous. **Silica** as 'sand' in some cells of chlorenchyma.

Culm apex surface

Epidermis: most cells with simple row of several papillae. **Stomata** as in lower part of culm.

Culm apex T.S.

Sectioned immediately below inflorescence.

Outline circular, ungrooved. **Vascular bundles** of 3 main types: (i) small, irregularly distributed, peripheral vb's, having poorly developed xylem and phloem; (ii) larger, individual vb's each with well-developed phloem and several mx vessels; tending to be amphivasal; (iii) fusion bundles composed of larger types, fused laterally. All bundles appearing to be in a state of anastomosis and re-segregation at this level in the culm. **Ground tissue** aerenchymatous, of lobed cells and small-celled parenchyma. **Silica:** none observed.

Root T.S.

Root-hairs frequent, arising from cells similar in size to remaining epidermal cells; each hair slightly widened at base. **Cortex** parenchymatous, up to 20 layers thick; cells thin-walled; intercellular spaces becoming progressively larger between cells of inner layers. **Endodermis** 1-layered, outer walls thin, inner and anticlinal walls moderately thickened; thin-walled passage cells frequent. **Pericycle** of 4–5 layers of thick-walled fibres, these rounded in outline. **Vascular system:** phloem strands numerous, peripheral, alternating with px poles; mx vessels solitary, or in groups of up to 3, scattered throughout parenchymatous central ground tissue.

Material Examined

Ecdeiocolea monostachya F. v. Muell.: (i) Drummond s.n.; Swan River, W. Australia (K). (ii) Pritzel, E. s.n.; between Rivers Moore and Murchison, Sept. 1901 (K). (iii) Steward, F. 522; Lederville (K). (iv) Koch, M. 2754; Merredin, 22.9.23 (K). (v) Blake, S .T. 18106; 10–20 miles N. of Northampton, 3 Sept. 1947 (K). (vi) Cheadle, V. I. CA458; Perth, 1960 (S).

Taxonomic Notes

The distinctive anatomy of *Ecdeiocolea* differs very widely from that of any member of the Restionaceae. The irregular arrangement of vb's in sections taken from immediately below the inflorescence does not resemble that in sections taken from any point of a typical restionaceous culm. So far no characters of leaf or culm have been found which indicate affinities between *Ecdeiocolea* and any other particular genus. *Ecdeiocolea* is anatomically dissimilar from members of the Xyridaceae, in spite of the fact that it is said to resemble this family in gross morphology.

The previous inclusion of *Ecdeiocolea* in the Restionaceae was probably mistaken, possibly because of its highly adapted form as a xeromorph, and the lack of easily recognizable morphologically distinctive characters.

The family Ecdeiocoleaceae shows general affinities with the Glumiflorae.

Bibliography for Ecdeiocoleaceae

ARBER, A. (1925) *Monocotyledons. A morphological study.* Cambridge University Press.

BORTENSCHLAGER, S., ERDTMAN, G., and PRAGLOWSKI, J. (1966) Pollenmorphologische Notizen über einige Blütenpflanzen incertae sedis. *Bot. Notiser* **119**, 160–8.

CUTLER, D. F. and SHAW, H. K. A. (1965) Anarthriaceae and Ecdeiocoleaceae: two new monocotyledonous families, separated from the Restionaceae. *Kew Bull.* **19**, 489–99.

GILG-BENEDICT, C. (1930) Restionaceae in Engler and Prantl, *Die natürlichen Pflanzenfamilien*, 2nd edn, Vol. 15a, p. 17.

TOMLINSON, P. B. (1969) Xyridaceae in *Anatomy of the monocotyledons*, Vol. III. *Commelinales-Zingiberales*. Clarendon Press, Oxford.

DESCRIPTION OF PLATES

I. Juncaceae. A, *Juncus inflexus*: culm T.S. ×33 (p. 30); B, *Luzula arcuata*: leaf T.S. ×96 (p. 54).

II. Juncaceae. *Marsippospermum grandiflorum*: A, part of leaf T.S. at abaxial groove ×180 (p. 60); B, part of culm T.S. ×180 (p. 61).

III. A. Juncaceae. *Prionium serratum*: part of leaf T.S. ×240 (p. 65); B, Thurniaceae. *Thurnia sphaerocephala*: vascular bundle pair from leaf T.S.; note inverted small vascular bundle, and stellate diaphragm cells. ×260 (p. 78).

IV. A, Juncaceae. *Juncus gerardii*: part of culm T.S. ×160 (p. 27); B, Restionaceae. *Restio gracilis*: part of culm T.S. ×200 (p. 255). Note that some peripheral vascular bundles are present to the outer side of the sclerenchyma cylinder in A, but all peripheral bundles are embedded in the sclerenchyma in B.

V. Restionaceae. A, *Chondropetalum paniculatum*: outer part of culm T.S. Note protective cells surrounding substomatal cavities, and the apertures between the protective cells. ×200 (p. 155); B, *Elegia vaginulata*, sector of culm T.S.; note the 2-layered epidermis and the protective cells. ×200 (p. 173).

VI. Restionaceae. A, *Cannomois acuminata*: part of culm T.S.; note short ribs from sclerenchyma sheath interrupting the parenchyma sheath. ×200 (p. 141); B, *Willdenowia striata*: sector of culm T.S.; note ribs from sclerenchyma sheath dividing chlorenchyma into sectors; note also protective cells. ×200 (p. 321).

VII. Restionaceae. A, *Leptocarpus simplex*: sector of culm T.S.; note single-layered, densely staining hairs closely applied to 1-layered epidermis; note also pillar cells dividing chlorenchyma into sectors. ×200 (p. 212); B, *Restio tetragonus* sector of culm T.S.; note rib of sclerenchyma; one of these occurs at each of the 4 corners of the culm. ×200 (p. 281).

VIII. Restionaceae. *Lepyrodia scariosa*: A, part of culm T.S. ×200; B, detail of outer part of culm T.S.; note silica-bodies in epidermal cells, also the difference between chlorenchyma cells of outer and inner layers. ×500 (p. 223).

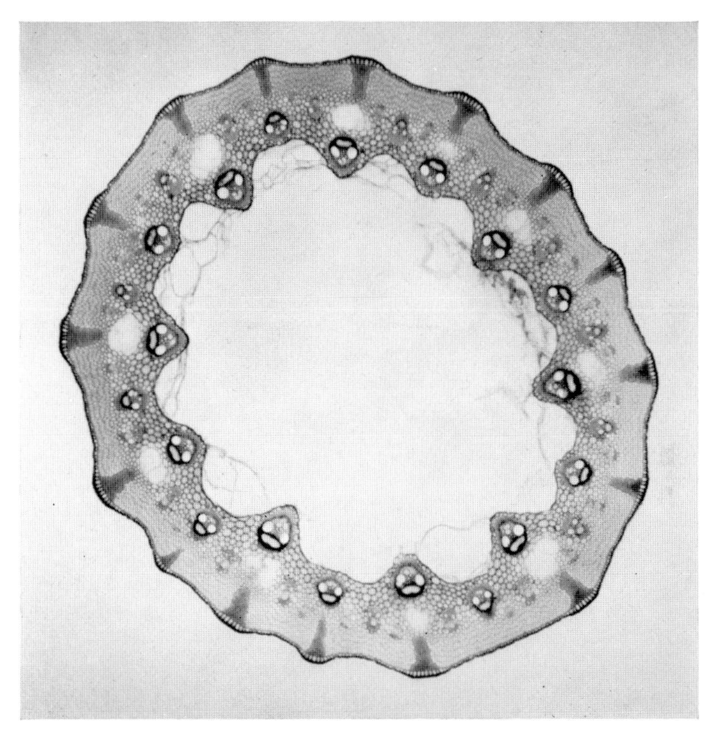

A. Juncus inflexus culm T.S. × 33

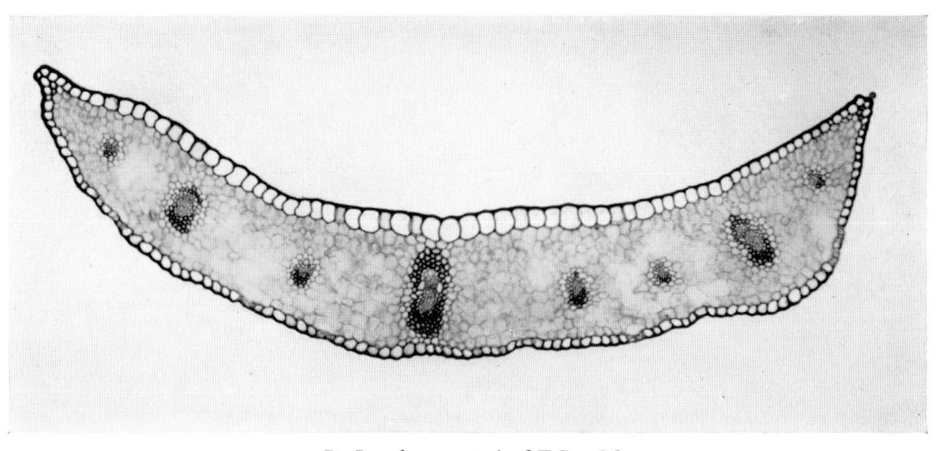

B. Luzula arcuata leaf T.S. × 96

PLATE II

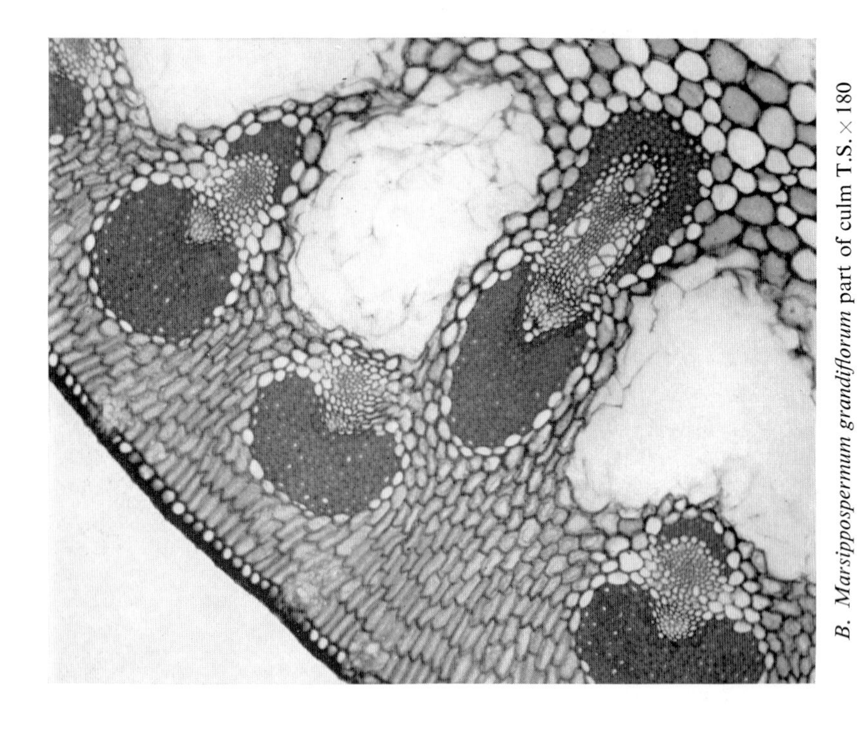

B. Marsippospermum grandiflorum part of culm T.S. × 180

A. Marsippospermum grandiflorum part of leaf T.S. × 180

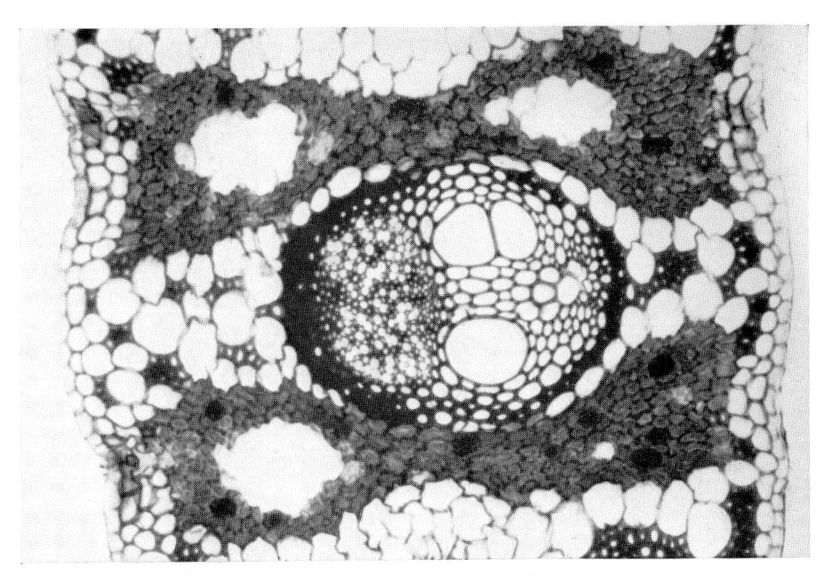

A. Prionium serratum leaf vb T.S. × 240

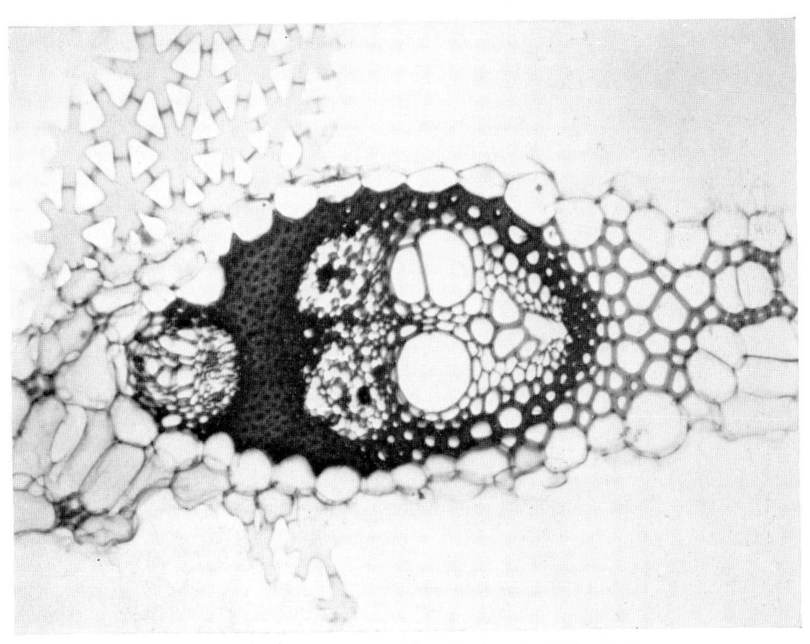

B. Thurnia sphaerocephala leaf vb pair, T.S. × 260

PLATE IV

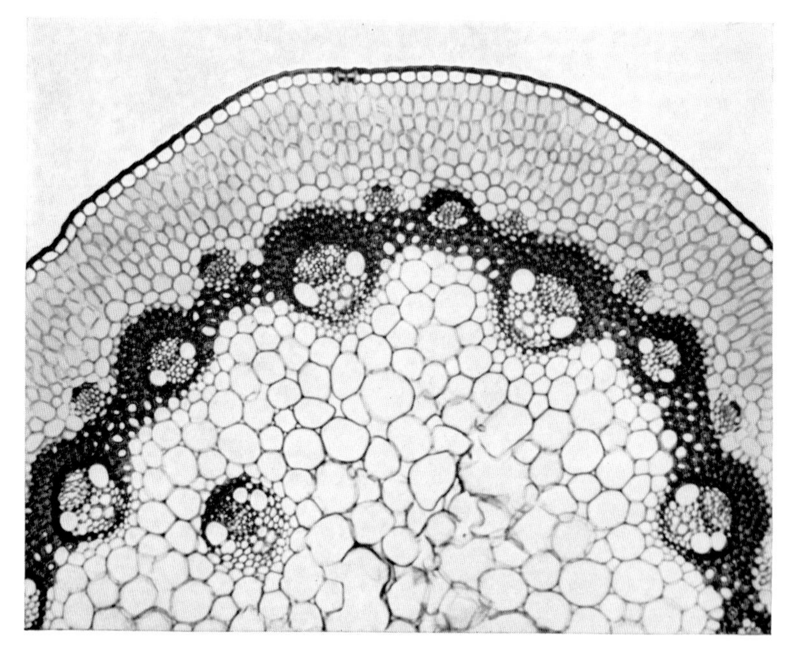

A. Juncus gerardii part of culm T.S. × 160

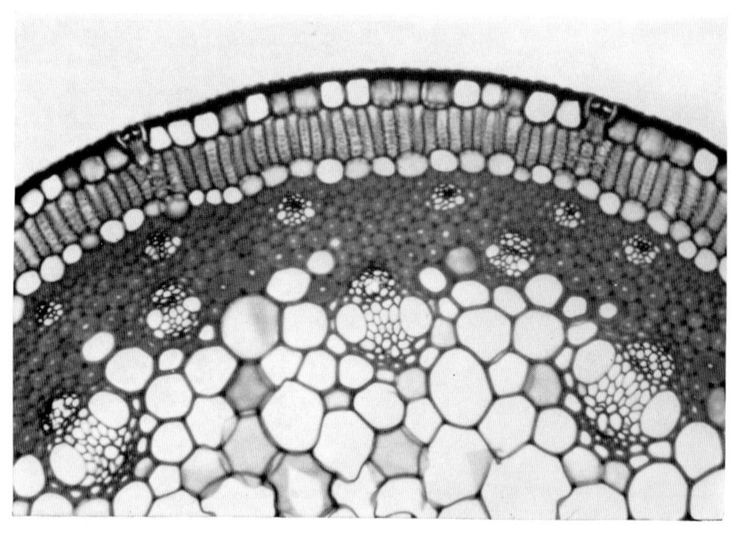

B. Restio gracilis part of culm T.S. × 200

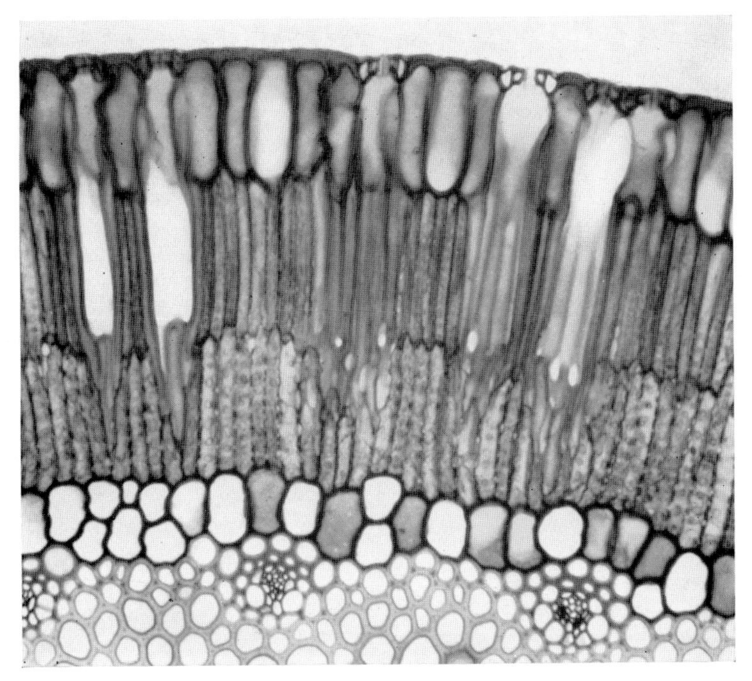

A. Chondropetalum paniculatum part of culm T.S. × 200

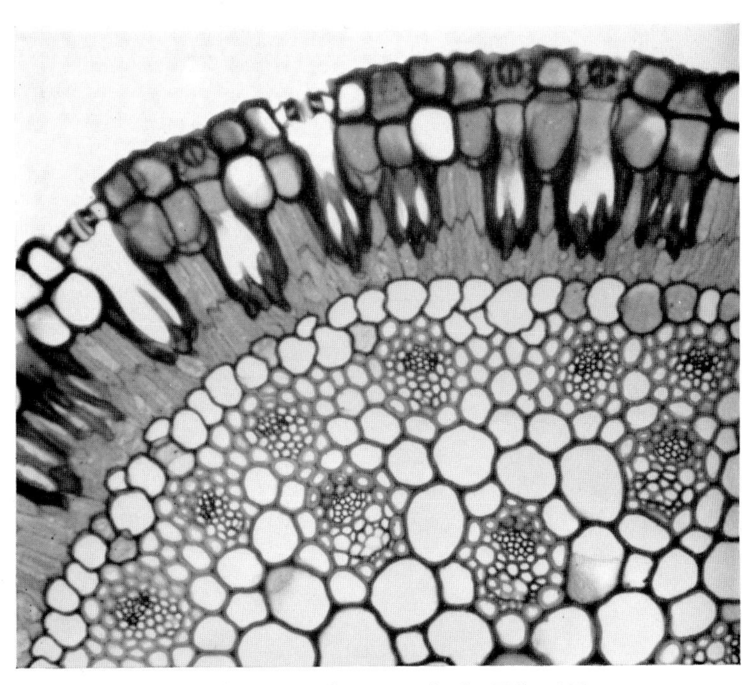

B. Elegia vaginulata part of culm T.S. × 200

PLATE VI

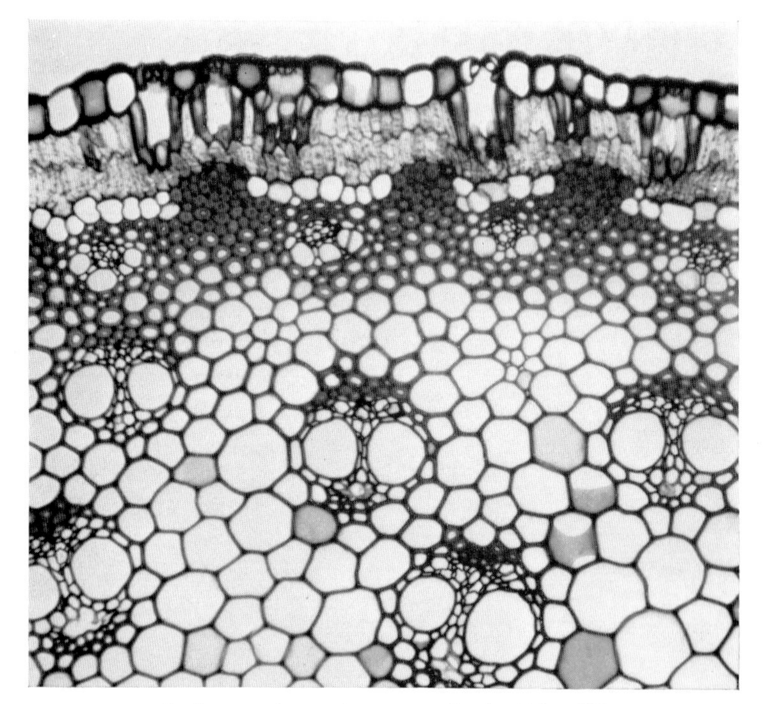

A. Cannomois acuminata part of culm T.S. × 200

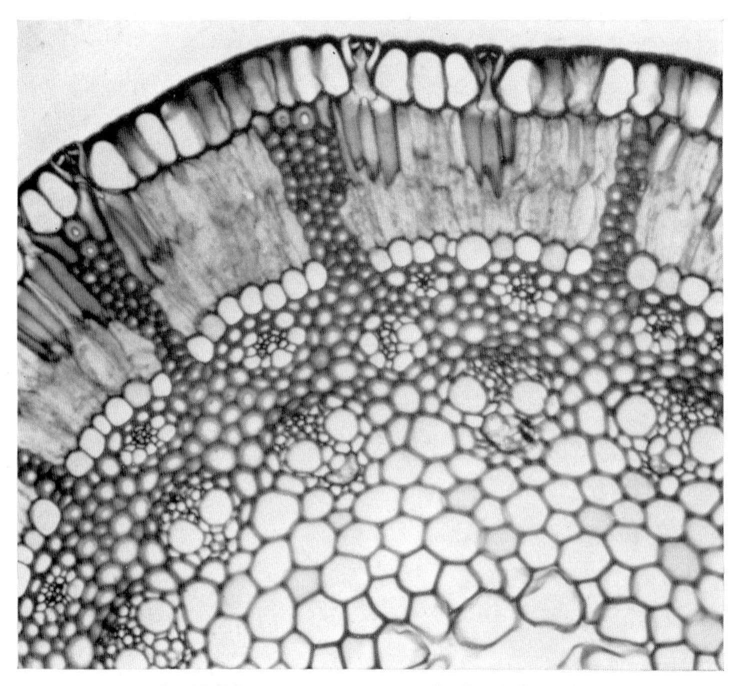

B. Willdenowia striata part of culm T.S. × 200

PLATE VII

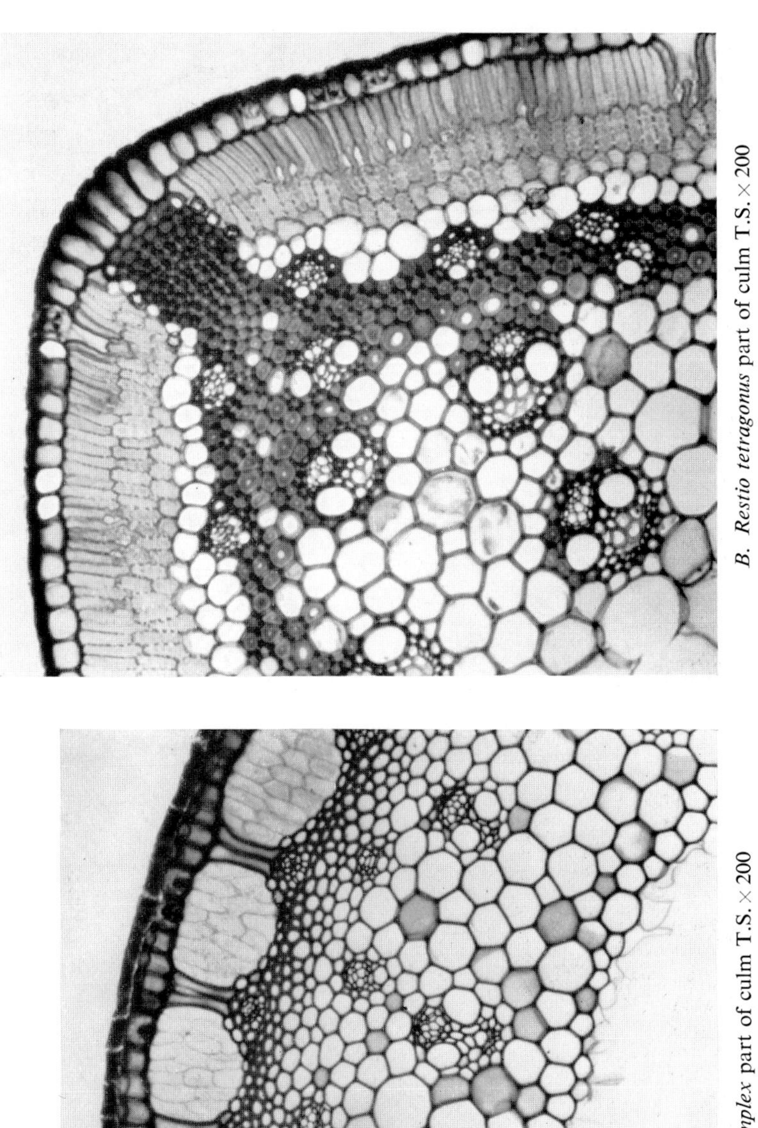

A. *Leptocarpus simplex* part of culm T.S. × 200

B. *Restio tetragonus* part of culm T.S. × 200

PLATE VIII

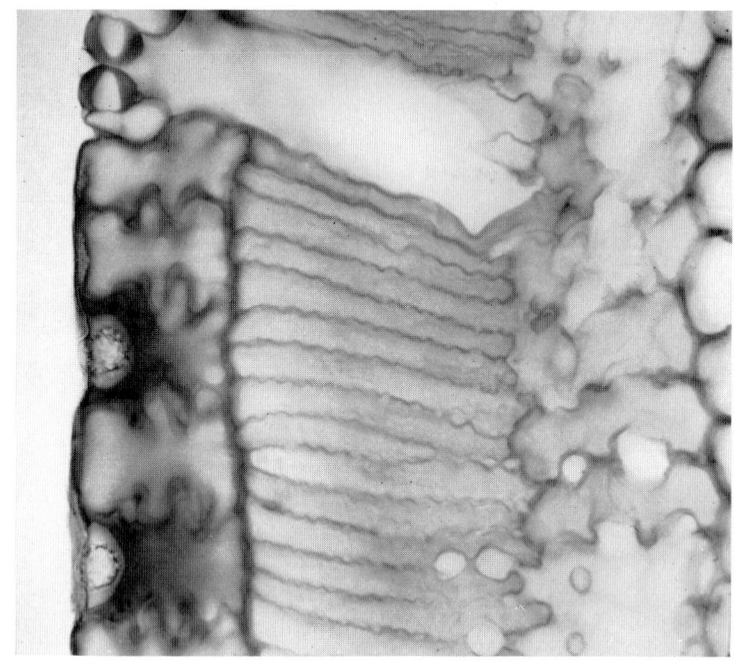

B. Lepyrodia scariosa H.P. detail of culm T.S. × 500

A. Lepyrodia scariosa part of culm T.S. × 200

AUTHOR INDEX

NOTE: *Page numbers in bold type refer to entries in the bibliographies*

SUBJECT INDEX

NOTE: *Where families have been specially studied, the page numbers are given in bold type. Italic page numbers are references to figures*

PRINTED IN GREAT BRITAIN
AT THE UNIVERSITY PRESS, OXFORD
BY VIVIAN RIDLER
PRINTER TO THE UNIVERSITY